Computers, Privacy and Data Protection: an Element of Choice

Serge Gutwirth · Yves Poullet · Paul De Hert ·
Ronald Leenes
Editors

Computers, Privacy and Data Protection: an Element of Choice

Springer

Editors
Serge Gutwirth
Vrije Universiteit Brussel
Center for Law, Science, Technology &
Society Studies (LSTS)
Pleinlaan 2
1050 Brussels
Belgium
serge.gutwirth@vub.ac.be

Prof. Yves Poullet
University of Namur
Research Centre for Information
Technology & Law
Rempart de la Vierge 5
5000 Namur
Belgium
yves.poullet@fundp.ac.be

Prof. Paul De Hert
Vrije Universiteit Brussel
Center for Law, Science, Technology &
Society Studies (LSTS)
Pleinlaan 2
1050 Brussels
Belgium
paul.de.hert@vub.ac.be

Ronald Leenes
Tilburg University
TILT
Warandelaan 2
5037 AB Tilburg
Netherlands
R.E.Leenes@uvt.nl

ISBN 978-94-007-0640-8 e-ISBN 978-94-007-0641-5
DOI 10.1007/978-94-007-0641-5
Springer Dordrecht Heidelberg London New York

Library of Congress Control Number: 2011922121

Printed on acid-free paper

Springer is part of Springer Science+Business Media (www.springer.com)

Preface

Privacy and data protection have never been static. On the contrary, the history of the last 40 years shows the reverse. Even if some important parts of the legislative framework – such as the EC Data Protection Directive of 1995 – have remained unaltered, new issues and challenges continue to emerge, requiring an ongoing process of interpreting their effect in terms of reach, objectives and their deeper significance. Such new issues do also trigger the elaboration of more specifically targeted legislative interventions that do not always fit seamlessly into the pre-existing framework. But this is not a surprising picture, since the developments, which resort under the denominators of privacy and data protection are generating and challenging the concerns of numerous different stakeholders who effectively use their voices and power to transform those concerns into political issues. Moreover, the number of factors and actors that sensibly impact upon privacy and data protection are manifold. Indeed, the consequences of technological applications due to unprecedented storage, processing and transmission capacities and by the possibilities of miniaturisation, convergence, interoperability and ubiquity, represent powerful triggers and challenges of emerging developments, but they are not the only determining factor. The current developments in the field are also linked to many other sources of action and change, such as business models, security policies, population management, police work and law enforcement, leisure, culture, health policies, practices in the "real" and in the "virtual" world and so on. Through its large variety of issues discussed, this book indeed evidences such complex cartography of issues related to privacy and data protection.

From the first edition on, in 2007, the annual international *Computers, Privacy and Data Protection* or "CPDP"-conferences[1] held in Brussels have taken these dynamic and multi-faceted features of privacy and data protection seriously and have upheld those at the core of their concept and organisation. The conferences are conceived as lively and interactive meetings where academics, practitioners, policy-makers, business representatives, data protection authorities, lawyers, activists and artists come together to exchange ideas and discuss emerging and/or hot issues related to privacy and data protection. Since the program is built up around an

[1] For more information about the CPDP-conferences: see http://www.cpdpconferences.org

important number of panels, sidetracks and side events with a high autonomy, the conference is a fertile soil for the prospering of debates and the taking form of new issues, questions and ideas, as is indeed evidenced by the rich collection of chapters that were published in the books made after the first and second editions of the conference, namely *Reinventing data protection?* (S. Gutwirth, Y. Poullet, P. De Hert, C. de Terwangne en S. Nouwt, Eds, Springer 2009) and *Data protection in a profiled word* (S. Gutwirth, Y. Poullet and P. De Hert, Eds, Springer 2010).

Like the two books we just mentioned, this book is again a sequel of a CPDP-conference. It has been made in the wake of its, again successful and inspiring, third edition, which was held in Brussels, in the famous *Kaaitheater*, immediately after "Privacy Day" on the 29th and 30th January of 2010.[2] This edition of the conference had *An element of choice* as title, and the editors decided to keep it as the title of the book, because the notion of "choice" expresses the wish to individual and societal control and channel evolutions and change in a direction that matches our both individual and collective projects and dreams. For sure, and as we suggested already, the developments and changes concerning data protection and privacy are the resultant of complex interplays and articulations of pertinent, but sometimes very heterogeneous, factors and actions, in which the moments and locations of choices and decisions are diverse, disseminated and difficult to identify. Hence, our focus upon "choice" as the overarching concept of this collection of peer reviewed chapters which focus on privacy and data protection from the perspective of conceptual changes (part 1), of counterforces and drawbacks (part 2), of inventive and singular practices (part 3) and of the emergence of the transversal technological development of cloud computing (part 4), must be read as a claim, or even stronger, as a reclaim.

In the face of the developments affecting privacy and data protection, "choice" unambiguously evokes both the need to collectively take responsibility and direct those developments in a desirable direction, providing the ambit to influence and steer the course of things in a way that matches our expectations related not only toward privacy and data protection, but also more broadly, to the kind of world we are building up. This challenge is not an easy one since all "big" policy choices we might be willing to make are conditioned by a myriad of "small" decisions and bifurcations that already have set many switches changers in irreversible positions. In one or another way, all the contributions to this book do express the complexity of

---

[2]The CPDP2010 was The CPDP2009 conference was organised by the Research Centre for *Law, Science, Technology & Society* (LSTS) at the Vrije Universiteit Brussel, the *Centre de Recherches Informatique et Droit* CRID at the University of Namur, the *Tilburg Institute for Law and Technology* at Tilburg University, the *Fraunhofer Institut für System- und Innovationsforschung* (ISI) in Karlsruhe, the *Institut National de Recherche en Informatique et en Automatique* (INRIA) in Grenoble, the Vlaams-Nederlands huis *deBuren* and the VUB *instituut voor PostAcademiche vorming* (iPAVUB). The Brussels Capital Region, the Fonds voor Wetenschappelijk Onderzoek (FWO), Microsoft, Google, the European Privacy Association, Hunton & Williams and Stibbe, further sponsored it. The conference was also supported by crucial players and stakeholders such as the *European Data Protection Supervisor*, the Dutch *College Bescherming Persoonsgegevens*, the Belgian *Commission for the protection of the private sphere* and both the Belgian Flemisch and French speaking Human Right Leagues.

the making of choices regarding to the issues of privacy and data protection which interest and concern their authors. May this book relay the interests, concerns and commitment of its authors to as many possible readers and may this, in turn, lead to the making of good choices with regards to matters we believe to be important for the future design of our societies. This is all the more relevant since the revision of the 1995 EC Data Protection Directive, the centrepiece of European data protection, is being revised as these sentences are written.

Brussels, Belgium	Serge Gutwirth
Namur, Belgium	Yves Poullet
Brussels, Belgium	Paul De Hert
Tilburg, The Netherlands	Ronald Leenes
October 21, 2010	

Contents

Contributors

Christel Backman Department of Sociology, University of Gothenburg, SE 405 30, Gothenburg, Sweden, christel.backman@sociology.gu.se

Rocco Bellanova Center for Law, Science, Technology and Society Studies (LSTS), Vrije Universiteit Brussel, Pleinlaan 2, 1050, Brussels, Belgium, rocco.bellanova@vub.ac.be

Paul Bernal Information Technology Law, University of East Anglia, Norwich, UK; London School of Economics and Political Science, London, UK, p.a.bernal@lse.ac.uk

Amos Bianchi Nuova Accademia di Belle Arti (NABA), Coordinamento Istituto Master e Bienni, Via Darwin 20, 20143 Milano, Italy, amos.bianchi@naba.it

Jean-François Blanchette Department of Information Studies, UCLA, Los Angeles, United States, blanchette@ucla.edu

Franziska Boehm University of Luxembourg, Luxembourg, franziska.boehm@uni.lu

John J. Borking Borking Consultancy, Wassenaar, The Netherlands, jborking@xs4all.nl

Ian Brown Oxford Internet Institute, The University of Oxford, Oxford, UK, ian.brown@oii.ox.ac.uk

Johann Čas Institute of Technology Assessment, Austrian Academy of Sciences, A-1030, Vienna, Austria, jcas@oeaw.ac.at

Valentina Casola Dipartimento di Informatica e Sistemistica, Università degli studi di Napoli Federico II, Napoli, Italy, casolav@unina.it

Norberto Nuno Gomes de Andrade Department of Law, European University Institute, Florence, Italy, norberto.andrade@eui.eu

Paul De Hert Center for Law, Science, Technology and Society Studies (LSTS), Vrije Universiteit Brussel, Pleinlaan 2, 1050, Brussels, Belgium, paul.de.hert@vub.ac.be

Katja de Vries Center for Law, Science, Technology and Society Studies (LSTS), Vrije Universiteit Brussel, Pleinlaan 2, 1050, Brussels, Belgium, edevries@vub.ac.be

Gurpreet Dhillon Virginia Commonwealth University, Richmond, VA, 23284-2512, USA, gdhillon@vcu.edu

Claire Gayrel Research Centre on IT and Law (CRID), University of Namur, Namur, Belgium, claire.gayrel@fundp.ac.be

Jacques Gérard Research Centre on IT and Law (CRID), University of Namur, Namur, Belgium, jacques.gerard@fundp.ac.be

Fosca Giannotti ISTI-CNR, Pisa, Italy, fosca.giannotti@isti.cnr.it

Serge Gutwirth Center for Law, Science, Technology and Society Studies (LSTS), Vrije Universiteit Brussel, Pleinlaan 2, 1050, Brussels, Belgium, serge.gutwirth@vub.ac.be

Wannes Heirman Department of Communication Studies, Research Group MIOS, University of Antwerp, Antwerpen, Belgium, Wannes.Heirman@ua.ac.be

Ella Kolkowska Orebro University, Orebro, Sweden, ella.kolkowska@oru.se

Laks V.S. Lakshmanan University of British Columbia, Vancouver, BC, Canada, laks@cs.ubc.ca

Ronald Leenes Tilburg Institute for Law, Technology and Society (TILT), Tilburg University, Tilburg, The Netherlands, R.E.Leenes@uvt.nl

Raffaele Lettiero Dipartimento di Ingegneria dell'Informazione, Seconda Università di Napoli, Naples, Italy, lettiero@gmail.com

Fadhila Mazanderani Oxford Internet Institute, The University of Oxford, Oxford, UK, mazanderani@gmail.com

Jean-Philippe Moiny Research Centre on IT and Law (CRID), University of Namur, Namur, Belgium, jean-philippe.moiny@fundp.ac.be

Anna Monreale University of Pisa, Pisa, Italy, annam@di.unipi.it

Dino Pedreschi University of Pisa, Pisa, Italy, pedre@di.unipi.it

Wolter Pieters University of Twente, Enschede, The Netherlands, w.pieters@utwente.nl

Yves Poullet Research Centre on IT and Law (CRID), University of Namur, Namur, Belgium, yves.poullet@fundp.ac.be

Nadezhda Purtova 2513BS Torenstraat 12, den Haag, The Netherlands, npurtova@yahoo.com

Massimiliano Rak Dipartimento di Ingegneria dell'Informazione, Seconda Università di Napoli, Naples, Italy, massimiliano.rak@unina2.it

Denis J. Roio Dyne.org Foundation, Nederlands Instituut voor Mediakunst, Keizersgracht 264, 1016 EV Amsterdam, The Netherlands, jaromil@dyne.org

Joep Ruiter Faculty of Sciences, VU University Amsterdam, The Netherlands, jrr260@few.vu.nl

Bibi van den Berg Tilburg Institute for Law, Technology and Society (TILT), Tilburg University, Tilburg, The Netherlands, bibi.vandenberg@uvt.nl

Jean-Marc Van Gyseghem Research Centre on IT and Law (CRID), University of Namur, Namur, Belgium, jean-marc.vangyseghem@fundp.ac.be

Umberto Villano Dipartimento di Ingegneria, Università del Sannio, Benevento, Italy, villano@unisannio.it

Michel Walrave Department of Communication Studies, Research Group MIOS, University of Antwerp, Sint Jacobstraat 2, BE 2000, Antwerpen, Belgium, michel.walrave@ua.ac.be

Hui (Wendy) Wang Stevens Institute of Technology, Hoboken, NJ, USA, hwang@cs.stevens.edu

Martijn Warnier Faculty of Technology, Policy and Management Delft University of Technology, M.E.Warnier@tudelft.nl

Arnd Weber Institute for Technology Assessment and Systems Analysis, Karlsruhe Institute of Technology, Hermann-von-Helmholtz-Platz 1, D-76344 Eggenstein-Leopoldshafen, Germany, arnd.weber@kit.edu

Dirk Weber Institute for Technology Assessment and Systems Analysis, Karlsruhe Institute of Technology, Hermann-von-Helmholtz-Platz 1, D-76344 Eggenstein-Leopoldshafen, Germany, daw@dawitcon.de

About the Authors

Norberto Nuno Gomes de Andrade is a legal researcher and PhD candidate at the Law Department of the European University Institute (EUI, Italy). He graduated in Law at the Faculty of Law of the University of Lisbon, and he holds a Master of Arts in International Relations and European Studies from Central European University (CEU, Hungary), as well as a Master of Research from the European University Institute. He has previously worked as a legal expert at the External Relations Department of the Portuguese Regulatory Authority for Communications (ANACOM, Portugal). His research interests are focused on law and technology (including biotechnology, neuroscience, artificial intelligence, genetics and genomics, digital environments, ambient intelligence), data protection and privacy law, intellectual property, philosophy of law and legal theory. In 2009 he co-edited and published "Law and Technology: Looking into the Future – Selected Essays".

Christel Backman (LL.M, BSc in Sociology) is currently completing her doctoral thesis 'Criminal Records: Law and Practice in Sweden' at the Department of Sociology, University of Gothenburg. The thesis examines the use of criminal background checks in hiring and the regulation of employer access to criminal history data from a socio-legal perspective. Christel Backman is a member of the Management Committee of COST Action IS0807 Living in Surveillance Society, and coordinator of Working Group 4 'Public Policy and the Regulation of Surveillance' in the same Action.

Rocco Bellanova holds a Research Master in political science and international relations (IEP-Paris) and graduated in international relations (University of Bologna). He has been visiting student at the University of Montreal and at the IEP in Lyon. After a trainee-experience at the European Parliament in the field of Justice and Home Affairs, he is currently doing his PhD on data protection applied to security and surveillance measures at both the Facultés universitaires Saint-Louis (FUSL) and the Vrije Universiteit Brussel (VUB). At FUSL, he also works as assistant in international relations, political science and contemporary political issues ("liberties and securities"). He collaborates in the organization of the Computers, Privacy and Data Protection (CPDP) conferences (2009, 2010 & 2011). In line with his research interests, Rocco is focusing on the powers and politics of data protection; as well as on the evolutions and intertwining of surveillance and security.

Paul Bernal's background is very varied. His initial degree was in mathematics at the University of Cambridge, before he trained and worked as a Chartered Accountant with Coopers & Lybrand (now PWC), qualifying in 1990. His subsequent work included working for Reuters, doing pioneering work in the internet in the early 1990s, including setting up and running the UK's first online real-time education system for children. In the 2000s, Paul was the financed director at a leading charity dealing with mental health and criminal justice, before returning to academia, completing a Masters in Human Rights at the LSE. He is currently in the process of completing his PhD at the LSE, under Professor Conor Gearty and Mr Andrew Murray, working in the field of human rights on the Internet, funded by a Doctoral Grant from the Arts and Humanities Research Council.

Amos Bianchi (1975) is coordinator of postgraduate programmes at NABA (Nuova Accademia di Belle Arti Milano, www.naba.it). At the same institution, he has been Assistant Professor of Theory and Method of Mass Media, and he is currently Lecturer of Research Methods. He is a Ph.D. researcher of the University of Plymouth, Planetary Collegium, at the M-Node, based in Milan.

Jean-François Blanchette is an Assistant Professor in Department of Information Studies, University of California, Los Angeles. He received a B.Sc. and a M.Sc. in Computer Science from the Université de Montréal in 1995 and 1997, and a Ph.D. in Social Studies of Science from Rensselaer Polytechnic Institute in 2002. Between 1999 and 2001, he was an invited researcher at the CNRS in Paris, where he investigated the definition of a new legal framework for recognizing the evidential value of electronic documents, including those produced by notaries, baillifs and judges. Between 2002 and 2004, he was a Post-Doctoral Fellow with the InterPARES project at the University of British Columbia. Professor Blanchette teaches in the area of electronic records management, digital preservation, social dimensions of computing, and systems design. His current research focuses on developing a theoretical framework for analysing the materiality of computing, and its implications for long-term preservation of digital objects.

Franziska Boehm is a research assistant at the University of Luxembourg where she is also preparing her PhD thesis in European data protection law. After the *Licence en Droit* in 2003 (University of Nice, France) and the German state exam in law in 2006, she specialized in European data protection law and obtained a Master in this field in 2007 (University of Gießen, Germany). Her research focuses on the data protection rights of individuals, in particular in a law enforcement context.

John J. Borking (1945), (Dutch Laws - University of Leyden 1973) (Dutch Nationality) is Director of Borking Consultancy, an one man SME consultancy firm on privacy protection and (e)-ADR. He was from January 1, 1994 till January 1,

2006 Privacy Commissioner and Board member and associate board member of the Dutch Data Protection Authority (CBP) in The Hague (The Netherlands) (www.cbpweb.nl), He is (e-) arbitrator / mediator of the Dutch Foundation for Alternative Dispute Resolution for ICT (SGOA) (www.sgoa.org). He is also since 2000 board member of the Netherlands Lottery and Gaming Control Board (www.toezichtkansspelen.nl/) is appointed as an expert on new IC technologies for the lotteries and gaming market and interactive gaming (casinos on-line). Since 1999 he is member and chair of the working party on data protection of CEN /ISSS WP/DPP in Brussels (www.cenorm.be.) He has been general manager of COSSO, the Dutch Trade Association for Information & Communication Technology Providers, Supervisory board member of Børsen Netherlands B.V (a publisher of newspapers and management books) and senior legal counsel and company secretary for Xerox Corp. in The Netherlands, France and UK (1974–1986).

Ian Brown is a senior research fellow at the Oxford Internet Institute, the University of Oxford. Dr. Brown's research is focused on public policy issues around information and the Internet, particularly privacy and copyright. He also works in the more technical fields of communications security and healthcare informatics. Since 1998 Dr Brown has variously been a trustee of Privacy International, the Open Rights Group and the Foundation for Information Policy Research and an adviser to Greenpeace, the Refugee Children's Consortium, Amnesty International and Creative Commons UK. He is a Fellow of the Royal Society of Arts, the International University of Japan and the British Computer Society, and a senior member of the ACM. He is currently the Principal Investigator of the Privacy Value Networks and Towards a Future Internet projects.

Johann Čas holds a degree in Communications Engineering from a Higher Technical College and a degree in economics from the University of Graz. He is a researcher at the Institute of Technology Assessment of the Austrian Academy of Sciences since 1988. He has worked on several aspects of the Information Society and on societal impacts of Information and Communication Technologies. Past foci of research include technological development programs, impact on employment and regional development, information systems for policy makers, regulatory issues of new telecommunication technologies and privacy. He has also been giving lectures on technology assessment at technical universities. His current research focus is on data protection and privacy in the information society, privacy enhancing technologies and their deployment within Ambient Intelligence, security technologies and health related applications. He was the coordinator of the PRISE EU-Project (http://prise.oeaw.ac.at/), which applied a participatory research approach to develop guidelines and criteria for privacy enhancing security technologies.

Valentina Casola is an assistant professor at the University of Naples Federico II. She received the Laurea degree in Electronic Engineering cum laude from the

University of Naples in 2001 and she received her PhD in Electronic and Computer Engineering from the Second University of Naples in 2004. Her research activities are both theoretical and experimental and are focused on security methodologies to design and evaluate distributed and secure infrastructures.

Paul De Hert is an international human rights expert. The bulk of his work is devoted, but not limited, to criminal law and technology & privacy law. At Brussels, Paul De Hert holds the chair of "Criminal Law", "International and European Criminal Law" and "Historical introduction to eight major constitutional systems". In the past he has held the chair of "Human Rights", "Legal theory" and "Constitutional criminal law". He is Director of the VUB-*Research group on Fundamental Rights and Constitutionalism* (FRC), Director of the Department of Interdisciplinary Studies of Law (Metajuridics) and core member of the internationally well-accepted VUB-*Research group Law Science Technology & Society* (LSTS) (see: www.vub.ac.be/LSTS). At Tilburg he holds a position as an associated-professor in the internationally renowned Institute of Law and Technology at the Tilburg University (http://www.tilburguniversity.nl/faculties/frw/departments/tilt/profile/). He is member of the editorial boards of several national and international scientific journals such as the *Inter-American and European Human Rights Journal* (Intersentia), *Criminal Law & Philosophy* (Springer). He is co-editor in chief of the *Supranational Criminal Law Series* (Intersentia) and of the New Journal of European Criminal law (Intersentia).

Katja de Vries is a PhD researcher in the interdisciplinary group on Law, Science, Technology and Society (LSTS) at the Vrije Universiteit Brussel (VUB) working on profiling technologies and data protection. She has degrees in Law, Psychology and Philosophy. De Vries has studied at Universiteit Leiden, SciencesPo in Paris and at Oxford University.

Gurpreet Dhillon is Professor of Information Systems at Virginia Commonwealth University, USA and a Guest Professor at ISEG, Universidade Téchnica De Lisboa, Portugal. He holds a Ph.D. from the London School of Economics and Political Science, UK. Gurpreet has authored seven books including *Principles of Information Systems Security: text and cases* (John Wiley, 2007) and over 100 research manuscripts. He is also the Editor-in-Chief of the *Journal of Information System Security*. Gurpreet's research has been featured in various academic and commercial publications and his expert comments have appeared in the *New York Times, USA Today, Business Week, NBC News*, among others. www.dhillon.us.com

Claire Gayrel graduated in European Law and political sciences. She worked a period for the European Regional Development Found (ERDF) in French Guyana, before joining the CRID in September 2008. Her research currently focuses on transborder data flows and the protection of personal data in the Justice, Liberty and Security area of the European Union.

Jacques Gérard received his Degree in Physics from the University of Namur (FUNDP) in 1986 and his Master of Computer Science from the University of Namur in 1988. Since September 1989, he is working at the Research Centre on IT & Law (CRID – Law Faculty of the University of Namur, FUNDP, Belgium) where he developed researches in legal expert systems. He teaches the fundamentals of computing to law students and explains computer concepts to legal researchers of the Centre.

Fosca Giannotti is a senior researcher at the Information Science and Technology Institute of the National Research Council at Pisa, Italy, where she leads the Knowledge Discovery and Data Mining Laboratory – KDD LAB – a joint research initiative with the University of Pisa, one of the earliest European research groups specifically targeted at data mining and knowledge discovery. Her recent research interests include data mining query languages, mining spatio-temporal and mobility data, privacy preserving data mining, and complex network analysis. She has been the coordinator of various European-wide research projects, including GeoPKDD: Geographic Privacy-aware Knowledge Discovery and Delivery. She is the author of more than one hundred publications and served as PC chair and PC member in the main conferences on Databases and Data Mining. She is the co-editor of the book "Mobility, Data Mining and Privacy", Springer, 2008.

Serge Gutwirth is a professor of human rights, legal theory, comparative law and legal research at the Faculty of Law and Criminology of the Vrije Universiteit Brussel.(VUB), where he studied law, criminology and also obtained a postgraduate degree in technology and science studies. Gutwirth founded and still chairs the VUB-research group *Law Science Technology & Society* (http://www.vub.ac.be/LSTS). He publishes widely in Dutch French and English. Amongst his recent co-edited publications are *Safeguards in a world of ambient intelligence* (Springer, 2008), *Profiling the European citizen* (Springer 2008), *Reinventing data protection?* (Springer 2009) and *Data protection in a profiled world* (Springer, 2010). Currently, Serge Gutwirth is particularly interested both in technical legal issues raised by technology (particularly in the field of data protection and privacy) and in more generic issues related to the articulation of law, sciences, technologies and societies.

Wannes Heirman is teaching assistant at the department of Communication Studies of the University of Antwerp. After obtaining his master's degree in Communication Sciences (University of Antwerp, 2006), he started doing research work for the research team MIOS. In the period 2008–2009 he worked together with other MIOS-colleagues on a research report concerning cyberbullying. This report was destined for the Belgian Internet Rights Observatory and ultimately resulted in a policy advice approved by the Observatory's members. Besides the topic of cyberbullying, the author has a broad interest for research into how information and communication technologies are impacting the daily lives of young children and teenagers. Currently he is conducting the fieldwork for his PhD project, which aims to examine how young children and teenagers deal with their privacy on the Internet.

Ella Kolkowska is a lecturer and PhD student at Swedish Business School at Orebro University in Sweden. Her research is about social and organizational aspects in information security, value conflicts as well as compliance with information security policies.

Laks V.S. Lakshmanan is a professor of computer science at the University of British Columbia, Vancouver, Canada. His research interests span a wide spectrum of topics in data management and mining, including: relational and object-oriented databases, advanced data models for novel applications, OLAP and data warehousing, database mining, data integration, semi-structured data and XML, large social and other networks, search, and recommender systems. He collaborates widely with both industry and academia the world over. He is a Research Fellow of the BC Advanced Systems Institute. Laks has served on the program committees of all top database and data mining conferences, chaired several, and has edited several special issues of the top journals. The latest of these, a special issue of the VLDB Journal on Data Management and Mining Issues on Social Networks and Social Media, co-edited with Klemens Boehm, will be published in Fall 2010.

Ronald Leenes is professor in Regulation by Technology at TILT, the Tilburg Institute for Law, Technology, and Society (Tilburg University). His primary research interests are privacy and identity management, regulation of, and by, technology. He is also involved in research in ID fraud, biometrics and Online Dispute Resolution. Ronald was work package leader in the EU FP6 PRIME project for socio-cultural aspects of privacy enhanced identity management. He is currently responsible for TILT's contribution to the FP7 project PrimeLife and leads the work package on social networks and collaborative workspaces. He has contributed to and edited various deliverables for the EU FP6 Network of Excellence 'Future of IDentity in the Information Society' (FIDIS).

Raffaele Lettiero is a young researcher at the Second University of Naples. He received the Laurea degree in Computer Science Engineering from the Second University of Naples in 2009. His current research interests are in the area of Grid Computing.

Fadhila Mazanderani is a doctoral student at the Oxford Internet Institute, the University of Oxford, where her research centres on how women living with HIV use the Internet in relation to their health. Before joining the Oxford Internet Institute Fadhila worked as a consultant specialising in telecommunications and new media technologies in South Africa. She has an undergraduate degree in Information Technology from the University of Pretoria and a masters degree in New Media, Information and Society from the London School of Economics and

Political Science. In addition to her doctoral research Fadhila works as project coordinator for the Privacy Value Networks project and is actively involved in volunteer work focussed on orphaned and vulnerable children in South Africa.

Jean-Philippe Moiny obtained his Law Degree from the University of Liège (ULg, Belgium) in 2008, and joined the Research Centre on IT & Law (CRID – Law Faculty of the University of Namur, FUNDP, Belgium) the same year as a junior researcher. He began studying two main research topics: access to and reuse of public sector information (as regards geographical data) and data protection (particularly in the context of Social Network Sites [SNSes]). Since February 2009, he is effective member of the Belgian Appeal Committee for the reuse of public sector information. And since October 2009, he is also research fellow for the F.R.S.-FNRS, in the CRID, pursuing a Ph.D. research related data protection and its interactions with other legal fields in the context of cloud-based SNSes.

Anna Monreale is a is a PhD student at Computer Science Department of the University of Pisa and a member of Knowledge Discovery and Data Mining Laboratory – KDD LAB – a joint research group with the ISTI – CNR of Pisa. Her research is in anonymity of complex forms of data, including sequences, trajectories of moving objects and complex networks, and privacy-preserving outsourcing of analytical tasks.

Dino Pedreschi is a full professor of Computer Science at the University of Pisa. He has been a visiting scholar at the University of Texas at Austin (1989/90), at CWI Amsterdam (1993) and at UCLA (1995). His current research interests are in data mining and logic in databases, and particularly in data analysis, in spatio-temporal data mining, and in privacy-preserving data mining. He is a member of the program committee of the main international conferences on data mining and knowledge discovery and an associate editor of the journal Knowledge and Information Systems. He served as the coordinator of the undergraduate studies in Computer Science at the University of Pisa, and as a vice-rector of the same university, with responsibility in teaching affairs. He has been granted a Google Research Award (2009) for his research on privacy-preserving data mining and anonymity-preserving data publishing. He is the co-editor of the book "Mobility, Data Mining and Privacy", Springer, 2008.

Wolter Pieters (1978) studied computer science and philosophy of science, technology and society at the University of Twente. From 2003 to 2007, he did his PhD research at the Radboud University Nijmegen, resulting in his interdisciplinary thesis "La volonté machinale: understanding the electronic voting controversy". After finishing his PhD, he worked for the Dutch Ministry of the Interior for one year, on electronic voting and electronic travel documents. Since September 2008 he is employed as a postdoc researcher in the VISPER project at the University of Twente. The project concentrates on de-perimeterisation, the disappearing of traditional boundaries in information security.

Yves Poullet, Ph.D. in Law and graduated in Philosophy, is full professor at the Faculty of Law at the University of Namur (FUNDP) and Liège (Ulg), Belgium. He teaches "Sources and Principles of the Law", "Internet Regulations", "International Commercial Law" and "Human Rights in the Information Society". Yves Poullet heads the CRID, since its creation in 1979. He conducts various researches in the field of new technologies with a special emphasis on privacy issues, individual and public freedom in the Information Society and Internet Governance. He is legal expert with the European Commission, the UNESCO and the Council of Europe. He has been during 12 years (1992–2004) member of the Belgian Commission on Data Protection. In addition, he was since its origin, member of Legal Advisory Board of European Commission. He has received the Franqui Chair in 2004. He also chaired the Belgian Computer Association ABDI (Association Belge de Droit de l'Informatique). Yves Poullet is an active member of the Editorial Board of various famous law reviews.He is a founder of the European Telecommunication Forum, ECLIP and FIRILITE. Recently (2009), he has been nominated as member of the Royal Belgian Academy and as Rector of the University of Namur

Nadezhda Purtova holds a law degree from Mari State University, Russia, an LLM degree from Central European University, Hungary, and an MSc in public administration from Leiden University, The Netherlands. At the time of submitting her contribution she is working towards completing her PhD research at Tilburg Institute for Law, Technology and Society. The goal of the PhD research is to develop a European perspective on property rights in personal data drawing on comparative European property law and privacy and data protection.

Massimiliano Rak is an assistant professor at Second University of Naples. He got his degree in Computer Science Engineering at the University of Naples Federico II in 1999. In November 2002 he obtained a PhD in Electronical Engineering at Second University of Naples. His scientific activity is mainly focused on the analysis and design of High Performance System Architectures and on methodologies and techniques for Distributed Software development.

Denis Roio aka Jaromil (1977) operates in research and development activities at NIMk in Amsterdam (Nederlands Instituut voor Mediakunst, www.nimk.nl). He is a software developer and artist, board member of Dyne.org and of the Free Culture Forum, Ph.D. candidate of the University of Plymouth (Planetary Collegium, M-Node), in 2009 and together with researcher Brian Holmes he was honoured with the Vilém Flusser Award.

Joep Ruiter recently finished the MSc. Information Sciences at the VU University Amsterdam. He will receive his graduation in the near future. Currently Joep participates in a second MSc. in Business Administration until August 2011, also at the VU University Amsterdam.

Bibi van den Berg has a postdoc position in philosophy of technology at the Tilburg Institute for Law, Technology and Society (TILT) of Tilburg University. Her main research interests are : (1) social, ethical and legal issues surrounding autonomous technologies and robotics, and (2) identity and privacy in online worlds.

Jean-Marc Van Gyseghem received his Law Degree from the University of Louvain (UCL, Louvain-la-Neuve, Belgium) in 1995 and his DESS in Medical Law from the University of Poitiers (France) in 1996. Since December 2001, he is working at the Research Center for Computer & Law (CRID – Law Faculty of Namur, FUNDP, Belgium) where he is now Head of the Unit "Liberties in the information society" (Medical Law (including liability), Insurance and Private Law, in Medical Data Protection and in eHealth Services and Products). He is also member of the Bar of Brussels (since 1997) and Partner at the Rawlings Giles law firm in Brussels. He is frequently invited to present communications on medical data protection and on eHealth and he is the author of several papers and book chapters in this domain. He's member of an ethical committee in a Belgian hospital and is deputy manager of the collection "Cahiers du Crid".

Umberto Villano is full professor at the University of Sannio at Benevento, Italy, where he is Director of the Department of Engineering. His major research interests concern performance prediction and analysis of parallel and distributed computer architectures, tools and environments for parallel programming and distributed algorithms. He received the Laurea degree in Electronic Engineering cum laude from the University of Naples in 1983.

Michel Walrave is an Associate Professor and head of the Department of Communication Studies of the University of Antwerp. He leads the research group MIOS (www.ua.ac.be/mios) that conducts research on, amongst others, young people's ICT uses. His research focuses on societal implications of ICT in general, and ICT-use related privacy risks in particular. He has conducted several research projects on, amongst others, e-marketing, social networking sites and privacy and on cyberbullying. He teaches societal implications of ICT, marketing communications and e-marketing at the University of Antwerp and as a guest lecturer in several other universities. He is co-supervisor of the research team of the Belgian Internet Rights Observatory and member of several European research networks. Michel Walrave holds a master's degree in Communication Studies, a Master in Information Science and a PhD in Social Sciences.

Hui Wang received the BS degree in computer science from Wuhan University in 1998, the MS degree in computer science from University of British Columbia in 2002, and the PhD degree in computer science from University of British Columbia in 2007. She has been an assistant professor in the Computer Science Department, Stevens Institute of Technology, since 2008. Her research interests include data management, database security, and data privacy.

Martijn Warnier received his PhD in Computer Science at the Radboud University Nijmegen, the Netherlands. After four years as a Postdoctoral Researcher in computer systems group at the VU University Amsterdam, he moved to Delft University of Technology where he currently holds the position of Assistant Professor.

Arnd Weber is an economist, with a PhD in sociology from the University of Frankfurt, Germany. He has been the project manager of several research projects on IT-related subjects, with the Institute for Technology Assessment and Systems Analysis of Karlsruhe Institute of Technology (ITAS, KIT, formerly Karlsruhe Research Centre), Institut für Sozialforschung (Frankfurt) and the University of Freiburg, Germany. He led the requirements and specifications work in several research projects (CAFE, SEMPER, OpenTC).

Dirk Weber is a certified SAP Technology Associate, Microsoft Certified Systems Engineer and Certified Novell Engineer and has worked as an IT consultant for several companies. He worked for the Institute for Technology Assessment and Systems Analysis of Karlsruhe Institute of Technology (ITAS, KIT, formerly Karlsruhe Research Centre), Germany, and edited several requirements and specifications documents within the OpenTC project.

Part I
Building and Rebuilding Legal Concepts for Privacy and Data Protection

Chapter 1
The German Constitutional Court Judgment on Data Retention: Proportionality Overrides Unlimited Surveillance (Doesn't It?)

Katja de Vries, Rocco Bellanova, Paul De Hert, and Serge Gutwirth

1.1 Introduction

On 15 March 2006, the Data Retention Directive, demanding the retention of telecommunications data for a period of 6 months up to 2 years, was adopted.[1] Since then, this seemingly straightforward directive has "generated" quite an impressive number of court judgments. They range from the European Court of Justice[2] (ECJ)

K. de Vries (✉)
Center for Law, Science, Technology and Society Studies (LSTS), Vrije Universiteit Brussel, Pleinlaan 2, 1050, Brussels, Belgium
e-mail: edevries@vub.ac.be

Earlier versions of this article have been published on the 23rd of March 2010 at the TILT Weblog for Law and Technology (http://vortex.uvt.nl/TILTblog/?p=118) and on the 18th of May 2010 in the CEPS Liberty and Security in Europe – publication series (http://www.ceps.eu/book/proportionality-overrides-unlimited-surveillance). The authors want to thank Patrick Breyer (AK Vorratsdatenspeicherung: German Working Group against Data Retention) and Caspar Bowden (Microsoft) for their salient comments.

[1] Directive 2006/24/EC of the European Parliament and of the Council of 15 March 2006 on the retention of data generated or processed in connection with the provision of publicly available electronic communications services or of public communications networks and amending Directive 2002/58/EC OJ L105, 13.04.2006. Hereinafter: Data Retention Directive. Available at: http://eur-lex.europa.eu/LexUriServ/LexUriServ.do?uri=OJ:L:2006:105:0054:0063:EN:PDF

[2] Judgment of the Court (Grand Chamber) of 10 February 2009 – Ireland v European Parliament, Council of the European Union (Case C-301/06) *(Action for annulment – Directive 2006/24/EC – Retention of data generated or processed in connection with the provision of electronic communications services – Choice of legal basis)*. Available at: http://curia.europa.eu/jurisp/cgi-bin/form.pl?lang=EN&Submit=rechercher&numaff=C-301/06

S. Gutwirth et al. (eds.), *Computers, Privacy and Data Protection: an Element of Choice*, DOI 10.1007/978-94-007-0641-5_1, © Springer Science+Business Media B.V. 2011

to the administrative (e.g. Germany[3] and Bulgaria[4]) and constitutional courts (e.g. Romania[5]) of some Member-States.

In particular, the judgment of the German Federal Constitutional Court,[6] delivered on 2 March 2010, has already caught the attention of several commentators, from civil society, lawyers, journalists and politicians (cf. infra, section 4). In the judgment, the Court annuls the German implementation laws of the Data Retention Directive.

This paper has two main goals. On the one side, it aims at offering a first critical overview of this important judgment, highlighting some of the key features of the ruling and its main similarities and divergences with other similar judgments. On the other side, given the relevance of the issues at stake, it aims at contextualizing the judgment in the wider framework of European data processing and protection debates, assuming a critical posture on the increasing emphasis on proportionality as the "golden criterion" to assess and limit surveillance practices.

1.2 The 2 March 2010 Judgment

1.2.1 Background

In its judgment of 2 March 2010 the German Federal Constitutional Court abrogated the national implementation of the data retention directive: Art.113a and 113b of the *Telekommunikationsgesetz*[7] (TKG), i.e., the Telecommunications Law, and Art. 100g, paragraph 1 sub 1, of the *Strafprozeßordnung*[8] (StPO), i.e., the Criminal Procedural Code, in combination with the aforementioned Art 113a TKG. This legislation, which was originally passed by the Bundestag on 9 November 2007 and entered into force on 1 January 2008, imposed the retention of information about all calls from mobile or landline phones for 6 months, including who called whom, from where and for how long. In 2009, the law was extended to include the data

[3]Administrative Court of Wiesbaden, 27 February 2009, file 6 K 1045/08.WI. See commentary in English: http://www.vorratsdatenspeicherung.de/content/view/301/79/lang,en/

[4]Decision no. 13627, Bulgarian Supreme Administrative Court ('Върховния административен съд'), 11 December 2008. Original text available at: http://www.econ.bg/law86421/enactments/article153902.html. Commentary in English: http://www.edri.org/edri-gram/number6.24/bulgarian-administrative-case-data-retention

[5]Decision no.1258, Romanian Constitutional Court, 8 October 2009. Published in the Romanian Official Monitor, no. 789, 23 November 2009. English translation (unofficial): http://www.legi-internet.ro/fileadmin/editor_folder/pdf/decision-constitutional-court-romania-data-retention.pdf

[6]*Vorratsdatenspeicherung* [Data retention] BVerfG 2 March 2010, 1 BvR 256/08. Available at: http://www.bundesverfassungsgericht.de/entscheidungen/rs20100302_1bvr025608.htmll. Hereinafter: the judgment or the German Court judgment.

[7]Available in German at the "Juristische Informationsdienst": http://dejure.org/gesetze/TKG/113a.htm, and http://dejure.org/gesetze/TKG/113b.html

[8]Available at http://dejure.org/gesetze/StPO/100 g.html, ibid.

surrounding e-mail communications as well. This being said, the law did forbid authorities from retaining the contents of either form of communication.

Since its adoption, the German national implementation law had met considerable resistance. On 31 December 2007 on the eve of its entry into force, the German privacy group *Arbeitskreis Vorratsdatenspeicherung* (AK Vorrat: Working group on data retention) filed a constitutional complaint with the German Federal Constitutional Court. The complaint was backed by more than 30,000 people, and requested, *inter alia*, the immediate suspension of the law.[9] The judgment of 2 March 2010 is the outcome of this complaint.

1.2.2 The Main Findings: A Proportionality Check

The case could have been tricky and threatening for EU law, but the German Court did not criticize the EU directive itself, arguing that the problem lay instead with how the German Parliament chose to interpret it. The German legislation was found to breach art. 10 paragraph 1 of the German Constitution (*Grundgesetz*[10]) which ensures the privacy[11] of correspondence and telecommunications (the so-called "Fernmeldegeheimnis" or "Telekommunikationsgeheimnis"). The text of the German Constitution protects communication in what might be termed an old-fashioned way. Article 10 of the German Basic Law seems to suggest that we still communicate by writing letters, but through the activity of the Court the protection goes well beyond the paper medium. All forms of (tele)communications are in fact protected, and this protection does not only cover the content of the communication, but it reaches also out to the data about this communication.[12] In the judgment of 2 March 2010, the Court stated that: "the protection of communication does not include only the content but also the secrecy of the circumstances of the communication, including especially if, when and how many times some person (. . .) contacted another or attempted to." (section 189)

Hence, the German Constitution also applies to the data that are the object of the retention measures. But this does not necessarily mean that the implementation law is unconstitutional. So how did the German Court come to the conclusion that

---

[9] Available in English at the website of the "Arbeitskreis Vorratsdatenspeicherung": http://www.vorratsdatenspeicherung.de/content/view/184/79/lang,en/

[10] Available in German at the website of the German Bundestag: http://www.bundestag.de/dokumente/rechtsgrundlagen/grundgesetz/gg_01.html

[11] Privacy is not mentioned in the German Constitution, but the German Court has developed a broad right to privacy and "informational self-determination" ("das Recht auf informationelle Selbstbestimmung") as tenets of the right to human dignity in Article 1 of the Constitution in its famous 1983 "Census Decision". BVerfG [Judgments of the Federal Constitutional Court] 15 December 1983, (*Volkszählung*), *BVerfGE* 65, 1. The plaintiffs in the German data retention case also claimed that the national implementation laws infringed both their right to informational self-determination and their privacy of telecommunication (art 10 GG), but the annulment of the Court was only based on the infringement upon the latter.

[12] This is indeed fully in line with the case law of the Strasbourg Court: ECrtHR, *Malone vs. UK*, 2 August 1984

the implementation law, doing no more than implementing EU legislation, breaches Article 10 of the Constitution?

As also remarked by Mohini, the Court bases its analysis on a "privacy test" similar to the one developed by the European Court of Human Rights.[13] From Strasbourg's point of view, the "privacy test" as contained in the second paragraph of Article 8 of the European Convention on Human Rights (ECHR),[14] not only requires a check of the quality of the legal basis,[15] but also of the legitimate aim and proportionality of the proposed initiative. We will see in the following that the German Court follows this scheme and carries out a check of the three requirements. It is however useful to observe that the European Court sees minimum safeguards with regard to data (e.g. safeguards on duration; storage conditions; usage, access by third parties and preserving the integrity of data) as being part of the first requirement (legality requirement),[16] whereas the German Court sees these safeguards as elements of the third requirement (proportionality). We will come back to this. Now let us turn to the privacy check by the German Court in the judgment of 2 March 2010.

As all the transposition laws were made with the proverbial German accuracy, the first requirement (legality) was not the problem. With regard to the second (legitimacy) the German Court found that a 6-month retention period can be legitimate

[13] Mohini, (2010), 'On the BVG ruling on Data Retention: "So lange" – here it goes again…', 13 April, available at http://afsj.wordpress.com/2010/03/05/so-lange-here-it-goes-again/.

[14] Article 8 of the European Convention on Human Rights states: "Everyone has the right to respect for his private and family life, his home and his correspondence" (first paragraph); "There shall be no interference by a public authority with the exercise of this right except such as is in accordance with the law and is necessary in a democratic society in the interests of national security, public safety or the economic well-being of the country, for the prevention of disorder or crime, for the protection of health or morals, or for the protection of the rights and freedoms of others" (second paragraph).

[15] The legality principle is expressly laid down in Articles 2, 5, 6 and in the second paragraphs of Articles 8 to 11. Interferences by the executive with the rights and freedoms of the individual should not be permitted unless there is a clear legal basis to do so. By the same token, individuals should be able to predict with reasonable certainty when and under what conditions such interferences may occur. Hence the need for a legal basis to be accessible and foreseeable are key features of the first requirement of the privacy check.

[16] The Court recalls in its well established case-law that the wording "in accordance with the law" requires the impugned measure both to have some basis in domestic law and to be compatible with the rule of law, which is expressly mentioned in the preamble to the Convention and inherent in the object and purpose of Article 8. The law must thus be adequately accessible and foreseeable, that is, formulated with sufficient precision to enable the individual – if need be with appropriate advice – to regulate his conduct. For domestic law to meet these requirements, it must afford adequate legal protection against arbitrariness and accordingly indicate with sufficient clarity the scope of discretion conferred on the competent authorities and the manner of its exercise" (ECtHR, *Case of S. and Marper versus the United Kingdom*, Application nos. 30562/04 and 30566/04, Strasbourg, 4 December 2008, § 95 with ref. to ECtHR *Malone v. the United Kingdom*, 2 August 1984, *Series A* no. 82, §§ 66–68; ECtHR *Rotaru v. Romania* [GC], no. 28341/95, *ECHR* 2000-V, § 55; and ECtHR, *Amann v. Switzerland* [GC], no. 27798/95, *ECHR* 2000-II, § 56).

in principle: firstly because under the current laws the data are stored in a dispersed manner by private actors (section 214). Secondly, such data retention is in accordance with the challenges posed by the current era:

> Storage of telecommunication traffic data for the period of six months is also not a measure which aims at the complete interception ("eine Totalerfassung") of the communication and activities of citizens as a whole. Much more it ties in, in a rather restrained manner, to the special significance of telecommunications in the modern world and it reacts to the specific potential danger which it brings along. The new means of telecommunication overcome time and space in a way which is incomparable to other forms of communication and basically exclude public observation. Thus these new means make it easier for criminals to communicate and act in a hidden way and enables dispersed groups of a few persons to find each other and effectively collaborate with each other [...] Thus precisely the reconstruction of connections by means of telecommunication is of special significance for effective criminal prosecution and the prevention of dangers. (section 216)

However, while the Court holds that the current legislation is in principle legitimate and not contrary to the German Constitution, it also notes that this would not be the case for more all-encompassing and intrusive legislation:

> In contrast the retention of telecommunication traffic data should not be understood as a stepping stone towards a legislation which aims at a potentially blanket measure of preventive data retention which stores all data which could be useful for the prosecution of crime or prevention of dangers. Such legislation would be, irrespective of the regulations concerning its usage, a priori incompatible with the Constitution. (section 218)

Yet, even though the Court deemed the contested national data retention laws not to be unconstitutional in principle, it did acknowledge that these measures do constitute a heavy infringement ("schwerwiegender Eingriff", section 212). Such measures:

> "largely increase the risk of citizens to be the subject of further investigations, although they did not do anything wrong. It is enough to be at a wrong time (...) contacted by a certain person (...) to be under an obligation to provide justifications", [and, further in the judgment, that the preventive collection of data] "can establish a feeling of permanent control" [and] "diffuse threat" ("diffuse Bedrohlichkeit"). (sections 212 and 242)

Because of the heavy infringements that such data retention can bring along, the major problem with the German implementation laws was that they did not satisfy the third requirement of proportionality. This requirement involves, at least in the German Court's understanding, that the transposition laws should contain regulations that are in accordance with all requirements of *legality* ("normenklare Regelungen"). Thus, contrary to Strasbourg's interpretation, proportionality is not directly discussed as part of the more sensitive criterion of *necessary in a democratic society*. Following the proportionality check the Court concludes that, while the idea underlying data retention is not "absolutely incompatible with Art.10 of the German Constitution (protecting the privacy of telecommunications)" (section 205), its application in national law did not meet the constitutional need for proportionality, which can be subdivided into four criteria:

(i) proportional data security standards. Given that data retention is a very heavy infringement, the threshold for those standards should be set very high;
(ii) proportional purpose limitation. When direct use of data is sought, and thus the possibility to create very detailed behavioral profiles is at stake, these standards should be very high (only in case of "schwerwiegende Straftaten", i.e. heavy crimes). However, the Court assesses indirect use (as in the case of requests to a service provider for the identifying information that belongs to an IP-address) as a less intrusive practice, and thus the standards concerning purpose limitation can be more lenient (no need for an exhaustive catalogue);
(c) transparency. This criterion aims at counter-acting the feeling of "diffuse threat" (discussed in section 242): using data without knowledge of the involved should only be allowed if the purpose of the investigation would become jeopardized otherwise, and if the involved people are at least notified afterwards. This criterion applies both to direct and indirect use of the data;
(d) judicial control and effective legal remedies. Proportionality requires that in case of direct use there should be judicial control, while in the case of indirect use this is not necessary.

None of these requirements were met. Seven out of eight judges (section 308) therefore agreed that the national transposition laws infringed upon art. 10, paragraph 1, of the German Constitution. After suspending the law several times during interim proceedings, the Court annulled[17] it in its final judgment. All data already collected by carriers and providers had to be deleted.

1.2.3 The German Court on Access and Use and the Role of Private Companies

According to the Constitutional Court it is important to distinguish between the mere retention and the actual access and use of data. In practice this difference is expressed by the fact that the data are not directly accessible as they are stored by a multiplicity of private companies (telecommunications services and providers).

---

[17] Judge Schluckebier wrote an extensive dissenting opinion in which he argues that the retention of mere location and traffic data, particularly when executed by private companies and not by the state itself, does not infringe upon art. 10,1, GG. According to Schluckebier data retention cannot be compared to truly intrusive infringements such as the acoustic surveillance of private premises or remote searches of information technical systems (section 314). Moreover he points at the need for judicial self-restraint in order to give the legislator more room to create regulations which it deems necessary. However, while the majority of the judges agreed that the transposition laws infringed upon the German Constitution, the question whether the law should be declared nullified (which implied that all stored data had to be erased immediately) or whether the legislator should get the opportunity to adapt the laws during a set period of time in which the data would be kept, was a harder question: with four out of eight judges in favor of the latter (section 309), it was a really close call that the transposition laws were completely nullified.

Although the complaints concerning the excessive economic burden of data retention on these companies were not accepted, their remarkable consolation prize was that the court assigned them the constitutionally pertinent and important role of incorporators of the distinction between storage and access. The private and dispersed nature of the collection and retention of data was thus welcomed by the German Federal Constitutional Court as something very positive. The fact that the obligation to retain data rests with private service-providers even became a "decisive element" for the assessment of the "non-unconstitutionality" of the principle of data retention. In fact, "when the data are stored, they are not gathered in one place, but they are scattered over many private companies and thus they are not at the State's disposal as a total collection. More importantly the State does not have (...) direct access to the data" (section 214 of the judgment).

Thus, while clearly stating that "the retention of telecommunication traffic data should not be understood as a step towards a legislation that aims at a potentially blanket measure of preventive data retention" (section 218), the Constitutional Court seems to identify a fundamental guarantee in the two-step procedure: a general but dispersed retention by private actors followed by a justified direct or indirect use by public actors. However, following up on the judgment of the German Federal Constitutional Court, the German Federal Commissioner for Data Protection, Peter Schaar, said in an interview with the *Focus* magazine that the data retention practised by private companies such as Google and Facebook should also be limited: "After all, private data collections of large companies, such as Google, are much more precise, extensive and more meaningful than that what is captured by a retention that was ordered by a state".[18] This raises not only the question of how large private actors can be without endangering the dispersed character of the retention, but also of the relativity of the notion of "dispersion" given the existence and availability of powerful data mining and aggregative software tools.

Another important elaboration by the German Federal Constitutional Court with regards to the use of the retained data is the distinction between "direct" and "indirect" use of data by law enforcement authorities and secret services. On the one hand, direct use is particularly sensitive and needs stronger safeguards, because it can lead to the construction of behavioural and mobility profiles. In particular, stricter rules have to apply to secret services. On the other hand, indirect use, namely the possibility for officials to request of service providers that they inform them of the holders of connections with specific IP addresses, requires "less strict guidelines". Because the Court deems the indirect use of data to be a relatively light infringement, the purpose limitation for such requests is proportionally light: "the production of such requests for information is independent of an exhaustive catalogue of legal interests or criminal offences, and can be allowed more widely than the request and the use of telecommunication traffic data themselves." (section 254)

[18]Online Focus (2010, 06.03.2010). Bundesdatenschutzbeauftragter: Google, Facebook & Co. Reglementieren. *Online Focus*, from http://www.focus.de/digital/internet/bundesdatenschutzbeauftragter-google-facebook-und-co-reglementieren_aid_487099.html

1.2.4 Other Important Findings

As widely discussed by journalists, the German Federal Constitutional Court stresses that what should be prevented at all costs is the creation of an opaque, blanket and centralised data retention that can engender a "feeling of unease" with the citizens. In the words of the Court:

> a preventive general retention of all telecommunications traffic data (...) is, among other reasons, also to be considered as such a heavy infringement because it can evoke a sense of being watched permanently (...). The individual does not know which state official knows what about him or her, but the individual does know that it is very possible that the official does know a lot, possibly also highly intimate matters about him or her (section 241).

This is why such a "diffuse threat" should be "counteract[ed] (...) by effective rules of transparency" (section 242). The Court's posture on "unease" is quite a strong official acknowledgment of the potential perverse effects of wide, even if soft, surveillance measures on individuals' lives.[19]

The Constitutional Court also underlines (section 238) that "as a product of the principle of proportionality" there has to be "a fundamental prohibition of transmission of data, at least for a narrowly defined group of telecommunications connections which rely on particular confidentiality".[20] The Court continues that these "might include, for example, connections to persons, authorities and organisations in the social or ecclesiastical fields which offer advice in situations of emotional or social need, completely or predominantly by telephone, to callers who normally remain anonymous, where these organisations themselves or their staff are subject to other obligations of confidentiality in this respect".

Notwithstanding the attempt of the German Constitutional Court to keep national and EC matters separate from each other (cf. infra, section 3.1), the judgment also provides some reflections that can give food for thought on the EC level. In particular this is the case with regard to the question of whether location and traffic data that have to be stored according to the Data Retention Directive (2006/24/EC) should be considered personal data as defined in Art. 2(a) of the Data Protection Directive 95/46/EC:

> "personal data" shall mean any information relating to an identified or identifiable natural person ("data subject"); an identifiable person is one who can be identified, directly or indirectly, in particular by reference to an identification number or to one or more factors specific to his physical, physiological, mental, economic, cultural or social identity.

Although the German Federal Constitutional Court does not make any explicit reference to the notion of personal data in the Data Retention Directive, it recognises that location and traffic data also deserve protection, because technologies

---

[19]For a critical overview of the shift towards a "soft surveillance" approach in law-enforcement, cf. Marx, G.T. "Soft Surveillance. The Growth of Mandatory Volunteerism in Collecting Personal Information – 'Hey Buddy Can You Spare a DNA'?." In *Surveillance and Security. Technological Politics and Power in Everyday Life,* edited by T. Monahan. New York London: Routledge.

[20]Press release in English: http://www.bverfg.de/pressemitteilungen/bvg10-011en.html

can extract from their processing important, and sometimes even sensitive, personal data. Because the Court was reluctant to pose a preliminary question to the ECJ and underlined the importance of Germany's constitutional identity, it also let the opportunity pass to take a stance with regard to how its judgment relates to similarly important questions within the EU directive. Even though it is understandable that the court did not want to get its fingers burned, it would have been interesting if the Court had taken the debates on the European level into consideration more explicitly. Thus, for instance, it could have been interesting if the Court would have taken into account the Working Party (WP) 29 Opinion (2007) on the definition of personal data. In this, not uncontested, opinion[21] the Working Party stated that dynamic IP addresses should be treated as personal data, unless the ISP can establish with "absolute certainty that the data correspond to users that cannot be identified": but in practice this is almost impossible to ascertain. Also, the Court did not take into account Directive 2002/58/EC, the so-called e-Privacy Directive, that provides for a distinctive protection of traffic and location data. The rationale of this protection is that these data can threaten privacy even if they are not personal data (which implies that "privacy" and "data protection" cannot be reduced one to the other, although they do surely overlap).[22]

1.3 The German Constitutional Court Judgment and Europe

1.3.1 Fundamental Rights and Data Retention

In order to get to the "core of the problem" the plaintiffs who addressed themselves to the German Federal Constitutional Court had hoped that the Court would pose a preliminary question about the constitutionality of the Data Retention Directive to the ECJ. However, the Constitutional Court did not deem such a preliminary question necessary. The questions we want to consider here are the following: When is the constitutionality of the data retention legislation part of the jurisdiction of the German Federal Constitutional Court and when is it part of the powers of ECJ? And what is the difference between mere retention and actual access to the data?

The German Court has on several occasions shown a reluctance to accept an unconditional and full supremacy of EC law. In the Solange II case[23] it famously stated that "as long as" ("so lange") the EC "ensured an effective protection of fundamental rights" that were "substantially similar" to that of the fundamental

---

[21] Article 29, Data Protection Working Party (2007). *Opinion 4/2007 on the concept of personal data*. Brussels. Available at http://ec.europa.eu/justice_home/fsj/privacy/docs/wpdocs/2007/wp-136_en.pdf

[22] Cf. Gonzales-Fuster, G., and S. Gutwirth. "Privacy 2.0 ?," *Revue du droit des Technologies de l'Information*, Doctrine 32 (2008): 349–359.

[23] *Solange II - Wünsche Handelsgesellschaft*, 22 October 1986, BVerfGE 73, 339, 2 BvR 197/83. English translation: *Wünsche Handelsgesellschaft* [1987] 3 CMLR 225.

rights safeguarded by the German Constitution, the German Court would "no longer exercise its jurisdiction to decide on the applicability of secondary Community legislation". Recently, in the complex and controversial Lisbon Judgment, the German Court took an even more outspoken stance and showed its constitutional teeth towards EC law.[24] In this judgment it held that the primacy of Community law could never infringe upon the constitutional identity of the Member-States (identity review, section 240) and should not transgress its competences (*ultra vires* review, section 240).[25] Even though it is difficult to say whether the judgment should be characterised as a triumph of nationalist euroscepticism or of constitutionalism, it has in any case become clear once more that the relationship between EC law and the German Constitutional Court is far from an unequivocal given.

If we keep this in mind, and return to the data retention judgment of 2 March 2010, it is noteworthy to stress how the Federal Constitutional Court avoids referring the case to the ECJ with a preliminary question. In sections 80–83 the Court briefly discusses the European legal context: it gives some bibliographical references to articles that raise doubts about the compatibility of Directive 2006/24 with European fundamental rights and refers to case C-301/06, 10 February 2009. In this case the ECJ rejected the claims that the Directive should be annulled because of its adoption within the first pillar (i.e., Art. 95 EC Treaty) instead of the more appropriate third pillar: according to the ECJ the first pillar is the correct legal basis. The way in which the German Court uses this judgment as an argument to avoid and circumvent a preliminary question to the ECJ is ingenious. After the general observation that Directive 2006/24 only ordains the storage of data for a period of at least 6 months, and does not give any prescriptions regarding the access and use of the data (section 186) it points out that this leaves a large margin of appreciation ("einen weiten Entscheidungsspielraum") to the national legislator. Looking at the ECJ judgment, this large margin of appreciation seems only natural to the German Court: after all, if the Directive has rightly been construed as a first pillar measure its main object is the establishment and functioning of the internal market, whereas its applicability with regards to the detection, investigation, and prosecution of crime has to be considered as the responsibility of individual Member-States. Henceforth, the regulations of the Directive do

> neither harmonise the question of access to data by the competent national law enforcement authorities nor the question of the use and exchange of this data between these authorities (cf. ECJ, C-301/06, 10 February 2009, section 83). Based on the minimal requirements of the Directive (Articles 7 and 13 of Directive 2006/24/EC), the Member States are the ones

---

[24] BVerfG 30 June 2009, 2 BvE 2/08 (*Lisbon*). A preliminary English translation: http://www.bundesverfassungsgericht.de/entscheidungen/es20090630_2bve000208en.html See also: Steinbach, A. "The Lisbon Judgment of the German Federal Constitutional Court – New Guidance on the Limits of European Integration?" *German Law Journal* 11, 4 (2010), 367–390; Lanza, E. "Core of State Sovereignty and Boundaries of European Union's Identity in the Lissabon – Urteil." *German Law Journal* 11, 4 (2010), 399–418.

[25] *Ultra vires* review is a concept that has already been around for a while in the case law of the German Court, the identity review was a new concept which was forwarded in the *Lisbon* judgment.

> who have to take the necessary measures to ensure data security, transparency and legal safeguards (section 186).

Even more telling is section 218 of the 2 March 2010 judgment, wherein the Court refers again to the notion of "constitutional identity" of its own Lisbon Judgment:

> That the free perception of the citizen may not be completely captured and subjected to registration, belongs to the constitutional identity of the Federal Republic of Germany (cf. on the constitutional proviso with regard to identity, Judgment of the second senate, 30 June 2009 - 2 BvE 2/08 etc. -, section 240) and the Federal Republic has to devote itself to guarantee this in a European and international context. By a preventive retention of telecommunications traffic data the room for other blanket data collections, also by means of the European Union, becomes considerably smaller.

Thus, especially when read together, the ECJ judgment of 10 February 2009 and the German judgment of 2 March 2010 seem to indicate the emergence of a very important demarcation within data retention: on the one hand there is the question of the storage and retention of data, which is regulated by Directive 2006/24/EC, and on the other there is the question of the use of and access to these data, which fall under the competency of the individual Member-States. It is striking that the UK Home Office uses the same distinction to brush aside the human rights concerns that the UK implementation law of the Data Retention Directive could lead to a disproportionately large "acquisition of communications data by the police, law enforcement agencies the security and intelligence agencies".[26] According to the Home Office, the critics overlook the difference between mere retention and access: "It is important to state that access to communications data is governed by the Regulation of Investigatory Powers Act 2000 (RIPA) and no changes to the safeguards set out in that Act are planned".[27]

In the judgment of the German Federal Constitutional Court this distinction between retention and access is further elaborated upon by the importance that is assigned to the fact that the retention is carried out by private companies instead of governmental organs and by the introduction of the notions of "direct" and "indirect use" (cf. supra, section 2.3).

1.3.2 Affinities and Differences Among Judgments

As said before, the German judgment is not the first to rule on the topic of data retention.[28] Apart from the ECJ ruling on the legal basis of the directive itself,

---

[26] Home Office (2009). *Government Response to the Public Consultation on the Transposition of Directive 2006/24/EC*. Available at http://www.homeoffice.gov.uk/documents/cons-2008-transposition-dir/cons-2008-transposition-response?view=Binary.

[27] Ibid., p. 27.

[28] The Bulgarian, Romanian and German judgments discussed in this section are not the only constitutional challenges which have been raised against the implementation of the Retention Directive. A decision regarding a constitutional complaint directed towards Hungarian Telecom Data Retention Regulations is still pending before the Hungarian Constitutional Court:

it is important to note that two other important judgments were formulated by the Romanian Constitutional Court,[29] on 8 October 2009, and by the Bulgarian Administrative Court,[30] on 11 December 2008. It is interesting to compare these two judgments, which are relatively concise, with the much more elaborated 2 March 2010 judgment of the German Federal Constitutional Court. Though certain similar elements can be discerned in the three judgments, in the Romanian case the differences are most striking, while in the Bulgarian case a focus on similarities is more enlightening.

First, we will take a closer look at the differences between the German and the Romanian decision. The question that differentiates these judgments is whether, given that there are enough legal and technological safeguards, constitutional data retention could be possible, or whether such an idea is a categorical contradiction in terms. Is "constitutional data retention" as unthinkable as a square circle? Both the German and the Romanian judgments subject the national implementation of Directive 2006/24 to similar tests, which concern the legality, the legitimate purpose, and proportionality of the measures. Yet, the criticisms forwarded by the German Court focus on the *use* and *access* of the data. It does not deem the data *retention* in itself, as required by the Directive, to be necessarily unconstitutional (section 205). On the other hand, the Romanian Court underlines that the *use* of data can be lawful and proportional in certain circumstances:

> the Constitutional Court does not deny [...] that there is an urgent need to ensure adequate and efficient legal tools, compatible with the continuous process of modernization and technical upgrading of the communication means, so that the crime phenomenon can be controlled and fought against. This is why the individual rights cannot be exercised *in absurdum*.

However, while there might be circumstances wherein the *use* may be justified, the Court considers the *blanket retention* of data to be disproportional by nature:

> The Constitutional Court underlines that the justified use, under the conditions regulated by law 298/2008, is not the one that in itself harms in an unacceptable way the exercise of the right to privacy or the freedom of expression, but rather the legal obligation with a continuous character, generally applicable, of data retention. This operation equally addresses all the law subjects, regardless of whether they have committed penal crimes or not or whether

---

http://tasz.hu/en/data-protection/constitutional-complaint-filed-hclu-against-hungarian-telecom-data-retention-regulat. In a similar case (Record No. 2006/3785P) pending before the High Court of Ireland the presiding judge decided on the 5th of May 2010 to refer the case to the ECJ. This means the ECJ will finally have to give a substantive decision on the constitutionality of Directive 2006/24/EC. We will return to this important development later in this paper.

[29] Decision no.1258, Romanian Constitutional Court, 8 October 2009. Published in the Romanian Official Monitor, no. 789, 23 November 2009. English translation (unofficial): http://www.legi-internet.ro/fileadmin/editor_folder/pdf/decision-constitutional-court-romania-data-retention.pdf

[30] Decision no. 13627, Bulgarian Supreme Administrative Court ('Върховния административен съд'), 11 December 2008. Original text available at: http://www.econ.bg/law86421/enactments/article153902.html. Commentary in English: http://www.edri.org/edri-gram/number6.24/bulgarian-administrative-case-data-retention

> they are the subject of a penal investigation or not, which is likely to overturn the presumption of innocence and to transform *a priori* all users of electronic communication services or public communication networks into people susceptible of committing terrorism crimes or other serious crimes.

Contrary to the German Court, the Romanian Court considers the use of the data to be a less radical threat than the blanket storage as such, as only the latter creates a situation where the infringement on "the right to private life and freedom of expression, as well as processing personal data" is no longer the exception but the rule:

> The legal obligation that foresees the continuous retention of personal data transforms though the exception from the principle of effective protection of privacy right and freedom of expression, into an absolute rule. The right appears as being regulated in a negative manner, its positive role losing its prevailing role.

Because the focus of the German Constitutional Court is on access and use, its criticisms are mainly aimed at the national implementation law. Moreover its criticisms are a matter of proportionality. Given the right safeguards not only retention, but also use and access can be constitutional. The Romanian focus, on the other hand, is on data retention as such and therefore the judgment is not only a frontal attack on national law 298/2008, but also on the Directive itself. Clearly, the Court considers ubiquitous and continuous retention for a period of 6 months to be intrinsically in opposition with Art 8 ECHR (right to respect for private and family life). Thus, the Romanian Court takes a particularly strong stance, and states that:

> the obligation to retain the data, established by Law 298/2008, as an exception or a derogation from the principle of personal data protection and their confidentiality, empties, through its nature, length and application domain, the content of this principle.

In Bulgaria the Supreme Administrative Court (judgment of 11 December 2008) annulled Art. 5 of *Regulation # 40 on the categories of data and the procedure under which they would be retained and disclosed by companies providing publicly available electronic communication networks and/or services for the needs of national security and crime investigation*, which partially transposed Directive 2006/EC, for being unconstitutional. Article 5 stated that "the data would be retained by the providers and a directorate within the Ministry of Interior (MoI) would have a direct access via a computer terminal"[31] and specified not only that the MoI would have "passive access through a computer terminal" but also that "security services and other law enforcement bodies" would have access "to all retained data by Internet and mobile communication providers"[32] without needing court permission. The constitutional aversion to centralised storage and direct access without any court control is very similar to the reasoning found in the German judgment. In 2009, the Bulgarian government tried to reintroduce a law that would give direct access to the

---

[31] Access to Information Programme (AIP) Foundation, available at http://www.aip-bg.org/documents/data_retention_231209eng.htm

[32] Digital Civil Rights in Europe, available at http://www.edri.org/edri-gram/number6.24/bulgarian-administrative-case-data-retention

Ministry of Internal Affairs to all data held by the providers, but the law was rejected by Bulgaria's Parliament. On 17 February, Parliament "approved the second reading of amendments to the Electronic Communications Act, but only after serious concessions".[33] One of the concessions made by the Ministry of Interior was that it had to renounce to its

> demand to have permanent, direct access to personal communication data. From now on, mobile phone and internet operators will have to supply requested communication data within 72 h and not, as Interior Minister Tsvetan Tsvetanov wanted, in 2 h. The Interior Minister, or his representative, would have the right to set a different deadline, shorter or longer, in exceptional cases and depending on the severity of the case.[34]

1.4 The Politics "Around" the Judgment of 2 March 2010

1.4.1 The Reactions to the German Judgment

It is noteworthy that the German judgment attracted much more attention than either the Bulgarian or Romanian one. This is probably due to a set of different reasons, among which are: the strong civil society participation behind the plaintiffs, namely 34,000 persons which were mostly mobilised by the *Arbeitskreis Vorratsdatenspeicherung*[35] (Working Group on Data Retention); and the timing of the very extensive and substantial judgment, just in the midst of EU debates on transatlantic data-sharing agreements.

In Germany, the reactions to the judgment came from three types of actors in particular: the privacy group that promoted and supported the complaint; the Federal Criminal Police and the government. It is particularly interesting that in the aftermath of the publication of the Court's decision, several international media focused on the contrast between the respective positions of the Justice Minister and the Interior Minister.[36] On the one side, the Justice Minister, an FDP party member of the opposition at the moment of the adoption of the German legislation and amongst the plaintiffs as a private citizen, publicly welcomed the judgment. On the other side, the Interior Minister, member of the CDU, expressed a thinly veiled criticism, and underlined the need for a quick redrafting of the law to fill the "legislative gap" created by the Court's judgment. A similar posture has been taken by the

---

[33]The Sofia Echo, available at http://sofiaecho.com/2010/02/17/860017_bulgarias-parliament-approves-eavesdropping-act

[34]The Sofia Echo, ibid.

[35]Stoppt die Vorratsdatensspeicherung! [Stop data retention!], available at http://www.vorratsdatenspeicherung.de/content/view/355/55/lang,en/

[36]See, among others: Q. Peel & S. Pignal (2010), "Germany's top court overturns EU data law", *Financial Times*, 2 March, available at http://www.ft.com/cms/s/0/563e0fc8-25f6-11df-b2fc-00144feabdc0.html; and H. Mahony (2010), "German court strikes blow against EU data-retention regime", *euobserver.com*, 3 March, available at http://euobserver.com/9/29595

Federal Criminal Police[37] which not only urged German politicians to come up with new legislation as soon as possible, but also sent out an open letter to Chancellor Angela Merkel wherein it reproaches the German Constitutional Court their naïve outlook.[38]

The reaction of the AK Vorrat deserves particular attention. First, they criticised the reasoning of the Court, and one of their members stated in a press release that:

> [the Court's] decision proclaiming the recording of the entire population's behaviour in the absence of any suspicion compatible with our fundamental rights is unacceptable and opens the gates to a surveillance state.[39]

Then, in the same press release, they already announced a double move: the continuation of the "legal fight" against data retention in Germany to avoid the re-enacting of the implementation law;[40] as well as a sort of "Europeanization" of their fight at the EU level, planning an EU-wide campaign based on the preparation of a European Citizens' Initiative concerning data retention.[41] This double move reflects their focus on the linkage between the national and the European (and even international) level. Indeed, they also invited the German government to refrain from agreeing to a new international agreement on data exchange, and they advised the Justice Minister to liaise at EU and international level with the EU Commissioner of Justice, Fundamental Rights and Citizenship and with the other Member-States that have not yet passed data retention implementation laws, in order to repeal data retention.

Finally, it is noteworthy that the telecom and internet providers, while playing such a crucial role in data retention, have not been a subject of much attention in the reactions of the first commentators. However, according to some news sources,

---

[37]Online Focus (2010, 02.03.2010). BKA will schnell ein neues Gesetz. *Online Focus*, from http://www.focus.de/politik/deutschland/vorratsdatenspeicherung-bka-will-schnell-ein-neues-gesetz_aid_486040.html

[38]Original text of the letter available at http://www.bdk.de/kommentar/artikel/vakuum-bei-der-kriminalitaetsbekaempfung-im-internet-ist-ein-hochrisiko-fuer-die-sicherheit-der-buerger-sonder-sitzung-der-imk-und-jumiko-zur-schadensbegrenzung-unverzichtbar/5920af02d045433601f31c9d0dde1180/?tx_ttnews[year]=2010&tx_ttnews[month]=03

[39]Arbeitskreis Vorratsdatenspeicherung (2010), After data retention ruling: Civil liberties activists call for political end to data retention. Available at http://www.vorratsdatenspeicherung.de/content/view/355/79/lang,en/

[40]Arbeitskreises Vorratsdatenspeicherung (2010). Kampagne: Stoppt die Vorratsdatenspeicherung 2.0! Retrieved 16.04.2010, http://www.vorratsdatenspeicherung.de/static/portal_de.html

[41]AK Vorratsdatenspeicherung is lobbying to get directive 2006/24/EC rejected or at least amended, so that Member-States can opt out of data retention: http://www.vorratsdatenspeicherung.de/content/view/362/79/lang,en/ and http://www.vorratsdatenspeicherung.de/images/antworten_kommission_vds_2009-11-13.pdf In a phone interview held on 30 April 2010, Patrick Breyer of the AK Vorrat told the authors that AK Vorrat was waiting for the adoption of the relevant European Citizens' Initiative legislation to launch their citizens' initiative campaign. The European Commission has already presented a first proposal: European Commission (2010), *Proposal for a regulation of the European Parliament and of the Council on the citizens' initiative*.

both Deutsche Telekom and Vodafone immediately complied with the German Constitutional Court's order to delete all already stored data.[42]

1.4.2 From the EU Perspective

As stated above, the interest and impact of the German judgment at European level are also due to the timing of the decision. Indeed, the judgment arrived in the midst of European and international debates on the next moves in data-sharing and protection, and, in particular, just weeks after the rejection of the so-called "SWIFT agreement by the European Parliament".[43] The judgment brought back emphasis on the issue of the implementation of the data retention directive. In fact, several Member-States have still not implemented the directive or are still in the course of passing the relative implementation law.[44] The slowness of the process is partly due to several and different layers of resistance (national political and juridical debates) and partly due to other less direct reasons (e.g. election schedules). Two months after the decision of the German Court, the High Court of Ireland has finally done what everybody has been hoping for: in its decision of the 5th of May 2010 (Record No. 2006/3785P) it refers the case to the ECJ. This is an important breakthrough because it means getting to the core of the matter, which is the constitutionality of Directive 2006/24/EC itself, rather than the constitutionality of the national implementation legislation.

At present, the most official reaction from the Commission has been the decision to schedule a "Proposal for a review of [the Data Retention] Directive" in the Commission Work Programme 2010.[45] Indeed, the official motivation of this decision states that:

> [f]ollowing an evaluation of the existing Data Retention Directive and recent judgments of MS constitutional courts, a review of the Directive is aimed at better matching data retention obligations with law enforcement needs, protection of personal data (right to privacy) and impacts on the functioning of the internal market (distortions).[46]

---

[42]Die Presse.com (2010, 04.03.2010). Deutsche Telekom vernichtet 19 Terabyte an Vorratsdaten. *Die Presse.com*, from http://diepresse.com/home/techscience/internet/544115/index.do?from=gl.home_tech

[43]Among the main reasons behind the massive rejection of the new "Swift Interim Agreement" were the European Parliament's requests for increased data protection guarantees and further inter-institutional cooperation to ensure proper parliamentary control. See European Parliament website: http://www.europarl.europa.eu/news/expert/background_page/019-68530-032-02-06-902-20100205BKG68527-01-02-2010-2010-false/default_en.htm

[44]In particular, Belgium and Luxembourg have not yet passed the implementation laws.

[45]European Commission (2010), *Communication from the Commission to the European Parliament, the Council, the European Economic and Social Committee and the Committee of the Regions, Commission Work Programme 2010 – Time to act.*

[46]Idem, p. 18 (annex).

In fact, the said evaluation was already planned in the very text of the Data Retention Directive itself.[47] According to the directive, such evaluation is supposed to be released to the public not later than 15 September 2010.[48] It has been already planned in the Action Plan Implementing the Stockholm Programme, which also mentions the possibility, if "necessary", of following the evaluation with a "proposal for revision".[49]

Apart from the issues concerning the future of the Data Retention Directive itself, the German judgment will probably prove to be very important in the numerous debates surrounding data protection and processing. The analysis of the German Constitutional Court judgment takes a position on important issues such as the definition of personal data; the recourse to commercial data for security purposes (and thus the relations with private entities, and the legal framework to adopt); the adoption of technological instruments to limit data use and abuse; the effects of diffuse surveillance on personal and social behaviour, even when surveillance takes the form, or relies, on the "mere" retention of data.

1.5 Provisional Conclusions

Even if it is still completely uncertain what the future will bring, and what will be the effective contribution of the German judgment to the evolution and solution of the current tensions and issues, it is already possible to advance some final considerations. In particular, it seems important to advance a more critical approach to the increasing emphasis on proportionality.

(i) The "proportionality check" approach of the German Constitutional Court confirms the relevance of this bundle of criteria in assessing the acceptability of privacy and data protection derogations for the benefit of security measures. It not only enriches the case-law on privacy and data protection, but also pays specific attention to the technological features of the measures and the need for adequate technological solutions (data security, control against misuse, encryption).

(ii) However, even an enhanced "proportionality test" of this kind does not substitute political and social choices concerning data retention, or data processing

---

[47] Art. 14(1) Data Retention Directive.

[48] A draft version of this document has recently been leaked (https://docs.google.com/fileview?id=0B2Rh7x7YpF3KNTZlNTU0NDAtZjgwMS00YzJkLWFiODktMDQwNTUxMjE3MTcz&hl=en). See also: Karlin Lillington "Leaked report reveals big surge in call data requests", *Irish Times*, 14 May 2010, online available at: http://www.irishtimes.com/newspaper/finance/2010/0514/1224270357547.html

[49] European Commission (2010), *Communication from the Commission to the European Parliament, the Council, the European Economic and Social Committee and the Committee of the Regions. Delivering an area of freedom, security and justice for Europe's citizens, Action Plan Implementing the Stockholm Programme*, p. 30.

for security purposes at large. The reaction of the AK Vorrat, as well as the tensions within the German government, seem to confirm the increasing request for having "politics" back into these debates, and not merely "around" them. The posture taken by the European Parliament in the discussions concerning transatlantic data sharing and processing could be partially read in this sense.

(iii) Moreover, there is no unanimous vision of what "the" proportionality test is, since the methods and criteria do not only vary from jurisdiction to jurisdiction, but also from case to case. The German Federal Constitutional Court, the European Court of Human Rights and the European Court of Justice, to name just these three, have a distinct understanding of what a proportionality test should comprise, and they all seem to apply the test in a strict and in a more lenient way, depending on the case. In his study on the use of the proportionality principle by the European Court on Human Rights, Sébastien van Drooghenbroeck deplores the lack of reflexivity from the side of the judges. There are no leading cases and very little can be distilled about the scope and impact of the requirement.[50] Nothing in European case law comes close to the three-tiered approach to scrutiny developed by the U.S. Supreme Court over recent decades under the Equal Protection Clause (rational basis review, middle-tier scrutiny and strict scrutiny).[51] It is clear that the European Court of Human Rights reaches a similar result through acknowledging to state authorities a "margin of appreciation". This margin and the standard of scrutiny will vary according to the context of the case.[52] However, there is no guidance in case law about this margin. Looking back at the Court's case law on security issues, one *can* observe that the Court is prepared to accept the legitimacy of the fight against crime and terrorism as well as to acknowledge the need to take effective measures. Without going as far as to say that the Court gives full discretion to Member States it is clear that almost always less strict scrutiny of the proportionality requirement is applied, especially when the bulk of the litigation is (only) on privacy, and not on other human rights enshrined in the Convention. This careful approach of sensitive issues by the European judges explains, so we believe, a tendency to concentrate on the first requirement

---

[50] van Drooghenbroeck, S. *La proportionalité dans de le droit de la convention européenne des droits de l'homme*. Brussels: Bruylant, 2001, 777.

[51] Henrard, K. *Mensenrechten vanuit international en nationaal* perspectief. The Hague: Boom, 2007, 258. See also Deverman, B.A. "Fourteenth Amendment - Equal Protection: The Supreme Court's Prohibition of Gender-Based Peremptory Challenges." *Journal of Criminal Law and Criminology* 85 (1995).

[52] On the nature of the Court's review see, e.g., ECtHR, Handyside, *Series A-24*, §§ 49–50 and ECtHR, Olsson, *Series* A-130, §§ 67–69 Relevant factors include the nature of the Convention right in issue, its importance for the individual and the nature of the activities concerned. If the Court finds that one or more of these factors are present, e.g. the right at stake is crucial to individual's effective enjoyment of intimate or key rights, then the state has a narrow margin of action. If they are not the state's action will be assessed against a wider margin of appreciation. See E. Guild, "Global Data Transfers: The Human Rights Implications", *Inex policy brief* no. 9, May 2009, 10p., (http://www.ceps.eu/ceps/download/3400)

(legality) of the privacy check.[53] This explains why the European Court studies the presence of safeguards to avoid abuse of data as elements of the legality requirement, rather than elements of the proportionality requirement, as the German Court in its judgment of 2 March 2010. There might be good reasons for both approaches. Like the German Court, Sébastien van Drooghenbroeck, seems to consider that safeguards against abuse are part of the proportionality requirement, but they are, and this deserves some emphasis, to be considered as the more formal aspects of this requirement. The other half of the requirement of proportionality, the substantive part, consists of balancing the interests at stake.[54] A fixation on the formal requirements of proportionality by the judges, might allow them to avoid the more sensitive, but necessary, substantive proportionality test. A bit of this is lurking in the German judgment and raises the question whether this judgement is really to be understood as a break-through in the European case law.

(iv) The foregoing shows that the existence as such of a proportionality test is not automatically a warrant for a strong protection of human rights and liberties. It all depends on the strictness of the test applied by the judges.[55] Will the judges address the substantive issues of the requirement or will they only concentrate on the formal issues? Even when they do address substantive questions regarding proportionality, it remains to be seen how this is done. A weak proportionality test, consisting of a mere balancing of a fundamental right and another interest – for example: privacy and crime control – does in fact not offer any guarantee for the preservation of that fundamental right, since the approach itself assumes that preserving the one per definition implies weakening the other, and vice versa. It excludes the possibility that both interests can be fostered and protected together. Such a proportionality test is doomed to weigh one interest *against* the other, and makes impossible the search of a *composition* in which the different interests at stake are all preserved in an optimal way. Such criticisms however do not apply to stronger proportionality tests that include the possibility to decide that some measures are inacceptable from a constitutional point of view – an exercise known to the Strasbourg court as the "necessary in a democratic state" test – since they encompass the possibility to refuse a measure because it harms the essence of a fundamental

---

[53] De Hert, P. "Balancing security and liberty within the European human rights framework. A critical reading of the Court's case law in the light of surveillance and criminal law enforcement strategies after 9/11." *Utrecht Law Review* 1, 1 (2005): 68–96 (http://www.utrechtlawreview.org/publish/articles/000005/article.pdf).

[54] van Drooghenbroeck, S. *La proportionalité dans de le droit de la convention européenne des droits de l'homme*. Brussels: Bruylant, 2001, 728.

[55] See on this more in detail: De Hert, P. "Balancing security and liberty within the European human rights framework. A critical reading of the Court's case law in the light of surveillance and criminal law enforcement strategies after 9/11." *Utrecht Law Review* 1, 1 (2005): 68–96 (http://www.utrechtlawreview.org/publish/articles/000005/article.pdf). See on the strict proportionality test in the *Marper* judgment: Guild, E. "Global Data Transfers: The Human Rights Implications" *Inex policy brief* 9 (May 2009): 10., (http://www.ceps.eu/ceps/download/3400)

right or of the constitutional order, even if it can be shown that this measure can effectively realise another legitimate interest. The issue at stake then is not a "balancing" between two values, but an answer to the questions "How much erosion of a fundamental right is compatible with the democratic constitutional state in which fundamental rights are a constitutive element?" or "In which society do we want to live?". Another aspect of a stronger proportionality test is indeed the obligation to explore if there are alternative measures that allow for the realisation of the legitimate interest in a way that does not affect the fundamental rights in the same way as the proposed measure. That is, in other words, answering the question: "Is there a way to protect and enforce both values without loss at the fundamental rights' side ?"

(v) Also noteworthy is the growing interest of national civil liberties groups to articulate their campaign at European level, and take advantage of the capacity to operate on different layers. This seemed to be mainly a prerogative of other actors, and in the field of security measures, of Interior Ministries and, to a certain degree, data protection authorities.[56]

(vi) In the context of a debate already underway on the possible revision of the Data Protection Directive, the German Constitutional Court judgment's concern for traffic and location data is particularly precious. In particular, the decision to assess the level of data protection on the base of data processing technology has to be welcomed. This should offer some guidance when discussing the possible, and most adequate, regulations for "data mining" and other "risk assessment" tools.

(vii) The German Constitutional Court judgment highlights the idea that even "mere" data retention is not a trivial measure, but a measure that has concrete consequences on societies and thus must undergo a severe check. This echoes the Strasbourg Court decision on the so-called Marper case, that criticized the "mere", but not time-limited, retention of personal data of acquitted or discharged people.[57] This posture is particularly important in the face of a continuous shift in the nature of security and surveillance measures, heading towards systems based on the "proactive" or random accumulation of commercial and non-commercial data of a great number of people.[58]

---

[56]Such ability of some Interior Ministries to operate along several layers to shape in a specific way security measures based on data exchange, and foster their adoption at European and international levels, was particularly evident in the case of the Prüm measure, dealing with exchange of DNA, fingerprints and vehicle registration data. For an analysis of the re-shaping of power relations, cf. Bellanova, R. "The 'Prüm Process': The Way Forward for Police Cooperation and Data Exchange?" In *Security vs. Justice? – Police and Judicial Cooperation in the European Union*, edited by E. Guild and F. Geyer, 203–221. Aldershot: Ashgate, 2008.

[57]European Court of Human Rights, *Case of S. and Marper versus the United Kingdom*, Application nos. 30562/04 and 30566/04, Strasbourg, 4 December 2008.

[58]Bellanova, R., and P. De Hert. "Le cas S. et Marper et les données personnelles: l'horloge de la stigmatisation stoppée par un arrêt européen." *Cultures & Conflits* 76 (2009): 101–114; De Beer, D., P. De Hert, G. Gonzalez Fuster, and S. Gutwirth. "Nouveaux éclairages de la notion de la notion de « donnée personnelle » et application audacieuse du critère de proportionnalité. Cour

(viii) Finally, the German Constitutional Court judgment takes an interesting stance on the role of private companies, praising their participation to data retention as an important guarantee against possible excess of state surveillance. However, the role and the responsibilities of private actors in the setting of security measures based on data processing is still far from being clear, or from achieving political consensus. The principle that crime fighting and guaranteeing public security by means of legitimate restrictions of fundamental rights and liberties is the exclusive prerogative of the democratic constitutional state certainly deserves to be reanimated during this debate. Given the aforementioned modifications to the nature of security systems, the issue of the "privatisation" of security and crime-fighting deserves crucial attention.

européenne des droits de l'homme Grande Chambre *S et Marper c. Royaume Uni*, 4 décembre 2008." *Revue Trimestrielle des Droits de l'Homme* 81 (2010): 141–161 and Gonzalez Fuster G., P. De Hert, E. Ellyne and S. Gutwirth (2010) *Huber, Marper and Others: Throwing new light on the shadows of suspicion*, INEX Policy Brief No. 11, 2010 Centre for European Policy Studies (CEPS), 9p. via http://www.ceps.eu/book/huber-marper-and-others-throwing-new-light-shadows-suspicion

Chapter 2
The Noise in the Archive: Oblivion in the Age of Total Recall

Jean-François Blanchette

2.1 Introduction

As the first electronic computers were deployed in the late 1940s to tackle business, scientific, and military problems, the scope of their intellectual capabilities generated intense debate. Was computation the same as thinking? How long before their computational prowess exceed that of the human brain? It is perhaps fitting that 60 years or so into the computing age, the conversation is shifting to concerns over the status of computers' memories. How well do they remember? Can they be made to forget? What is the relation between human, institutional, and machine memory?

These questions arise in the context of what has been described as the coming "data deluge": in addition to the highly granular data collection that is an intrinsic feature of online environments, the marketplace is soon to be flooded with inexpensive sensors (smart phones already integrate cameras, microphones, GPS, and accelerometers) that can collect a wide variety of data in digital form — e.g., continuous video and sound of one's daily activities, sport performance, body weight, or sleep patterns. The data may then be automatically geolocated and uploaded to websites providing statistical and visualization tools for sharing, comparing, and forecasting.[1] Furthermore, for the first time in history, these different types of records — images, quantitative data, audio recordings, written documents, etc. — are available for transmission, storage, indexation, analysis, retrieval, and visualization in a single media.

This convergence of pervasive forms of data collection, widespread digitization of analog records, economics of infinite storage, and subsumption of all media into the digital format is, according to the majority of commentators, inexorably leading us into an age of "perfect remembering," enabling individuals to "google their past," recalling at will individual events in full multimedia richness, identifying trends in

J.-F. Blanchette (✉)
Department of Information Studies, UCLA, Los Angeles, United States
e-mail: blanchette@ucla.edu

[1] See for example Nike's *Nike+ Sportband*, Withing's *WiFi Body Scale*, and Zeo's *Personal Sleep Coach*.

S. Gutwirth et al. (eds.), *Computers, Privacy and Data Protection: an Element of Choice*, DOI 10.1007/978-94-007-0641-5_2, © Springer Science+Business Media B.V. 2011

personal health, work activities, and lifestyle. Indeed, if one factors in the gradual elimination of paper in favor of digital forms for commercial transactions, communication, documentation, etc., and the continually plummeting costs of digital storage, the picture of a world where "everyone is on the record all the time" does not seem far-fetched.

This paper critically examines the trope of "perfect remembering" by confronting two of its most widely circulated articulations, Gordon Bell's *Total Recall,* and Viktor Mayer-Schönberger's *Delete.*[2] According to both authors, forgetting will soon become a thing of the past, a quirky feature of a bygone technological age, like cars that need cranking and very large cell phones. But while *Total Recall* rejoices in the innumerable benefits—from increased productivity to immortality—the "e-memory revolution" will bring, *Delete* is concerned that "perfect remembering" will cast a chilling shadow on individuals' ability to think and act in the present. Published within weeks of each other in the fall of 2009, the two books usefully encapsulate antagonistic positions on the changing state of memory.

I begin by summarizing Bell & Gemmel's arguments for why the coming "e-memory revolution" will reap untold riches for mankind, and, in any case, can hardly be averted, as the book's dust jacket loudly proclaims; I then move to Mayer-Schönberger's concerns for the potentially oppressive nature of a world drained of its capacity for oblivion, and conclude with my own critical evaluation of the assumptions shared by proponents and foes alike.

2.2 Total Recall

Gordon Bell is a computer engineer with a long and distinguished career in industry, notably with DEC.[3] After joining Microsoft Research in 1995, Bell embarked on a quest to become history's first "paperless man." The experiment, dubbed "MyLifeBits," was profiled in the *New Yorker,*[4] *Scientific American,*[5] and *Fast Company*[6] (where Bell was photographed with an external hard drive plugged into his forehead) and its insights broadly inform *Total Recall*'s vision. With endorsements from names like Gates, Negroponte, Myrhvold, Shirky, and Drexler generously sprinkled on the book jacket, *Total Recall* is Bell's bid to accede to the elite group of visionaries that have defined much of the information technology public imaginary and the terms under which it is debated.

---

[2]Bell, G., and J. Gemmel. *Total Recall: How the E-Memory Revolution will Change Everything.* New York, NY: Dutton, 2009; Mayer-Schönberger, V. *DELETE: The Virtue of Forgetting in the Digital Age.* Princeton, NJ: Princeton University Press, 2009.

[3]I abide by the authors' own convention of using Bell's as the sole voice for their text.

[4]Wilkinson, A. "Remember This?" *The New Yorker* (May 28, 2007): 38–44.

[5]Bell, G., and J. Gemmel. "A Digital Life." *Scientific American* (March 2007): 58–65.

[6]Thompson, C. A "Head for Detail." *Fast Company* 73-79 (November 2006): 110–112.

The prose is light, long on examples and anecdotes from Bell's personal life; the tone, decisive, with few if any concessions made to contrary arguments — either ignored altogether, or quickly expedited with swift rhetorical blows.[7] Bell resorts liberally to the "e-" prefix as well as to short lyrical scenarios of what our future lives *will* look like: "Imagine Dan, a blueberry farmer. ... He loves to sit down with his e-memories and a cup of tea to contemplate how he might make his farm better." (p. 133) The argument is organized in roughly three parts: first, an overall description of the Total Recall vision and its origin (Vannevar Bush's *Memex* makes a mandatory appearance); second, more detailed excursions into the potential benefits and consequences of Total Recall in the spheres of work, health, learning, and everyday/afterlife, and finally, material on how to adapt to, and get started with Total Recall, including an annotated guide to relevant research and literature.

In the workplace, Total Recall will be simultaneous with emancipation from paper and the mental fatigue that too often accompany it. The paperless office will be "pleasant", "calming", and provide everyone with "an incredible sense of freedom." (p. 73) Job training will also radically improve, as new hires tap into their predecessor's data holdings to access the tacit knowledge that is the first casualty of employee turnover. If concerns for liability have in the past shaped institutional record keeping policies, Bell confides that, in the digital age, "I don't see how corporate e-memory destruction policies can continue." (p. 90)

With regard to both health and learning, similar patterns will obtain. Instead of relying on patients' vague account of their ailments, doctors will finally have access to "minutely detailed chronicles of vital signs, behavior, diet, and exercise, along with physician' diagnoses, prescriptions, advice, and test results." (p. 94) Similarly, usage data, collected as students peruse, annotate, and share their electronic textbooks, recorded lectures, and online resources, will eliminate the fuzziness of current (paper-based) evaluation methods, while increasing self-knowledge for both students and teachers. Better information gathering will also foster self-motivation for learning and health, aided by algorithmic assistants that will mine our data stores to identify correlations and trends and issue recommendations that may prolong our lives and supercharge our learning.

It is in the realm of the personal that Bell takes his most extreme stand, when he recommends discarding commemorative artifacts in favor of digital surrogates (mostly photographs). While he recognizes the evocative power of material objects, he contends that "most people's physical mementos gather dust in an attic—if they even have them." (p. 118) With regard to leisure, the e-memory revolution will not only result in fantastically detailed travelogues "that might even exceed the actual trip experience" (p. 142), but also eventually leave our descendants with much more than slide shows to remember us by: Bell suggests that we may in the future endow

[7] For example, "They ask: Do we really want to know all this stuff? Liam Bannon, writing in favor of forgetting, offers up the inarguable: "More data does not imply better-quality decisions." Of course, that's true—but flawed human resources do not imply quality decisions either." (p. 165)

digital avatars with our lifetime store of data, achieving in the process a certain kind of immortality.[8]

Bell is well aware that in spite of its transformative potential, changes as profound as those implied by Total Recall will require individual adaptation, collective choices, as well as technical innovation. He foresees some difficulties with regard to data loss and decay (will your data be readable 50 years from now?), data entanglement (how can we separate work data from personal data?), adapting to more self-knowledge (how much truth about ourselves can we handle?), adapting to being recorded (how will consent be negotiated?), adapting in court (could your memories be used against you in court?). For those concerned with the Big Brotherian potential of *Total Recall*, Bell points out that a society in which "the recording equipment is not controlled by a single central authority, but by millions of individuals and private entities" (p. 14) is a *democratic* surveillance society.

What forces will propel this adaptation forward? Quite simply, technology itself: "I am a technologist, not a Luddite, so I'll leave abstract discussions about whether we should turn back the clock to others. Total Recall is inevitable regardless of such discussions." (p. 159) Bell is understanding that some might consider curtailing their participation in such a movement, but reminds them that "Total Recall, like the automobile, is rejected only at the price of giving up great advantages." (p. 174) How long before the tidal wave washes upon us? When it comes to prediction, Bell cannot quite resist the decisive statements that signal the supremely confident visionary: "It is absolutely clear that by 2020 these streams of technology will have matured to give the complete Total Recall experience" (p. 24) or the instant classic: "It's impossible to know exactly how long it will take for lifelogging to become common practice, but it's almost a sure bet that it will do so within a decade." (p. 21)

2.3 Delete

Viktor Mayer-Schönberger, formerly of the Kennedy School of Government at Harvard, and now a professor at the National University of Singapore, provides a somewhat more nuanced analysis of the consequences of the "e-memory revolution." At less than 200 pages, with relatively few notes and a short bibliography, *Delete* is directed at a broader public than mere academics, with the goal of stimulating policy debates in information governance. Like Bell, Mayer-Schönberger's premise is that mass digitization, cheap storage, improved retrieval techniques and

[8]There is more than a passing acquaintance here with the "singularity" movement, something Bell believes will eventually occur, although not in the form of machine consciousness—see Nordmann, A. "Singular Simplicity," *IEEE Spectrum* 45 (June 2008): 60–63. On the ambiguous promise of memorials for effecting remembrance, see also Bindman, D. "Bribing The Vote Of Fame: Eighteenth-Century Monuments And The Futility Of Commemoration." In *The Art of Forgetting,* edited by A. Forty and S. Küchler, 93. Oxford: Berg, 1999: "What could be more forlon than a grand and costly tomb, put up to prolong a reputation beyond the grave, but now long neglected and falling into ruin, with the name of the deceased completely forgotten."

the global reach of computer networks are making it easier today to remember than to forget, "because it no longer requires a conscientious act, a tiny bit of time, energy or money that we need to expend to commit information to digital memory." (p. 169) Unlike Bell however, Mayer-Schönberger sees this shift as deeply problematic, as the individual and collective unlearning of "one of the most fundamental behavioral mechanisms of humankind." (p. 92)

Mayer-Schönberger's analysis of the issue proceeds along two main dimensions, power and time. In the first case, he sees the demise of forgetting as leading to enormous asymmetries between individuals and the institutions that collect their personal information. The accessibility, durability, and comprehensiveness of digital information will provide an almost overwhelming incentive for individuals to censor themselves, as they contemplate the potential damage that even the mildest forms of deviant or oppositional behavior may inflict on their reputation long into the future — Mayer-Schönberger can in fact already point to such frightening cases. In the second case, he points to how the continuous availability of an instantly available and detailed representation of past events will have deleterious effects on our ability to act in the present. The selective processes of forgetting, he argues, are not so much flaws as they are the necessary foundations of our ability to generalize, and thus, to rise above the particular. By gradually undermining cognitive processes rooted in millennia of human evolution, we may in effect leave ourselves vulnerable to impaired individual and collective decision-making.

Mayer-Schönberger then proceeds to the analysis of three pairs of possible responses to the power/time dimensions of the forgetting crisis, operating respectively on the level of individual behavior, law, and technology. Individuals may redress the power imbalance by simply refraining from sharing personal information, the practice of "digital abstinence." (p. 128) It is also possible they may cope with the continuous intrusion of the past by disregarding it and focusing on an individual's most recent actions, forms of "cognitive adjustment." (p. 154) Laws may also address the power imbalance by further defining information privacy rights that grant individuals the power to restrict access to their information, and may also expand existing mechanisms that mandate sealing or deletion of certain types of information (e.g., juvenile crime records, bankruptcy in credit reports). And technology may also join in the fight, either through "digital privacy rights infrastructure" that could enforce policies for the retention and disposal of personal information, or through providing the "perfect contextualization" that would situate each piece of information within its full historical context. (p. 163)

While each of these approaches does contribute something to restoring a certain balance, each has also significant drawbacks. Practitioners of digital abstinence must systematically forego the various benefits service providers offer in exchange of release of personal information; privacy rights have historically enjoyed limited successes in the US, and policies for automatic negotiation of privacy settings between information sharing devices are notoriously complicated for dedicated experts, let alone for casual users.

Mayer-Schönberger's own proposal aims to "flip the default back" to forgetting by attaching a new type of metadata — an expiration date — to each piece of information. When saving a file, for example, users would be forced to specify a retention period, in the same manner they specify the file's name and location. A search query could similarly include an additional parameter specifying its retention period. Mayer-Schönberger's proposal draws its inspiration from the warnings triggered by Web sites attempting to install cookies:

> the core goal of expiration dates for information is precisely not to push the problem of digital memory off our consciousness by delegating it to technology, but rather the opposite: to make humans aware of the value and importance of forgetting. (p. 185)

While well-aware that a combination of measures will likely prove necessary to restore the ecology of remembrance and forgetting, Mayer-Schönberger believes that confronting users with choices of suitable retention periods performs an important function, by reminding them that "the value of information is not timeless." (p. 173)

2.4 Noisy Bits

The confrontation of these two antagonistic theses immediately leads to two questions. First, are Bell and Mayer-Schönberger right? Is it indeed the case that we have switched from default forgetting to wholesale remembering? Second, if they are right, which conclusion is the correct one? Will *Total Recall* usher in a golden age of enhanced self-learning and objective historical truth for humanity, or on the contrary, a perpetual dark age of domination by an omnipresent past?

2.4.1 Digital Decay

Bell-Gemmel's and Mayer-Schönberger's analysis posits a historical progression of technologies for remembering, along an axis where the unreliabilities of "biological memories" are gradually supplemented by stronger and stronger grades of "external memories" inscribed on various media—from writing, paintings, books, and movies, to information's supreme incarnation in the digital. For Mayer-Schönberger, digital information is superior to all previous analog forms of remembering "because it lacks the noise problem" (p. 57), that is, it does not decay with use, reproduction, or time. For Bell, in contrast to its biological counterpart, "digital memory is objective, dispassionate, prosaic, and unforgivingly accurate." (p. 56)

This characterization is dependent on the pervasive Western analogy that identifies human memory with recording technologies that function through imprinting (and erasing) of traces on substrates of variable malleability, from wax to stone.[9]

[9]Forty, A. "Introduction." In *The Art of Forgetting*, edited by A. Forty and S. Küchler, 2. Oxford: Berg, 1999; Burton, J. Bergson's "Non-Archival Theory of Memory." *Memory Studies* 1 (2008): 322; Carruthers, M.J. *The Book of Memory: A Study of Memory in Medieval Culture.* Cambridge: Cambridge University Press, 1992, 16. The Roman punishment of *damnatio memoriae* involved

In such a framework, material decay is akin to forgetting and, conversely, "perfect remembering" a direct consequence of the noiselessness of digital media, of its imperviousness to decay.

The ascription of such improbable qualities to any system for recording information is best understood as another manifestation of the historical association of electric communication with transcendence of the material properties of physical media,[10] of the sublimity of overcoming the ordinary limitations of space, time, and energy through technology,[11] and of the pervasive metaphors within computer science that "minimizes our sense of representations as material things."[12]

The transcendental properties of information technology, its seemingly mysterious ability to both exist within the physical plane and yet escape its most fundamental law (decay) have been recently questioned by scholars trained in methods of bibliographic analysis, focusing on the material context of production, expression, and interpretation.[13] For our purposes, such analysis have yielded two observations of particular importance: first, and despite how pervasive the distinction in ordinary discourse, "form is constitutive of information, not its transparent representation";[14] second, the specific material instantiations of informational artifacts are "always undergoing changes, aging, crumbling, acquiring or resisting wear." (p. 142) How then do Bell and Mayer-Schönberger account for the wear and tear of digital media, given its purported noiselessness and unforgiving accuracy?

Driven by the pragmatic constraints of the "My LifeBits" experiment, much of *Total Recall* can in fact be read as a list of contradictory footnotes to Bell's vision of seamless and perfect remembering, each highlighting a different dimension of the

---

erasing all traces of a person, including those written in stone—see Hedrick, C.W. Jr. *History and Silence: Purge and Rehabilitation of Memory in Late Antiquity.* Austin: University of Texas Press, 2000.

[10] Rosenheim, S.J. *The Cryptographic Imagination: Secret Writing From Edgar Poe to the Internet.* Baltimore, MD: The Johns Hopkins University Press, 1997.

[11] Nye, D. *American Technological Sublime.* Cambridge, MA: The MIT Press, 1994.

[12] Agre, P. "Beyond The Mirror World: Privacy and The Representational Practices of Computing." In *Technology and Privacy: The New Landscape*, edited by P. Agre and M. Rotenberg, 29-61. Cambridge, MA: The MIT Press, 1997. It should be noted that Bell's *Total Recall* vision essentially reiterates one articulated almost 20 years ago by David Gelernter, who argued that ubiquitous sensors will inevitably lead to a distributed computer running a real-time simulation of the physical world, faithfully mirroring reality, yet augmenting it with software capabilities. Gelernter's *Mirror World* is real-time and concerned exclusively with public spaces, while *Total Recall* is about stored data and confined to the personal sphere, but in every other respect, they share the same basic assumptions about computers' ability to substitute for reality. See Gelernter, D. *Mirror worlds: Or the day software puts the universe in a shoebox . . . How it will happen and what it will mean.* New York, NY: Oxford University Press, 1991.

[13] Kirschenbaum, M.G. *Mechanisms: New Media and the Forensic Imagination.* Cambridge, MA: The MIT Press, 2008; Gitelman, L. *Always Already New: Media, History, and the Data of Culture.* Cambridge, MA: The MIT Press, 2006.

[14] Drucker, J. *SpecLab: Digital Aesthestics and Projects in Speculative Computing.* Chicago, IL: University of Chicago Press, 2009, 139.

materiality of digital information. A first category relates to the reformatting, degradation, or unavailability of data arising from the various incompatibilities exhibited by file formats, hardware platforms, and software applications, as they ceaselessly morph into their next market-driven incarnations. Despite our limited collective experience with digital preservation, it already appears inevitable that in order to stave off obsolescence, data will require some kind of continuous process of re-instantiation into (newer) formats, each one dictating in effect new formal conditions of production, expression, and interpretation. Despite the relative ease with which a string of characters may migrate from one format to another, plain text is not XML is not Word is not PDF is not TIFF, and no single format will ever transcend the entirely different conceptions of what a text is each embodies. The problem is already well upon us with Web-based documents (Bell recommends printing them to PDF) and gets only worse when one considers Flash-based sites or social media — Bell suggests such sites should "wise up . . . and release our data from captivity." (p. 201)

More radical forms of loss also loom large, from hard drive crash to improper backups and licensing of content that constrain the availability of documents. All in all, as Bell admits, "1000-year preservation is a matter shrouded in uncertainty" (p. 224), but much more worrisome, so is that of 100-year preservation. Despite this, Bell suggest that in general, one may confidently destroy original artefacts, unless their qualities as material objects justify keeping them, as in the case perhaps of a photo album, which you may keep and enjoy "until it falls apart and fades — in any case, you should rest assured that you have the digital version forever." (p. 187)

A second type of material impingment on the transparent manifestation of digital information falls under the heading of "representation": resolution, classification, and description bring specific kinds of constraints to the digital archive: pictures and video offer more or less detail, and only ever from specific viewpoints, and for recall to occur at all, they must be properly described, whether through authoritative descriptors, social tagging, or automatic analysis. As Bell remarks, "all this takes work" (p. 197), involving considerations of intellectual and physical labor, time, space, energy, and value, considerations which induce in turn a certain stratification of the total archive, in terms of its discoverability by search algorithms.

Yet another set of constraints to perfect remembering include those related to selection and appraisal, that is, what gets included or excluded in the archive. Exclusion may occur, among other things, through technical processes (e.g., the possible resolution of a given measurement), appraisal policy (e.g., "my goal is to record everything I actually read, not what others send me. It's my choice, not their, that counts." [p. 32]), cultural and legal norms (e.g., requiring consent before recording), intellectual property agreements (Bell recounts how, after Jim Gray's disapearance at sea, Microsoft returned his laptop to his wife, meticulously expunged of documents susceptible of leaking trade secrets), not to mention individuals' desire to bequeath posterity a favorable image of themselves.

Mayer-Schönberger's investigation also forces him to eventually work through the dichotomy of a transcendent yet materially instantiated media. After an analysis of media history that concludes with the inherent superiority of digital media over all previous analog technologies, he points to our collective faith in this superiority —the dazzling speed and comprehensiveness of our digital stores — as the very danger that threatens our appreciation for forgetting. But as he points out, the digital archive is necessarily "biased against information that is not captured in digital form and not fed into digital memory." (p. 123) Furthermore, this collective faith is even more dangerous because, contra to analog media, we simply lack experience with evaluating this systematic bias (one we already experience insofar as if it can't be found with Google, it doesn't exist), or the kind of forgeries digital media will fall prey to.

At the same time he unveils these shortcomings, Mayer-Schönberger cannot shake off his intellectual commitment to technologies of representation that may 1 day provide for objective, unmediated, comprehensive remembering. When he recommends against unconditionally trusting the digital archive, it is on the basis that it "can be modified after the fact, and thus does not necessarily represent an accurate rendition of a past event." (p. 120) And when he critiques the "perfect contextualization" approach, it is on the basis that it would require an investment in "the technical means for true digital remembering." (p. 165) Digital information thus retains its superior status as transparent representation, an ontological circle that is fully closed when he concludes that until "our internal thoughts are remembered, digital memory will remain fundamentally incomplete." (p. 166)

2.4.2 Can You Handle the Truth?

What if, in addition to media's intrinsic ambiguity as representation, suffused with noise, decay, degrees of resolution, etc., records themselves figured in an ambiguous relationship to our psychological need for remembering and forgetting? What if humans create commemorative artifacts not only because, as Mayer-Schönberger puts it, they "yearn to remember" (p. 93), but because they yearn just as much to deny, repress, forget?

Bell's vision for *Total Recall* has little room to entertain such academic sophistry: if digital memory stands as objective and accurate capture of reality, the only question remaining is whether we are ready to confront it: "successful people don't shy away from the honest record. . . . In court, we ask for the truth, the *whole* truth, and nothing but the truth. It might be painful, but I believe better memory is really better." (pp. 166–167) Freud would have understood Bell's Total Recall as not merely painful, but in fact, profoundly alien to the dynamics of the psyche, dynamics in which forgetting plays an active and fundamental role in the constitution of the self. Bell's characterization is also inaccurate: court proceedings are ruled by elaborate rules governing the admissibility and evaluation of evidence, and the most cursory examination of these rules cannot fail to point to the fact that courts have, thanks to the adversarial process, a sophisticated understanding of the technological mediation of evidence.

Instead of mere prosthesis to aid our failing biological mechanisms, one might thus understand commemorative artifacts as mediating the conflicting demands of self-building, whether individual or collective. At the most basic level, this mediation operates simply by causing "only certain things to be remembered, and by exclusion, cause others to be forgotten."[15] An archive is necessarily a condensation of a larger whole, and is thus founded on a fundamental process of exclusion that defines the boundary between its inside and outside. On another level, commemorative artifacts, by serving as focal points for remembering, may enable more pervasive forms of forgetting to take place. This ambiguity has been explored in particular by the artists and architects who have created memorials to humanity's most systematic attempt at consignment of a people to oblivion, the Holocaust. Several of them have explicitly attempted to eschew the traditional figure of the memorial as a durable imprint of an historical event — for example, Rachel Whiteread's "Memorial to the Victims of the Holocaust" in Vienna, seeks to actively provoke the audience, by presenting them with an hermetically-sealed library of nameless books "causing us to try to remember what remains permanently out of reach, and inaccessible to us."[16]

The *Total Recall* vision is thus problematic not only on account of the purported imperviousness of digital technologies to decay, but also in its implication of a direct correspondence between records and remembering. It relies on a number of hypotheses about digital media that prove difficult to maintain in any sustained encounter with the practical constraints of digital information capture, storage and curation. Like any other media, digital media brings to the table its own dialectics of objectivity and subjectivity, signal and noise, integrity and decay, authenticity and forgery, transparency and censorship, remembrance and repression. These dialectics are never fully determined by the material characteristics of the technology, as Bell would have us (at least partially) believe. José van Dijck captures the point succinctly:

> Media are not confined to private and public areas, and neither do they store or distort the past in relation to the present or future. Like memories, media's dynamic nature constitutes constantly evolving relations between self and others, private and public, past and future.[17]

If there *is* something unique about digital media, it is to be found in the powerful association between computers and mathematics that endows digital information with a special cultural authority, that which has historically accrued to mathematics as pure symbolic expression of natural laws. Unlike mathematics however, computers are thoroughly physical devices (for those in doubt, the electrical cord is a dead giveaway). How is it then that otherwise scientifically minded individuals align themselves with conceptions of memory that manage to abstract away the basic laws of decay? One answer is provided by historian of computing Michael Mahoney, who notes:

---

[15]Forty, "Introduction," 9.

[16]Forty, "Introduction," 13.

[17]van Dijck, J. *Mediated Memories in the Digital Age.* Stanford, CA: Stanford University Press, 2007, 26.

> The dual nature of the computer is reflected in its dual origins: hardware in the sequence of devices that stretches from the Pascaline to the ENIAC, software in the series of investigations that reaches from Leibniz's combinatorics to Turing's abstract machines. Until the two strands come together in the computer, they belong to different histories, the electronic calculator to the history of technology, the logic machine to the history of mathematics, and they can be unfolded separately without significant loss of fullness or texture. Though they come together in the computer, they do not unite. The computer remains an amalgam of technological device and mathematical concept, which retain separate identities despite their influence on one another.[18]

The computer's split personality problem leads information age pundits to almost unfailingly focus on its *logical* dimension, happily ignoring its *material* dimension, the mechanical components that compute, exchange, and store bits. From John Perry Barlow's "there is no matter here"[19] to Nicolas Negroponte's "from atom to bits,"[20] the material dimension of computing consistently gets short thrift. Of course, no one explicitely denies that digital information is dependent on physical hardware for its existence, yet, this material dimension is largely understood as merely providing a support system for the processing, transport, and storage of immaterial bits. Yet, as the looming digital preservation crisis signals, the messy materiality of computing will become increasingly harder to ignore.

2.5 Conclusion

Even if we put aside the easy rhetoric of the coming "e-memory revolution," there is little question that we are facing something rare and exceptional, nothing less than a "new regime of memory practices,"[21] a sweeping changing of the guard in the modes of production, expression, and reception of commemorative artifacts. While this essay has argued that characterizing such a shift as a revolution obscures rather than illuminates, the transition to a new dominant media offers special opportunities for analysis and critique:

> There is a moment, before the material means and the conceptual modes of new media have become fixed, when such media are not yet accepted as natural, when their own meanings are in flux. At such a moment, we might say that new media briefly acknowledge and

---

[18] Mahoney, M.S. "The History of Computing in the History of Technology." *Annals of the History of Computing* 10 (1988): 113–125.

[19] "Governments of the Industrial World, you weary giants of flesh and steel, I come from Cyberspace, the new home of Mind. ... Your legal concepts of property, expression, identity, movement, and context do not apply to us. They are all based on matter, and there is no matter here. ... Our identities have no bodies, so, unlike you, we cannot obtain order by physical coercion. ... In our world, whatever the human mind may create can be reproduced and distributed infinitely at no cost. The global conveyance of thought no longer requires your factories to accomplish." John Perry Barlow, "A Declaration of the Independence of Cyberspace," https://projects.eff.org/~barlow/Declaration-Final.html.

[20] Negroponte, N. *Being Digital.* New York, NY: Vintage, 1996.

[21] Bowker, G.C. *Memory Practices in the Sciences.* Cambridge, MA: The MIT Press, 2005.

> question the mythic character and the ritualized conventions of existing media, while they are are themselves defined within a perceptual and semiotic economy that they then help to transform.[22]

In conclusion, I want to offer a brief probe into one such ritualized convention of remembrance and oblivion, with a far-reaching impact on the historical record, and how it might be successfully challenged during our current time of transition.

Mayer-Schönberger argues that "information ecologies" (i.e., regulatory constraints on personal information collection and storage) approaches have held limited purchase so far on the demise of forgetting and stand to erode even further. The legal safeguards built around, for example, credit reports and the sealing of juvenile crime records[23] seem quaint and desperately out of touch with reality at a time where Google seeks to make available "the world's information" and transparency has become synonymous with effective government.

Yet, few are questioning the effectiveness of such constraints when it comes to one social actor, the corporate person. An extensive regulatory apparatus, including an array of governmental agencies (e.g., SEC, FDA, EPA), trained professionals (counsels, records managers) and comprehensive body of rules operates with single-minded devotion to the precise delimitation of corporate accountability and liability through the creation, preservation, and destruction of records, as defined by the retention periods established in various sectors of corporate activity. In his ground-breaking analysis of the political economy of personal information, Oscar Gandy pondered why similar limits seem unthinkable for individuals:

> Corporations, unlike individuals, can be rather easily dissolved and formed anew on action of their boards of directors. Why should corporations as fictional persons already have rights that natural persons still long to enjoy?[24]

Driven by a concern for the vast gap in the historical record such a regulatory framework induces, business historian David Kirsch has begun questioning this remarkable discrepancy. While every year in the United States, more businesses are created than marriages celebrated, historians will have little to rely on when the time comes to document the extraordinary surge of entrepreneurship that is synonymous with the rise of the Internet. In 2007, the records of a bankrupt Silicon Valley law firm, Brobeck, Phleger, & Harrison, were assigned to a liquidation committee. Comprised of several millions records and one and half terabyte of data, the collection constitutes an extraordinarily rich and unique historical archive of thousands of dot com ventures. Yet, as private business records, issues of legal privilege and confidentiality prevent their use as primary sources for historical research.

---

[22] Pringree, G.B., and L. Gitelman. "Introduction: What's new about new media." In *New Media 1740-1915*, edited by L. Gitelman and G.B. Pringree, xii. Cambridge, MA: The MIT Press, 2003.

[23] Blanchette, J.-F., and D. Johnson. "Data Retention and the Panoptic Society: The Social Benefits of Forgetfulness." *The Information Society* 18 (2002): 1–13.

[24] Gandy, O.H. *The Panoptic Sort: A Political Economy of Personal Information.* Boulder, CO: Westview Press, 1993, 225.

With help from the Library of Congress, Kirsch's efforts have led to the creation of the Brobeck Closed Archive, operating under a set of innovative guidelines that seek to reconcile the interest of Brobeck's clients with a public interest in private records. As Kirsch notes,

> even if the bulk of the Brobeck Archive would need to remain off-limits to historians, potentially into perpetuity, the scale and breadth of the collection could support social science research to answer a host of interesting questions without requiring that specific confidential information be disclosed.[25]

The case of the Brobeck Closed Archive suggests it is indeed possible to develop "information ecology" approaches that negotiate in innovative ways the trade-offs between liability, accountability, and the collective interest in a comprehensive historical record. Rules constraining the use of information relative to personal bankruptcy and juvenile crime records were developed with a similar concern in balancing individual rights with the collective interest in ensuring that individuals eventually participate again in economic and social life. If such regulatory constraints appear today a lost cause, it is not on account of the impossibility of effectively regulating digital information, but of the enormous economic value of the data points that constitute the finely grained fabric of our online personas.

Thus, in evaluating the various claims for both risks and benefits of digital technologies for memory, we should remain mindful that "remembering," whether perfect or fallible, is always the remembering of specific social actors, with varying degrees of access and exposures to these risks and benefits. The ability to be forgotten is thus a privilege likely to remain unevenly distributed among these social actors and, for some of them at least, the best years for oblivion are probably still to come.

References

Agre, P. "Beyond The Mirror World: Privacy and The Representational Practices of Computing." In *Technology and Privacy: The New Landscape*, edited by P. Agre and M. Rotenberg, 29–61. Cambridge, MA: The MIT Press, 1997.

Barlow, J.P. "A Declaration of the Independence of Cyberspace." https://projects.eff.org/~barlow/Declaration-Final.html (accessed May 28, 2010).

Bell, G., and J. Gemmel. *Total Recall: How the E-Memory Revolution will Change Everything.* New York, NY: Dutton, 2009.

Bell, G., and J. Gemmel. "A Digital Life." *Scientific American* (March 2007): 58–65.

Bindman, D. "Bribing The Vote Of Fame: Eighteenth-Century Monuments And The Futility Of Commemoration." In *The Art of Forgetting,* edited by A. Forty and S. Küchler, 93–105. Oxford: Berg, 1999.

Blanchette, J.-F., and D. Johnson. "Data Retention and the Panoptic Society: The Social Benefits of Forgetfulness." *The Information Society* 18 (2002): 1–13.

Bowker, G.C. *Memory Practices in the Sciences.* Cambridge, MA: The MIT Press, 2005.

Burton, J. "Bergson's Non-Archival Theory of Memory." *Memory Studies* 1 (2008): 321–339.

[25] Kirsch, D.A. "The Record of Business and the Future of Business History: Establishing a Public Interest in Private Business Records." *Library Trends* 57 (2009), 362.

Carruthers, M.J. *The Book of Memory: A Study of Memory in Medieval Culture.* Cambridge: Cambridge University Press, 1992.

Dijck, J.v. *Mediated Memories in the Digital Age.* Stanford, CA: Stanford University Press, 2007.

Drucker, J. *SpecLab: Digital Aesthestics and Projects in Speculative Computing.* Chicago, IL: University of Chicago Press, 2009.

Forty, A. "Introduction." In *The Art of Forgetting,* edited by A. Forty and S. Küchler, 1–18. Oxford: Berg, 1999.

Gandy, O.H. *The Panoptic Sort: A Political Economy of Personal Information.* Boulder, CO: Westview Press, 1993.

Gelernter, D. *Mirror Worlds: Or the Day Software Puts the Universe in a Shoebox ... How It Will Happen and What It Will Mean.* New York, NY: Oxford University Press, 1991.

Gitelman, L. *Always Already New: Media, History, and the Data of Culture.* Cambridge, MA: The MIT Press, 2006.

Hedrick, C.W. *History and Silence: Purge and Rehabilitation of Memory in Late Antiquity.* Austin, TX: University of Texas Press, 2000.

Kirsch, D.A. "The Record of Business and the Future of Business History: Establishing a Public Interest in Private Business Records." *Library Trends* 57 (2009): 352–370.

Kirschenbaum, M.G. *Mechanisms: New Media and the Forensic Imagination.* Cambridge, MA: The MIT Press, 2008.

Mahoney, M.S. "The History of Computing in the History of Technology." *Annals of the History of Computing* 10 (1988): 113–125.

Mayer-Schönberger, V. *DELETE: The Virtue of Forgetting in the Digital Age.* Princeton, NJ: Princeton University Press, 2009.

Negroponte, N. *Being Digital.* New York, NY: Vintage, 1996.

Nordmann, A. "Singular Simplicity." *IEEE Spectrum* 45, 6 (June 2008): 60–63.

Nye, D. *American Technological Sublime.* Cambridge, MA: The MIT Press, 1994.

Pringree, G.B., and L. Gitelman. "Introduction: What's New About New Media." In *New Media 1740-1915*, edited by L. Gitelman and G.B. Pringree, xi–xxii. Cambridge, MA: The MIT Press, 2003.

Rosenheim, S.J. *The Cryptographic Imagination: Secret Writing from Edgar Poe to the Internet.* Baltimore, MD: The Johns Hopkins University Press, 1997.

Thompson, C. "A Head for Detail." *Fast Company* 73–79(November 2006): 110–112.

Wilkinson, A. "Remember This?" *The New Yorker* (May 28, 2007): 38–44.

Chapter 3
Property in Personal Data: Second Life of an Old Idea in the Age of Cloud Computing, Chain Informatisation, and Ambient Intelligence

Nadezhda Purtova

3.1 Introduction

This contribution considers a familiar idea of property rights in personal data, in light of some new developments in information technology that have not yet been accounted for in the existing debate. Indeed, proposals to introduce property rights in personal data emerged in the US in the 1970s[1] and were still developing until the early 2000s. Propertisation then was considered as a possible and a better way to achieve the goals of privacy and data protection.[2] The argument was made from various perspectives. It is not the aim of this paper to give a detailed description of all propertisation theories.[3] It suffices only to mention that for some, formal propertisation was only a *post factum* recognition of the *de facto* commodification of personal data; the property regime was argued to be able to respect the interests of the information industries and, somehow, channel otherwise the uncontrolled process of trade in personal data and even empower individuals. For others, propertisation was a solution to the data protection problem associated with growing private and government databases. One of the most discussed dimensions of the problem was the deprivation of people of control over their personal information; treating this information as property would arguably give that control back.[4] Other property

N. Purtova (✉)
2513BS Torenstraat 12, den Haag, The Netherlands
e-mail: npurtova@yahoo.com

[1]See, e.g., Westin, A.F. *Privacy and Freedom.* London, Sydney, Toronto: the Bodley Head, 1967.

[2]Despite the lack of agreement on the content of those goals.

[3]For a detailed outline and analysis of the US debate on propertisation of personal data see Purtova, N. "Property Rights in Personal Data: Learning from the American Discourse." *Computer Law & Security Report* 25, 6 (2009).

[4]E.g., Westin, *Privacy and Freedom.*, 7; Solove, D.J. "Privacy and Power: Computer Databases and Metaphors for Information Privacy." *Stanford Law Review* 53 (2001): 1428

S. Gutwirth et al. (eds.), *Computers, Privacy and Data Protection: an Element of Choice*, DOI 10.1007/978-94-007-0641-5_3, © Springer Science+Business Media B.V. 2011

claims derived from the natural rights theory,[5] alleged the rhetorical value of property talks,[6] economic analysis of law, and finally from the shortcomings specific to the US legal and political systems in general and its data protection law in particular. The idea was mainly discussed by the US scholars in the context of their domestic law and received little attention of the European authors.[7] The reason possibly lies in the fact that the approaches to property on the two continents are quite different: the US view on property, at least in academic debate also spreading to the law books and practice, is under the influence of the discipline of economic analysis of law and therefore, arguably, more flexible to believe in a new application of the old idea; whereas Europeans may be more conservative to accept such an unconventional use of the traditional property regime; one may blame such conservatism on a relatively unbending character of the classical model of property rights in (predominantly) Continental European tradition. Whatever the reason, ever since its culmination in the 1990s – early 2000s inter alia in the debate between Lessig[8] and Rotenberg,[9] The idea of propertisation has gradually lost its academic attractiveness. One of the last academic publications devoted entirely to the idea of property in personal data was the 2004 article "Property, Privacy, and Personal Data" by Paul Schwartz in which he powerfully argued in favour of the hybrid model of information privacy that combines property approach and regulation.[10]

The idea of propertisation of personal data was not implemented in law then and has been nearly forgotten. As a result, developments in technology and information practices over the last 5–6 years have received virtually no reflection in the propertisation discourse calling to revive the abandoned debate, but this time from

---

[5]Solove, "Privacy and Power: Computer Databases and Metaphors for Information Privacy.", 1446 (although he does not develop the natural law argument further); Vera Bergelson, "It's Personal, but Is It Mine? Toward Property Rights in Personal Information," *UC Davis Law Review* 37 (2003): 430; according to Radin, there is a certain inherent connection between an individual and data pertaining to him. This connection arguably justifies property status of personal data. Margaret Jane Radin. "Property and Personhood." *Stanford Law Review* 34, 5 (1982): 959.

[6]"Property talk is just how we talk about matters of great importance" (Lawrence Lessig, "Privacy as Property." *Social Research: An International Quarterly of Social Sciences* 69, 1 (2002): 247); "If you could get people (in America, at this point in history) to see certain resource as property, then you are 90% to your protective goal." (Lessig, "Privacy as Property.")

[7]Among few European authors commenting on the issue of property in personal data see J.E.J. Prins, "When Personal Data, Behavior and Virtual Identities Become a Commodity: Would a Property Rights Approach Matter?," *SCRIPT-ed* 3, 4 (2006).; on the possibility of private law solutions in data protection, including propertisation, see Colette Cuijpers, "A Private Law Approach to Privacy: Mandatory Law Obliged?," *SCRIPT-ed* 4, 4 (2007).; for a dignitarian argument against market solutions in data protection see Yves Poullet, "Data Protection Legislation: What Is at Stake for Our Society and Democracy?", *Computer Law & Security Report* 25 (2009).

[8]Lawrence Lessig, *Code and Other Laws of Cyberspace* (New York, NY: Basic Books, 1999).; Lawrence Lessig, "The Law of the Horse: What Cyberlaw Might Teach," *Harvard Law Review* 113 (1999).; Lessig, "Privacy as Property."

[9]Marc Rotenberg, "Fair Information Practices and the Architecture of Privacy (What Larry Doesn't Get)," *Stan. Tech. L. Rev.* 1 (2001).

[10]Paul M. Schwartz, "Property, Privacy, and Personal Data," *Harv. L. Rev.* 117, 7 (2004).

the perspective of those developments. This paper is written in response to that call; it reviews the recent evolution of information technology and practices and argues that in the age of cloud computing, chain informatisation, and ambient intelligence, a property regime, combined with regulation, not only deserves a second look but might capture, and hence channel, new and otherwise difficult to control relationships with regard to personal data.

This analysis is formulated in several steps. It will start by introducing concepts central to the argument. Namely, Section 3.2 will describe the notions of chain informatisation, cloud computing, and ambient intelligence and explain how they have altered the personal data problem in a way that the existing approach is insufficient to cope with the changes. Section 3.3 will briefly address the notion of property in general and as used in the earlier debate on propertisation of personal data. It will be argued that the concept of property may be flexible enough to include a range of new objects like personal data and achieve desired regulatory goals also in more conservative continental Europe. Section 3.4 will first offer blueprints of a model of fragmented property rights in personal data, and then elaborate on its benefits and drawbacks. Section 3.4 will present the summary of the argument and main conclusions.

3.2 New Challenges for the Information Society

A body of law known in Europe as the law of data protection has constantly been evolving in reaction to new developments of information technologies and practices.[11] Each subsequent generation of data protection was an attempt to capture and control changing relationships between the participants of data processing – individuals as data subjects and the actors on the receiving end of the data flow. This section will show that some recent and (not so distant) future developments in information technology and practices already have and will continue to reshape the relationships within the personal data flow including shift in accountability of data processing actors (Section 3.2.1). The current data protection regime is not adequately responding to the new challenges (Section 3.2.2).

3.2.1 New Structure of Relationships

Let us take a very brief glance at the evolution of the relationships in the data flow and data protection legislation so far. Data protection law has always had to conform to both quantitative and qualitative changes of relationships within the data flow. Quantitative because the number of actors collecting, analysing, and using

---

[11]For an overview of the evolution of data protection up to the 1990s see Viktor Mayer-Schőnberger, Data Protection in Europe, In *Technology and Privacy: The New Landscape*, edited by P.E. Agre and M. Rotenberg (Cambridge: The MIT Press, 1997).

personal data has been constantly growing, as did the number of relationships. Qualitative because the relationships were becoming more complex. For instance, at the beginning of the Information Revolution because the computers were expensive and available only to a small number of actors, there were expected to be only few databanks. As a result, the first-generation data protection norms targeted those few databases individually and did not contain generally applicable data protection rights.[12] As computers became easily available and the number of actors processing personal data grew and counted thousands, second-generation data protection shifted to the generally applicable negative – non-disclosure – rights of the citizens so that they could protect their interests.[13] Later on, as data protection relationships grew beyond mere collection of data, third-generation data protection turned to find a balance between privacy and participation in the information society. It did so by matching non-disclosure with more participatory positive rights to control subsequent data use.[14] Finally, to address another complexity of the data flow, *i.e.* inequality of negotiating powers of weak data subjects and powerful information industries, the fourth-generation data protection laws – among those the 1995 Data Protection Directive – by means of regulation have established some ground rules of the game – principles of data processing.[15] However, the number of actors involved in chain informatisation, cloud computing, and (potentially) ambient intelligence is so high, and the relationships between them are so intertwined and complex that they are difficult to capture in data protection measures.

3.2.1.1 Chain Informatisation

Chain informatisation is a part of a phenomenon of organisational cooperation and refers to automated sharing of information both between private sector organisations and government agencies. It is argued that it aids more speedy, smooth and customer-friendly provision of services. In practice it means that many small databases are effectively merged into one big database. For instance, when an individual refers to a state agency or a private entity, he/she does not need to supply these agencies with documented proof of the facts so that a certain decision can be made; the relevant entity already has access to all necessary data which has been supplied via the chain of databases of other entities. Multiple actors are involved in the operation of that database: some actors collect authentic personal data, others process, and others use it. The actors who collected the information do not always end up using it, and the ones making decisions on the basis of that information are not the ones who originally collected it. Besides customer convenience, chain informatisation is said to improve cooperation between various private and public

[12]Ibid. 225

[13]Ibid. 227–228

[14]Ibid. 229–232

[15]Ibid. 232–235

agencies to address complex situations, like e.g. child welfare[16] or prevention of child abuse. The complexity of a real-life situation is dealt with by breaking it into separate segments; each of these segments is then dealt with by a separate body or authority. Each authority collects or needs data to do its share of the work. This leads to multiple actors possessing relevant information and exchanging this information. For instance, in the autumn of 2003, in the Green Paper *Every Child Matters* the UK government published a plan to introduce local databases containing "a list of all children living in the area" and other "basic details," the latter including not only the child's name, address, details on parents, carers and education, but also "any cause of concern in relation to a child."[17] Such tagging is proposed "*for preventive purposes*, without the consent of the child or their carers. We would also welcome views on whether warning signs should reflect factors within the family such as imprisonment, domestic violence, mental health or substance misuse problems amongst parents and carers."[18] When the data forming that database comes from or is accessible through other government agencies or private organisations, this is chain informatisation in action.

The bigger the chain the more actors it includes; the more actors are involved, the higher the likelihood that something will go wrong in the process. For instance, there is a danger of incorrect records getting into the chain and, though used for lawful purposes, it could result in harmful consequences for a citizen or consumer. The Dutch ombudsman cites an example of an entrepreneur who was mistakenly "given" a criminal record and was experiencing consequences of it for 13 years.[19] The possibility of removing children "by mistake" is as well not excluded.

3.2.1.2 Cloud Computing

The term "cloud computing" refers to the body of web-based – as opposed to on-premises – services: storage capacity and applications including customer records, healthcare records, employee databases management.[20] Cloud computing – similar to chain informatisation – is often presented to businesses as a cheaper way of delivering IT services. Instead of maintaining an expensive complete IT infrastructure required for on-premises execution of the relevant information processes, customers of cloud computing vendors[21] pay only for the services they consume.

---

[16] "De Burger in De Ketens: Verslag Van Nationale Ombudsman over 2008," (Dutch National Ombudsman, 2008).

[17] Paul Michael Garret, "Social Work's 'Electronic Turn': Notes on the Deployment of Information and Communication Technologies in Social Work with Children and Families," *Critical Social Policy* 25, 4 (2005). 536

[18] Chief Secretary to the Treasury, 2003: 53-4, cited in Ibid.538, emphases added by Garrett.

[19] "De Burger in De Ketens: Verslag Van Nationale Ombudsman over 2008."

[20] For more details on cloud computing see, e.g. Richard Martin, J., Hoover, Nicholas, "Guide to Cloud Computing," In *Information Week: the business value of technology* (2008).

[21] Vendors of cloud computing services include Amazon Web Services, Google App Engine, Salesforce, etc.

Cloud computing is also widely available for private use in the form of web-based email services, photo storing services, online backup services, file transfer services such as YouSendIt, online medical records storage such as Microsoft's HealthVault, and applications associated with social networking sites.[22] When customers store their data with the vendors' hardware, they lose visibility and a large share of control over the fate of that data, including its protection from hacker attacks and transfers to marketing industry and government agencies.[23]

3.2.1.3 Ambient Intelligence

Ambient intelligence (AmI) – or Internet of things – refers to an architecture where computers "melt invisibly into the fabric of our [. . .] life."[24] From a technological point of view, ambient intelligence is enabled by data communication tools, e.g. RFIDs, "planted" into various items: household objects, clothes, personal communication devices, goods, etc,[25] which, as a result, become "smart" and communicate information about or around themselves and "act" in accordance with this information. For example, this technology can be used to monitor supply of goods and provide for their immediate delivery.[26] It also can be used to monitor and identify people, "since all possible everyday objects will be part of a network."[27] Various "intelligent" objects have been marketed already, including a Japanese "intelligent bathroom" where one's blood pressure, weight, and sugar level are measured, urine analyzed; the test results are transferred to a home network and displayed on a computer spreadsheet, followed by advice on diet and exercise, and all without any human intervention.[28] But imagine that all these data are transferred to one's GP. In fact, "smart" wrist bands have already been used to monitor from a distance the condition of chronic patients and report them to a hospital if they had a seizure. The idea of full-scale ambient intelligence responsive to every need of an individual may sound like something out of Science Fiction, however, one gets a sense of how close this future is after checking, e.g. the Phillips research web-site which

---

[22] Privacy Rights Clearinghouse, "The Privacy Implications of Cloud Computing".

[23] The personal data related concerns resulting from cloud computing will be addresses in more detail further on in this Chapter. Meanwhile, see e.g. Ann Cavoukian, "Privacy in the Clouds - a White Paper on Privacy and Digital Identity: Implications for the Internet " (Information and Privacy Commissioner of Ontario, 2008).; Robert Gellman, Privacy in the Clouds: Risks to Privacy and Confidentiality from Cloud Computing, (The World Privacy Forum, 2009).

[24] Paul De Hert, "A Right to Identity to Face the Internet of Things?."

[25] H. Rolf Weber, "Internet of Things - New Security and Privacy Challenges," *Computer Law & Security Report* 26, 1 (2010). at 23

[26] Ibid.

[27] De Hert, "A Right to Identity to Face the Internet of Things?."

[28] "Health Checks from Your Doctor Could Be Replaced by Visits to the Bathroom, Thanks to a Smart Toilet Developed by a Japanese Company.," *CNN.com* 2005, no. June 28 (2005), http://www.cnn.com/2005/TECH/06/28/spark.toilet/index.html.

reports progress in this area.[29] Wikipedia predicts the AmI will become a reality in 2010–2020.[30]

At the moment ambient intelligent is not yet such a contemporary factor as chain informatisation and cloud computing. Yet, it presents similar features: increased number of "smart objects" collecting personal information not only in the home but also on the streets, supermarkets, etc. being present in ever more human life matters, connected into a network controlled by various and multiple actors – goods- and service providers, soft- and hardware maintenance services and an individual himself (the computer controlling the "smart bathroom" is located in one's home).

3.2.1.4 The Challenges

The common factor present in all three phenomena and distinguishing them from earlier developments in data processing is the growing number of actors involved and the relationships between them. On the one hand, this is a step along the line of the earlier tendency for the information technology to become more widely used and an increase in a number of the data processing actors. On the other hand, if the presently-in-force fourth generation of data protection is aimed to regulate relatively simple sequences of relationships between those actors, relationships currently characteristic of chain informatization, cloud computing, and in future, ambient intelligence are of a different scale of complexity and clearly are the case when quantity of the actors and relationships had impact on their quality.

More specifically, although the data flow in the 1990s already involved more and more participants, it was relatively easy to map. After collection, personal data was retained by the initial collector for his needs or transferred to several other actors for processing on the order of the collector or for other use. Despite a growing number of transfers, the data flow remained relatively linear with a few branched lines. With the advance in information technology and practices of the 2000s, especially, developments in the Internet use enabling data clouds and chains, and the Internet of things, the number of actors involved in the data flow has multiplied in geometric progression, so have the number of relationships between them, going beyond simple chains to form a massive three-dimensional spider's web. In fact, research revealed that the paths that packets of information take as they travel across the Internet form a dandelion-like structure.[31]

A data subject is in the centre of the web. Each node in the structure represents a data processing actor that is connected with the actors in the same chain of data flow but also, by means of cloud computing, interconnected with other actors not in

---

[29] http://www.research.philips.com/newscenter/pictures/systsoft-ambintel.html

[30] http://en.wikipedia.org/wiki/Ambient_intelligence

[31] Daniel Kane, "Digital Dandelions: The Flowering of Network Research," *USCD News Center*, no. August, 31 (2007). Available at <http://ucsdnews.ucsd.edu/newsrel/science/08-07DigitalDandelionsDK-.asp>

one but in multiple chains. The links connecting the nodes stand for the paths data pertaining to the data subject take – relationships within the data flow.

The dandelion represents not several independent databases but effectively one large database where a piece of data can move from actor X to Y by taking a multiplicity of various shorter or longer paths, with a smaller or bigger number of steps and, where a smaller or bigger number of actors are involved.

This new complexity of relationships within the data flow reinforces old and raises new data protection concerns, in particular, those of transparency and accountability. Even more so, the lack of transparency in the data flow makes accountability for data protection violations a virtually unattainable goal. First, the paths personal data may take within the web are extremely entangled and difficult to trace or predict and therefore also to regulate; second, within the multiplicity of the intertwined information chains, it is not clear how the burden of accountability for data protection is distributed among all involved actors since the identity of the participating actors, as well as their exact contribution to the entire process are not clear.

For instance, when a mistake or a data security breach occur in the context of chain informatisation, it is difficult to name a single responsible government agency supplying, retaining or analysing data, for it is not always clear how the data at hand moved from point A to point B. Failures are blamed on the system, i.e. on its complexity. Moreover, it takes a long time to correct a mistake: first, the mistake has to be reported to the agency which used data in question, then the original database from where the authentic data was retrieved has to be notified, has to look into the mistake, verify the data, and then let the subsequent link of the chain know if the data was indeed false and share corrected data. The organisation receiving new data has to make sure the mistake is corrected in its database. In the meantime, a citizen suffers the consequences of "bad" informatisation. In addition, because different actors possessing better or poorer data management resources are in charge of them, databases of different scales and quality are merged together and inherently difficult to control, and protect against data security breaches. It is also difficult to make sure that all the actors who copied the false piece of data into their systems subsequently corrected it.[32]

Cloud computing and, in future, ambient intelligence open access to personal data to the third parties – contractors providing data storage, management, and analysis services – therefore represent similar quantity-grown-into-quality dangers. Briefly, the bigger the number of data transfers between the actors the higher the likelihood of errors, data loss, security breaches, etc., and the lower the chances of identifying the responsible actors. Finally, especially on the Internet, the facts of collection, analysis and implementation of one's personal information are not obvious to a lay individual: although the knowledge that some information is being collected can be expected, which information that is will not be obvious, just as who collected it, what algorithms have been used to analyse it, and who, how and when if at all

[32]"De Burger in De Ketens: Verslag Van Nationale Ombudsman over 2008."

they will be using it. The next section considers how the current European data protection mechanism copes with the challenges presented by the modern structure of the data flow.

3.2.2 Shortcomings of the Current Approach

This section will show that the data protection mechanisms established by the 1995 Data Protection Directive do not adequately grasp the new structure of relationships within data flow and therefore, are not able to control modern processing of personal data. In particular, they are not able to cope with the lack of transparency in the modern data flow; as a result, the structure of accountability established in the Directive is difficult to enforce.

By accountability this paper means the system of judicial remedies, liability and sanctions prescribed in Chapter III of the Directive, as well as the answerability to the supervisory authority under, e.g. Art. 18 etc. The Directive imposes the entire burden of accountability on the data controllers – only one certain type of actors involved in data processing. A data controller is defined as a person or entity who determined the purposes and means of processing of personal data (Art. 2(d)). At the same time, the Directive itself lists three other types of actors who can be potentially involved: data processor (Art. 2(e)), third parties (Art. 2(g)), and a recipient of data (Art. 2(f)).[33] Hypothetically, it is the controller who will be liable although the actual fault may lay with any other actor in a chain where the controller is only a link, or even in a totally different segment of the information dandelion as long as that segment is connected by a single link. Relationships between a controller and processor are governed by a contract (Art. 17(3)) giving no rights to the data subject. In case of a violation, to determine if a certain actor is a controller and not a processor, a third party or a receiver may present a difficult task since the business models regarding data processing are so variable that in a given data transaction they do not always fit into the rigid definitions of the actors given in the Directive.[34]

However, the current system relying on the liability of controllers does not motivate the actors who cannot be unambiguously classified as controllers to take steps to ensure proper level of data protection since there is no immediate possibility of action but a delayed contractual liability. It may be argued that assigning liability to the controller in all cases aids data subjects by strengthening their position; indeed, if that is a processor who is at fault for a breach, in the data protection dispute it faces not weak individual but its peers – controllers who are also large companies just like their opponents. It arguably serves to protect a data subject – a weaker party,

[33]Lee A. Bygrave, *Data Protection Law: Approaching Its Rationale, Logic and Limits*, vol. 10, Information Law Series (Kluwer Law International, 2002). at 21

[34]The problem of distinguishing a data controller from a processor is quite common in the data protection literature. See, e.g., Christopher Kuner, *European Data Protection Law: Corporate Compliance and Regulation*, 2nd ed. (New York, NY: Oxford University Press, 2007). 70–71

and reassign the burden of dispute resolution to the controller – a stronger corporation. However, before that happens, an individual first has to face that same strong corporation and get his way with it. Besides, whether or not to pursue the processor is on the discretion of controllers and, especially in cases of individual and other low-profile breaches, not to pursue is a cheaper option.

However, a bigger problem of accountability in the modern data flow has nothing to do with an inadequacy of formal classifications, but with a lack of certainty regarding a *specific actor* that has to be accountable for a data protection violation. Indeed, in the information chains every actor may be considered as a controller since every actor determines their own purposes and means of personal data processing. A clear example would be the system of sharing of medical files between general practitioners, medical specialists, and hospitals. They all have their own specific interest in the data of a patient. The same may be said about the actors in data clouds. The reality of business models is such that the providers of the on-line services may act not only as processors, but also process personal data for purposes and in ways they need. For instance, Rebeca Wong has argued that on social networking sites such as Twitter and Facebook individual users who post information of their friends should be regarded as controllers just like Facebook itself.[35] It seems valid to continue that Facebook applications constitute controllers when they provide the individual with access to games, quizzes and other services in exchange for access to the data of the user and his friends. This well illustrates the problem of accountability: when something goes wrong in the data flow of the social network site and personal data is abused, corrupted, or disclosed to third parties as a result of a security breach, without ambiguity as to who a controller is, it is still hard to establish with certainty within which fragment of the cloud the breach occurred and which specific controller is liable.

The complexity of the modern data flow and resulting lack of transparency makes government monitoring and enforcement of data protection even more difficult than it used to be. As Bergkamp points out, even "in the past, business could survive under European privacy legislation only because enforcement was extremely lax and the government could grant ad-hoc privileges in any event. Even in member states that have had data protection laws on the books for more than a decade, the number of sanctions imposed for violations of the legal standards is very small."[36] In the age of chain informatisation, cloud computing, and the advent of ambient intelligence, when the number of controllers to supervise has exploded it is unreasonable to think that the supervisory authorities who are limited in time and resources, will cope with data protection enforcement better than before. Professor Bergkamp, also a practicing lawyer, pessimistically concludes that "as a result, regulated entities do

[35]Rebecca Wong, "Social Networking: Anybody Is a Data Controller?," *Social Science Research Network* (2008).

[36]Lucas Bergkamp, "Eu Data Protection Policy the Privacy Fallacy: Adverse Effects of Europe's Data Protection Policy in an Information-Driven Economy," *Computer Law & Security Report* 18, 1 (2002). at 37

not have appropriate incentives to comply with the law."[37] Otter observes that data protection is on the bottom of the list of priorities for the IT companies as well.[38]

Hence, in light of the new complexity of the relationships within the data flow, there is an apparent need to transform the system of accountability for data protection violations in a way that reflects the reality of dandelion structure of the relationships between the actors in the modern data flow and their matching responsibilities. The rest of the paper will be dedicated to show that propertisation of personal data – combined with regulation – may be the way forward.

3.3 Introduction into the Propertisation Debate

The goal of this section is to make some basic ground statements concerning property vital for further analysis of propertisation of personal data. It will establish that, although the idea has nearly lost its attractiveness for an academic debate, to speak of property in personal data still makes more sense when the participants of the debate agree on the perspective of analysis. The perspective of choice of this contribution is the one of law (3.1). Moreover, the legal concept of property, at least, in theory, is fluid, i.e. able to cover a wide range of objects and various scopes of rights, enough to achieve multiple regulatory goals (3.2.1) not only in the common law discourse, but also in continental Europe (3.2.2).

3.3.1 Agreeing on Terms

A remarkable trait of the body of literature on propertisation of personal data is a puzzling and persistent presence of contradictory statements about what property is and is not, what it will or will not achieve, and whether it is moral or not. One of the reasons for such confusion is that arguments for and against propertisation have often been made and criticised from different perspectives, and the participants of debate have not spent much time specifying from which perspective they approached the subject, *i.e.* whether they advocate for or object against property as understood in economics, a certain philosophical theory or elsewhere. As a result, when engaging in the debate about property in personal data, its participants derive from their economic or philosophical background to draw some initial assumptions about property and as a foundation for the position they take. The problem, however, is that the assumptions about property in different intellectual areas may not be the same. Not acknowledging those differences provides for a debate without a constructive outcome.

---

[37] Ibid.

[38] Thomas Otter, "Data Protection Law: The Cinderella of the Software Indudtry?," *Computer Law & Security Report* 23 (2007).

It is quite common in private law literature on the concept of property to accentuate that the meaning of property in economic theory is substantially different from the one in law,[39] and moral theories only focus on moral justifications of property rights, not their content like the law does.[40]

However, in the literature on property in personal data, moral arguments are still often used in a legal debate, and legal property rights are still equated with property rights in economic theory.[41] In the meantime, it is overlooked that the concept of property in law as well carries a meaning, in a large part independent of the meanings assigned to property elsewhere. Therefore, for the consistency of the position defended in this contribution it is essential to note that the analysis here relies on a *legal* concept of property. That means that the idea of property in personal data is considered viable not because it is just for an individual to have property rights over the data pertaining to him, as normative theories would command, and not because it makes economic sense; the concept of property in personal data is considered here solely based on the implications – or consequences – of its use in law. The next section elaborates on the meaning of property in law and its implications.

3.3.2 Possibility of Propertisation of Personal Data

This section focuses on the possibility of property in personal data, namely, that this legal concept is flexible enough to stretch and embrace new objects and rights of a scope needed to achieve various regulatory goals (3.2.1) both in common and, theoretically, continental law (3.2.2).

3.3.2.1 Fluid Nature of the Concept of Property in Law

Development of the legal institution of property is a good example of pragmatism in law: while philosophers of law are occupied with normative justifications of existence of property, the actual property rights address certain practical needs

---

[39]E.g., Yoram Barzel, *Economic Analysis of Property Rights*, 2nd ed. (Cambridge: Cambridge University Press, 1997). p. 3: "Property rights in economics are "the individual's ability, in expected terms, to consume the good (or the services of the asset) directly or to consume it indirectly through exchange. [...] Legal rights are the rights recognized and enforced by the government. These rights, as a rule, enhance economic rights, but the former are neither necessary nor sufficient for the existence of the latter. A major function of legal rights is to accommodate third-party adjudication and enforcement. In the absence of these safeguards, rights may still be valued, but assets and their exchange must then be self-enforced."

[40]E.g. James Gordley, *Foundations of Private Law : Property, Tort, Contract, Unjust Enrichment* (Oxford [etc.]: Oxford University Press, 2006).

[41]For examples of this terminological confusion in the US discourse see Purtova, "Property Rights in Personal Data: Learning from the American Discourse."

that emerged in a given society.[42] Those needs change, as does the society in question. Therefore a noticeable characteristic of the concept of property is its fluidity. The commentators speak of the evolution of property, its flexibility and dynamism regarding different objects, scope of rights varying across time and space and determined by socio-economic reality.[43]

The range of objects open to property rights is not static and may well include personal data. Indeed, as Gray points out, "I may have "property" in a resource today, but not tomorrow."[44] Equally, the fact that no property rights in an object are recognized at the moment does not necessarily mean that this will not change in future. To name only few examples of exclusion and inclusion of the objects of property rights, human beings themselves stopped being an object of property rights relatively recently;[45] and early in the twentieth century Canadian and US laws laid down that "no property rights were to exist in alcoholic beverages."[46] Regular air traffic as a consequence of technological developments has led to the "shrinking" of the object of rights in land in English law; if before the advance of aeronautics the holder of the rights in land had, optimistically stated, a prima facie ownership "of everything reaching up to the very heavens and down to the depth of the earth,"[47] to allow air traffic over England the landowners' property rights over the airspace had to be limited to the "lower stratum" control over which is essential to the enjoyment of the piece of land itself.[48] Advances in medicine have routinely resulted into (at least attempts of) property claims in human bodies and bodily parts: eggs, donor organs and tissue, etc.[49] The example of intellectual property indicates another direction in the evolution of property – its dephysicalization. As a consequence of dephysicalization property rights shifted from being the rights with regard to solely tangible things to intangible, including information such as

---

[42]Gordley, *Foundations of Private Law : Property, Tort, Contract, Unjust Enrichment.*

[43]J.W. Bruce, Ely, James W. Jr., *Cases and Materials on Modern Property Law*, 6th ed. (Thomson West). p. 19; Remigius N. Nwabueze, *Biotechnology and the Challenge of Property*, edited by Sheila McLean, Medical Law and Ethics (Aldershot – Burlington: Ashgate, 2007).; Kevin Gray, "Property in Thin Air," *Cambridge Law Journal* 50, 2 (1991); Roy Vogt, *Whose Property? The Deepening Conflict between Private Property and Democracy in Canada* (Toronto, ON: University of Toronto Press, 1999). Etc.

[44]Gray, "Property in Thin Air." P.296

[45]In the US the XIII Amendment to the Constitution abolished slavery in 1865; in Eastern Europe slavery started gradually to disappear in the 15th century but formally seized to exist in Russia in 1861 (see "Slavery." Wikipedia, at http://en.wikipedia.org/wiki/Slavery).

[46]Arnold S. Weinrib, "Information and Property," *University of Toronto Law Journal* 38 (1988). At 121

[47]Gray, "Property in Thin Air." p. 253

[48]Ibid. p. 254

[49]A recent example of such attempts is Donna Dickenson, *Property in the Body: Feminist Perspective* (Oxford: Oxford University Press, 2007).

trade secrets,[50] ideas and claims,[51] and bits of data composing virtual objects in on-line worlds.[52]

Ultimately, debates on propertisation of new objects often present a struggle to find a new regulatory solution in a certain area rather than a mere debate on whether certain objects may or may not be objects of property rights. Whether participants of those debates realize it or not, they talk of property as a legal means to achieve regulatory goals. For instance, when anatomy became a standard medical practice, dead bodies have suddenly gained economic value; in the absence of legitimate institutional arrangements, the initial source of supply was a group of people known as "body-snatchers." They stole newly buried bodies from their graves, but the absence of the common-law property in corpses did not allow to charge them with theft. In response the government adopted anatomy legislation; Nwabueze suggests that "part of the solution [...] is to consider corpses as limited property."[53]

Objects of property rights vary not just across time, but also across jurisdictions; same things may be treated as property in one country but not in the other. A good illustration is the so-called "virtual property" – commodities in cyberspace including on-line equivalents of the real world things, as well as e-mail addresses, domain names, social network site accounts, etc.[54]. At present the on-line resources are explicitly given property protection in the Republic of Korea and China's Taiwan and Hong Kong;[55] in the US or Europe, however, the recognition by law of virtual property is only debated.[56]

Similar to the objects of legal property rights, their structure and scope, too, "differ from one society to another, and within the same society from one period to another, because they are historically determined."[57] The scope of property rights in a given country constantly adapts to current needs – read "regulatory goals" – of the jurisdiction in question;[58] one of the common examples of such adaptation

---

[50] US case; in French civil law, new objects of property also have been developed, such as a business enterprise, and information - trade secrets Sjef Van Erp, "Security Interests: A Secure Start for the Development of European Property Law," *Maastricht University Faculty of Law Working Papers* (2008). P. 18 citing Libchaber, *La Recodification Du Droit Des Biens*.

[51] E.g. business goodwill in common and civil law (see Van Erp, "Security Interests: A Secure Start for the Development of European Property Law.")

[52] see Joshua Fairfield, "Virtual Property," *Boston University Law Review* 85 (2005).

[53] Nwabueze, *Biotechnology and the Challenge of Property*. P. 17

[54] See

[55] All these jurisdictions passed relevant laws and set precedents in giving criminal sentences to those infringing upon others' virtual property. (e.g., see Fairfield, "Virtual Property.")

[56] Ibid.

[57] Vogt, *Whose Property? The Deepening Conflict between Private Property and Democracy in Canada*. p. 17

[58] Although the rules of the civil law property model are characterised as "hard" and "inflexible," the commentators of the continental European property law observe that it as well "undergoes an evolutionary and thus gradual change, caused by changing social, economic, cultural and political conditions." (in Van Erp, "Security Interests: A Secure Start for the Development of European Property Law.", p. 16)

is increasing state regulation of property.[59] Besides, moral limits on property may differ across space: two different societies may operate under different normative convictions shaping the two sets of ownership interests differently;[60] those were considerations of public policy that prevented the court from vesting in Mr Moor a property right in his spleen,[61] but a court of another country could have operated under a different policy leading to a different outcome of the case and a property right in a part of a human body. As long as property rights are enforced by the state, their scope and objects are political and therefore depend on the political environment in a particular state.[62] To sum up, the concept of property in law is flexible; therefore, nothing in the nature of property in law prevents it from transforming to include personal data as one of its objects, provided it serves current needs of a jurisdiction in question, and there is political will to transform property.

3.3.2.2 Possibility of the Common-Law Debate in Continental Europe

In the preceding section a reader may have noticed that the examples of flexibility of the concept of property primarily come from the common law jurisdictions. Arguments in favour of propertisation of personal data are also made mostly by the common law scholars. One of the possible explanations is that, as noted by many commentators of property law,[63] common law property is much more flexible and open to new objects and scope of property rights. Continental law property based on the classical model is more conservative. However, research on the modern European law of property shows that some continental law systems slowly adopt some elements of common law property and therefore become more open to the idea of propertisation of personal data. This section will explain what quality of property in common law is the key of its flexibility and show that the same quality is slowly entering the continental Europe.

As it has been mentioned in the previous section, what property is varies a lot depending on whether a particular country, especially on whether that country belongs to the Anglo-Saxon legal tradition or has a continental legal system. To say that X has a property right in his house in a country within a continental legal system would most likely mean that X has a full ownership of the house, i.e.

---

[59]Think of, e.g. the gun laws, changing registration requirements in land law, etc.

[60]Human rights considerations may serve as moral limits on property rights: "'Property' in a resource stops where the infringement of more basic human rights and freedoms begins." (Gray, "Property in Thin Air." At p. 294)

[61]*Moore v. Regents of the University of California* (51 Cal. 3d 120; 271 Cal. Rptr. 146; 793 P.2d 479)

[62]The idea of property as a political institution appears in Jeremy Bentham, "Security and Equality of Property," In *Property: Mainstream and Critical Positions*, edited by C.B. Macpherson (Toronto, ON: University of Toronto Press, 1978), Gray, "Property in Thin Air."; Nwabueze talks about property rights reflecting expectations of the members of a given society as expressed by its political system. Nwabueze, *Biotechnology and the Challenge of Property*. E.g. p. 25

[63]E.g. Nwabueze, *Biotechnology and the Challenge of Property*. p. 9

with some limitations, he can possess it, enjoy it by living there himself or renting it out, and finally sell or otherwise alienate it. To an ear more familiar with the Anglo-Saxon legal lexicon, the same statement would not convey the same message. First, in English law the term "land law" rather than property law is used with regard to realty.[64] Second, and more importantly, a characteristic trait of the Anglo-Saxon system of property, especially, the land law, is the so-called "fragmentation of property rights" which means that, next to the ownership in the fullest sense – "fee simple" in the English land law vocabulary – other "smaller" property rights can exist in the same object, such as the rights of a tenant, leases of land, etc.[65] Property in the Anglo-Saxon legal tradition "can involve very different combinations of [the] constituent parts."[66] Such system of property rights is often described by a metaphor of a "bundle of rights." Roughly, the complete bundle represents full ownership and each "stick" in the bundle represents one of many "fragments" composing full ownership: right to use a resource, right to use it for a fixed period of time, conditioned upon fulfilment of an obligation or unconditionally, etc. Fragmentation of property is the key to the fluid character of property,[67] as well as to its ability to achieve various regulatory goals.

The phenomenon of fragmentation makes property flexible, first, because it enables transfer of resources without the necessity for an original proprietor to completely surrender all control over the resource which may be not desirable regarding some resources; second, fragmentation implies that property does not always mean the fullest control over a resource in one hands not desirable regarding some objects; in common law property rights of scope narrower than the full ownership receive the same protection against third parties. Simultaneously, giving property rights in an object in the meaning of the classical model always implies the fullest possible control over the resource; hence, propertisation of any object, especially as unconventional as personal data, has greater implications and is harder to accept in continental than in common law.

Each "stick" can be kept in the bundle or held independently. As a result, there may be more than one person holding different property rights towards the same object. By assigning property rights of various scope (and corresponding obligations to respect those rights) it is often possible to create a regulatory regime – a system of desired rights and responsibilities – with regard to a certain resource including personal data.[68] For instance, tenant-landlord relationships were given a (partial) property status when it became clear that the purely contractual nature of the tenants' rights did not provide them with desired protection.[69]

---

[64] F.H. Lawson, Rudden, B., *The Law of Property*, 3rd ed., Clarendon Law Series (Oxford, Oxford University Press, 2002).

[65] Ibid. starting on p. 90

[66] Weinrib, "Information and Property." At 121

[67] Nwabueze, *Biotechnology and the Challenge of Property*. P. 9

[68] Section 3.4 will elaborate on how it is possible with regard to personal data.

[69] Lawson, *The Law of Property*.

The boundaries of property in the Anglo-Saxon legal tradition – possible objects and content of property rights – "are still to be explored."[70] That does not mean, however, that the boundaries of what can be a property right both in the continental and common law systems do not exist. Under a so-called principle of *numerus clausus*, parties are not free to create previously non-existing property rights at will.[71] The application of this principle in the continental legal systems is quite strict, although the degree of rigour varies from country to country.[72] In English law, although the property law is not completely inclusive and the *numerus clausus* principle applies, the courts are more willing to recognise new property rights than their counterparts in Continental Europe.[73] The Continental model of property law is based on a so-called "classical model" coming from Roman law and implying a closed system of undividable ownership rights.[74]

Nevertheless, globalisation of modern economy has led to the need of laws in different countries to accommodate international trade practices, including first steps towards convergence of property laws; the fragmentation of property rights has touched property institutions in the continental Europe. Especially French law shows signs of openness to fragmentation of property law;[75] the process may become European-wide under the influence of the EU legislation and ECJ case-law (which already recognises claims and social security rights as property).[76] Finally, the ECJ decisions of the *Cassis de Dijon* line promote further harmonization of European property law when establish that when an object is tradable in one country, it has to be tradable to the same extent throughout the common market.[77]

In one way or another, the idea of fragmentation of property rights has entered continental Europe. That means that, even if only in theory, the continental legal thought became more open to recognize rights "lesser" than absolute dominion as property rights and, as a result, the idea of property rights in unconventional objects like personal data.

---

[70]See Charles A. Reich, "The New Property," *Yale L.J.* 73 (1964).; common law property framework is used for analysis of many relationships, also unconventional objects of property such as race, social security entitlements, etc. Nwabueze, *Biotechnology and the Challenge of Property*.

[71]Bram Akkermans, *The Principle of Numerus Clausus in European Property Law* (Antwerp - Oxford - Portland: Intersentia, 2008). P. 19

[72]Ibid.

[73]Ibid. 389 et seq.

[74]Sjef Van Erp, "From 'Classical' To Modern European Property Law?," *Maastricht University Faculty of Law Working Papers* (2009).

[75]Van Erp, "Security Interests: A Secure Start for the Development of European Property Law."

[76]K. Lenaerts, Vanvoorden, K., "The Right to Property in the Case Law of the Court of Justice of the European Communities," in *Property and Human Rights*, edited by H. Vandenberghe (Bruylant, 2006).

[77]Prof. van Erp develops this point in Van Erp, "Security Interests: A Secure Start for the Development of European Property Law."

3.4 Property Rights As a Regulatory Framework for the Modern Data Flow

The previous section established the possibility of extending the legal institution of property to include new objects and rights. This section will explain why propertisation of personal data is not only possible but also makes sense in light of the new challenges of the modern information flow (4.1). Section 3.4.2 addresses the most common criticism of the idea of propertisation and explains that introduction of property rights in personal data does not mean free trade in that data.

3.4.1 What Property Rights Have to Offer

It has been shown earlier in this paper that the current data protection mechanism enshrined in the 1995 Directive does not account for a new complexity of relationships within the modern data flow. Namely, the obligations of data protection effectively lie only with and are enforceable against data controllers. This approach disregards two main characteristics of the current information flow: first, that the paths personal data take are extremely entangled and difficult to trace and hence it is difficult to know where the wrongfully disclosed or used data came from and who the relevant controller is; second, even when it is known where the data came from, it is not clear how the functions of a controller or processor are distributed among all involved actors since the number and identity of the participating actors, as well as their exact contribution to the entire process are often not clear. Further analysis shows that due to its *erga omnes* effect and a possibility of fragmentation, property is able to deal with those complexities.

The *erga omnes* effect is a feature that distinguished property ("real") rights from personal rights.[78] This holds both for the common law and continental legal systems. The *erga omnes* effect entails that property rights have effect against all persons by creating negative obligations for them without their consent.[79] Transforming rights towards personal data into property rights and attributing the *erga omnes* effect to them would mean the elimination of differences in responsibilities between the data controllers and data processors; if the legislators chooses to keep this classification at all, both categories of actors would be bound by a negative obligation to respect the rights of a data subject towards data pertaining to him. This change would be consistent with the uncertainty of the roles of the actors in the modern data processing described earlier in this paper.

---

[78] Personal rights create obligations only for the parties of a contract. Steven Bartels, et al., *Content of Real Rights* (Nijmegen: Wolf Legal Publishers, 2004).; Michael J. Milo, "Property and Real Rights," in *Elgar Encyclopedia of Comparative Law*, edited by Jan M. Smith (Edward Elgar, 2006).; Van Erp, "From 'Classical' To Modern European Property Law?."; Gray, "Property in Thin Air."

[79] Van Erp, "From 'Classical' To Modern European Property Law?."

Whatever the position of any given actor is within an information chain of any degree of complexity, that actor will be expected to make sure his actions are not crossing the borders of property rights in personal data. As a result of propertisation, the burden of finding the right actor to bring an action against are is removed from a data subject. Namely, when a violation of data protection principles is discovered, the choice of a data subject is not limited to any particular actor in a given information chain; an action could be brought against any actor if, however, it is not clear where exactly the data regime was violated and what actor at whose disposal the personal data in question was at the time when the data protection breach was discovered, or against the actor who was "caught" using personal data in question without proper authorization. The burden to ensure that data transfers occurred without violations would lie on each and every actor "in the cloud" or "in the chain", so would the liability. After paying damages, the actor in question would have a chance to look further "down the chain" for the source of the violation.

Special attention has to be paid here to the scope of property rights in personal data. To gain a better insight into how property may grasp the complex relationships vis-à-vis personal data and form regulatory framework of the data flow it will be helpful to look at the English land law system governing, what a continental lawyer would call "property rights in immovables". Similar to personal data, land is a valuable resource which is also put to many uses. To accommodate those, and also grant interests in land special protection, modern land law developed into a pyramid-like system of rights and interests with a right of a widest scope – fee simple – at the bottom, and leases – property rights of a narrower scope.[80] Skipping the details,[81] let us just make some basic observations relevant for the present discussion. The content of those rights has been tailored to account for the most popular uses of land, and, according to the principle of *numerus clausus*, no other rights in land except the ones on the list receive the *erga omnes* protection.[82] The transfer of the leases – "smaller" rights in a piece of land does not undermine, although limits, the "bigger" right of fee simple. But at all time, until fee simple is transferred in full, its holder retains some control over his property, e.g. the right of access to maintain the property in a proper state, etc.

In search for that quality – retention of control after transfer – a similar system of property rights could be built around personal data. An individual – data subject – may be said to have the widest, albeit not unlimited, property right possible

---

[80]Next to the common law rights in land (property in law), there are rights in equity developed by the courts within English system but of a different jurisdiction (e.g., covenants prohibiting a certain use of land for future buyers). It is not the purpose of this paper to go into details of the English land law.

[81]For more information on the matter see, e.g. Alison Clarke, Kohler, Paul, *Property Law: Commentary and Materials*, edited by William Twining, McCrudden, Christopher, Law in Context (Cambridge: Cambridge University Press, 2005).; Akkermans, *The Principle of Numerus Clausus in European Property Law*.

[82]As Akkermans explains, there is a slim chance of inclusion of a new right into the list of property interests. Akkermans, *The Principle of Numerus Clausus in European Property Law*.

including a right to transfer his personal data for remuneration (in common language known as a right to "sell"). The most important limitation on the possible scope of this right would be prohibition of waiver of the data protection guarantees, e.g. consent, etc.[83] Therefore, this biggest right would never have been able to be completely alienated. The part of the rights concerning personal data transferred from the data subject to data controllers and processors is comparable to the leases in land law; the alienable "leases" in personal data may be tailored to reflect most common practices with regard to personal data and vary in types, depending, for instance, on the duration and purpose limitations of the "lease," e.g. excluding the use of the data for profiling. The "leases," similar to the ones in land law, may be transferable as well; this way the introduction of the "leases" would be a response to the calls of the information industry to protect their investments in collecting data by recognising their property rights; the system of "leases" will protect the investments (by granting *erga omnes* protection, also against data security breaches). Moreover, recognising pursuant to the principle of *numerus clausus* only a closed list of "smaller" property rights in personal data would be one step closer to ensuring that individuals are not forced, as it often happens, into giving away unlimited range of their control over personal information by giving them a choice either to provide data or not to be able to use services which can be more or less difficult to do without, e.g. an email account or a plane ticket.[84]

Further transfers of personal data within a cloud or a chain may also take shape of transfer of – even "smaller" – property rights or contractual relationships. The benefits and drawbacks of the two options may be the subject of another publication. Whichever that will be, vesting the "biggest" *erga omnes* property right with a data subject and the possibility of fragmented property rights in personal data alone are able to achieve a desired goal to secure that a data subject retains control over his personal data regardless of the complexities of the modern data flow.

3.4.2 Market vs Non-Market Meaning of Property: Rebuttal to One Objection Against Property in Personal Data

A traditional objection to propertisation of personal data is that it would encourage free market in personal information, not control it. Since the present analysis rested on a core of the concept of property in law – *inter alia* its *erga omnes* effect – it is essential to explain here that some features – like free market alienability – often attributed to property in lay debate are not defining. This section will demonstrate that it is a misperception to link property rights and free market, and that modern

[83] For more on limitations of alienability of personal data see Nadezhda Purtova, "Private Law Solutions in European Data Protection: Relationship to Privacy, and Waiver of Data Protection Rights," *NHRQ* (2010).

[84] see Roger Brownsworth pointing at such a shortcoming of a consent requirement.

property law is increasingly relied on to exercise its protective rather than its market function.

A number of the commentators generally see commodification (and propertisation as a legitimized commodification) of certain goods including personal data as a problem. This is a "public good" argument which generally implies that information privacy has value not only for an individual, but also for a wider society. The market is unable to account for the latter. For instance, Katrin Schatz Byford submits that regarding "privacy as an item of trade ... values privacy only to the extent it is considered to be of personal worth by the individual who claims it."[85] Pamela Samuelson argues that propertisation of information privacy as a civil liberty might be considered "morally obnoxious."[86] "If information privacy is a civil liberty, it may make no more sense to propertize personal data than to commodify voting rights."[87, 88] One of the main points made in this contribution in defence of propertisation of personal data is that, if property rights are structured in a certain way, even after transfer of some control also by "selling" a fraction of rights an individual would always retain essential control over his personal data, e.g. allowing and defining the goals of data processing. It is more appropriate to define this function of property as regulatory or protective of data protection rights rather than serving free market. This partially addresses the "public good" objection to propertisation. However, further follow some additional points rejecting the idea of a purely market nature of property and supporting its use as a protective or regulatory tool.

First, unlimited market alienability is a myth. Alienability of any object of some public significance, e.g. food, medication, children's products, homes with minors as residents etc. – is heavily regulated by the state. Free alienability inherent in and necessary for the market function of property may be limited since property is never absolute.

Second, although the main property function is rightly said to be to protect "value" from third parties, that value may be both material and immaterial.[89] The latter is in no way linked to the free market. For instance, in English law, as Lawson points out, in the twentieth century the concept of property changed from a means of securing and investing wealth to "a direct denial of the general commercial thesis that every physical thing can be adequately replaced by its price in money."[90] This new vision has especially influenced tenant-landlord relationships where the value of a home for a family began to substitute the value of wealth invested in a house.[91]

[85] Katrin Schatz Byford, *Privacy in Cyberspace: Constructing a Model of Privacy for the Electronic Communications Environment*, 24 Rutgers Computer & Tech. L.J. 1 (1998)

[86] Samuelson 2000, p. 1143

[87] Samuelson 2000, p. 1143

[88] Antoinette Rouvroy, Poullet, Yves, "The Right to Information Self-Determination and the Value of Self-Development: Reassessing the Importance of Privacy for Democracy," In *Reinventing Data Protection?*, edited by Serge Gutwirth, et al. (Berlin: Springer, 2009).

[89] Nwabueze, *Biotechnology and the Challenge of Property*.

[90] Lawson, *The Law of Property*. p. 198

[91] Ibid.

Third, the concept of property used in criminal, constitutional and human rights law also serves a protective (against theft, damage, or state taking) rather than market function. For instance, virtual property in the webherlands was granted criminal law protection sooner than property protection in private law.[92]

Fourth, it seems that the notion of property is so closely related to free market only in the western legal thought. Nwambueze quotes examples of some aboriginal societies familiar with property but not with sale.[93] Gray explains that a large proportion of the western scholarly writing focuses on the market side of property since formation of the common law property coincided with "the age of bargain and exchange."[94] One of the points of dispute in the modern common law property debate is whether alienability is a necessary element of property rights, with a sound position against it.[95]

Finally, as has been mentioned in Section 3.3.2.1, property and its meaning are inherently political. As long as property rights are enforced by the state, their scope and objects are governed by political goals of a given society. Hence, if there is a need in property as a protective or regulatory tool and a respective political will, the meaning of property will be shaped accordingly.

3.4.3 Limitations of Property: Necessity of Regulation

Several times in the course of the argument, a disclaimer has been made that property complimented with regulation will be able to achieve the desired control over the modern data flow. This section explains why property alone is not sufficient, and regulation is necessary.

A minor reservation with regard to propertisation is that it is a tool aimed to provide an individual with better control. However, part of the modern personal data problem is that, thanks to profiling and excessive availability of personal data, one does not need to reveal his/her personal information to be subjected to personal data related treatment, like prise discrimination. As long as there is enough data about people like the individual in question to build a profile, a very small bit of data like an IP address is enough to identify a citizen or a consume with a group and treat him/her accordingly.[96] Regulation may be a better fitting tool to address the collective dimension of the personal data problem.

---

[92] A.C. Lagemaat, Boonk, M.L., Briet, M., "Vermogensrechtelijke Aspecten," In *Recht in Een Virtuele Wereld: Juridische Aspecten Van Massive Multiplayer Online Role Playing Games.* (Elsevier, 2007).

[93] Nwabueze, *Biotechnology and the Challenge of Property.*

[94] Gray, "Property in Thin Air." P. 294

[95] According to Gray, "the criterion of 'excludability' gets us much closer to the core of 'property' than does the conventional legal emphasis on alienability or enforceability of benefits." Ibid.

[96] Lawrence Lessig, *Code 2.0* (New York, NY: Basic Books, 2006). at 217

Finally, the main limitation of the idea of propertisation is that it only bears negative obligations whereas positive obligations, are a vital part of data protection[97] and should therefore be introduced by regulation.

3.5 Conclusions

This contribution proposed to re-examine a familiar idea of property rights in personal data in view of the recent developments in information technology and practices. It has been shown that, as a result of chain informatisation, cloud computing, and the advent of ambient intelligence, the number of actors involved in processing of personal data and relationships and the connections between them have grown and will keep growing in geometrical progression. The resulting structure of the data flow is too complex for the existing data protection approach to grasp; namely, the paths taken by personal data and participation of individual actors are difficult to trace and, hence, to regulate. Property, with some limitations resolved by regulation, due to its *erga omnes* effect and fragmentation of property rights, has the potential to reflect and control this complexity of relationships. This may be considered an instance of property exercising its protective rather than market function; it aims at making sure that even after transfer of a fraction of rights, a data subject always retains basic control over his personal information.

A disclaimer should be made at this point. It is not the aim of this contribution to argue that propertisation is the best or only possible way to create a certain regulatory regime for personal data. Arguably, the same results can also be achieved by, *inter alia*, eliminating the distinction between the data controller and the data processor concerning data protection obligations under 1995 Directive. As Nwambueze notes when defending his "remedial framework" of property in dead bodies, body parts, and reproductive materials, "the choice of legal categories [property vs regulation – N.P.] is strategic and there is nothing in one category that makes it inherently better than the other."[98] Consistent with the earlier statements about the pragmatic nature of law, "the regime of property is adopted on the basis of its practical utility compared to the other frameworks."[99] Provided a political will is present, propertisation and regulation may achieve roughly the same results.

Among the competitive advantages of propertisation, proponents of Lessig would name the rhetoric effect of the word "property." Possibly, propertisation will raise more interest among people concerning their data protection rights than the

[97] Paul De Hert, Gutwirth, Serge "Making Sense of Privacy and Data Protection: A Prospective Overview in the Light of the Future of Identity, Location-Based Services and Virtual Residence in the Institute for Prospective Technological Studies: Report Eur 20823 En," *Security and Privacy for the citizen in the post-September 11 digital age: a Prospective overview* (2003).

[98] Jennifer Nedelsky, Property in Potential Life? at 44 cited in Nwabueze, *Biotechnology and the Challenge of Property*. 39–40

[99] Jane Churchill, Patenting Humanity at 281 cited in Ibid. 40–41

regulatory approach managed up to now. However, no conclusions can be made without some empirical research into the influence of property rhetoric.

Among the factors making propertisation less practical is the possible unwillingness of national governments, especially, in continental Europe, to change their traditional property law, even more so under international or supranational pressure. This unwillingness has been demonstrated during debates on Art. 295 TEU and Art. 1 Protocol 1 ECHR. Nevertheless, the goal of this contribution was merely to take a second look at the idea of property rights in personal data in view of the new challenges of information technology.

References

Akkermans, B. *The Principle of Numerus Clausus in European Property Law.* Antwerp, Oxford, Portland: Intersentia, 2008.

Bartels, S., et al. *Content of Real Rights.* Nijmegen: Wolf Legal Publishers, 2004.

Barzel, Y. *Economic Analysis of Property Rights*, 2nd ed. Cambridge: Cambridge University Press, 1997.

Bentham, J. "Security and Equality of Property." In *Property: Mainstream and Critical Positions*, edited by C.B. Macpherson. Toronto, ON: University of Toronto Press, 1978.

Bergelson, V. "It's Personal, but Is It Mine? Toward Property Rights in Personal Information." *UC Davis Law Review* 37 (2003): 379.

Bergkamp, L. "EU Data Protection Policy the Privacy Fallacy: Adverse Effects of Europe's Data Protection Policy in an Information-Driven Economy." *Computer Law & Security Report* 18, 1 (2002): 31.

Bruce, J.W., and J.W. Ely Jr. *Cases and Materials on Modern Property Law*, 6th ed. Thomson West, 2007.

Bygrave, L.A. *Data Protection Law: Approaching Its Rationale, Logic and Limits,* Vol. 10. Information Law Series. The Hague: Kluwer Law International, 2002.

Cavoukian, A. *Privacy in the Clouds – a White Paper on Privacy and Digital Identity: Implications for the Internet*: Information and Privacy Commissioner of Ontario, 2008.

Clarke, A., and P. Kohler. In *Property Law: Commentary and Materials*, edited by W. Twining and C. McCrudden, Law in Context. Cambridge: Cambridge University Press, 2005.

Cuijpers, C. "A Private Law Approach to Privacy: Mandatory Law Obliged?." *SCRIPT-ed* 4, 4 (2007): 304–318.

"De Burger in De Ketens: Verslag Van Nationale Ombudsman over 2008." Dutch National Ombudsman, 2008.

De Hert, P., and S. Gutwirth. "Making Sense of Privacy and Data Protection: A Prospective Overview in the Light of the Future of Identity, Location-Based Services and Virtual Residence in the Institute for Prospective Technological Studies: Report Eur 20823 En." *Security and Privacy for the citizen in the post-September 11 digital age: a Prospective overview* (2003).

Dickenson, D. *Property in the Body: Feminist Perspective.* Oxford: Oxford University Press, 2007.

Fairfield, J. "Virtual Property." *Boston University Law Review* 85 (2005).

Garret, P. "Social Work's 'Electronic Turn': Notes on the Deployment of Information and Communication Technologies in Social Work with Children and Families." *Critical Social Policy* 25, 4 (2005): 529–553.

Gellman, R. "Privacy in the Clouds: Risks to Privacy and Confidentiality from Cloud Computing." *The World Privacy Forum* (2009).

Gordley, J. *Foundations of Private Law: Property, Tort, Contract, Unjust Enrichment.* Oxford [etc.]: Oxford University Press, 2006.

Gray, K. "Property in Thin Air." *Cambridge Law Journal* 50, 2 (1991): 252–307.
"Health Checks from Your Doctor Could Be Replaced by Visits to the Bathroom, Thanks to a Smart Toilet Developed by a Japanese Company.". CNN.com, no. June 28 (2005), http://www.cnn.com/2005/TECH/06/28/spark.toilet/index.html.
Kane, D. "Digital Dandelions: The Flowering of Network Research." *USCD News Center* no. August, 31 (2007).
Kuner, C. *European Data Protection Law: Corporate Compliance and Regulation*, 2nd ed. New York, NY: Oxford University Press, 2007.
Lagemaat, A.C., M.L. Boonk, and M. Briet. "Vermogensrechtelijke Aspecten." In *Recht in Een Virtuele Wereld: Juridische Aspecten Van Massive Multiplayer Online Role Playing Games*, 21–40. Elsevier, 2007.
Lawson, F.H., and B. Rudden. *The Law of Property*, 3rd ed. Clarendon Law Series. Oxford: Oxford University Press, 2002.
Lenaerts, K., and K. Vanvoorden. "The Right to Property in the Case Law of the Court of Justice of the European Communities." In *Property and Human Rights*, edited by H. Vandenberghe, 195–241. Bruylant, Brussles, 2006.
Lessig, L. *Code 2.0.* New York, NY: Basic Books, 2006.
Lessig, L. *Code and Other Laws of Cyberspace*. New York, NY: Basic Books, 1999.
Lessig, L. "Privacy as Property." *Social Research: An International Quarterly of Social Sciences* 69, 1 (2002): 247–269.
Lessig, L. "The Law of the Horse: What Cyberlaw Might Teach." *Harvard Law Review* 113 (1999): 501.
Libchaber. *La Recodification Du Droit Des Biens*.
Martin, R., and J.N. Hoover. "Guide to Cloud Computing." In *Information Week: the business value of technology*, 2008, http://www.informationweek.com/news/services/hosted_apps/showArticle.jhtml?articleID=208700713.
Mayer-Schőnberger, V. "Data Protection in Europe." In *Technology and Privacy: The New Landscape*, edited by P.E. Agre and M. Rotenberg, 219–243. Cambridge, MA: The MIT Press, 1997.
Milo, M.J. "Property and Real Rights." In *Elgar Encyclopedia of Comparative Law*, edited by J.M. Smith, 587–602. Edward Elgar, 2006.
Nwabueze, R.N. In *Biotechnology and the Challenge of Property,* edited by S. McLean. Medical Law and Ethics. Burlington: Ashgate, 2007.
Otter, T. "Data Protection Law: The Cinderella of the Software Indudtry?" *Computer Law & Security Report* 23 (2007): 67–72.
Poullet, Y. "Data Protection Legislation: What Is at Stake for Our Society and Democracy?" *Computer Law & Security Report* 25 (2009): 211–226.
Prins, J.E.J. "When Personal Data, Behavior and Virtual Identities Become a Commodity: Would a Property Rights Approach Matter?" *SCRIPT-ed* 3, 4 (2006): 270–303.
PrivacyRightsClearinghouse. "The Privacy Implications of Cloud Computing ".
Purtova, N. "Private Law Solutions in European Data Protection: Relationship to Privacy, and Waiver of Data Protection Rights." *NHRQ* (2010).
Purtova, N. "Property Rights in Personal Data: Learning from the American Discourse." *Computer Law & Security Report* 25, 6 (2009): 507–521.
Radin, M.J. "Property and Personhood." *Stanford Law Review* 34, 5 (1982): 957–1015.
Reich, C.A. "The New Property." *Yale Law Journal* 73 (1964).
Rotenberg, M. "Fair Information Practices and the Architecture of Privacy (What Larry Doesn't Get)." *Stanford Technology Law Review* 1 (2001).
Rouvroy, A., and Y. Poullet. "The Right to Information Self-Determination and the Value of Self-Development: Reassessing the Importance of Privacy for Democracy." In *Reinventing Data Protection*?, edited by S. Gutwirth et al., 45–77. Berlin: Springer, 2009.
Schwartz, P.M. "Property, Privacy, and Personal Data." *Harvard Law Review* 117, 7 (2004): 2055–2128.

Solove, D.J. "Privacy and Power: Computer Databases and Metaphors for Information Privacy." *Stanford Law Review* 53 (2001): 1393.

Van Erp, S. "From 'Classical' To Modern European Property Law?" *Maastricht University Faculty of Law Working Papers* (2009).

Van Erp, S. "Security Interests: A Secure Start for the Development of European Property Law." *Maastricht University Faculty of Law Working Papers* (2008).

Vogt, R. *Whose Property? The Deepening Conflict between Private Property and Democracy in Canada.* Toronto, ON: University of Toronto Press, 1999.

Weber, H.R. "Internet of Things – New Security and Privacy Challenges." *Computer Law & Security Report* 26, 1 (2010): 23–30.

Weinrib, A.S. "Information and Property." *University of Toronto Law Journal* 38 (1988).

Westin, A.F. *Privacy and Freedom.* London, Sydney, Toronto: the Bodley Head, 1967.

Wong, R. "Social Networking: Anybody Is a Data Controller?" *Social Science Research Network* (2008), http://ssrn.com/abstract=1271668.

Chapter 4
Right to Personal Identity: The Challenges of Ambient Intelligence and the Need for a New Legal Conceptualization

Norberto Nuno Gomes de Andrade

4.1 Introduction

Personal identity is, unquestionably, one of the most complex, multifarious and polysemic concepts permeating the intellectual history of mankind. Such primordial concept has been the focus of many studies and analyses, pervading a myriad of different disciplines and fields of study, ranging from philosophy and anthropology, to psychology, biology and medicine, among many others. Such concept is not only intrinsically multidisciplinary and multifaceted, but also – and incessantly – dynamic, being subject to constant evolution. Such characteristics render the task of defining the exact meaning and scope of identity an extremely difficult one, if not ultimately impossible.

Through this article I will attempt to circumvent those difficulties and tackle such task in two steps. Firstly, I will narrow the analysis of the subject of identity to a strictly legal dimension, focussing upon the so-called right to personal identity. Its analysis shall include a brief account of its legal evolution (Section 4.2), followed by an examination of the main challenges brought to the right to personal identity by new and forthcoming technological developments, namely the ones posed by the vision of ambient intelligence (Section 4.6). Secondly, and while acknowledging those challenges, I will strive to demonstrate the pressing need to rethink the right to personal identity. As such, and after a critical analysis of the configuration of the right to identity made by the Italian jurisprudence, I will propose a renewed conceptualization of this legal disposition. Within such proposal, I will present a list of different "sub-rights" which should be accommodated under the umbrella of the right to identity: the right to be forgotten and the right to multiple identities (Section 4.7).

N.N.G. de Andrade (✉)
Department of Law, European University Institute, Florence, Italy
e-mail: norberto.andrade@eui.eu

S. Gutwirth et al. (eds.), *Computers, Privacy and Data Protection: an Element of Choice*, DOI 10.1007/978-94-007-0641-5_4, © Springer Science+Business Media B.V. 2011

4.2 The Right to Personal Identity

The right to identity is a rather intriguing and puzzling right. Before entering into any legal technicalities and theoretical constructions, the right to identity – on a first approach – could be described as a right designed to protect aspects and elements of my own identity. In other words, and in an even more direct and linear fashion, I could bluntly define the right to identity as the right through which I protect "who I am." Immediately, two important questions arise. First, what do we mean by "who am I", that is, what makes me *me* and you *you*? Second, what contributes to and constitutes my personal identity in law? Such notion of identity remits us to think upon the elements and properties that make me unique as an individual and different from all the others. But then, two further questions arise. First, what elements and properties could those be? Am I my physical appearance, my external image, my body, my name, my memories, my intellectual creations, my psychological traits, my actions, my life experiences, my ideals, my beliefs? Second, is the right to identity suppose to protect "who am I" according to myself (how a person perceives herself), or according to others (how this person is perceived and represented by others, or in other words, the image that this person provides to her environment)? [1] It seems that while such right to personal identity should involve the way in which I see myself, it cannot – nonetheless – live only in my head (and apart from society), requiring thus the recognition of other's perceptions of myself.[2] Another question then comes up; how can a balance be struck between these two dimensions (my self-perception and the perception that others have of me)?[3]

---

[1]Such distinction has been advanced in philosophy by Paul Ricoeur (Paul Ricœur, *Oneself as Another* (Chicago, IL: University of Chicago Press, 1992). More recently, and within the field of legal studies, such distinction has been re-captured by Mireille Hildebrandt (Mireille Hildebrandt, Privacy and Identity, In *Privacy and the Criminal Law*, edited by Erik Claes, Antony Duff, and Serge Gutwirth (Antwerpen: Intersentia ; Oxford : Hart Pub. [distributor], 2006). According to the latter, "[i]dem (sameness) stands for the third person, objectified observer's perspective of identity as a set of attributes that allows comparison between people, as well as a unique identification, whereas ipse (self) stands for the first person perspective constituting a 'sense of self'. Their intersection provides for the construction of a person's identity" Mireille Hildebrandt, Profiling and Ami, In *The Future of Identity in the Information Society : Challenges and Opportunities*, edited by Kai Rannenberg, Denis Royer, and André Deuker (Berlin ; London: Springer, 2009), 274.

[2]As we shall see in the following, an important aspect of the right to personal identity is the correct recognition and representation of oneself in the eyes of others. In this sense, the right to personal identity has operated (in some legal systems, at least) under the assumption that a lack of recognition or misrepresentation by others weakens a person's sense of identity, by projecting an erroneous and flawed image of that person.

[3]There is, moreover, a kind of symbiotic and reinforcing relationship between these two different dimensions. As explained by Marshall, "the other's recognition of a person's identity makes the person aware of their specificity and difference from all others on an ongoing dynamic basis thus forging a stronger sense of identity" (Jill Marshall, *Personal Freedom through Human Rights Law?: Autonomy, Identity and Integrity under the European Convention on Human Rights*, International Studies in Human Rights, V. 98. (Leiden ; Boston, MA: Martinus Nijhoff Publishers, 2009), 96.

In order to find an answer to those questions, and to many others related, one should delve into the meanders of legal history and ascertain how did the right to identity came into existence and evolved. Its history describes how different aspects and elements of identity were progressively broadened and taken into consideration by different branches of law and legal instruments. Furthermore, the history of the evolution of the right to personal identity is also the history of jurisprudential creation and doctrinal innovation, marked by important theoretical constructions. As we shall see, and as I will attempt to demonstrate, the history of the continuous evolution of the right to personal identity is far from being over. In fact, present and prospective technological developments demand further evolutionary (and revolutionary) steps in the legal conceptualization of the right to personal identity. Before peeking into the future and examining those further steps, I shall first look into the past.

The notion of the right to personal identity is inexorably fluid and necessarily changeable in time.[4] More than changeable, the right to personal identity is a continuously growing legal figure. With the passing of time, and due to the incessant pace of technological progress,[5] it has been conceptually expanded in order to include more and more elements deemed as constitutive and representative of one's personal identity.[6] The history documenting the evolution of this legal right is, as such, the history of this expansive process. Within such process, I will underline how the initial scarcity and simplicity of the right to personal identity gave place afterwards to an increasing level of complexity and sophistication.

Looking into the history of the creation and evolution of the right to identity in Western Europe, it is curious to note that the concept of personal identity – at the legal level – is a modern concept. Hence, before the twentieth century there was no legal protection of "personal identity" as such, and neither a legal recognition of a proper right to identity. During the ancient and for most of the medieval periods in history, people were ruled by the law according to their geographical place of origin, family ancestry, tribal and religious affiliations. A person's identity was assimilated to the groups and communities to which she pertained, and not according to her own individual characteristics and features. Nevertheless, the 'legal' characterization of individuals according to their geographical provenance, class or guild started to change in the late Middle Ages and the Renaissance.[7]

[4]Sergio Niger, "Il Diritto All'identità Personale," In *Diritto All'anonimato. Anonimato, Nome E Identità Personale*, edited by Giusella Finocchiaro (Padova: Cedam, 2008), 116.

[5]The technological progress and its implications to the theme of identity can be illustrated by a multiplicity of different advancements, such as the establishment and widespread of the internet, the developments observed in the scientific areas of genetics and genomics, as well as the discoveries made in the field of neuroscience.

[6]The enshrinement of the concept and definition of 'genetic identity' in Human Rights Law is a good example of such trend.

[7]In fact, law already made use of a series of personal identity features during the medieval period. For an overview of the early modern European history of identification practices and identity insignia (such as seals, stamps, portraits, badges, clothes, signatures and coats of arms), see

As such, from the sixteenth century onwards, with the development of centralized administrations and bureaucratic apparatuses, a person's identity began to be perceived as personal. The consolidation of law in the hands of the state, essential to the process of state-building, required the implementation of an administrative machinery capable of individuating individuals. The individuation and differentiation of persons was crucial for the state to oversee tax collection, law enforcement, judging, etc. In other words, the emergence of identity corresponded to the administrative effort of regulating this new society.[8] The State had to identify in order to regulate.

In this process, it was imperative to know, evaluate and tabulate the people subject to this new form of administration. In order to capture this dense and growing mass of people, the solution was to select a number of criteria through which people could be described. These criteria, duly noted down on paper, formed a set of references through which the administrative system could trace back to each individual. As a result, those observable identifying features constituted, once collected and registered through administrative procedures, our so-called paper identity, enabling our identification as distinct individuals. At this time, and "[i]n a rather obsolete conception, personal identity was understood as the whole of the official personal data resulting in public records, and important mainly for the public purpose of making the citizen identifiable by the public administration."[9]

Such initial (and narrowed) definition of personal identity was merely instrumental, as it served the exclusive purpose of identification, assisting law in the task of individuating one person from another. Personal identity was thus used for governmental purposes, encompassing the sum of elements strictly necessary for the newly formed nation states of the nineteenth century to individualize their subjects. Such elements included the name, parentage, nationality, domicile, birth date and gender of each individual. It was based on such elements that the state could individuate each person in order to, for example, grant citizenship and impose taxes.

As a result, the early version of what would afterwards constitute the right to personal identity was associated with the collection and stipulation of a set of distinctive external signs which could describe and individualize each person. Accordingly, the antecedent of the right to personal identity referred solely to a person's external and official descriptive elements attributed by the state, encompassing only a number of

---

Valentin Groebner, *Who Are You? : Identification, Deception, and Surveillance in Early Modern Europe* (Brooklyn, NY: Zone Books, 2007).

[8] Jean-Claude Kaufmann, *L'invention De Soi : Une Théorie De L'identité*, Collection Individu Et Société. (Paris: Armand Colin, 2004).

[9] Giorgio Pino, "The Right to Personal Identity in Italian Private Law: Constitutional Interpretation and Judge-Made Rights," In *The Harmonization of Private Law in Europe*, edited by M. Van Hoecke and F. Ost (Oxford: Hart Publishing, 2000), 225.

limited features. The focus was upon the state-imposed distinctive characteristics and identifying elements of an individual, rendering her distinct from everybody else.

In this way, law only privileged a heteronymous dimension of identity, captured according to a third observer's standpoint. In other words, the right to identity only protected "what we were" in the view and perspective of others. Such standpoint encompassed a restricted list of external features representative of one's identity, which could then be checked and confirmed by the competent registries and administrative authorities. In that way, "[t]he emphasis thus placed is on the way in which society, rather than the individual, is made up."[10] The personal identity guaranteed by law did not protect any truly intrinsic elements of our identity, but a set of artificial and externally imposed identifying elements, crystallized on paper, and used by external authorities to identify and distinguish one person from any other.[11]

Nevertheless, such notion of identity ensured by law represented an important step in the configuration of a right to personal identity, as it conveyed the idea that each person had the legal right to be identified and distinguished from other subjects. With the further development and consolidation of centralized administrative structures, people began to be correlated with their own distinctive characters and signs. Through such elements, an individual was now able to distinguish herself and to be distinguished from others. For the first time, any given person could be recognised, in a formal and institutionalized way, through individual characteristics and registered data. Furthermore, it was in the context of the so-called paper-identity (characterized by the issue of the identity cards) that the first rights connected to identity were born and enshrined into law. In fact, it was during the nineteenth and twentieth centuries that a qualitative step in the theoretical conceptualization of the right to personal identity was taken. Both these legislative and doctrinal innovations took place within the auspices of the so-called personality rights.

[10]Hélène Boussard, "Individual Human Rights in Genetic Research: Blurring the Line between Collective and Individual Interests," In *New Technologies and Human Rights*, edited by Thérèse Murphy (Oxford ; New York, NY: Oxford University Press, 2009), 250.

[11]This aspect of the right to identity – the body of rules governing the means by which one can be identified through different elements pertaining to one's identity – is still (and increasingly more) important nowadays. With the incessant pace of technological improvement, the ways through which any person can be identified have grown exponentially (photo, camera surveillance, electronic data, biometrics, etc). In fact, the corner stone of the Data Protection legal regime, the concept of personal data, is precisely defined according to the possibility of identifying a given person, that is, as "any information relating to an identified or identifiable natural person" (article 2 of Directive 95/46/EC).

4.3 Personality Rights and the Right to Identity

The rights of personality[12] (*diritti della personalitá, droits de la personalité, persönlichkeitsrecht*), a general category of rights which has for centuries[13] protected aspects related to our personality,[14] "recognise a person as a physical and spiritual-moral being and guarantee his enjoyment of his own sense of existence."[15] Emerging from the need to safeguard and protect the value of human dignity, the rights of personality protect juridical interests and values deeply related to the human person, such as life, physical and moral integrity, honour, reputation and privacy.

Under the label of rights of personality, many European countries established in their legal systems the so-called right to personal identity.[16] Such particular personality right, oriented towards the protection of human dignity, reaffirmed in written law what resulted in practice from the establishment of the paper identity, that is, the right of being individuated and identified. Along this perspective, a subject earned the right to possess, control and impose a set of particular characteristics and features which individualized and distinguished her from all the others. Moreover, such capability was in line with the administrative and bureaucratic task of the state of individuating persons, attributing them the right to such individuation.

More importantly, the enshrinement of the right to identity as a personality right equated to the recognition of identity as a particular and autonomous personality interest. Such particular conceptualization – the legal recognition of identity as a personality interest worthy of legal protection – led the way to a succession of theoretical constructions and jurisprudential interpretations. Hence, the exercise consisted in ascertaining the meaning and the scope of such particular personality interest. In other words, the scope was to determine how a personality interest, like identity, differed from similar ones, such as privacy. According to Neethling,

---

[12]The conceptualization of personality rights as a separate group of private rights is firmly established in Europe (on the European Continent), being present also in other jurisdictions, as for example in the USA and South Africa.

[13]Personality rights, in fact, can be traced back to Roman Law, namely to the *actio iniuriarum*, which provided a detailed scheme of personality protection, including the rights to *corpus* (physical mental integrity), *libertas* (physical freedom) and *fama* (reputation). For more details, see J. Neethling, J. M. Potgieter, and P. J. Visser, *Neethling's Law of Personality* (Durban: Butterworths, 1996), 3–4.

[14]Personality rights "are private law (subjective) rights which are by nature non-patrimonial and highly personal in the sense that they cannot exist independently of a person since they are inseparably bound up with his personality" Johann Neethling, "Personality Rights: A Comparative Overview," *Comparative and International Law Journal of Southern Africa* 38, 2 (2005): 223. For more details on the nature of personality rights, see Neethling, "Personality Rights: A Comparative Overview", 223 ss.

[15]Neethling, "Personality Rights: A Comparative Overview," 210.

[16]In fact, the right to identity has been recognized *eo nomine* in countries such as Italy, France, Switzerland and South Africa.

> *Identity as an interest of personality can be defined as a person's uniqueness or individuality which identifies or individualises him as a particular person and thus distinguishes him from others. Identity is manifested in various indicia by which that particular person can be recognised; in other words, facets of his personality which are characteristic of or unique to him, such as his life history, his character, his name, his creditworthiness, his voice, his handwriting, his appearance (physical image), etcetera. A person has a definite interest in the uniqueness of his being and conduct being respected by outsiders*[17]

It is precisely the idea of the uniqueness and singularity of the human being that has conferred to the right to personal identity its own conceptual autonomy within the group of personality rights to which it is attached. Beneath the aspect of control over the several indicia of one's identity (nowadays increasingly translated and dispersed in electronic processed data and information), the right to identity presupposes an "inalienable interest in the uniqueness of his being."[18]

What started as a right restricted to a few elements imposed by the state, for the strict purpose of identification of its citizens, evolved into a mature and complex right, underlining a definite interest in the uniqueness of the being. Furthermore, the right to personal identity began including elements (or *indicia*) that went clearly beyond the name, domicile and nationality, such as life history, voice and appearance. At the present time, it is conceptualized as not only a right granting protection through the imposition of negative obligations upon others (duties not to act), but also as a right which demands positive obligations to be performed by third parties (duties to act). The right to personal identity is, as such, both a negative and positive right.[19]

The right to personal identity was first enshrined at the level of private law, in a multiplicity of European civil codes, under the label of personality rights. Afterwards, and along the so-called movement of *constitutionalization* of private law, it was enshrined as a fundamental right at the level of constitutional law.[20] At such level, the right to personal identity was, in some cases, enshrined explicitly in law, while in others it was derived indirectly from jurisprudential interpretation. Finally, and more recently, the right to personal identity has climbed another stair, reaching the level of international public law, entering the realm of human rights. Here, such right has also been enshrined explicitly in legislation (namely in the United Nations Convention on the Rights of the Child [UNCRC], article 5), as well as derived from jurisprudential interpretation (namely through article 8 of the European Convention on Human Rights [ECHR].

---

[17]Neethling, Potgieter, and Visser, *Neethling's Law of Personality*, 39.

[18]Ibid.

[19]This aspect, as we shall see in continuation, has been further developed and elaborated by the international human rights legal framework.

[20]Moreover, as Werro notes, "it is now recognized that defining the scope of personality rights is no longer a question of private law only, but also, and perhaps primarily, a question of constitutional and European law" (Franz Werro, "The Right to Inform V. The Right to Be Forgotten: A Transatlantic Clash," In *Liability in the Third Millenium; Georgetown Public Law Research Paper No. 2*, ed. Aurelia Colombi Ciacchi, et al. (Baden-Baden: F.R.G., 2009), 289).

The ascension of personality rights into higher rankings of the legal hierarchy is obviously linked to the considerable importance that individuals tend to attach to the protection of the facets and aspects of their personalities. In this light, Neethling affirms that all legal systems should not only strive towards, but indeed have an obligation to provide for comprehensive personality protection. In fact, and as I shall explain in continuation, both constitutional law and human rights play an important role not only in the protection of personality rights in various legal systems, but also in the development of the theoretical conceptualization and construction of those rights. The right to personal identity is no exception to this trend.

4.4 Right to Personal Identity and Constitutional Law: Jurisprudential Creation and Doctrinal Innovation

A paradigmatically case and excellent example of further and rich legal conceptualization of the right to personal identity can be learned from its doctrinal theorization[21] and judicial recognition in Italy.[22] The Italian jurisprudence, during the last quarter of the twentieth century, literally created the right to personal identity.[23] For that reason, in Italy, the right to personal identity is essentially a judge-created right.[24] Through jurisprudential interpretation of the Italian Constitution (namely from Art. 2),[25] Supreme Court magistrates and other judges created and endowed the right to personal identity with a clear and precise autonomous meaning and scope within the group of personality rights. Considered as a new personal right

---

[21]The contribution of the Italian doctrine for the development of the right to personal identity precedes, as a matter of fact, its juridical recognition. In this respect, the work of De Cupis on the autonomization of this legal right is particularly relevant. See Adriano De Cupis, *Il Diritto All' Identità Personale* (Milano: A. Giuffrè, 1949).

[22]For a detailed overview of the construction of the right to identity by the Italian jurisprudence, see Giorgio Pino, *Il Diritto All'identità Personale : Interpretazione Costituzionale E Creatività Giurisprudenziale*, Ricerca; (Bologna: Il mulino, 2003).

[23]The landmark case that introduced the right to identity in Italy is the Pretura Roma 6-5-1974 (Pangrazi and Silvetti v. Comitato Referendum). Pino summarizes the case in the following manner: "The facts of the case are quite interesting. In the days of the referendum propaganda about the abrogation of divorce in Italy, the anti-divorce committee, for the purpose of its campaign, used the picture of a man and a woman working in the country. The picture was meant to evoke a 'traditionalist' atmosphere (and old-style family) and was of course associated with an anti-divorce message. The problem was that, first of all, the picture was taken without the consent of the people portrayed; in the second place, they were not married; finally, and what is more, they were in favour of the existing divorce legislation" (Pino, "The Right to Personal Identity in Italian Private Law: Constitutional Interpretation and Judge-Made Rights," 234).

[24]In this regard, Pino observes that "it is possible to regard the introduction of the right to personal identity as an interesting example of a common-law technique in a civil-law system, such as that in Italy" (Ibid.).

[25]"[A]ccording to which Italy "acknowledges and protects the fundamental rights of the human being, both as an individual and as a member of social groups" (Ibid., 228).

in that country, the right to personal identity acquired novel and different contours, adding to the ones I have previously analysed.

The right to personal identity, framed according to the Italian legal culture, could then be described "as the right everybody has to appear and to be represented in social life (especially by the mass media) in a way that fits with, or at least does not falsify or distort, his or her personal identity."[26]

This new legal conceptualization incorporated two new key elements. On the one hand, it emphasized the inherent relational and social character of the right to personal identity, while, on the other, introduced an important idealistic component. In the words of Pino, such new concept of personal identity "adds to the traditional concept a sort of 'abstract', 'moral' or 'ideal' feature that can be expressed as the interest everybody has to be represented with his/her real identity, i.e., with the identity that appears in concrete and unequivocal circumstances of social life."[27] Along these lines, the Italian jurisprudence moved the right to personal identity to the context of social life, claiming that the infringement of that right should be asserted in the light of the subject's interest of being (ideally) represented in accordance with her real identity, that is, with the identity which emerges (and can be verified) from the way one lives and acts in society. Pursuant to this logic, the Italian jurisprudence introduced the idea of "real" identity as the projection of oneself in society.

This qualitative step in the construction of the right to personal identity acknowledged identity as something more (and fundamentally different) than a mere individuation and identification of a given person, based on registered elements ascertained by public powers. Identity, as such, was conceived as the demand of affirming the intrinsic quality of the subject in the life of social relations. This conception, moreover, entailed a comprehensive assessment of the spiritual values that each person holds and carries with her.[28] Such identity equated, thus, to the social projection of the being.[29] The person's identity was no longer (or not only) subsumed to a simple record or a registry in the shelf of a public archive, it encompassed the quality of that subject, the notion of someone with her own set of values and ideas, who should be protected from having such inner quality distorted or misrepresented. In this way, the right to identity translated "the claim that one's (ascertainable) cultural, professional, religious, political, social experiences should not be distorted, misrepresented, falsified, confused, contested, or the like, by means of the ascription of false (even if not necessarily defamatory) statements or acts."[30]

---

[26]Ibid., 225.

[27]Ibid., 226.

[28]Niger, "Il Diritto All'identità Personale," 116.

[29]Rafaelle Tommasini, "L'identitá Dei Sogetti Tra Apparenza E Realtà: Aspetti Di Una Ulteriore Ipotesi Di Tutela Della Persona," In *Il Diritto Alla Identità Personale* edited by Guido Alpa, Luca Boneschi, and Mario Bessone (Padova: CEDAM, 1981), 82–83.

[30]Pino, "The Right to Personal Identity in Italian Private Law: Constitutional Interpretation and Judge-Made Rights," 226. According to the cited scholar, such quote corresponds to the commonest definition of the right to personal identity, which can be read in the decision of the

In this regard, most of the cases dealt with by the Italian jurisprudence concerned the protection of the person from misrepresentations conveyed by the mass media.

Furthermore, the Italian courts established a mechanism to assert the potential violations of the right to personal identity. In that way, its infringement would be evaluated and ascertained according to someone's real identity, that is, according to her projection in the social. In order to assert such identity, the exercise consisted in the analysis of the objective facts related to and acts developed by the individual in society. In this manner, the real identity of a person could be verified from the analysis of these facts and actions. Such fictional "real" identity of the subject (an identity that "ideally" corresponds to the "truthful" identity of the person), contextualized at the level of social life and subject to objective and factual scrutiny, brought to the right to personal identity an added layer of complexity and thoroughness.

As a result, and in addition to the relational and idealistic elements analysed above, the Italian jurisprudence made another two contributions to the conceptualization of the right to identity, rendering it conceptually autonomous and functionally "flexible." Regarding its autonomy, the Italian case law distinguished the right to personal identity from other rights, such as the right to someone's likeness, reputation and privacy, providing such right with an "identity" of its own. Concerning the flexibility of the right to personal identity, its application "shows a considerable 'elasticity', because it covers a wide, and not a priori determinable, range of assaults on the integrity of one's personal history, perpetrated by the mass media."[31] In other words, the right to personal identity proved to be adaptable to a myriad of different situations and cases which configured a violation of one's identity. As I shall note in Section 4.7, the plasticity and adaptability of the right to personal identity constitutes a central attribute of the latter in order to advocate and legitimize its application to novel situations deriving from prospective technological developments.

4.5 Human Rights and the Right to Personal Identity

The tendency to expand the protection of personality rights[32] can be observed not only in national jurisprudence, but also in the international one, namely in the field of human rights. Such phenomenon, in truth, should be anything but surprising. In fact, both human rights and personality rights rely upon the same premises, that is, the protection of the intrinsic value of the human person from acts coercing and undermining their dignity and personal freedom. In view of that, it is noticeable that

---

Corte di Cassazione, I, 22.6.1985, n. 3769, in Nuova giurisprudenza civile commentate, I, 1985, pp.647–654.

[31] Ibid., 235.

[32] As Pino observed, "the growing interest in the legal protection of the various aspects of human personality ... characterizes the evolution of almost every single European legal system in the second half of this century", (Ibid., 226).

many human rights relate to interests of personality. As Neethling observes, "[a] survey of different charters of human rights reveals that it is common to find protection of the following personality rights or interests: life, liberty, dignity, privacy, religion and freedom of movement."[33] A more recent addition to such list is the right to identity. Beyond its explicit reference in the UN Convention on the Rights of Child, the right to identity has also been recurrently invoked in the case law of the European Court of Human Rights in Strasbourg (ECtHR). In this respect, the ECtHR has derived a right to identity through jurisprudential interpretation of the right for respect of one's private life, as set out in Article 8 of the European Convention on Human Rights (ECHR).[34] The Convention on the Rights of Child (which is based, inter alia, on the Universal Declaration of Human Rights) expressly recognises the right to identity. Despite the fact that the Rights of Child Convention does not define "identity", article 8(1)[35] stipulates "the right of the child to preserve his or her identity, including nationality, name and family relations as recognised by law without unlawful interference." In the words of Sullivan, "[a]rticle 8 is unique in that specifically refers to a right to identity and articulates three components of identity."[36] As such, "[i]t clearly contemplates a concept of identity made up of elements which include 'name and family relations' and 'nationality'."[37]

The right to identity has also been recently invoked in the jurisprudence of the European Court of Human Rights. Despite not being specifically mentioned in any of the articles of the European Convention on Human Rights (ECHR), the Court in Strasbourg has derived a right to identity from the "right to one's private life", enshrined in article 8 of the ECHR (in a similar way as to the right to privacy). In the view of some authors, the lack of any explicit and literal reference to the right to identity does not undermine the latter. On the contrary, as Sullivan reiterates:

> *It is clear that like the Universal Declaration, the ECHR is based on recognised, intrinsic human rights like dignity and autonomy in the sense of self-determination, which form the basis for the right to identity. Although the right to identity is not specifically covered by*

---

[33] Neethling, Potgieter, and Visser, *Neethling's Law of Personality*, 19.

[34] The full text of the Article is as follows:

1. Everyone has the right to respect for his private life and family life, his home and his correspondence.
2. There shall be no interference by a public authority with the exercise of this right except such as in accordance with the law and is necessary in a democratic society in the interests of national security, public safety or the economic well-being of the country, for the prevention of disorder or crime, for the protection of health and morals, or for the protection of the rights and freedoms of others.

[35] For further details on the inclusion of Article 8 of the Rights of Child Convention, as a result of a proposal by Argentina, see Sharon Detrick, editors. *The United Nations Convention on the Rights of the Child. A Guide to The "Travaux Préparatoires"* (Dordrecht, Boston, MA, London: Martinus Nijhoff Publishers,1992).

[36] Clare Sullivan, "Privacy or Identity?," *Int. J. Intellectual Property Management* 2, 3 (2008): 296.

[37] Ibid.

> *any particular article in the ECHR, it nevertheless is a fundamental human right which underpins the ECHR.*[38]

Further to being an implicit fundamental human right underlining the ECHR, it is important to note that the ECtHR has acknowledged a right to identity from article 8. Referring to such article, the Court stated that "[t]he Article also protects a right to identity and personal development, and the right to establish and develop relationships with other human beings and the outside world."[39]

Accordingly, the jurisprudence of Strasbourg has underlined (along side the Italian jurisprudence, but in different terms and with different nuances)[40] the link between the social context and identity. Such link expresses not only the role of social factors in creating and shaping individual identities, but also the importance of social recognition for valuing and promoting such personal identity. In this regard, the ECtHR has broadened the scope (and also the *rational*) of the right to respect for one's private life. Such court stressed not only the negative, protective and traditional side of the right to respect for one's private life (protection against unwanted intrusions into people's private lives in the classical sense of keeping a person's private space intact), but also underlined a more promotional and positive aspect of such right. In fact, the developing jurisprudence of Strasbourg has included a right to develop one's personality within the right to respect for one's private life,[41] in the sense of "such personality being developed not only 'alone' but also in our relationship with others and the outside world."[42] Such important inclusion "stresses the importance of social conditions and relationships between human beings in creating and developing ... human personality: a social context is not only needed for this personality to thrive but also for it to form."[43]

Such approach to the right to personal identity is based upon an existential and developmental meaning of identity. In this case, identity is not looked at as a sum of different elements, representative of one's identity and subject of being misrepresented and falsified. Identity is understood as a narrative, an individual inner story that each individual needs to build, develop and rewrite over time in order to

---

[38]Ibid., 297.

[39](2003) EMLR 15 ECHR, 57 citing P.G. and J.H v United Kingdom ECHR 2001-IX

[40]While both underline the importance of the social context in which identity must thrive, for the Italian jurisprudence the "social" was the stage where identity was projected and could be ascertained, whereas for the ECtHR jurisprudence, the social is the main factor creating and shaping identity itself.

[41]In theory, and taking into account the scope and the latitude of these rights, it should be the other way around, that is, the more specific right to private life included within the more general and transversal right to free development of personality. Furthermore, it is somewhat paradoxical that the emphasis on the social context is done through a right to private life. In fact, the way in which the concept of private life has been constantly broadened reveals a number of problems (which, nevertheless, go beyond the scope of this article).

[42]Marshall, *Personal Freedom through Human Rights Law? : Autonomy, Identity and Integrity under the European Convention on Human Rights*, 3.

[43]Ibid.

define the meaning of their lives. Moreover, such "autobiographical" characterization of identity is also in clear contrast with the previously described administrative and static meaning of identity, configured by private law. While the latter had the state as point of departure, setting up the identity elements that would characterize and individuate each individual, now, and within the human rights perspective, it is exactly the contrary: the individual as starting point, stipulating and deciding which identity elements she wants to be described and recognized by.

It is the idea of identity as a narrative process, a life-time exercise of creating and recreating our own story, which explains the Court's open and progressive stance in establishing sexual identity rights[44] (namely in the cases of sex change). It also explains the Court's decision of allowing access to information about one's origins and past experiences (also considered as important aspects of one's identity). Remaining within the "narrative" metaphor, human rights protect not only the right to read the initial and introductory chapters of one's story, the right to know its beginning (that is, our origins – the right to access information relating to one's birth and childhood existence);[45] but also the right to write the subsequent chapters, that is, the right to write one's own story (illustrated, for example, by the developing jurisprudence in sexual identity cases). With regard to the reading of the first chapters, Jackson has pointed out that the "right to identity is a right not to be deceived about one's true origins,"[46] amounting thus to the right of the children to know their biological identity.

Another important aspect of the new legal construction of the right to identity relates to its effective enforcement, that is, to the measures and impositions which guarantee its application. As such, the right to identity encompasses also the right to be provided with the necessary conditions for identity to be shaped and developed, allowing not only for identity recognition but also (and especially) for identity formation. Further to its negative dimension, as a right to be protected from intrusions or infringements on one's identity, the right to identity also encompasses a positive obligation for third parties (namely the states), that is, the right to be endowed with the indispensable means to forge, develop and sustain one's identity. This improvement in the construction of the right to identity is connected to the Court's development of "positive" obligations, which – in our particular case – affirms the state's duty of assisting and contributing to the creation and formation of one's identity. At the level of the human rights framework, the obligation imposed to the states of creating conditions (either social, technical or legal) to enable the flourishing of

[44]For a more profound view on this particular issue, see Ibid., chapter 7.

[45]Stressing the same idea, the Glover Report for the European Commission has argued that "a life where the biological parents are unknown is like a novel with the first chapter missing" (Jonathan Glover, *Ethics of New Reproductive Technologies : The Glover Report to the European Commission*, Studies in Biomedical Policy (DeKalb: Northern Illinois University Press, 1989), 37). See also Emily M. Jackson, *Regulating Reproduction : Law, Technology and Autonomy* (Oxford ; Portland, OR: Hart, 2001).

[46]Jackson, *Regulating Reproduction : Law, Technology and Autonomy*, 214–215.

one's identity has undoubtedly transformed the right to identity into a much more efficient and compelling one.

Furthermore, the international legal framework of human rights has also had the merit of engaging into a re-conceptualization of the right to personal identity which takes into account new technological developments, namely in the field of genetics.[47] A paradigmatical example is the reference to the concept of genetic identity, which – once again – advances significantly in the conceptualisation of the right to identity.

In light of the knowledge attained in the scientific disciplines of genetics and genomics, the human rights legal framework protects the human person not only as an individual per se, but also as a depositary of the genetic heritage of the human species. Thereby, "the protection of genetic identity appears as the new rationale of the right to identity (the genetic identity of the individual) and the right to uniqueness of the individual (the genetic identity of the human species)."[48] In the conceptualization of the right to identity endorsed by this perspective, and given the collective dimension of the human genome, it is interesting to note that the right to identity no longer protects personal identity in the sense of "who am I", but in the sense of "who will my descendants be." The rational behind this particular manifestation of the right to identity is no longer shaped by the need to protect one's self-perception (who am I to myself) or the perception of myself forged by a third person (who am I to others), but "who am I" according to the genetic characteristics of the human species. Here, the individual is protected as part of the whole, as representative of an important community, that is, humanity. As such, "the individual is protected as a depositary of the genetic heritage of the species and the right to genetic identity protects the interests of future individuals and of the human species over time."[49]

Nevertheless, it is crucially important to observe that the relevance and weight attributed to the concept of genetic identity (as the right to identity encompassing genetic characteristics) are not extrapolated. Genetic attributes are seen as only a part (albeit important) of someone's identity, not equating to the whole of the identity itself. Such appreciated moderation and balance in dealing with this concept prevents the risk of engaging into any sort of "genetic essentialism", thus reducing the human person to a mere expression of her genetic architecture. Consequently, the definition of personal identity professed by the human rights approach masterly balances the influence of genetics alongside the role of the environment and social conditioning in the formation of one's identity. In this regard, and despite not being a legally binding instrument, the International Declaration on Human Genetic Data

[47] Human genetics, in fact, can be seen "as the 'new' engine of the modern construction of human rights" (Boussard, "Individual Human Rights in Genetic Research: Blurring the Line between Collective and Individual Interests," 246).

[48] Ibid., 249.

[49] Ibid., 259.

is particularly relevant. In its article 3, "person's identity" is referred to as encompassing not only the genetic components of each individual, but also the "complex educational, environmental and personal factors and emotional, social, spiritual and cultural bonds with others."

4.6 Ambient Intelligence and the Challenges to the Right to Personal Identity

In his detailed study of personality rights, Neethling lists the main factors which paved the way for the development of the modern doctrine of the law of personality in the early nineteenth century, pointing as the first one "the technological and industrial revolution where inventions and dramatic developments in the press and photography, increased production and commercial competition gave rise to new forms of infringement of the personality of others, endangering personality interests as never before."[50] Faced with a new wave of technological advancements, history seems to repeat itself, as the prospective technological revolution (already well underway) promises to carry new and sophisticated means of harming our personality interests, namely our identity. Those technological developments, in truth, do not only bring new threats and perils, but also new ways and instruments of expressing and developing one's identity. It is because of both these perils and opportunities that, once again, a further re-conceptualisation of the right to personal identity is needed. In fact, only through a re-theorization of the right to personal identity can the latter accompany the new technological threats and opportunities that will be posed to a person's identity.

In this manner, and interestingly enough, one observes that the permanent dialogue between technological progress and re-theorization of law (in particular of the rights of personality) assumes a cyclical character. The first "wave" took place with the industrial revolution and its unprecedented technical developments at the level of the printing press and photography. Such technologies, by increasing exponentially the creation and distribution of news and various contents, gave birth to novel ways of infringing one's personality. The "second" wave is now about to start. But, this time, the technological revolution in hands covers a much wider range of technologies, encompassing the convergence of areas such as information technology, nanotechnology, biotechnology, cognitive and neuroscience, robotics and artificial intelligence. The concatenation of all these technologies will, inexorably and once again, provide newer and stronger threats to the protection of personality. In light of this historical observation one could say that technology cyclically

[50] Neethling, Potgieter, and Visser, *Neethling's Law of Personality*, 6.

provides a renewed, and perhaps more justified, sense to personality rights.[51] This is undoubtedly the case for the right to identity.

The need to reconceptualise the right to identity emerges from a myriad of different technological challenges and threats posed to the protection of one's identity. The scope of my analysis will be restricted to a number of specific challenges, namely those related to the protection of our personal identity posed by the so-called vision of ambient intelligence (AmI).

The ambient intelligence scenario, also denominated by the terms internet of things, or ubiquitous, pervasive, proactive and autonomic computing, constitutes a vision for the future, a technological ecosystem in which "people will be surrounded by intelligent and intuitive interfaces embedded in everyday objects around us and an environment recognising and responding to the presence of individuals in an invisible way."[52] Such new paradigm forms a complex technological environment, requiring little deliberate human intervention and encompassing a wide array of different emerging technologies, such as mobile sensors, radio frequency identification (RFID) tags, software agents, brain computer interfaces, ICT implants, affective computing and nanotechnology.[53] Furthermore, the ambient intelligence scenario builds upon automated profiling practices and human-centric computer interaction design, dispersing and integrating networked devices into the environment by attaching them to everyday objects. Such technologies and devices, moreover, are endowed with the capability of recognizing a given person and her situational contexts, adapting to the users' needs, and anticipating their desires without conscious mediation. The AmI will thus be characterized, on the one hand, by its invisibility, discretion and unobtrusiveness and, on the other, by its sensitivity, interactivity and responsiveness to the human person.[54]

[51] Stefano Rodotà sustains, in this regard and in a somewhat humorous manner, that private law has been saved by technology (Stefano Rodotà, "Lo Specchio Di Stendhal: Riflessioni Sulle Riflessioni Dei Privatisti," *Rivista critica del diritto privato* 15(1997): 5).

[52] ISTAG report 1999, in which the term "ambient intelligence" was, for the first time, coined (Information Society Technologies Advisory Group (ISTAG), "Orientations for Workprogramme 2000 and Beyond," (1999).

[53] For further details on the vision of Ambient Intelligence, see Emile Aarts, Rick Harwig, and Martin Schuurmans, "Ambient Intelligence," In *The Invisible Future : The Seamless Integration of Technology into Everyday Life*, edited by Peter J. Denning (New York, NY: McGraw-Hill, 2001). E. H. L. Aarts and Stefano Marzano, *The New Everyday : Views on Ambient Intelligence* (Rotterdam: 010 Publishers, 2003). Werner Weber, Jan M. Rabaey, and E. H. L. Aarts, *Ambient Intelligence*, 1st ed. (Berlin ; New York, NY: Springer, 2005). Giuseppe Ph D. Riva, *Ambient Intelligence : The Evolution of Technology, Communication and Cognition Towards the Future of Human-Computer Interaction*, Emerging Communication, 1566–7677 (Amsterdam ; Oxford: IOS Press, 2005). *Handbook of Ambient Intelligence and Smart Environments*, 1st ed. (New York, NY: Springer, 2009). Norberto Nuno Gomes de Andrade, "Technology and Metaphors: From Cyberspace to Ambient Intelligence," *Observatorio (OBS*) Journal* 4, 1 (2010).

[54] Summarising this group of features, Hildebrandt describes AmI as an adaptive, smart environment which "should always be one step ahead of the user, like a butler who unobtrusively

In this way, ambient intelligence, along with its innovative profiling techniques, embedded sensors and sophisticated autonomous software agents, promises to revolutionize the way we live and interact in society. Moreover, such revolution will (and already is) putting into question the classical and static ideas we have about ourselves and about our own identities. But how exactly will the AmI scenario affect and change our sense and definition of personal identity? In the following, I shall analyse the fundamental changes operated by new ambient technologies, giving particular attention to those that exert a greater impact in the classical constructions and concepts of the right to identity.

The AmI scenario will carry a number of important transformations to the way a person's identity is captured, represented and disseminated.[55] Such important changes will derive from a number of new characteristics and tendencies present in the future world of AmI.

In the first place, there will be a radical increase in the production, creation, circulation and exchange of personal information. The AmI scenario will increment and accentuate the continuous digitisation of personal information, generating, collecting, analysing, processing and storing massive amounts of personal data.[56] It is thus expected an explosive boost of digitisation of personal characteristics and personal information. Such fact will bear important changes on future identification processes, as well as upon the ways in which individuals' identities will be represented and used. The increase of personal information will, moreover, derive from both the embedded smart objects, as well as from people themselves. In this way, electronic systems, sensors and other objects distributed throughout the physical world – via the constant monitoring of our actions and behaviour – will, themselves, generate and produce massive amounts of personal data and information concerning our identity and behaviour.[57] The boost of personal information will also originate from the users themselves, as people will be able to create, describe and define their identities through a greater number of instruments and platforms.

anticipates his master's wishes even before the master becomes aware of them" (Hildebrandt, "Profiling and Ami," 287).

[55] It is important to stress that many of these changes are already under motion (through the web 2.0. mobile applications, augmented realities and location-based services), having AmI an accelerating (and aggravating) effect.

[56] The Future Group Report (2008), written by the Informal High Level Advisory Group on the Future of European Home Affairs Policy, has used the expression "tsunami" of data to illustrate the massive amounts of data expected to be produced by RFID systems and sensor technologies (Hildebrandt, "Profiling and Ami," 274).

[57] Many times, and to some extent, people will not even know about or be aware of the collection and processing of such information.

Second, AmI technologies will blur the distinction between the physical and the digital worlds.[58] In this regard, the frontiers that demarcate the physical territory from the digital one will become increasingly difficult to distinguish, as both spaces will tend to converge in "one seamless environment of computing, advanced networking technology and specific interfaces."[59] With the slow and relentless "disappearing of the frontier between the offline and the online world, the new identity that will emerge will bring in the physical world many of the characteristics present in the online worlds, such as increased transparency and massive tracking and profiling."[60] In this regard, Rodotà speaks of "networked persons – persons who are permanently on the net, configured little by little in order to transmit and receive signals that allow tracking and profiling movements, habits, contacts and thereby modify the meaning and contents of individual's autonomy"[61] and, I would add, identity. In this manner, it goes without saying that this "new" traceable, profiled and networked identity will be more easily subject to personality infringements.

Third, this novel and technological environment will favor the multiplication of identities. In view of that, the tendency to multiply and polarize various and distinct identities from a single one[62] will only tend to increase. Such tendency, in particular, will consist of a more intense virtualization and multiplication of distinct identities, with virtual and partial identities being created for the most different purposes and reasons, such as for security, business, convenience or entertainment. In addition, and within the AmI world, "[t]hese virtual and multiple identities and the paradigms behind them are feeding back into the 'physical' world, offering a mix of physical

[58] Alongside this change there will also be another important one: the blurring between private and public spaces, as both the public and the private spheres will increasingly become ever more entangled and intertwined. In this regard, and along the same lines, Nissenbaum has argued that the digitalisation of our environment has blurred the borders between the private and the public spheres, while also decreasing the anonymity traditionally associated with many public spaces (Helen Nissenbaum, Privacy as Contextual Integrity, *Washington Law Review* 79, 1 (2004).

[59] Information Society Advisory Group (ISTAG), "Ambient Intelligence: From Vision to Reality" (2003), 8. An illustrative example of the symbiosis between the physical and the digital world is given by emerging notions of "virtual residence" and "digital territories" developed by the European Commission's Institute for Prospective Technological Studies (IPTS). Regarding the concept of digital territories, "the underlying premise is that citizens should be empowered to create, shift, and sustain borders in order to develop and sustain their personal identity" (Hildebrandt, "Profiling and Ami," 302).

[60] Kai Rannenberg, Denis Royer, and André Deuker, *The Future of Identity in the Information Society : Challenges and Opportunities* (Berlin ; London: Springer, 2009), 23. As I shall point out afterwards, it is the spill-over of typical features pertaining to digital and virtual identities (such as multiplicity and permanent availability) that justifies a re-conceptualization of the right to personal identity, namely through the incorporation of the right to multiple identities and the right to be forgotten.

[61] S. Rodotà, "Data Protection as a Fundamental Right," In *Reinventing Data Protection?*, edited by Serge Gutwirth, et al. (Dordrecht ; London: Springer, 2009), 81.

[62] Phenomena recurrently observed in the Internet and its paraphernalia of communication and interaction platforms: social networks, virtual worlds, blogs – spaces which offer different "lives" and "existences".

and virtual plural identities and processes to deal with them."[63] As a result, further to becoming increasingly profiled and networked, identity will be fragmented into different partial and virtual identities.

Fourth, the AmI environment will also display a more encompassing and sophisticated capacity of identifying, distinguishing and classifying each human being. In the "Age of Identification", as Hildebrandt has qualified it, both public administrations and private entities, through automated profiling technologies, biometrics, monitoring and location technologies, will have at their disposal a set of advanced and sophisticated instruments to identify, track and monitor their citizens or (potential) customers. Such new technological apparatus will render the various elements and aspects (indicia) of one's identity, protected under the right to personal identity (such as voice, physical appearance and psychological traits) more easily detected and, what is worst, more easily reproduced and replicated in the AmI scenario. In this regard, and taking into account the technologies involved in the AmI vision, along with its main characteristics and purposes (briefly exposed above), it is perfectly possible to imagine a new generation of electronic agents programmed to act on our own behalf. In order to perform such task, those agents (in an autonomous and invisible fashion) would incorporate and replicate several determining and constitutive elements of one's identity (observed and collected through their constant monitoring process), acting afterwards in the same manner as the correspondent subject would act.

To sum up, the concept of identity in an AmI world will be essentially characterized by its multiple facets and ubiquity.[64]

Regarding its ubiquitous character, traces of one's identity will become dispersed, decentred and permanently registered. In the first place, ubiquitous identity presupposes that traces of our identity will be dispersed in the environment, scattered throughout smart objects, intelligent interfaces, databases and networks located everywhere. The merge between on-line and off-line spheres constituting this future environment, along with the "proliferation of communications, exchanges of personal user data, and identity information, and their storage by means of numerous types of technologies, sensors and devices" will definitely imply "the omnipresence of identity information."[65] Secondly, ubiquitous identity will also mean that the traces of one's identity, further to being dispersed, will also become decentred, that is, outside oneself sphere of command. In other words, our identity will inexorably escape our control, as it already does whenever reconstructed and represented in the eyes of other people. But with AmI, such trend will only tend to aggravate, as our personal identity will not only be susceptible of being (mis)represented by other people, but also by machines and autonomous

[63] Rannenberg, Royer, and Deuker, *The Future of Identity in the Information Society : Challenges and Opportunities*, 1.

[64] Thierry Nabeth, "Identity of Identity," In *The Future of Identity in the Information Society : Challenges and Opportunities*, edited by Kai Rannenberg, Denis Royer, and André Deuker (Berlin; London: Springer, 2009), 53.

[65] Ibid., 54.

agents. The "idem" aspect of one's identity (how am I perceived and represented by others) will not only encompass the perceptions that other people have of our own identities, but also the profiles and the representations of one's identity constructed by those AmI machines and agents (namely through profiling automated processes).[66] [67] Thirdly, the traces of one's identity, besides existing everywhere and existing outside oneself, will also tend to exist perpetually. Such traces will not only be spread out in the physical-digital world of AmI (hybrid space), but they will also be permanently stored and registered (as it already happens with the Internet). This tendentiously eternal character of one's identity elements draws our attention to the need of incorporating the so-called right to be forgotten within the umbrella of the right to personal identity.

Concerning the multifaceted aspect of identity, it is worth underlining that a series of technical developments observed in the Internet[68] (which will only tend to be aggravated with the development of the AmI vision) pose a serious challenge to the traditional understanding of identity. Such understanding, as I have briefly exposed above, tends to correlate identity to a single person, advocating a classical, strict and unequivocal identity bound to a certain person as "a one-to-one link."[69] In the AmI world, and on the contrary, the connection between "one person – one identity" will no longer apply, as identity will be increasingly fluid, undetermined, variable and fragmented. This phenomenon can already be seen today, through different cases and examples. As such, people nowadays manage different and simultaneous identities through their email accounts and social networks, or in online forums and virtual worlds. The reverse case also occurs quite frequently, with single identities being shared and managed by several persons (as it is, for example, the case of an email account of a given institution shared by its members).

Another case proving how obsolete the classical link one person – one identity is becoming are given by those identities that, despite being able to engage into communication and interaction with persons, do not correspond to an actual person,

---

[66]"Profiling is the process of 'discovering' correlations between data in databases that can be used to identify or represent a human or nonhuman subject (individual or group) and / or the application of profiles (sets of correlated data) to individuate and represent an individual subject or to identify a subject as a member of a group or a category." Mireille Hildebrandt and Serge Gutwirth, *Profiling the European Citizen : Cross-Disciplinary Perspectives* (New York, NY: Springer, 2008), 19.

[67]I shall come back to this important aspect when evaluating critically the conceptualization of the "Italian" right to personal identity in light of the new technological developments brought by the AmI scenario.

[68]People nowadays (and many times unconsciously) generate multiple identities. Besides the different identities one may have and develop in the physical world (according to the context in which one is: at work, at home, with family, etc), people are increasingly undertaking different identities (virtual and partial) through their email accounts, online forums, social networks and virtual worlds.

[69]David-Olivier Jaquet-Chiffelle et al., "Virtual Persons and Identities," In *The Future of Identity in the Information Society : Challenges and Opportunities*, edited by Kai Rannenberg, Denis Royer, and André Deuker (Berlin ; London: Springer, 2009), 76.

but rather to "artificial (intelligent) agents moving avatars in video games, and expert systems administering forums or dealing on the stock exchange."[70]

Identity is thus becoming an increasingly complex phenomenon, inherently multifaceted and mutable. In this light, the idea of a unique and stable identity assumes pre-historical contours, as people tend to present, more and more, different identities, dissociated from one another and, many times, constructed upon profound incongruences. In this sense, "[p]eople are not only different to one another, but they are also different within themselves. They are object of incessant variations. They have dissociated identities build upon internal contradictions and opposing forces. The person exists but is not unified and it would therefore be very problematic to encapsulate him or her in fixed screenplays that do not take into account this fluid and complex dialectic."[71] In order to account for this progressive trend, it will be proposed the right to multiple identities as an important element of the right to identity and as a derivation of the right to the free development of personality.

The next section provides a number of hints for a re-conceptualization of the right to personal identity, proposing the incorporation of the right to be forgotten and the right to multiple identities, along with a critical analysis of the theorization of the right to identity made by the Italian jurisprudence. Furthermore, such renewed conceptualization of the right to personal identity derives from the technological developments and the vision of the Ambient Intelligence scenario previously examined.

4.7 Broadening the Scope of the Right to Personal Identity: Critical Analysis of the Italian Jurisprudence in Light of the AmI Scenario

The conception of the "real" identity of the subject, as the projection of oneself in the social, constituted the Italian attempt to struck a reasonable and satisfactory balance between what we are for ourselves and what we are for others, that is, between the ipse- and idem-identity. The idea was to reach a middle-way point between those two conceptions in order to catch sight of the real and objective identity of the subject. The idea behind this particular conceptualization, moreover, seems to be based upon the conviction that someone's real identity can be assessed through her

---

[70] Ibid.

[71] Patrick Boumard, Georges Lapassade, and Michel Lobrot, *Le Mythe De L'identité. Apologie De La Dissociation* (Paris: Economica-Anthropos, 2006)., cited in P. De Hert, "A Right to Identity to Face the Internet of Things," (at http://portal.unesco.org/ci/fr/files/25857/12021328273de_Hert-Paul.pdf/de%2BHert-Paul.pdf. Also on the CD of Commission Nationale française pour l'Unesco, Ethique et droits de l'homme dans la societé d'information. Actes, synthèse et recommandations, 13–14 septembre 2007, Strasbourg., 2008), 14.

public behaviour in society, or (borrowing again Pino's expression) in the "concrete and unequivocal circumstances of social life."[72]

Nevertheless, and bearing in mind the expected AmI features and transformations previously analysed, this vision is now obsolete, as it presupposes the "one person – one identity" link. According to such old assumption, we have only an ascertainable unique identity, subject of being objectively assessed by our actions in the public stage. However, society, nowadays (and even more with the concretization of the AmI vision), is no longer clearly divided into public and private spaces. Accordingly, any given individual's personal identity, instead, lives in the limbo of this blurry and hybrid space and, as such, cannot be simply and objectively inferred from actions performed in such utopian public space. In addition, the current society and the future AmI world allow us to express, develop and live simultaneously different identities (one or several in the Internet – virtual worlds, social networking sites, another one in the physical world, etc), fact which will put into question the vision of this single and ascertainable identity enshrined in legislation and professed in courts.

Having said that, the right to personal identity in the forthcoming AmI world cannot be restricted to the protection of personal identity against false representations engendered by the mass media and exposed to the public eye (as the Italian jurisprudence has constructed this figure). In my opinion, the right to personal identity has to go further. It has to cover also misrepresentations engendered by machines and agents which do not come out to the public eye but, on the basis of which, decisions affecting people are made, questioning their autonomy. This important aspect of the AmI vision has been captured by Hildebrandt and Gutwirth, in their study of profiling technologies.[73] The problem identified in this regard is that the sophisticated personalised profiles constructed by those technologies, which define an individual in many detailed aspects of her social, private and public life, do not only run the risk of misrepresenting that individual, but also (and mainly) of influencing a person's sense of self. Such technologies threaten to affect the very process of identity construction (and this without people being even aware of). In other words, the influence of third person's perspective on a subject's identity may not only originate from other's people perceptions and the mass media, but also (and ever more) from a whole apparatus of silent and invisible machines, devices, sensors and intelligent agents. Such aspect, furthermore, draws our attention to the need of developing the notion of identity as a narrative, that is, as a constructive process. In this way, the right to personal identity, learning from the human rights experience, should devote its attention not only to the aspect of identity representation, but also to identity construction.

Despite the advanced conceptual evolution of the right to personal identity observed in the Italian case, the construction of this right – when confronted with the challenges posed by the AmI technological developments – needs to be brought

[72]Pino, "The Right to Personal Identity in Italian Private Law: Constitutional Interpretation and Judge-Made Rights," 226.

[73]Hildebrandt and Gutwirth, *Profiling the European Citizen : Cross-Disciplinary Perspectives*.

up to date in order to protect new types of personal identity infringement. Such updating imperative can be better understood by underlining the problems and insufficiencies of the particular features of the right to personal identity, as listed by Pino.[74]

The first problem concerns the fact that, according to the Italian jurisprudence, the protection granted by the right to personal identity can only be invoked if a false representation of the personality has been conveyed to the public eye. In this light, and bearing in mind the mechanics of the AmI vision, what will happen if a false representation of someone's identity is constructed by machines and autonomous agents? And what if such representation is done without extrapolating to the public eye but solely addressing the individual, thus influencing her behaviour and undermining her autonomy and self-determination? Should the right to personal identity also not work in those cases?

The second problem relates to the concept of personal history, referred by the right to personal identity, and the requirement to interpret it only in a factually-objective way. In other words, "[t]his means that this right does not cover all those personal ideas and thoughts that have never been publicly exposed or actually revealed in concrete acts."[75] Such restrictive scope of the right to personal identity may turn out to be problematic in an AmI world. As such, imagine that those personal ideas and thoughts are not publicly exposed nor revealed in concrete acts but are still detected by AmI technologies, namely by automated profiling techniques? Should the right to personal identity not be also invoked in those cases?

To sum up, there are essentially two main problems affecting the proposed and current right to personal identity, at least as framed in Italy. The first problem consists in the excessively restricted context in which such right can be applied, that is, in cases of public exposure by the mass media. The second problem refers to the strict conceptualization of the right itself, that is, as social projection and image. With the inexorable advent of the AmI world, both that context and conceptualization bear the risk of becoming obsolete and reductive. The society in the making will no longer (or, at least, not only) be composed of people reading newspapers and watching TV news, but it will also be populated by innumerable machines, sensors and agents that will monitor and track us permanently, creating their own representations of ourselves. The perils of misrepresentation and distortion of our identities will go much beyond the ones eventually perpetrated by the mass media. In addition, the assessment of our identity according to the one projected in society will be difficult to operate, taking into account the fragmentation of our identity into diverse ones and the overload of information (in many cases contradictory) concerning those very same identities. It is exactly the fragmentation of identity that leads us to another important aspect of the re-conceptualization of the right to personal identity: the right to multiple identities.

[74]Pino, "The Right to Personal Identity in Italian Private Law: Constitutional Interpretation and Judge-Made Rights," 234–235.

[75]Ibid., 235.

4.8 The Right to Multiple Identities

In "The Multiple Self",[76] Elster begins by stating that "[t]he idea that the individual person may be seen as – or actually is – a set of sub-individual, relatively autonomous 'selves' has a long history." In fact, the cited book encompasses a number of contributions that elaborate on the conceptual strategies that have been used to make sense of the notion of "several selves." In this regard, the work of the American economist and Nobel Laureate Thomas Schelling is worth remembering. Departing from his work on conflicts between nation-states, particularly those with nuclear weapons, Schelling went beyond the military paradigm, applying a similar analysis to individuals' internal struggles. The problem tackled by Schelling, and explained by Vedral, "is that pretty much everybody suffers from a split personality on various issues ... [b]ut they do not exist at the same time and which side wins depends on the strategies that the two personalities use."[77] Furthermore, and as Schelling suggested:

> the human being is not best modelled as a speculative individual but as several alternates according to the contemporary body chemistry. Tuning in and tuning out perceptual and cognitive and affective characteristics is like choosing which "individual" will occupy this body and nervous systems[78]

With the advancement of technology and the eruption of the internet, the possibilities for identity construction and the ways to accommodate the "several selves" seem to have surpassed the constraints of body chemistry. Furthermore, and with the progressive implementation of the AmI scenario, several identities can now be formed across the mixed environment of physical and digital dimensions. Having previously seen how the traditional link "one person – one identity" has become obsolete, it is important to note that technological developments in the internet and, in the future, in the AmI environments are forging "new forms of identities that have been created partially separated from the original, unique identity of the person."[79] We are thus moving towards a deep fragmentation of personal identity, shattered into multiple and different concepts of partial and virtual identities,[80] such as avatars, pseudonyms, categories, profiles, etc. Now it seems to be possible to choose between different individuals using the same body and nervous systems.

---

[76]Jon Elster, *The Multiple Self*, Studies in Rationality and Social Change. (Cambridge [Cambridgeshire] ; New York, NY: Cambridge University Press, 1986).

[77]Vlatko Vedral, *Decoding Reality : The Universe as Quantum Information* (Oxford ; New York, NY: Oxford University Press, 2010), 92–93.

[78]Thomas C. Schelling, "Ethics, Law, and the Exercise of Self-Command," In *The Tanner Lectures on Human Values Iv*, edited by M. Sterling McMurrin (Salt Lake City: University of Utah Press, 1983), cited in Elster, *The Multiple Self*.

[79]Jaquet-Chiffelle et al., "Virtual Persons and Identities," 77.

[80]For a profound analysis of these new forms of identity, see David-Olivier Jaquet-Chiffelle, Emmanuel Benoist, Rolf Haenni, Florent Wenger, and Harald Zwingelberg "Virtual Persons and Identities" in "The Future of Identity in the Information Society", p.75–117

Furthermore, and as Nabeth has observed, identity intervenes very concretely in many facets of people's life, spanning from their private, family and biological lives, to their social, work, citizen and entrepreneurial ones. In this regard, "as individuals take on many different roles in the course of their life, different sets of characteristics, corresponding to these different roles, are used to represent their identity."[81]

Bearing that in mind, what if the right to personal identity could refer to a very specific set of personal characteristics, granting the possibility to encapsulate them in a distinct and isolated partial identity? Behind this idea lies the proposal for a right to multiple identities.

Departing from such hypothesis, and taking into account the previous considerations on the fragmentation and multiplication of identity emerging from the AmI scenario, a right to multiple identities seems absolutely fundamental in order to capture and regulate the increasingly complex and dissociated character of personal identity.

Rather than a schizophrenic exercise, the right to multiple identities addresses the need of every individual to have, according to the context in which one would act, her partial identities (both digital and physical) recognized by law. Such recognition entails, moreover, that every partial identity (that is, the sum of particular elements describing and representing that person's partial identity) would only be subject to identification according to those specific elements, preventing that the latter could in anyway be linked to any other elements and, thus, to other partial identities. Such important aspect of the right to personal identity, here proposed, would not only serve the privacy interests of the subject (by keeping important aspects of one's private life concealed, allowing the subject to act with only a restricted representation of his or her identity), but would also be in line with one of the data protection legal regime imperatives, that is, the minimization of personal data disclosure. In addition, the right to multiple identities would allow for the possibility to create different representations of oneself, keeping them separate from one another.

Elliot has described modern selfhood as "flexible, fractured, fragmented, decentred and brittle."[82] Legislation, nonetheless, treats identity as if it were a unique item.[83] My proposal for a right to multiple identities intends to alter such state-of-affairs, calling the attention to the need to adapt our current legal framework to the upcoming technological world, where different and simultaneous identities can be easily created and commanded.

This is clearly a right *in statu nascendi* and more work is needed in order to consolidate this legal figure, namely studies covering the possible connections between the right to multiple identities, on the one hand, and the rights to anonymity and the legal protection of pseudonyms, on the other.

---

[81] Nabeth, "Identity of Identity," 38.

[82] Elliott, *Concepts of the Self*, 8.

[83] M. Wigan, "Owning Identity – One or Many – Do We Have a Choice?," *Technology and Society Magazine, IEEE* 29, 2 (2010).

4.9 The Right to Be Forgotten

In a much more developed stage is the so-called "right to be forgotten," also known as the right to oblivion, "droit à l'oubli" (French) or "diritto al'oblio" (Italian). This legal figure has been formulated in the French[84] and Italian[85] law (although implicitly) and jurisprudence. Defined as "the right to silence on past events in life that are no longer occurring,"[86] the debate around this right has been recently resumed in Europe[87] with the question of whether internet users should be entitled to erase personal information stored in the internet. Such question becomes particularly pressing when one realizes that the internet tends to record everything and forget nothing. In practical terms, the issue of the right to be forgotten revolves around the question of granting (or not) to internet users the possibility of deleting personal data (such as images, texts, opinions, official documents, certificates and any other type of personal data describing past behaviours and actions, etc) from the list of results promoted by search engines, websites, social networks, blogs, etc. Taking into account what was previously said about the expected proliferation of personal data in the ambient intelligence scenario, the question of the admissibility of a right to be forgotten will only tend to become more relevant.

Being traditionally connected to the right to privacy, the issue of forgetfulness also bears an important association with the right to identity. Recurring again to the idea and metaphor of personal identity as narrative, the question that lies beneath the right to oblivion is the possibility of having parts of our identity narrative erased, preventing them from being accessed and acknowledged by the larger public. Here, the main issue meriting careful discussion is the appropriateness of broadening the scope of the right to personal identity, in order to cover not only the entitlement to construct one's future identity story, but also to erase her past one.

In this discussion, one should take into consideration that, contrarily to other rights of personality, personal identity changes with the evolution and the ageing of

---

[84]The right to be forgotten is implicitly recognized in a number of different legislative acts and legal instruments in France, namely in article 40 of the "Loi nº 78-17 du 6 janvier 1978 relative à l' informatique, aux fichiers et aux libertés," as well as in article 226-20 of the French Penal Code.

[85]The right to oblivion (*il diritto all'oblio*) is deemed to be implicitly enshrined in articles 7 and 11 of the Personal Data Protection Code (Legislative Decree n.196 dated 30 June 2003), as well as in a number of jurisprudential decisions (Italian Supreme Court [*Corte Suprema di Cassazione*], 18 October 1984, n.5259; Court of Rome [*Tribunale di Roma*], 27 November 1996, etc).

[86]Pino, "The Right to Personal Identity in Italian Private Law: Constitutional Interpretation and Judge-Made Rights," 237.

[87]It is interesting to note that the right to be forgotten is not protected in the United States, being clearly overshadowed by the right to inform and the right to free speech. Such fact results from the ever-broadening view of the First Amendment's protection of a free press and a clear preference for the latter over the privacy interests of individuals. For more details on the clash between Europe and the US concerning the tension between the right to inform and the right to be forgotten, see Werro, "The Right to Inform V. The Right to Be Forgotten: A Transatlantic Clash."

the person.[88] Adding to this particularity one of the main rationales of the right to personal identity, that is, the right not to have one's identity misrepresented or falsified, it seems that the changeable and the variable characters of personal identity demand the right to have our most recent and actual identity recognized and ascertained by others.[89] This implies, conversely, the right to have past traces of one's identity (that either go against the actual and current identity or not) erased, that is, to have older facts and actions representative of past identities deleted (if one so wishes). As it is only by forgetting the past ones that the actual identity can prevail, the right to be forgotten may develop an extremely important role in allowing an individual to reconstruct her identity's narrative, having the certainty that the past ones will not undermine such process.

In this sense, the right to be forgotten – as part of the right to personal identity – is intimately connected to the ability to reinvent oneself, to have a second chance to start-over and present a renewed identity to the world. As such, the right to oblivion seems to find an appropriate normative root in the right to personal identity.

Regarding the articulation between the right to oblivion and other "competing" rights, it is important to bear in mind that the (private) interest and right to be forgotten needs to be balanced with other important rights and interests. This is the case of the public and social interest to access information (the right to information), the right to freedom of speech and the need to preserve a collective and historical memory.[90] In this way, the right to oblivion needs to be carefully shaped in order to strike an adequate balance between the right to forgetfulness and the right to memory. Furthermore, and in particular cases, it is important to acknowledge that the right to be forgotten should not always prevail. As such, and as Werro explains, "[w]hen information about the past is needed to protect the public today, there will be no right to be forgotten. This could be the case, for example, when a person who has abused his managerial position to gain financial advantages in the past seeks

---

[88]Niger, "Il Diritto All'identità Personale," 125.

[89]Following this point of view, Niger observes that the need to protect one's projection in the reality of society, taking into account what one is and expresses through her present social presence, assumes enormous importance. The past of a person, as long as not necessary to define someone's actual and current social presence, should remain in oblivion, namely when its remembrance may alter her present position (Sergio Niger, "Il Diritto All' Oblio," In *Diritto All'anonimato: Anonimato, Nome E Identitá Personale*, edited by Giusella Finocchiaro (Padova: Cedam, 2008) – author's translation.

[90]A paradigmatically example of the preservation of a collective memory associated with new technologies (in this case within the so-called web 2.0) is reflected in the recent announcement made by the US Library of Congress. The world's largest library has announced that it will archive digitally every public tweet since Twitter's inception, back in March 2006 (Christopher Bean, Posterity. How Future Historians Will Use the Twitter Archives, http://www.slate.com/id/2251429/., accessed October 2, 2004). Twitter is a website which offers a social networking and microblogging service, enabling its users to send and read other user's messages called tweets (Wikipedia contributors, "Twitter," Wikipedia, The Free Encyclopedia, http://en.wikipedia.org/wiki/Twitter (accessed October 2, 2010).

employment in a comparable position."[91] The right to oblivion will also face difficulties regarding certain members of society (politicians, public figures) whose transparency is important from a point of view of e.g. democracy.

Leaving aside these particular cases, the move towards the substantiation of an implicit right to be forgotten has, increasingly and for quite some time, been supported by legislative modifications in European data protection legal regimes. In fact, and since the late 1960s, such legal systems have recurrently imposed stricter specific retention periods on data storage,[92] enshrining (indirectly) a true right to oblivion. In this light, and by specifying the periods beyond which certain categories of information should no longer be kept or used, data protection regulatory frameworks have, although not explicitly, recognized the right to be forgotten. More recently, and due to the enormous risks associated with the proclaimed total and eternal memory of the Internet, the right to oblivion has been proposed as an explicit right to be enshrined in specific legislation. In this matter, both France and Italy have presented legislative proposals in this sense.[93, 94] Moreover, Alex Türk, the French Data Protection Commissioner, has called for a "constitutional right to oblivion" that would allow citizens to maintain a greater degree of anonymity online and in public places.[95]

Academicians have also presented original suggestions that form interesting modalities of a right to oblivion. This is the case of the idea of "reputation bankruptcy" authored by Zittrain. According to the Harvard Professor:

> Like personal financial bankruptcy, or the way in which a state often seals a juvenile criminal record and gives a child a "fresh start" as an adult, we ought to consider how to implement the idea of a second or third chance into our digital spaces. People ought to

---

[91] Werro, "The Right to Inform V. The Right to Be Forgotten: A Transatlantic Clash," 291.

[92] In fact, and within European data-protection laws, one of the fundamental principles orienting the protection of individuals in this area has been the increasing limitations and restrictions upon data retention. For more details on the issue of data retention and the right to oblivion, see Jeremy Warner, "The Right to Oblivion: Data Retention from Canada to Europe in Three Backward Steps," *University of Ottawa Law & Technology Journal* 2, 1 (2005).

[93] In this sense, French Senators Yves Détraigne and Anne-Marie Escoffier have recently put forward a draft proposal for a "Law to Better Guarantee the Right to Privacy in the Digital Age" which explicitly recommends the establishment of a "droit à l'oubli numérique." Such legislative proposal is designed to regulate the storage of data, establishing a maximum retention period for personal data and including a right to delete information, guaranteed free of charge. "Proposition de loi sénatoriale visant à mieux garantir le droit à la vie privée à l'heure numérique", recorded by the Senate Presidency on Nov. 6, 2009, is available, in French, at http://www.senat.fr/leg/ppl09-093.html

[94] In Italy, the MEP Carolina Lussana has presented in Parliament a controversial draft proposal for a law that regulates the right to oblivion. Such proposal would prevent the storage and availability of information on the Internet concerning people already under investigation or facing charges in a criminal process. The proposed bill (*proposta di legge n.2455: nuove disposizione per la tutela del diritto all'oblio su internet in favore delle persone già sottoposte a indagini o imputate in un processo penale*), submitted to the Italian Parliament on May 20, 2009 is available, in Italian, at http://www.camera.it/_dati/leg16/lavori/stampati/pdf/16PDL0025880.pdf

[95] Jeffrey Rosen, "The Web Means the End of Forgetting," *New York Times*, July 21 2010.

be able to express a choice to deemphasize if not entirely delete older information that has been generated about them[96]

People would thus be allowed to declare "reputation bankruptcy" every 10 years or so, wiping their reputation slates clean (through the deletion of certain categories of ratings or sensitive information) and start over.

At the technological level, interesting proposals to enforce a true right to be forgotten have also been put forward. This is the case of Mayer-Schönberger who, in his recent book – "Delete – The Virtue of Forgetting in the Digital Age"[97], – argues that digital technology and global networks are eroding our natural capability to forget, proposing thus the establishment of expiration dates on information. According to the scholar,

> One possible way we can mimic human forgetting in the digital realm is by associating information we store in digital memory with expiration dates that users set. Our digital storage devices would be made to automatically delete information that has reached or exceeded its expiry date[98]

In fact, research is already being done to attain such objective and materialize a right to oblivion. As an example, researchers at the University of Washington are developing a technology called Vanish that makes electronic data "self-destruct" after a specified period of time.[99]

As Rosen observes, "[t]hese approaches share the common goal of reconstructing a form of control over our identities: the ability to reinvent ourselves, to escape our pasts and to improve the selves that we present to the world."[100]

The association between the right to be forgotten and the right to personal identity[101] that I hereby propose provides a stronger case for the emergence and consolidation of the right to oblivion. The latter, in this way, should not only be seen from a privacy point of view, but also from an identity standpoint. It is thus important to acknowledge not only the immediate consequences of the application of the right to be forgotten, that is, the possibility to conceal a number of past facts and actions from the public knowledge (privacy perspective); but to bear also in mind the more profound implications of the application of such right, that is, what the right to oblivion allows us to do afterwards. From an identity perspective, the right to be forgotten equates to the right to new beginnings, the right to start over from a clean slate, the right to self-definition. The right to be forgotten is thus an

---

[96] Jonathan Zittrain, *The Future of the Internet and How to Stop It* (New Haven, CT: Yale University Press, 2008), 229.

[97] Viktor Mayer-Schönberger, *Delete : The Virtue of Forgetting in the Digital Age* (Princeton, NJ: Princeton University Press, 2009).

[98] Ibid., 171.

[99] Rosen, "The Web Means the End of Forgetting."

[100] Ibid.

[101] In this context, Niger affirms that the right to identity represents the foundational reference of the right to oblivion ("Il diritto all'identità personale rappresenta, quindi, la matrice prima del diritto all'oblio", in Niger, "Il Diritto All' Oblio," 67)

important legal instrument to reconstruct one's identity, to have the opportunity to re-create oneself, exerting a better control over one's *idem* identity.

4.10 Conclusion

In an era of pervasive traceability, surveillance and profiling, personal identity is facing the risk of becoming less and less personal. Taking into account the challenges that lie ahead and the numerous new possibilities of infringing the right to one's identity, I believe that it is time to add and enrich the right to personal identity with new perspectives and insights. This is what this article sought to accomplish.

AmI technologies will promote a series of changes to our notions and feelings of personal identity. Those changes include the dispersion of one's identity traces and features across the physical and digital worlds, the potentially perpetual registration of information and knowledge about oneself, and the fragmentation of one's identity into partial segments. As such, the challenges posed by AmI technologies upon the protection of personal identity demand a further re-conceptualisation of this right.

In this regard, I have embarked in such re-theorization by taking into account the multifaceted definitions of personal identity that have been crafted throughout the evolution of this right. I have thus used the plasticity and flexibility of the right to personal identity[102] to cover new and upcoming situations, granting legal protection to the new ways and processes through which our identities can be reconstructed and fragmented (and, also, infringed and violated). Going beyond the proclaimed interest in the uniqueness of the human being, I have strived to go further in the conceptualization of the right to identity. In this way, I have formulated a proposal for a new right to identity by looking at the various challenges posed by new and prospective AmI technologies. In this context, and taking into account how personal identity traces will become dispersed, decentred and permanently registered, I underlined two aspects that will assume greater relevance for the control over our own identities: the possibility of self-reinvention and the option to feed and develop different (partial) identities within the construction of one's personality. In this light, and using the flexibility of this legal institute, I have used the right to personal identity as a normative root and anchor to affirm and consolidate two other legal rights: the right to be forgotten, through which the possibility of self-reinvention can be secured; and the right to multiple identities, through which the segmentation of one's identity into partial ones can come to fruition.

To finalize I would like to stress that new and emerging technologies provide both the obstacles and the solutions for the respect of the right to personal identity. In the same way that technologies trace, register and store personal information, they can also erase and delete. In the same way that they gather one's identity traces

[102] Pino stresses the same idea by qualifying the right to personal identity as a multiform, adaptable right (Ibid.)

and features into one single converging platform, they may also allow us to segment our identities into different pseudonyms, avatars and digital *personas*, enabling us to command and live different "selves." While it is up to us to maximize the benefits of technology in the construction and affirmation of our own personal identities, it is up to law to enclose such technological potentialities within an identity-friendly framework. A legal framework that enables every human person to freely construct, de-construct and re-construct their own identities. The conceptualization of the right to personal identity here proposed attempts to be an important step in such direction.

References

Aarts, E.H.L., and S. Marzano. *The New Everyday : Views on Ambient Intelligence*. Rotterdam: 010 Publishers, 2003.

Aarts, E., R. Harwig, and M. Schuurmans. "Ambient Intelligence." In *The Invisible Future : The Seamless Integration of Technology into Everyday Life*, edited by P.J. Denning, 235–250. New York, NY: McGraw-Hill, 2001.

Andrade, N.N.G. de. "Technology and Metaphors: From Cyberspace to Ambient Intelligence." *Observatorio (OBS*) Journal* 4, 1 (2010): 121–146.

Bean, C. "Posterity. How Future Historians Will Use the Twitter Archives." http://www.slate.com/id/2251429/.

Boumard, P., G. Lapassade, and M. Lobrot. *Le Mythe De L'identité. Apologie De La Dissociation*. Paris: Economica-Anthropos, 2006.

Boussard, H. "Individual Human Rights in Genetic Research: Blurring the Line between Collective and Individual Interests." In *New Technologies and Human Rights*, edited by T. Murphy. Oxford, New York, NY: Oxford University Press, 2009.

De Cupis, A. *Il Diritto All' Identità Personale*. Milano: A. Giuffrè, 1949.

De Hert, P. "A Right to Identity to Face the Internet of Things." 21 p.: at http://portal.unesco.org/ci/fr/files/25857/12021328273de_Hert-Paul.pdf/de%2BHert-Paul.pdf. Also on the CD of Commission Nationale française pour l'Unesco, Ethique et droits de l'homme dans la societé d'information. Actes, synthèse et recommandations, 13–14 septembre 2007, Strasbourg., 2008.

Detrick, S., editors, *The United Nations Convention on the Rights of the Child. A Guide to The "Travaux Préparatoires"*. Dordrecht; Boston, MA; London: Martinus Nijhoff Publishers, 1992.

Elliott, A. *Concepts of the Self*, 2nd ed. Key Concepts. Cambridge: Polity, 2007.

Elster, J. *The Multiple Self.* Studies in Rationality and Social Change. Cambridge [Cambridgeshire]; New York, NY: Cambridge University Press, 1986.

Glover, J. *Ethics of New Reproductive Technologies : The Glover Report to the European Commission*. Studies in Biomedical Policy. DeKalb: Northern Illinois University Press, 1989.

Groebner, V. *Who Are You? : Identification, Deception, and Surveillance in Early Modern Europe*. Brooklyn, NY: Zone Books, 2007.

Hildebrandt, M. "Privacy and Identity." In *Privacy and the Criminal Law*, edited by E. Claes, A. Duff and S. Gutwirth, 43–57. Antwerpen: Intersentia; Oxford: Hart Publications [distributor], 2006.

Hildebrandt, M. Profiling and Ami. In *The Future of Identity in the Information Society : Challenges and Opportunities*, edited by K. Rannenberg, D. Royer and A. Deuker, 273–313. Berlin, London: Springer, 2009.

Hildebrandt, M., and S. Gutwirth. *Profiling the European Citizen : Cross-Disciplinary Perspectives*. New York, NY: Springer, 2008.

Information Society Advisory Group (ISTAG). "Ambient Intelligence: From Vision to Reality", 31, 2003.

Information Society Technologies Advisory Group (ISTAG). "Orientations for Workprogramme 2000 and Beyond." 7, 1999.

Jackson, E.M. *Regulating Reproduction : Law, Technology and Autonomy*. Oxford ;Portland, OR: Hart, 2001.

Jaquet-Chiffelle, D.-O., E. Benoist, R. Haenni, F. Wenger, and H. Zwingelberg. "Virtual Persons and Identities." In *The Future of Identity in the Information Society : Challenges and Opportunities*, edited by K. Rannenberg, D. Royer and A. Deuker, 75–122. Berlin ;London: Springer, 2009.

Kaufmann, J.-C. *L'invention De Soi : Une Théorie De L'identité*, Collection Individu Et Société. Paris: Armand Colin, 2004.

LeDoux, J.E. *Synaptic Self : How Our Brains Become Who We Are*. New York, NY: Viking, 2002.

Marshall, J. *Personal Freedom through Human Rights Law? : Autonomy, Identity and Integrity under the European Convention on Human Rights*, International Studies in Human Rights, V. 98. Leiden ;Boston, MA: Martinus Nijhoff Publishers, 2009.

Mayer-Schönberger, V. *Delete : The Virtue of Forgetting in the Digital Age*. Princeton, NJ: Princeton University Press, 2009.

Nabeth, T. "Identity of Identity." In *The Future of Identity in the Information Society : Challenges and Opportunities*, edited by K. Rannenberg, D. Royer and A. Deuker, 19–69. Berlin, London: Springer, 2009.

Nakashima, H., Aghajan, H., and J.C. Augusto. *Handbook of Ambient Intelligence and Smart Environments,* 1st ed. New York, NY: Springer, 2009.

Neethling, J., J.M. Potgieter, and P.J. Visser. *Neethling's Law of Personality*. Durban: Butterworths, 1996.

Neethling, J. "Personality Rights: A Comparative Overview." *Comparative and International Law Journal of Southern Africa* 38, 2 (2005): 210–245.

Niger, S. "Il Diritto All' Oblio." In *Diritto All'anonimato: Anonimato, Nome E Identitá Personale*, edited by G. Finocchiaro, 59–73. Padova: Cedam, 2008.

Niger, S. "Il Diritto All'identità Personale." In *Diritto All'anonimato. Anonimato, Nome E Identità Personale*, edited by G. Finocchiaro, 113–129. Padova: Cedam, 2008.

Nissenbaum, H. "Privacy as Contextual Integrity." *Washington Law Review* 79, 1 (2004): 119–158.

Pino, G. *Il Diritto All'identità Personale : Interpretazione Costituzionale E Creatività Giurisprudenziale*. Ricerca, Bologna: Il mulino, 2003.

Pino, G. "The Right to Personal Identity in Italian Private Law: Constitutional Interpretation and Judge-Made Rights." In *The Harmonization of Private Law in Europe*, edited by M. Van Hoecke and F. Ost, 225–237. Oxford: Hart Publishing, 2000.

Rannenberg, K., D. Royer, and A. Deuker. *The Future of Identity in the Information Society : Challenges and Opportunities*. Berlin, London: Springer, 2009.

Ricœur, P. *Oneself as Another*. Chicago, IL: University of Chicago Press, 1992.

Riva, G. Ph D. *Ambient Intelligence : The Evolution of Technology, Communication and Cognition Towards the Future of Human-Computer Interaction*, Emerging Communication, 1566–7677. Amsterdam, Oxford: IOS Press, 2005.

Rodotá, S. "Data Protection as a Fundamental Right." In *Reinventing Data Protection?*, edited by S. Gutwirth, Y. Poullet, P. De Hert, C. Terwangne and S. Nouwt, 77–82. Dordrecht, London: Springer, 2009.

Rodotá, S. "Lo Specchio Di Stendhal: Riflessioni Sulle Riflessioni Dei Privatisti." *Rivista critica del diritto privato* 15 (1997).

Rosen, J. "The Web Means the End of Forgetting." *New York Times*, July 21 2010.

Schelling, T.C. "Ethics, Law, and the Exercise of Self-Command." In *The Tanner Lectures on Human Values Iv*, edited by M. Sterling McMurrin, 43–79. Salt Lake City, UT: University of Utah Press, 1983.

Sullivan, C. "Privacy or Identity?" In *Int. J. Intellectual Property Management* 2, 3 (2008): 289–324.

Tommasini, R. "L'identitá Dei Sogetti Tra Apparenza E Realtà: Aspetti Di Una Ulteriore Ipotesi Di Tutela Della Persona." In *Il Diritto Alla Identità Personale,* edited by G. Alpa, L. Boneschi and M. Bessone, 78–91. Padova: CEDAM, 1981.

Vedral, V. *Decoding Reality : The Universe as Quantum Information*. Oxford, New York, NY: Oxford University Press, 2010.

Warner, J. "The Right to Oblivion: Data Retention from Canada to Europe in Three Backward Steps." *University of Ottawa Law & Technology Journal* 2, 1 (2005): 75–104.

Weber, W., J.M. Rabaey, and E.H.L. Aarts. *Ambient Intelligence*, 1st ed. Berlin, New York, NY: Springer, 2005.

Werro, F. "The Right to Inform V. The Right to Be Forgotten: A Transatlantic Clash." In *Liability in the Third Millenium; Georgetown Public Law Research Paper No. 2*, edited by A. Colombi Ciacchi, C. Godt, P. Rott and L.J. Smith, 285–300. Baden-Baden: F.R.G., 2009.

Wigan, M. "Owning Identity – One or Many – Do We Have a Choice?" *Technology and Society Magazine, IEEE* 29, 2 (2010): 33–38.

Zittrain, J. *The Future of the Internet and How to Stop It*. New Haven, CT: Yale University Press, 2008.

Part II
The Dark Side: Suspicions, Distrust and Surveillance

Chapter 5
Frames from the Life and Death of Jean Charles de Menezes

Amos Bianchi and Denis J. Roio

5.1 Premise

The process of subjectivation/desubjectivation is a very effective method of analysis for the contemporary world, which we'll adopt in this paper.

This method enables one to put the processes of intentional subjectivation/desubjectivation in a system of semantic apparatuses in order to investigate – retrospectively – if there are entirely new processes of subjectivity at work. Or, in other words: how, on 22 July 2005, were the contemporary, hyper- mediatic and hyper-technologic governmental control apparatuses able to reduce de Menezes' identity to his bare life? Seven keywords will suggest a partial, not-exhaustive answer.

5.2 Apparatus

This research is underpinned by the concept of apparatus, outlined by Michel Foucault as follows

> [it] is, firstly, a thoroughly heterogenous ensemble consisting of discourses, institutions, architectural forms, regulatory decisions, laws, administrative measures, scientific statements, philosophical, moral and philanthropic propositions – in short, the said as much as the unsaid. . . . Secondly, what I am trying to identify in this apparatus is precisely the nature of the connection that can exist between these heterogenous elements. . . Thirdly, I understand by the term "apparatus" a sort of – shall we say – formation which has as its major function at a given historical moment that of responding to an urgent need. The apparatus thus has a dominant strategic function (Foucault 2001, 299–300).

Foucault's concern was to understand how the subject and object position themselves and give shape to experience by the process of subjectivation: 'C'est . . . le sujet, qui constitue le thème général de mes recherces' (Foucault 2001, 1042). The

A. Bianchi (✉)
Nuova Accademia di Belle Arti (NABA), Coordinamento Istituto Master e Bienni, Via Darwin 20, 20143 Milano, Italy
e-mail: amos.bianchi@naba.it

S. Gutwirth et al. (eds.), *Computers, Privacy and Data Protection: an Element of Choice*, DOI 10.1007/978-94-007-0641-5_5,

history of humanity concerns the ways in which the human race has thus created the fields of objectification, and, in this way, has created itself as subject.

5.3 Desubjectivation

From a perusal of Foucault, Agamben derives a further strand: the division of everything into two macro-categories: living beings and the apparatus (the historical element). In the following passage the two macro-categories are merged together to create the concept of subjectivity: "The apparatus is, above all, a machine that produces subjectivity because it is also a machine of government" (Agamben 2006, 29). But, in another step, the Italian writer allies government apparatus with the most powerful catalyst of apparatuses of the contemporary era, capitalism: "In the present phase of capitalism, an apparatus does not operate through the construction of a subject, but through the processes of desubjectivation" (Agamben 2006, 30). And later:

> What is happening now is that the processes of subjectivation and the processes of desubjectivation seem to have become reciprocally indifferent and do not give rise to the reconstruction of a new subject, if not in a phantom form and, so to speak, spectral. In the non-truth of the subject, there is no truth in any form (Agamben 2006, 30–31).

In order to identify processes of desubjectivation, and what they consist of, it is necessary to verify in what ways capitalism reveals its apparatus.

5.4 Space

From Guy Debord, The Society of Spectacle, paragraph 169:

> The society that reshapes its entire surroundings has evolved its own special technique for molding the very territory that constitutes the material underpinning for all the facets of this project. Urbanism – "city planning" – is capitalism's method for taking over the natural and human environment. Following its logical development toward total domination, capitalism

Fig. 5.1 A CCTV still showing Jean Charles de Menezes at Stockwell Tube station – Photo: PNS

> now can and must refashion the totality of space into its own particular décor (Debord 2002, 45).

The city is the proper place for the murder of de Menezes, and a key concept to understand it. Urbanism, city planning, urban space management: by this element of total domination, capitalism takes possession of the space to build its own scenario. City becomes the ideal stage on which capitalism displays its power. But if the city has become the stage of capitalism, the question is: what is performed on it? And what instruments are working?

5.5 Body

In this scenario of a city the body itself is subjected to a radical transformation, as Agamben states in another writing

> the process of technologization, instead of materially investing the body, was aimed at the construction of a separate sphere that had practically no point of contact with it: What was technologized was not the body, but its image. . .To appropriate the historic transformations of human nature that capitalism wants to limit to the spectacle, to link together image and body in a space where they can no longer be separated, and thus to forge the whatever body, whose φύσις is resemblance – this is the good that humanity must learn now to wrest from commodities in their decline. Advertising and pornography, which escort the commodity to the grave like hired mourners, are the unknowing midwives of this new body of humanity (Agamben 1993, 50).

The world suggested by Agamben is a divided world, in which the apparatus of subjectivation (including capitalism) works on the identity of humans until, according to Debord, they become pure image and, ultimately, their bodies conceived as residual. If the liberal governamentality, theorized by Foucault, still requires the physical existence of the human beings, do the contemporary dataflows and databases open the possibility to the emergence of a new kind of governamentality, by which the relations of power can leave aside the material bodies, the *flesh*? Or could the present "whatever singularity" be linked to the double-binded practice of the "just do it", whose fake and faible disembodied communities are global-scale figures, compulsively clicking the button "Add to the basket" in order to buy the spectral image of themselves?

If the body is residual and reduced to pure spectacle, then the exterior image of the human being is the one that survives during the formation of the identity of the individual; and what happens within the space of the city is that the spectacle forges new human identities, reduced to exterior images by its technological apparatus.

5.6 Imago

Imago is the Latin word for image (italian: "immagine"; french: "image"). It has the common meaning of image, but its etymology is interesting: "imago" was the mortuary mask, made with wax, that was fixed on the face of dead men to reproduce

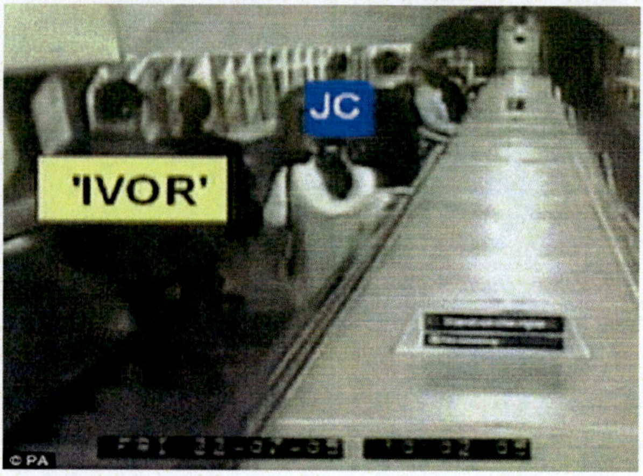

Fig. 5.2 The surveillance officer codenamed Ivor is filmed on CCTV following Jean Charles de Menezes ("JC") towards the Tube platform

human shapes during ancient pagan rites. Therefore image is, *ab origine*, related to death; it should be said that image returns to presence after death: it is the *representation of an absence*. As Agamben outlines

> the absurdity of individual existence, inherited from the sub-base of nihilism, has become in the meantime so senseless that it has lost all pathos and been transformed, brought out into the open, into an everyday exhibition: Nothing resembles the life of this new humanity more than advertising footage from which every trace of the advertised product has been wiped out. . . The fact is that the senselessness of their existence runs up against a final absurdity, against which all advertising runs around: death itself. In death the petty bourgeois confront the ultimate expropriation, the ultimate frustration of individuality: life in all its nakedness, the pure incommunicable, where their shame can finally rest in peace. Thus they use death to cover the secret that they must resign themselves to acknowledging: that even life in its nakedness is, in truth, improper and purely exterior to them, that for them there is no shelter on earth (Agamben 1993, 64–65).

The footage of de Menezes is advertising footage from which every trace of the advertised product has been wiped out. The division of his individuality into two poles, pure image (subjected to the CCTV control) and bare life, ultimately cancels out his life.

5.7 Media

In *Medienkultur* (Flusser 1997), Vilèm Flusser theorizes that writing has been the act detaching man from magic and opening doors to science. In a further step, he underlines that the modern era is living through a new transition in which contemporary techno-codes move humanity away from texts, because they transform concepts into images. Flusser's discourse is about photography, which, in this sense, is not the representation of a situation, but the representation of a series of concepts elaborated by the photographer in relation to a situation. The imagination of the photographer

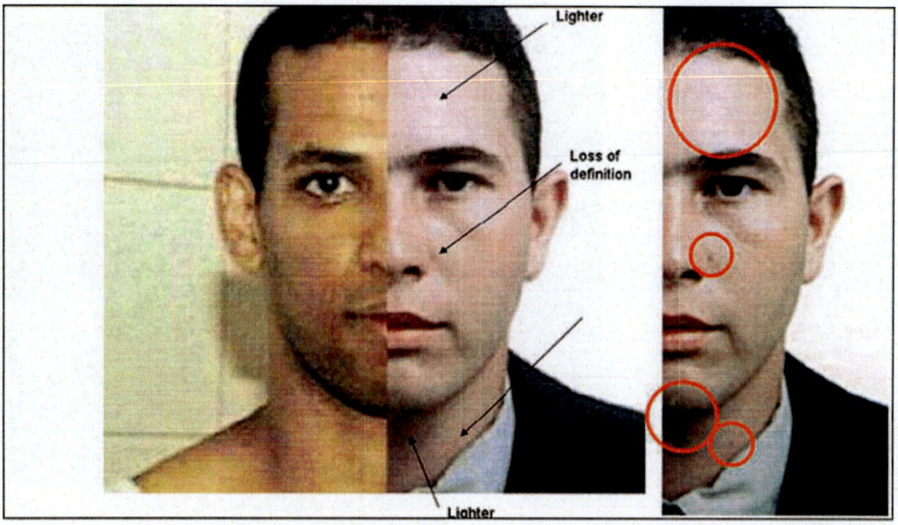

Fig. 5.3 The face shown on the ID of de Menezes is aired on media cut in half, besides that of a known terrorist, to demonstrate their resemblance

is mediated by a series of texts, by which it can exist as techno-image. The conversion of texts in techno-images generates a crisis, because the text has to mediate the techno-image without being a programme anymore.

CCTV technology mainly exasperates this condition of absence of control. The apparatus generating techno-images supplies the text, allowing them to be used. CCTV itself is a powerful apparatus of control.

We might ask: if photography is the image of a series of concepts owned by the photographer in relation to a situation, what is the role of a "CCTV photographer" if the scene shown is the whole world, a singular situation is all the situations, and a single concept is all the concepts? In this context the human being, both at the beginning and at the end of each apparatus, who tries to live in a condition of homeostasis, cannot do anything to control his creation.

The apparatus of control becomes apparatus out of control. The text that gives birth to the apparatus remains in the background, as ideology, and the will to panopticon becomes absence of vision.

It follows that the image of human being, transformed into techno- image, is incommunicable, and what remains is just his bare life, for which there is no shelter on earth, as Debord stated:

> The images detached from every aspect of life merge into a common stream in which the unity of life can no longer be recovered. Fragmented views of reality regroup themselves into a new unity as a separate pseudo-world that can only be looked at. The specialization of images of the world evolves into a world of autonomized images where even the deceivers are deceived. The spectacle is a concrete inversion of life, an autonomous movement of the nonliving (Debord 2002, 6).

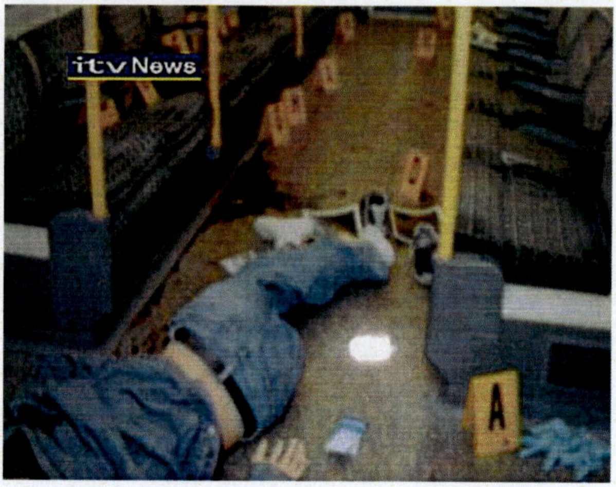

Fig. 5.4 The murder scene July 22, 2005

5.8 False Positives (Addendum)

In 2008 the Guardian UK reports plans of MI5, The Security Service of Great Britain, to datamine information out of public transportation: all the information about private traffic flow in England is made available for computational analysis conducted by law enforcement agencies in order to find out hints about terrorist activities.

This might seem just a quantitative change in terms of information processed, but what really happens in this case is that the power of the apparatus increases *qualitatively*: from a situation where, in obtaining private data about a specific citizen, is necessary a legal mandate issued by a judge (or by extraordinary anti-terrorism regulations), to a situation where the whole data about every citizen moving with public transport across a city is automatically examined by national law enforcement apparata. Such a large pool of generic information is not only available for immediate consultation when needed, but constitutes a flow of real life samples which can be constantly analysed in search for deviance patterns and relationships to suspicious individuals.

The scenario opened by this qualitative change transforms the apparatus in a high speed mechanism that, consequently, also leads to a quantitative change in the order of several magnitudes. At the origin of the new inquiries conducted by the security devices investigating reality are not humans anymore, but computer algorithms synthesising vast amounts of information on human behaviours whose results are eventually reviewed by humans.

In mathematics the errors grow exponentially when we increase the number of dimensions. The standard mean error formula resumes this concept:

$$SE_x = \frac{s}{\sqrt{n}} \quad (1)$$

Fig. 5.5 Brucker-Cohen, Policestate (2003)

While contemporary security research emphasizes on automatic pattern recognition in human behaviour, in a close future mass-analysis can be exercised on the totality of data available; relational networks around suspicious nodes will then be traced through evaluations that will multiply suspicion at a speed that was never seen before.

Illustrative analogies are offered by analysts of the socio-economic meltdown, a contemporary condition that offers vast case scenarios on the total commodification operated by financial algorithms: its causes are popularly ascribed to the financialization (Marazzi 2010) of global economies, while phenomenons described as computationalism (Golumbia 2009) and informationalism start to acquire importance, helping further comprehension. As Hakken recently stated:

> Notions like "the wisdom of crowds" (Suroweicki 2004) are virtually pure statements of informationalism. In its naive presumption that, as more information is added to the model – that is, the more computable it becomes – its accuracy tends to increase, informationalism becomes an aspect of computationalism. In short, reliance on prestigious, information filled, but systematicity presuming and therefore blind to systemic crisis, and thus deeply flawed, computationalist computing also caused the crisis. In this way, the models contained too much information, on the one hand, while ignoring crucial information (i.e., regarding systemic risk) on the other, because it was not easily quantified (Hakken 2010).

Algorithmic models fail to incorporate the risks of systemic failure: they presume systematicity, being generally incapable of incorporating "black swan/long tail" risks, like that of general failure.

Let us conclude with what Richard Stallman, a popular developer and independent thinker on civil liberties, responded to our article: he provided this quote of Chögyam Trungpa from Tibet (with foreword by Marco Pallis):

It is not only such obvious means of intimidation as machine guns and concentration camps that count; such a petty product of the printing press as an identity card, by making it easy for the authorities to keep constant watch on everybody's movements, represents in the long run a more effective curb on liberty. In Tibet, for instance, the introduction of such a system by the Chinese Communists, following the abortive rising of 1959, and its application to food rationing has been one of the principal means of keeping the whole population in subjection and compelling them to do the work decreed by their foreign overlords.

References

Agamben, G. *The Coming Community. Theory Out of Bounds*, Vol. 1. University of Minnesota Press, 1993.

Agamben, G. *Che cos'è un dispositivo? /. nottetempo + Debord, Guy. 2002. /The Society of the Spectacle*. Treason Press, 2006.

Flusser, V. *Medienkultur*. Fischer Taschenbuch Verlag, 1997.

Hakken, D. *Computing and the Current Crisis*. TripleC, 2010.

Guardian UK. 16 March `MI5 seeks powers to trawl records in new terror hunt` 2008.

http://en.wikipedia.org/wiki/Death_of_Jean_Charles_de_Menezes

http://news.bbc.co.uk/2/hi/in_pictures/7038430.stm

http://news.bbc.co.uk/2/hi/in_depth/629/629/7073125.stm

http://www.stockwellinquest.org.uk/

http://www.timesonline.co.uk/tol/news/uk/article556227.ece

http://www.guardian.co.uk/uk/menezes

http://www.guardian.co.uk/uk/2008/mar/16/uksecurity.terrorism

Chapter 6
Regulating Privacy: Vocabularies of Motive in Legislating Right of Access to Criminal Records in Sweden

Christel Backman

6.1 Introduction

Ever since the setting up of the Criminal Records Registry in Sweden in 1901, ability to access information on individuals' criminal record has been viewed as a potential hazard by Swedish state authorities, concerned about the sensitive and stigmatizing nature of that information. Restricting access to it has historically been an important strategy in managing the perceived risks. A fundamental change of approach, however, came in 1989 in response to recommendations of the Council of Europe. For the first time, full subject access disclosure was made possible for those wishing to view their own criminal record information.

Sweden has a history of keeping large databases about its population, created in the name of welfare promotion. Because of these, and because of the major role that the principle of public access to official records has played in the organization of its social and political life, the country has been called "the model surveillance society in the Western world".[1] The principle of public access to official records has a long history in the Swedish juridical tradition.[2] It was codified already in 1766 as a means to guarantee that citizens could exercise control over public bodies and prevent abuse of power. One consequence of this principle, however, is that citizens can obtain information not only on authorities and civil servants but also on other citizens. The fact that it makes tax and earnings information available to anyone is often brought forward as an example of how adherence to it affects people's privacy. To reduce the possibility for interference with the fundamental right of privacy, an exception has therefore always been made in the case of criminal records, which

C. Backman (✉)
Department of Sociology, University of Gothenburg, SE 405 30 Gothenburg, Sweden
e-mail: christel.backman@sociology.gu.se

[1] David H. Flaherty. *Protecting Privacy in Surveillance Societies.* Chapel Hill: The University of North Carolina Press, 4. 1989.

[2] See Cecilia Magnusson Sjöberg. "Constitutional Rights and New Technologies in Sweden." In *Constitutional Rights and New Technologies: A Comparative Study*, edited by R. Leenes et al. The Hague: T.M.C. Asser Press, 199–224, 2008

S. Gutwirth et al. (eds.), *Computers, Privacy and Data Protection: an Element of Choice*, DOI 10.1007/978-94-007-0641-5_6, © Springer Science+Business Media B.V. 2011

have been kept secret.[3] As will be shown below, the creation of a criminal records registry had an impact on how the principle of public access to records was implemented, prompting an amendment to the country's constitution that specifically left the criminal records out of its scope. The fact that access to criminal records was thus restricted from the start represents a rather special case, making the country unique in international comparison.

The way government databases are handled in Sweden changed with the computer revolution and the data protection debates of the 1960s and the 1970s.[4] Given its long history, the Swedish Criminal Records Registry, in all its subsequent transformations, offers a good glimpse into the "pre" and "post" periods in this evolution, making it possible to trace historical shifts in the regulation of an issue shaping up as a societally significant clash of interests between the need for privacy and the need for information.

Information about previous convictions has always been of interest to employers as well as to the state and its criminal justice system, and the likelihood of a conflict between the latter's preventive interests and the interest of individuals' in keeping that information private is thus obvious. Clashes like this provide a good opportunity to analyze how privacy has been understood, shaped, and defined at different historical conjunctions. Any study of such developments, however, will of necessity be more sociological than philosophical or legal in nature. The problem, at least in the present case, is not how privacy ought to be defined, but how it has actually been defined in relation to the use of criminal records; and this may be best examined by analyzing the vocabularies of motive used in the justification of the different positions.[5] In this chapter, I analyze government documents as examples of the kind of instructive "cultural stories" that shed light on how ruling interests in society influence or attempt to influence the understanding of how that society's core values and principles should best be balanced.

In what follows, I will concentrate on five transformative moments at which the Swedish criminal records legislation was amended in regard to its subject access provisions. My argument centers around three distinct vocabularies of motive that crystallized and were put to use during these moments, and on the observation that the way subject access has been regulated depends on the position in which these vocabularies place the individual.

In the early part of the twentieth century a "protective vocabulary" dominated, with people's possibility to participate in the labor market and thus provide for themselves high up on the legislators' agenda and the criminal justice system facing

---

[3]See Freedom of the Press Act (SFS 1949:105), Chapter 2, Article 2, and Public Access to Information and Secrecy Act (SFS 2009:400), Chapter 35, Article 3, for current regulations.

[4]Lars Ilshammar. *Offentlighetens nya rum – Teknik och politik* (Örebro: Universitetsbiblioteket, 2002).

[5]Kenneth Burke. *Permanence and Change.* Berkeley: University of California Press, 1984 [1935]); C. Wright Mills. "Situated Actions and Vocabularies of Motive." *American Sociological Review* 5, 6 (1940).

a need for a more permanent record keeping. Around mid-century, a "rehabilitative vocabulary" replaced the protective vocabulary. At this point, the criminal justice system's use of criminal records for sentencing purposes was seen as almost as harmful for recidivist offenders' rehabilitation prospects as other actors' potential access to that information. In both of these vocabularies, the individual was posited as someone needing protection from the state, which was best arranged by restricting subject access right to prevent the possibility of enforced subject access where third parties like employers force job applicants to make access requests to obtain their criminal records. During the second part of the century, another vocabulary, focusing on the rights of individuals, then took shape, centered on the notion that individuals needed protection from a too protective state. Full subject access became a strategy for addressing this power imbalance. The early years of the current century then saw the consequences of unrestricted subject access, especially the sheer number of enforced subject access requests being made, become increasingly problematic, opening up a possibility for a partial return of the rehabilitative vocabulary.

To my knowledge, the history and evolution of the Swedish criminal records registry legislation has never been reconstructed in this fashion before. In Britain, the equivalent UK law and how it has changed over the years has been analyzed by Grier and Thomas.[6] Thomas has identified two reasons for why the British registry was set up in 1869. First of all, there was a need for information in the criminal justice system to allow determination of whether an offender was a "habitual offender" or not. Secondly, the abandonment of capital punishment and transportation to colonies like Australia in favor of penal servitude as a prison sentence meant that there was now a greater interest in knowing where former prisoners were and what they were up to.[7] Over the years, the use of criminal records has then widened in the UK, resulting in a situation where both authorities outside the criminal justice system as well as private employers are given access to criminal records.

In the United States and Australia, several studies report on an increased demand for criminal records, mainly from employers.[8] While the regulations in these two

---

[6]Angela Grier and Terry Thomas. "The Employment of Ex-Offenders and the UK's New Criminal Record Bureau." *European Journal on Criminal Policy and Research* 9, 4 (2001); Terry Thomas *Criminal Records: A Database for the Criminal Justice System and Beyond* Basingstoke: Palgrave Macmillan, 2007.

[7]Thomas, *Criminal Records: A Database for the Criminal Justice System and Beyond*, 7ff.

[8]Bronwyn Naylor, Moira Paterson, and Marilyn Pittard. "In the Shadow of a Criminal Record: Proposing a Just Model of Criminal Record Employment Checks." *Melbourne University Law Review* 32, 1 (2008); Harry J. Holzer, Steven Raphael, and Michael A. Stoll. "Will Employers Hire Former Offenders? Employer Preferences, Background Checks, and Their Determinants." In *Imprisoning America: Employer Preferences, Background Checks, and Their Determinants*, edited by M. Pattillo, D. F. Weiman, B. Western. New York: Russell Sage Foundation, 2004; Harry J. Holzer, Steven Raphael, Michael A. Stoll. "The Effect of an Applicant's Criminal History on Employer Hiring Decisions and Screening Practises: Evidence from Los Angeles." In *Barriers to Reentry? The Labor Market for Released Prisoners in Post-Industrial America*, edited by S. Bushway, M. A. Stoll, D. F. Weiman New York: Russell Sage Foundation, 2007.

countries may somewhat vary from state to state, in general they have not restricted access to criminal records in a manner comparable to Sweden. In them, the issue has been approached from two opposing angles: on the one hand, those critical of the present system have developed proposals for regulation based on more restricted access, while, on the other hand, a more positive stance and broader utilization of the access opportunity among employers has been encouraged by mainly those coming from a management perspective.[9] Within the European Union, the regulation of criminal records differs between the countries, with the databases maintained by bodies such as the police, the ministry of justice, or the ministry of the interior. A 1994 overview of the legislation of the (then) twelve member states of the European Union from the perspective of ex-offenders' labor market opportunities noted that even though access to criminal records databases was restricted in many of the countries considered, the practice of enforced subject access was widespread, with several countries also providing legal means to access criminal record information through so-called certificates of conduct.[10]

6.1.1 Regulating Privacy Through Opacity Tools and Transparency Tools

The sensitivity of the information contained in criminal records, and especially the fact that this information was all stored in one central national registry, was an issue tackled by the Swedish legislation even long before data protection became a known concept. In more current terminology, the legislative history of the country's criminal records regulation can be described as a shifting dynamic between what De Hert and Gutwirth have called "opacity tools", such as right to privacy, and "transparency tools", such as the right of access to information.[11] As shown by other studies and also below, framing the issue of data storing and registers as a matter of transparency

---

[9]For examples of the critical approach, see Patricia M. Harris and Kimberly S. Keller. "Ex-Offenders Need Not Apply: The Criminal Background Check in Hiring Decisions." *Journal of Contemporary Criminal Justice* 21, 1 (2005); Helen Lam and Mark Harcourt. "The Use of Criminal Record in Employment Decisions: The Rights of Ex-Offenders, Employers and the Public." *Journal of Business Ethics* 47, 3 (2003); Naylor, Paterson, Pittard. "In the Shadow of a Criminal Record: Proposing a Just Model of Criminal Record Employment Checks." For examples of the more positive stance, see Mary L. Connerly, Richard D. Arvey, Charles J. Bernardy. "Criminal Background Checks for Prospective and Current Emploees: Current Practices among Municipal Agencies." *Public Personnel Management* 30, 2 (2001); Agnes Lam and Brian H. Kleiner. "Criminal Background Checks of Prospective Employees: Why and How Should It Be Done?" *Managerial Law* 43, 1/2 (2001); Jund-Ming Wang Brian H. Kleiner. "Effective Employment Screening Practices." *Management Research News* 23, 5/6 (2000).

[10]Nancy Louks, Olwen Lyner, Tom Sullivan. "The Employment of People with Criminal Records in the European Union." *European Journal on Criminal Policy and Research* 6, 2 (1998).

[11]Paul De Hert Serge Gutwirth. Privacy, "Data Protection Law and Law Enforcement: Opacity of the Individual and Transparency of Power." In *Privacy and the Criminal Law*, edited by E. Claes, A. Duff, S. Gutwirth Antwerpen and Oxford: Intersentia, 2006.

became topical only in the 1960s. Nevertheless, to look for traces of something that had yet to become actual can, paradoxical as it may sound, provide important clues to understand its origins and evolution as an issue and why it materialized the way it did.

A democratic constitutional state can be said to have two sets of different but complementary legal tools for regulating and limiting power. Opacity tools are used to limit or stop power and to guarantee noninterference in individual matters by the state and other private actors. Transparency tools, on the other hand, help channel power and guarantee accountability of actors in powerful positions. In contemporary legislation, the right to informational privacy serves as an example of opacity tools, with data protection laws functioning as transparency tools. Since transparency tools channel power, either through offering ways of holding, for example, those who process data accountable for their actions, or by regulating the way data may be processed, they must build on the achievement of the opacity tools. Opacity tools such as privacy rights set limits to the reach of power and thus serve to protect individuals from the illegitimate uses of power; in this sense, they can be said to be "normative" in nature. It is not until such limits are in place that transparency tools can be used to regulate legitimate power relations and "channel the normatively accepted exercise of power".[12]

De Hert and Gutwirth categorize data protection legislation as a transparency tool. Yet, as they have pointed out, when it concerns sensitive data and the prohibition of processing such data, data protection is also a tool of opacity. Bygrave has approached the issue from the opposite direction, reminding us that data protection is not just about privacy and that regulation also aims at making data processing possible.[13] This ambiguity – data protection as a question both of privacy and of enabling the use of personal information – became visible for instance in the debates that preceded Sweden's first Data Act in 1973. In those debates, the risks of data processing were contrasted with the societal benefits of possessing large databases of information about citizens.[14]

As stressed by De Hert and Gutwirth, a piece of legislation can only seldom be classified as either an opacity or a transparency tool from the outset. The subsequent uses and interpretations of the regulation will decide whether or not it comes to function as a power limiting or a power channeling tool. Below, I will argue that the regulation of criminal records in Sweden has shifted from being opacity-based to becoming more of a question of increasing transparency, in line with De Hert and Gutwirth's analyses of the European rights law. I will argue, moreover, that normative choices are not limited to opacity tools alone but are involved in the use of transparency tools as well, and that transparency tools can have an impact on the degree of opacity, thereby affecting the boundaries of privacy.

---

[12]Ibid., 70.

[13]Lee A. Bygrave. "The Place of Privacy in Data Protection." *University of New South Wales Law Journal* 24, 1 (2001).

[14]Åsa Söderlind. *Personlig integritet som informationspolitik* Borås and Göteborg: Valfrid, 2009.

The concepts of opacity and transparency may thus be useful in understanding and analyzing current legislation as well as the way laws are interpreted by courts and other actors. At the same time, these very notions themselves, and the tools they form, remain vaguely defined. While much work has already been done to clarify the concept of privacy, in this chapter I nonetheless approach the issue as a sociological rather than a philosophical or legal question.[15] As De Hert and Gutwirth remind us, "privacy is a relational, contextual and per se social notion which only acquires substance when it clashes with other private or public interests."[16] The theoretical concepts of "opacity" and "transparency" provide a framework for the analysis of the variable values that have influenced and shaped the legislation at different moments in the history of the criminal records registry. Hence, rather than asking how privacy ought to be defined, my guiding question is how privacy has been constructed and understood in shifting social realities and in legislative practice.

When discussing "privacy" and "opacity" below, it is only in relation to the use of criminal records, or of personal information from public registers. Privacy, to be sure, covers much more as a concept, and can relate to any number of activities in addition to the processing of data.[17] In the same manner, I will only analyze data protection to the extent that it has influenced the regulation of criminal record checks.

6.1.2 A Brief History of Swedish Crime Policy

Regulating criminal records is part of crime policy. Sweden, known for its social-democratic welfare system[18], exemplifies "Scandinavian exceptionalism" in this particular area as well, aiming at a more "humane" penal system and low prison rates.[19] The country's crime policies throughout the twentieth century manifest this orientation, even if a small deviation from the rule can be detected towards the end of the period. Although Pratt and Ericsson maintain that Sweden has kept intact its fundamental features as a welfare state, especially in comparison to the policy changes in the Anglo-Saxon countries (and others imitating them), other researchers

---

[15]For an overview of definitions of privacy, see Bygrave. "The Place of Privacy in Data Protection."

[16]De Hert and Gutwirth. "Privacy, Data Protection Law and Law Enforcement: Opacity of the Individual and Transparency of Power." 75.

[17]Cf. Flaherty. *Protecting Privacy in Surveillance Societies.*

[18]See, for example, Gøsta Esping-Andersen. *The Three Worlds of Welfare Capitalism* Cambridge, UK: Polity Press, 1990.

[19]John Pratt. "Scandinavian Exceptionalism in an Era of Penal Excess: Part I: The Nature and Roots of Scandinavian Exceptionalism." *British Journal of Criminology* 48, 2 (2008). and "Scandinavian Exceptionalism in an Era of Penal Excess: Part II: Does Scandinavian Exceptionalism Have a Future?" *British Journal of Criminology* 48, 3 (2008).

have put a greater focus on the country's transformation from a social-democratic welfare state to a more liberal one.[20]

The time period around 1900, when the Swedish Criminal Records Registry was established, was dominated by an ongoing struggle to reform the country's criminal law. The old retributive system was understood to be ineffective in making criminals not to reoffend, and a new system that could better cope with this task was deemed necessary.[21] Individual legal experts, often combining a legal or an academic career with a political one, had a central role in the reform work that followed. Although, on occasion, there were strong disagreements between the experts regarding the theoretical ground on which a criminal justice system should be built, an understanding was nonetheless shared that it was the state's responsibility to protect its citizens from crime. The idea of rehabilitating offenders and of adjusting the criminal justice system to better serve that purpose grew more and more influential, even when notions centered on retributive justice remained very much alive. In the aftermath of the First World War (during which the country, however, remained neutral), a movement called "the movement for social protection" (*socialskyddsrörelsen*) formed around demands to set limits on the means that the state could employ in its efforts to prevent crimes.[22] The movement gained momentum in the 1940s and the 1950s, and, as we will see later, the changes to the criminal records law in the 1960s can be understood as a result or a reflection of precisely such concerns about state power as were crystallized in this movement.

The support for the welfare state project continued unabated among all the political parties until the 1960s, with very few debates or political arguments about crime policy coming to the surface.[23] As in most other Western countries, however, the increasing crime rates of the post-World War II period nonetheless brought with them questions about the legitimacy of the rehabilitative approach.[24] Experts were now no longer seen as authorities in the field and the issue became politicized, making its way into the manifestos of political parties.[25]

In the 1970s, crime policy was then made the target of a broad-ranging critique of the rehabilitative ideal. The state's rehabilitative and reformative ambitions had

---

[20] Robert Andersson. *Kriminalpolitikens väsen*. Stockholm: Stockholm University, Department of Criminology, 2002; Robert Andersson and Roddy Nilsson, *Svensk kriminalpolitik* Malmö: Liber, 2009; John Pratt and Anna Eriksson. "Den skandinaviska exceptionalismen i kriminalpolitiken." *Nordisk Tidsskrift for Kriminalvidenskab* 96, 2 (2009).

[21] Christian Häthén. *Straffrättsvetenskap och kriminalpolitik. De europeiska straffteorierna och deras betydelse för svensk strafflagstiftning, 1906–1931*. Lund: Lund University Press, 1990.

[22] Christian Häthén. *Stat och straff. Rättshistoriska perspektiv* Lund: Studentlitteratur, 2004.

[23] Lena Lexbro. "Konflikt eller konsensus? Kriminalpolitiken och riksdagen 1946–1965." *Nordisk Tidskrift for Kriminalvedenskab* 87, 1 (2000); Henrik Tham. "From Treatment to Just Deserts in a Changing Welfare State." In *Beware of Punishment: On the Utility and Futility of Criminal Law*. edited by A. Snare Oslo: Pax Forlag, 1995.

[24] Cf. David Garland. *The Culture of Control* Chicago: The University of Chicago Press, 2001.

[25] Marie Demker and Göran Duus-Otterström. "Realigning Criminal Policy: Offender and Victim in the Swedish Party System over Time." *International Review of Sociology* 19, 2 (2009); Tham, "From Treatment to Just Deserts in a Changing Welfare State."

come to be seen as misuse of state power that contradicted with legal principles such as proportionality, equality, and foreseeability.[26] In the 1980s, a new penal system was introduced, based on the principle of just deserts and a stronger notion on individuals' rights vis-à-vis the state. At around the same time, the focus of crime policy shifted from conceptualizing crime as a societal failure to framing it as a matter of protecting honest citizens; and with this, penal welfarism made room for a more punitive, more victim-centered approach.[27] It is against this background, then, that the regulation of criminal records in general, and subject access in particular, must be examined in the Swedish case, something that I will attempt in the following.

6.2 Criminal Records Legislation and Subject Access, 1901–2009

In Sweden, bills submitted to parliament take the form of a proposal in which the government must account for its reasons for proposing the new legislation. The submission documents will subsequently be used by the courts, authorities, and other parties in interpreting the new law and in establishing guidelines for how certain aspects of it should be applied. For this study, I examined all the government proposals and commission reports connected to the legislative work around criminal records registry, data and record keeping, and access to records in the time period covered. The analyzed documents also included other materials and documents related to the government proposals to enact or amend the country's criminal records legislation regulating subject access, as well as reports by various commissions appointed by the government to analyze more specific issues related to the proposed legislation.[28] The perspectives emerging from this material can be interpreted as "cultural stories", or instructions about how to understand society, as "told from the point of view of the ruling interest and the normative order".[29] In practice, cultural stories can be analyzed by looking at the way they frame a political issue; and "[t]o study the framing of a political issue is to focus on the metaphors, symbols, imagery, catchwords and historical examples through which a position is elaborated...to create consensus".[30] In investigating this process, I will focus on what can be considered as "typical examples" provided by this documentation,

---

[26] Andersson. *Kriminalpolitikens väsen*; Andersson and Nilsson. *Svensk kriminalpolitik*; Raimo Lahti. "Towards a Rational and Humane Criminal Policy: Trends in Scandinavian Penal Thinking." *Journal of Scandinavian Studies in Criminology and Crime Prevention* 1, 2 (2000).

[27] Andersson. *Kriminalpolitikens väsen*; Demker and Duus-Otterström. "Realigning Criminal Policy: Offender and Victim in the Swedish Party System over Time."

[28] In the analysis, an inductive approach was used to identify the various thematic categories drawn upon in the discussion below.

[29] Laurel Richardson. "Narrative and Sociology." *Journal of Contemporary Ethnography* 19, 1 (1990): 128.

[30] Bengt Larsson. "Auditor Regulation and Economic Crime Policy in Sweden, 1965–2000" *Accounting, Organizations and Society* 30, 2 (2005): 129.

to illustrate how criminal records have been framed in relation to individuals and subject access at different points of time.

Even if I recognize the importance of studying "conflicts" between different actors before and during the preparation of a new act, such conflicts are not in the focus of this article. Instead, the analysis looks at the vocabularies that, at particular conjunctions, became the dominant ones in framing the issue. While highly important in itself, the question of whether, or to what extent, such vocabularies were then accepted by all the parties affected by the new legislation remains beyond the scope of this research.

6.2.1 1901: The Creation of a National Criminal Records Registry

When the national Criminal Records Registry was set up in 1901, the emphasis was on creating opacity not just between (convicted) individuals and other private actors, but also between individuals and the state church. The vocabulary used was centered on the notion of "protection", in that the intent was to shield the individual from any potentially negative consequences in the event that information about their status as ex-convicts were to become public. This, however, was so to a certain extent only: the need of courts to have access to this information was never questioned or even openly discussed.

The preparation of the legislation for the establishment of the registry was a somewhat lengthy process that involved, among other things, a change in the constitution. An 1892 commission report first outlined the reasons why a criminal records registry should be set up to begin with, and how it should be organized. Next came a proposal by the government, in 1896, to change the constitution. The proposal, however, was rejected by the parliament, on the basis that it would have caused the register to be governed by an ordinance and not a law, which in turn would have allowed the government to make changes to it without the parliament's approval. In 1899, a new proposal was thus submitted to the parliament, this time to set up a Criminal Records Registry regulated by laws passed by the parliamentary instead of ordinances. The proposal was subsequently adopted and the register set up with no more that minor deviations from the original 1892 plan.

6.2.1.1 A Protective Vocabulary

In 1753, it became mandatory for the courts to obtain a certificate from the defendant's parish containing information about the person's religious convictions and "repute", including whether she or he had been convicted of any crimes.[31]

[31] The courts had a duty to inform parishes about all sentences. This information was used by the parish priest e.g. to rule who could and who could not receive Holy Communion. See Gösta Lext, *Studier i svensk kyrkobokföring 1600–1946* (Gothenburg: University of Gothenburg, School of Business, Economics and Law, Department of Economic History, 1984), 210.

Sentencing rules demanded that the judge take into consideration "the need for severer sentencing to rectify the law-breaking mindset".[32] There were also some crime categories to which specific "repeat sentencing" guidelines were applicable. If a person, for example, was convicted for shoplifting for the third time, the offence was to be recorded as a theft.

As a first attempt to address the problem of unauthorized parties' gaining access to information from the parish registers, it was decided in 1864 that only those who were sentenced to the newly introduced penalty of "loss of civil rights" (*förlust av medborgerligt förtroende*) should be noted in the register. The sentencing in this case meant that the individuals in question would, for a certain period of time, lose their right to vote, to be employed as a civil servant, and so forth. However, this way limiting the information entered on the parish certificates also reduced the certificates' usefulness to the courts, which therefore led to a proposal to set up a separate national criminal records registry.

6.2.1.2 Creating Opacity

To create the intended opacity through this new initiative, however, it was necessary to first amend the country's constitution. As noted above, the principle of public access to official records was enshrined in the country's constitution already in 1766, and, accordingly, any information stored in official records needed to be made accessible to the citizenry upon request. To avoid a situation where criminal records could be accessed by any citizen, the constitutional right of access to public documents thus needed to be restricted in their case. A comparable exception concerning the parish registers had already been accommodated, although it was formulated in such a way that the information held on the registers only remained confidential if its disclosure was likely to cause harm to the persons in question. A similar provision concerning criminal records, however, would be undesirable, it was proposed, since it would make it possible for employers to exploit a loophole in the regulations by obtaining "criminal conviction certificates" (*straffrihetsintyg*) on job seekers: merely disclosing that a person had no criminal record, they could be claimed to cause no risk of harm to anyone, and could thus be legally issued.[33] Were the criminal records to be regulated in the same way as the parish registers, the thinking went, employers could then simply seek information from the register and, if that was not available, they would know that the person in question had a conviction on her or his record. To prevent this possibility, full restriction was imposed upon the register and all the information contained in it was classified as secret. In the "typical example" of this line of thinking below, the aim of increased opacity between private actors is taken for granted, even if it is not discussed in terms of

[32] Commission Report, *Förslag till förordning angående straffregister* (Stockholm: Kungliga Boktryckeriet and P.A. Norstedt & söner, 1892), 24. All translations from the original Swedish mine.

[33] Ibid., 62.

"privacy", a concept that made its way into the official rhetoric concerning criminal records only much later:

> As concerns those who, when applying for employment or otherwise asked to do so, are unable to produce such a ["No Convictions"] certificate, it could, again, be concluded that these persons have been convicted of a crime; and what this would amount to in practice would be the creation of just another kind of certificate of good conduct, which, in no insignificant manner, would then undermine the intended purpose of limiting access to the information on the parish certificates.[34]

As a second strategy to protect people with a criminal record, precautions were taken to minimize the risk of having the information end up in unintended hands, where it could "bring everlasting harm to those who have made themselves guilty of the offences" noted on the record.[35] It was, for example, considered that one central registry was a safer solution than several local ones, and clearly a more commendable option than publishing the information in internal police journals, as it would be difficult to prevent at least some of the copies in circulation from ending up in the hands of unauthorized persons.

6.2.1.3 Subject Access

The 1892 commission report, the first of its kind, made no proposals towards increasing "transparency" in the form of subject access provisions, and did not distinguish between subject access and access in general. In its proposal of 1899, the government argued for a system in which individuals would be entitled to a copy from the register if the king gave permission to this.[36] This proposal was met with criticism from members of the country's Supreme Court, who considered this deviation from the original commission report to not be in line with the other considerations put forward in the discussion. In their view, subject access would undermine the opacity that was strived for:

> [S]ince, as the report notes, it has been deemed to be of particular importance that the regulations regarding the cases in which access to the criminal records register ought to be allowed be codified in civil law, the scenario in which not only access would be provided to public authorities, as the commission report proposes, but that individuals, too, would have the right to access the criminal records register. . .cannot be considered compatible with this viewpoint.[37]

The criticism caused the government to modify its proposal so that only those individuals whose "right may be dependent upon access to register information" were to gain access to it upon permission of the king.[38] Nothing more was stated about

---

[34] Ibid.

[35] Ibid., 39.

[36] The king's power had gradually diminished following the new constitution of 1809, and the term 'king' was at this point used to refer to both the actual person of the king and his office.

[37] Proposal 1899 No. 18, "Kongl. Maj:ts nådiga proposition till riksdagen med förslag till lag om straffregister; given Stockholms slott den 22 december 1899," 14.

[38] Ibid., 15.

whom this might concern exactly or what this "right" was that was in question, although from later documents it becomes apparent that it referred to the possibility to travel or migrate to certain countries like Switzerland and the United States.[39]

Subject access, in other words, was not discussed in terms of "channeling power" or as a way to keep the state accountable for the content of the register. There is nothing in the documents involved to suggest that these early attempts at regulation were thought of in terms of "transparency" the way we understood the issue today. In the latter part of the nineteenth century, the only concern was to create opacity, and it was deemed necessary to limit individuals' access to their criminal records in order to uphold the kind of privacy protection that was created through that opacity.

In the years that followed, the new law then underwent minor revisions, mainly to bring it up to date with new sanctions that needed to be registered. As concerns subject access, the issue was not raised again until 1948, when a process ultimately leading to a proposal for a complete revision of the criminal records act in 1963 was begun.

6.2.2 1963: Rehabilitation and Access Restrictions

If, in the preparation of the first version of the Swedish criminal records act at the end of the nineteenth century, much attention was devoted to protecting individuals with a criminal history, the intent to do the same was even more pronounced during the process of rewriting the act in 1963. This time, however, there was a much clearer conception of what type of "permanent harm" unintended use of the records could bring, with rehabilitation of ex-offenders as the express aim of the penal law. It is thus the latter that allows one to examine the two periods through the lens of two different vocabularies. Although at first glance similar to the "protective vocabulary" in the way individuals' need for protection was conceived, the "rehabilitative vocabulary" was nonetheless based on a critique of how the previous legislation had failed to take into consideration the need to rehabilitate offenders. In the rehabilitative vocabulary, opacity became a matter of regulating the relation not just between individuals and other private actors, but also between individuals and the state as represented by the criminal justice system. In addition, there was also an element of the just desert thinking coming to surface that would then come to dominate the discussions over the next two decades.

The amendment process of the 1960s was initiated already in 1948, when, through two parliamentary motions, the government was called on to appoint a committee to investigate what alterations were necessary in the Swedish law to protect individuals from "an excessive system of records of crimes and offences".[40] This was, then, the first sign of a sense of concern about government data banks in the

---

[39]Proposal 1963:39, "Kungl. Maj:ts proposition till riksdagen med förslag till lag om allmänt kriminalregister m.m.," 54.

[40]Ibid., 7.

country and about the way in which they might affect individuals' privacy. Usually, the emergence of this kind of concerns about the state's increasing capacity to collect and store information about its citizens is dated to the late 1960s and the early 1970s.[41] Although more areas would need to be examined before any firm conclusions can be made, it may then well be that criminal records were the first type of state-stored information that became questioned from a privacy point of view.

As a result of the parliamentary motions, a penal code committee was appointed in 1949 to, among other things, review the criminal records act. In its report, submitted in 1953, the committee proposed restrictions to how long the information could be held in the records, but also referred its conclusions to an expert group set up by the Justice Department that had not completed its work yet.[42] Eventually, an expert investigator was appointed in 1956, and the report on which the government then based its proposal was issued on the last day of 1960.[43]

6.2.2.1 The Rehabilitative Vocabulary

In the government proposal of 1963, the negative effects potentially arising from criminal records keeping were described in great detail. Three distinct ways in which the existence and use of such records could threaten the aim of rehabilitation and cause harm to individuals were considered in particular:

> The mere knowledge that there exists information recorded about her- or himself that might generally be regarded as unfavorable can in itself constitute a burden to the individual. Maintenance of such records can also give rise to a situation where that information is retrieved in a connection in which the individual whom that information concerns has a strong interest in not being presented before authorities or other parties in an unfavorable light, and where that information – which may relate to events and circumstances long since passed – is more or less immaterial or otherwise unnecessary. Neither can the risk that such information becomes accessible to unauthorized third parties ever be completely eliminated. These kinds of situations can, in adverse conditions, negatively impact the reintegration of offenders at which the society's correctional measures are aimed.[44]

Firstly, "the mere knowledge" that personal information records exist was described as a potential burden for individuals. Secondly, since these records are kept for a very long time, individuals may, for instance, be denied travel opportunities to certain countries because of the information held about them on record; this was seen as an unjust consequence. Whether or not the use of information from criminal records might be called for, and whether the consequences would be just or unjust,

---

[41] Ilshammar, *Offentlighetens nya rum – Teknik och politik*; Sten Markgren, *Datainspektionen och skyddet av den personliga identiteten* (Lund: Studentlitteratur, 1984); Söderlind, *Personlig integritet som informationspolitik*.

[42] Swedish Government Official Reports, *Enhetligt frihetsstraff* (Stockholm: Department of Justice, SOU 1953:17).

[43] Swedish Government Offical Reports, *Den allmänna brottsregistreringen* (Stockholm: Department of Justice, SOU 1961:11).

[44] Proposal 1963:1939, "Kungl. Maj:ts proposition till riksdagen med förslag till lag om allmänt kriminalregister m.m." 17.

was now conceived as something that depended on the passing of time. Thirdly, it was considered not possible to wholly eliminate the chance that the information on record would fall into the wrong hands, which meant that the mere existence of an information register presented a threat to individuals' integrity.

To limit the negative consequences, so-called "rehabilitation rules" were therefore introduced. Information in the records would no longer be disclosed after a certain time period had elapsed. Up until this point, the information had been kept until the person in question had died, or a maximum of ninety years. In a 1961 report commissioned by the government, it was suggested that the new rehabilitation rules apply to individuals' access but not that of courts and other authorities.[45] In its response, however, the government stated that the report had underestimated the "impact of knowing that old information on record will automatically be retrieved even after years of living as a law-abiding citizen, if the person ever again ends up standing before the court".[46] Together with the estimation that old records were likely to have no more than limited significance for the courts, this led the government to conclude that the ability of the courts to access the information in the record could be restricted as well. Yet, a distinction was made between the criminal justice system and individuals, in that different "rehabilitation rules" were to be applied in each case. For criminal records provided to the courts, a rehabilitation time of ten years was set, and for records availed to individuals a rehabilitation time of five years. No grounds, however, were provided for this decision, beyond the mere statement that it was important to create a "simple" principle and that for individuals a substantially shorter time span would be applicable.[47]

The vocabulary used in the proposal left little room for considerations other than those related to rehabilitation and reintegration of ex-offenders. Nonetheless, it is important to note that this did not mean that the concern for the individual trumped all other interests. The question, for instance, of whether this kind of register was, ultimately, necessary at all if it indeed posed a burden and a serious risk of harm to individuals, was never really brought up in the text. Moreover, although the courts' need for the information on that register can be said to have been somewhat called into question (considering the restrictions brought by the new rehabilitation rule), part of the justification given for the new rule was simply that old records were of no

[45] Swedish Government Offical Reports, *Den allmänna brottsregistreringen,* SOU 1961:11.

[46] Proposal 1963:1939, "Kungl. Maj:ts proposition till riksdagen med förslag till lag om allmänt kriminalregister m.m." 45.

[47] Ibid., 46. This kind of 'practical reasoning' has influenced the way in which the various issues have been dealt with at different conjunctions. For example, records of any fines were left out of the 1901 register, and not only because the information was deemed irrelevant to the courts, but also because including them would make the register too cumbersome and difficult to handle. The opposite approach was chosen in 1997 when all convictions, regardless of the type of punishment, were included; at this point, it was considered too demanding for the courts to have to sort out the sentences that should be registered in the system from those that should not. The impact of technological development and automated data processing is obvious here.

use to courts if there had been no reoffending during the intervening time period.[48] This inability or reluctance to challenge fundamental assumptions was evident in text passages specifically addressing the courts' use of criminal records. The notion that repeat offenders should be punished differently from first-time offenders was strong, and the frequent cultivation of phrases such as "it is necessary," "need to be taken into account," and "cannot be avoided" serve as indications of the kind of taken-for-granted elements in the dominant conception that continued to live on:

> In sentencing, it is necessary to take into consideration whether it is a case of a first-time offence or a repeat offence. Also, when deciding on the penalty, the history of societal reactions to previous cases of similar nature forms one of the factors that need to be taken into account.[49]

One of the fundamental assumptions at play here was that knowledge about past convictions plays a pivotal role in the attainment of the rehabilitation goals. What was thereby created, however, was a differentiation between, on the one hand, other public authorities' and private actors' use of criminal records, which was viewed as a threat to rehabilitation goals, and, on the other hand, criminal justice authorities' use of these records, which was seen as vital for the successful rehabilitation of ex-offenders. The fact that rehabilitation rules were also made applicable to the criminal justice system can in this context be seen as an acknowledgment of the need to restrict state power, which the "social protection movement" had advocated for many years already, and as a reflection of the dominant position that labeling theories had come to occupy in Sweden and in most Western countries during this period.[50]

6.2.2.2 Subject Access

In the process of rewriting the country's criminal records legislation in the late 1950s and early 1960s, subject access was discussed in relation to the so-called certificates of conduct (*vandelsintyg*) issued by local police authorities. Prior to 1965, there was no national police organization in Sweden, and local police authorities could exercise a considerable degree of self-determination when deciding on how to run their business. This meant that, besides the national criminal records registry, there were also several local police registers containing information about both suspected and convicted offenders. When employers and job seekers were unable to obtain information from the criminal records registry, they turned to local police for it. Just like the previous parish certificates, the practice of issuing certificates of conduct was viewed by the government as having potentially harmful effects for individuals, and it was for this reason deemed undesirable. Tackling the issue, however, was not possible within the scope of the current criminal records act, but when a national

---

[48] Ibid., 44.

[49] Ibid.

[50] Cf. Andersson and Nilsson, *Svensk kriminalpolitik*; Häthén, *Stat och straff. Rättshistoriska perspektiv*.

police authority was created in 1965 it also made it possible to replace local registers with one national registry with restricted access to the information it held.

What happened during the first part of the 1960s was almost identical to the developments at the turn of the century, although this time it was the police registers that were given critical attention and not parish registers. The solution, however, was the same: creating one national registry to store the information instead of several locally kept registers, combined with access restrictions. The aim, once more, was to increase opacity between individuals and third-party actors, in order to protect ex-offenders from what was seen as unfair consequences. Establishing the national Criminal Records Registry in 1901 had turned out to be ineffective from this perspective, given the availability of information from the police registers, but the reorganization of the police force and the creation of one national police register then made it possible to maintain the opacity achieved up until 1989 when full subject access was implemented.

6.2.3 1987: Data Protection and Transparency

When the Swedish criminal records legislation was rewritten in 1963, the computer revolution was still to come. Computerization and the enhanced capacity to maintain and manage large databases that it meant, along with the existence of a national police authority database containing qualitatively different kinds of information (on both suspects and convicts), thus provided the motivation for revising the act in the years that followed. The work started out with some amendments to it in 1989, and was completed with the passing of the country's new Criminal Records Act in 1998.

In 1989, the existing law was amended to provide for full subject access rights to individuals, in response to the need to bring the access procedure on par with the new standards for data protection. Various Council of Europe recommendations clearly had an impact on the decision of the Swedish legislators to embark upon this work, as shown, for example, by the government's proposal text in which the formerly dominant vocabularies centering on protection and rehabilitation had now been replaced with a new vocabulary revolving around individuals' rights vis-à-vis the state.[51] Two major concerns were brought forward in the proposal. First, the criminal records registry had been transferred from under the authority of the National Prison and Probation Administration to the National Police Board in 1971, and subsequently merged with the police register into a new database called PBR (*Person- och belastningsregistret*), a register holding information on persons and criminal offences. Given the various recommendations and conventions of the Council of

[51] Proposal 1987/1988: 122, "Förslag till lag om ändring i lagen (1963: 197) om allmänt lriminalregister."

Europe,[52] this was seen as problematic, since the new database contained information about both suspected and convicted offenders. Second, the right of individuals to access the information on their record and the fact that the country's existing data protection laws fell short of the recommendations of the Council of Europe were brought up, having both been first addressed in several official reports preceding the actual government proposal. The computerization and the new media for storing information complicated the question of how the principle on public access to official records should be implemented, with the issue tackled in various governmental reports from the late 1960s onward after the previous amendments to the criminal records legislation had entered into force.[53] Access to criminal records was one aspect of this larger issue, and in the proposal for a new Secrecy Act the question of subject access in connection with police records was brought up.[54] It was, however, decided that the records kept by the police should be dealt with separately, and the issue was transferred to the Justice Department that subsequently addressed it in a 1980 memorandum and in two reports issued in 1981 and 1985.[55]

Regarding its first concern, about several registers sharing one common database that was regulated through a number of individual acts and ordinances, the government concluded that further inquiry was necessary, and that until such time as that inquiry could be completed, the country's criminal records legislation should be amended only minimally.[56] In consequence, the only revision made to the legislation in 1989 concerned the implementation of full subject access rights. This, to be sure, represented no small change, given the past orientation based on limiting subject access rights.

In 1997, the PBR was then reorganized and the criminal records database separated from the police records. The registry itself was physically relocated from Stockholm, the capital city, to Kiruna in the northern part of the country, and the act was rewritten so as to bring it up to date with the technical developments and the new European Union Data Protection Directive 95/46/EC.[57] It is worth mentioning that the 1997 government proposal was the first one to couch its discussion of criminal records in terms of the right to privacy.

---

[52] Convention for the Protection of Individuals with Regard to Automatic Processing of Personal Data (1981); Recommendation No. R (84) 10 on the Criminal Record and Rehabilitation of Convicted Persons; and Recommendation R (87) 15 regulating the use of personal data in the police sector.

[53] Cf. Swedish Government Offical Reports. *Offentlighet och sekretess* Stockholm: Department of Justice, SOU 1966: 60 and *Data och integritet* Stockholm: Department of Justice, SOU 1972: 47.

[54] Proposal 1979/1980: 2, Förslag till ny sekretesslag.

[55] Ministry Publication Series, *Lag om brottsregister m.m.* (Stockholm: Department of Justice, Ds Ju 1985: 8); Ministry Publications Series, *De registrerades rätt till insyn i kriminal- och polisregistren* (Stockholm: Department of Justice, Ds Ju 1981: 6); Department of Justice, "Promemoria 5 of February" (Record Number Dnr 301–380, 1980).

[56] Proposal 1987/1988: 122, "Förslag till lag om ändring i lagen (1963: 197) om allmänt kriminalregister," 13.

[57] Proposal 1997/1998: 97, "Polisens register."

6.2.3.1 Subject Access

In the text of the 1987 proposal, very little could be seen of the previous influence of the protection and rehabilitation vocabularies. Instead, a new vocabulary, centered on the rights of individuals, had emerged, putting transparency on the agenda. In the "typical example" below, nothing is said about negative consequences for individuals or about employers; rather, the stress is on how difficult it must be for a person with a criminal record to understand that it is kept secret:

> For the individual with a criminal record, the current prohibition against accessing information on one's own record must be difficult to understand in all those cases where the real and only reason for keeping that information inaccessible is to protect that individual's own best interest.[58]

The right of individuals to access information on their own record was now taken as something self-evident, and, unlike the picture painted by the other vocabularies, this was not a "sympathy story" about individuals facing potential harm. If anything was there to be felt sorry about it was in fact the existing legislation, which, even if only implicitly, was depicted as somewhat out of date and, perhaps, showing a misguided concern for the individual. In the proposal, it was then stated that the preparation works of the law (various commission reports and comments from relevant bodies) had clearly showed there to be a common agreement that individuals ought to have the right to review their own records.[59] The document also makes reference to other existing rules and regulations, including those incorporated in the country's constitution itself, the Data Act of 1973, and Article 8 of the Council of Europe recommendation concerning automatic processing of personal data.[60] Including a full subject access provision in the criminal records act, the argument went, would bring the act into line with all this other legislation and make it possible for individuals to verify that the information on them was correct and, where necessary, initiate rectification of any errors.[61] Similarly, it was proposed that individuals should have the right to access the same information that other legally authorized parties could retrieve from the register. In consequence, the 1963 decision to apply stricter rehabilitation rules to individuals accessing the register was to be revoked.[62] However, there was no discussion of the possible negative consequences the new decision might have from the perspective of offender rehabilitation.

---

[58] Proposal 1987/1988:122, "Förslag till lag om ändring i lagen (1963: 197) om allmänt kriminalregister," 17.

[59] Ibid.

[60] Regulating subject access on a general level had been part of the Swedish legislation since the enactment of the country's first Data Act in 1973. In case of contradiction with other laws, however, such as the criminal records legislation that did not permit individuals to access their personal data, the provisions of the latter nonetheless prevailed.

[61] Proposal 1987/1988: 122, "Förslag till lag om ändring i lagen (1963: 197) om allmänt kriminalregister," 18.

[62] Ibid., 20.

When it came to deciding how subject access requests would be handled in practice, a 1980 memorandum by the Department of Justice suggested that individuals living in Sweden be offered the possibility of reviewing their record on a police station computer screen with no printouts involved, so as to prevent employers from gaining access to this information by forcing job seekers to exercise their access rights.[63] The inspiration for this solution came from Norway, where, instead of obtaining extracts from their criminal record, individuals had the right to be told what information was kept about them on records. This way, a certain level of opacity could be created between individuals. The suggestion, however, met with a negative response from various quarters, and in the subsequent reports as well as in the government proposal itself the proposed solution was ruled out as too expensive and unfeasible in practice. Sending individuals a copy of their criminal record upon request was then deemed the best way to proceed. Yet, a concern lingered that the information might indeed end up in wrong hands, prompting a decision that the extracts be sent to their recipients via registered mail.[64]

For a solution with copies to be sent out by mail to appear realistic, however, the potential risk arising from employers' gaining access to the criminal records information needed to be downplayed. The government declared in its proposal that there was no reason to believe that employers would misuse the new regulations, stating that while "behind the current regulations concerning subject access there is an apprehension that [the opportunity availed through unrestricted subject access] might be taken advantage of by employers and landlords…the grounds for not allowing individuals access to their own record are no longer solid".[65] In this way, the new rule bringing more transparency was constructed as posing no threat to the degree of opacity between individuals and other private actors. Why the grounds for not allowing data subjects access to their criminal record information were "no longer solid", and why there was "no reason" to doubt employers' good faith in complying with the intent of the law, were not specified in the document. That assumptions like these could be left unaccounted for indicates the extent to which there must have been a common agreement on these points. Despite the way in which it was formulated in the government proposal, the 1989 amendment needs therefore to be understood in relation to the discussions about individuals' right to be protected from the state, which had been on the political agenda since the 1970s.[66] Indeed, it seems that the central issue was not the likelihood of employers' looking for and gaining access to this information, but rather the line of reasoning that culminated in the adoption of the 1963 act: it was seen to reflect the interests of the same "rehabilitative" state with too much influence over individuals' private

---

[63] Department of Justice, "Promemoria 5 of February."

[64] Proposal 1987/1988: 122, "Förslag till lag om ändring i lagen (1963: 197) om allmänt kriminalregister," 20.

[65] Ibid., 18.

[66] Andersson, *Kriminalpolitikens väsen*; Colin J. Bennett. *The Privacy Advocates*. Massachusetts: MIT Press, 2008.

lives that, among other things, also kept people incarcerated for indefinite periods of time. In dealing with such negative effects of the welfare state, the state could no longer portray itself as the *paterfamilias* that could infringe on individuals' rights in order to protect its citizens. For this reason, transparency had now to be constructed as something that would not affect opacity. Decreased opacity was thus presented not as a threat but rather as an unlikely outcome, one, moreover, that could be successfully managed if need arose, preferably by the labor market organizations and, as a last resort, through direct interventions of the state.[67]

The issue of subject access was discussed again in the 1997 government proposal concerning the reorganization of the police registers.[68] In the proposal, a Council of Europe recommendation[69] and the newly ratified Europol Convention[70] were used as justifications for the need to enact enhanced transparency and subject access. The risk of employers taking advantage of the opportunities offered by the new subject access rights was mentioned in passing, by referring to the fact that this risk had, in previous estimations, been seen as low and that for the time being there was no reason to reassess that conclusion:

> In the recently ratified Europol Convention. . .all citizens are given the right of access to data relating to them, either directly through accessing this data themselves or indirectly by having such data checked on their behalf for legality and accuracy (Article 19). It seems natural that what is stated in the Europol Convention should also apply to our Swedish regulations regarding disclosure of information from police registers in our own country. Neither should there be any doubts whatsoever that we shall also implement the recommendations of the Council of Europe regarding the use of personal data by the police. Therefore, individuals should always have a right of access to personal information about themselves that is held in police files. What this also means, however, is that third parties such as employers and landlords may misuse this access right by demanding job seekers and apartment seekers to prove their clean record. . .. The right to be able to verify the information on one's own record should be considered of vital importance and interest to all citizens. To date, there is no indication of such misuse extensive enough to require the government to consider implementing the right of individuals to access their personal information held in records in another way.[71]

The vocabulary centered on the rights of individuals still dominates the language of this proposal, although almost nothing is said in it about the benefits that subject access offered for the individuals. Instead, the government continued to dissociate itself from the earlier protective strategy. As noted above, Swedish crime policy has in many respects continued to be developed under rehabilitative and "humane" ideals, while increasingly incorporating neoliberal values and an individual-centered viewpoint. Demker and Duus-Otterström have argued that the punitive turn in

---

[67]Ministry Publications Series. De registrerades rätt till insyn i kriminal- och polisregistren.

[68]Proposal 1997/1998: 97, "Polisens register."

[69]Recommendation R (87) 15 regulating the use of personal data in the police sector.

[70]Council Act of 26 July 1995 drawing up the Convention on the establishment of a European Police Office.

[71]Proposal 1997/1998: 97, "Polisens register," 44f.

Swedish crime policy during this time period came about as a reflection of a more individualized society that puts crime victims and their rights at center stage.[72] It would thus seem possible to interpret the legislative developments in the regulation of criminal history record information in the same way: the growing individualization of society has brought with it an increasing emphasis on individuals' rights vis-à-vis the state, making "transparency" in turn look like a solution just as much in the interest of the individual as the safeguarding of privacy was previously.

6.2.4 2009: A Mixture of Vocabularies

Since 2000, the number of subject access requests from individuals in Sweden has been on a steady increase (see Table 7.1 below for an overview). In 2004, the National Police Board, which administers the criminal records database, submitted a report to the government in which it estimated that approximately seventy-five percent of all the requests made were so-called enforced subject access requests, or requests prompted by employers wanting to see a copy of the job applicant's criminal record.[73] The report pointed out that this was not in keeping with the intent of the new Criminal Records Act, and in fact formed a threat to the individuals' right to privacy.

Table 7.1 Subject access requests in 2000–2008

Year	1995[a]	2000	2002	2004	2006	2008
Number of requests	10,000[a]	43,400	38,600	52,108	86,838	115,815

[a]Estimations. During the relocation of the Registry in 1998–2000, all statistics dating from before 2000 were lost. A parliamentary report dating from 1997, however, makes a reference to 10,000 requests per year.[74]

The report prompted the government to appoint an expert commission to inquire into the need for revision of the criminal records legislation and of any other legislation regulating what was now called the "personal privacy in working life".[75] The commission completed its work in spring 2009. What is interesting to note about its

[72]Demker and Duus-Otterström, "Realigning Criminal Policy: Offender and Victim in the Swedish Party System over Time."

[73]The National Police Board, "Framställning om en översyn av den enskildes rätt att ta del av uppgifter om sig själv i belastningsregistret" (The National Police Board, Legal Secretariat, record number RÄS 442-3960/04, 2004).

[74]Ibid.; Swedish Government Official Reports, *Polisens register. Slutbetänkande* (Stockholm: Fritze, SOU 1997: 65); Swedish Government Official Reports, *Integritetsskydd i arbetslivet* (Stockholm: Fritzes, SOU 2009: 44).

[75]Terms of Reference, "Personlig integritet i arbetslivet" (Dir. 2006: 55).

subsequent report is that the rehabilitative vocabulary that was absent in the government proposals of 1987 and 1997 was now resorted to once more.[76] The government directives to the expert commission had clearly stated that a guiding principle for the Swedish society was that those who have served their sentences should be allowed to participate in the community on the same terms as everyone else.[77] This led the commission to stress that information about previous convictions was sensitive and that dispersal of such information might obstruct ex-offenders' rehabilitation.[78] Parallels were also drawn between the practice of enforced subject access and earlier forms of control such as the certificates of conduct, which had been eliminated in 1965 as inappropriate.[79] The need to increase the opacity between individuals had thus clearly returned to the center of policy discourse. Accordingly, the commission report spoke at considerable length of the vulnerable position of individuals vis-à-vis the demands of their prospective employers:

> The need for the proposed regulation [as a means to prevent enforced subject access] appears pressing, considering the precarious position in which employment seekers frequently find themselves. The applicant usually has no choice but to agree to the employer's request for access to criminal records if she or he wishes to be considered for the position.[80]

Indeed, the report went on to declare that the earlier fears about subject access leading to employers' demands to review criminal history records of job applicants had proved to be well founded, and that the Criminal Records Act needed to be amended to prevent the misuse.[81] New restrictions on subject access rights were, however, not deemed possible due to the international recommendations and agreements that the government considered itself bound by. Instead, it was proposed that employers be held liable to civil penalty if they compelled access to job applicants' criminal records.[82]

This development can be viewed as either a new shift in the society and crime policy or evidence that the vocabulary of individual rights had never really gained a foothold. There is some support for the conclusion that Swedish crime policy is becoming more and more under the influence of rehabilitative ideals again. For example, as concerns criminal records, new rehabilitation rules for juveniles will be introduced in 2010, with the aim of facilitating young offenders' reintegration into society.[83] In its session where the matter was decided, the parliament, however, also passed an amendment to the country's criminal law that increased the

---

[76] Swedish Government Official Reports, *Integritetsskydd i arbetslivet,* SOU 2009: 44.

[77] Terms of Reference, "Personlig integritet i arbetslivet." Dir. 2006: 55

[78] Swedish Government Official Reports, *Integritetsskydd i arbetslivet*, SOU 2009: 44, 275.

[79] Ibid., 281.

[80] Ibid.

[81] Ibid., 279.

[82] Ibid., 280.

[83] Proposal 2009/2010: 191, Gallring ur belastningsregistret av ippgifter om unga lagöverträdare.

minimum sentences for several existing offences.[84] Thus, what we see might instead be a somewhat confused mixture showing the influence of several different things, such as the critique of the welfare state, neoliberal values that have gained dominance in countries lacking similar social-democratic welfare state traditions, and surviving notions concerning rehabilitation and welfare. When neoliberal values are incorporated into a social-democratic welfare state context, the picture easily becomes complicated if the old ideals are not overthrown in the process. As a result, there may be an attempt to simultaneously both stress ex-offenders' right to privacy, by maintaining a certain level of opacity, and comply with ideals focused on transparency. This is also what Demker and Duss-Otterström discovered in their study of political party manifestos in Sweden.[85] Even if all the parties in the country can today be said to be in favor of a punitive approach to crime policy and general deterrence as the best way of pursuing the policy goals, they were still found to adhere to the ideal of rehabilitation.

It remains to be seen whether the proposals of the expert commission will be heeded by the government,[86] and if so, whether employers' liability for damages will be sufficient to bring the practice of enforced access to an end.

6.3 Concluding Remarks

In this article, I have analyzed three different vocabularies – a protective vocabulary, a rehabilitative vocabulary, and a vocabulary centered on the rights of individuals – that were actualized and drawn upon in legislative work to change the regulations governing access to criminal history record information and the way the question of subject access was handled in Sweden. At the turn of the twentieth century, restricted access, or a low degree of transparency, was seen as necessary to prevent harm to individuals. In the "rehabilitative vocabulary" of the mid-century, even criminal justice authorities' use of criminal records was seen as a potential source of harm to individuals, and rehabilitation rules were enacted to make the system more "just". Although the discussion of a "just" penal system is normally associated with the critique of the rehabilitative ideal that developed only in the 1970s and the 1980s, the same logic also informed the discussion of unjust consequences when the 1963 government proposal was being prepared. The state's use of criminal records has always been seen as a matter of necessity, and, except for the implementation of rehabilitation rules, the rationale behind it has never been called into question – not

[84]Preliminary Report of the Parliament Proceedings. "Onsdagen den 19 maj, kl 09.00–16.37", 2009/2010: 121.

[85]Demker and Duus-Otterström, "Realigning Criminal Policy: Offender and Victim in the Swedish Party System over Time."

[86]The proposals put forth by the expert commission have been referred for consideration to relevant bodies. The deadline for comments and feedback has expired but, at the time of this writing (September 2010), no proposals have come from the government yet.

even when the vocabulary emphasizing the rights of individuals became dominant, calling attention not so much to any actual need of citizens to control the state as to the country's need to act on various Europe-wide recommendations and conventions, and to the need to leave behind the legacy of the paternalistic state and the rehabilitative ideal.

As concerns data protection, four main findings emerged from this study. The first is the fact that limitations placed on free subject access to records have been used as a means to implement privacy rights, or "opacity" in individuals' relations with parties other than the justice authorities. Second, the state's attempts to protect the citizens' right to privacy by increasing the opacity between individuals and other private actors recurred at approximately half-century intervals, owing to new practices of employers. Third, the implementation of access rights, or "transparency" between the state and the citizens, in the 1980s seemed to undermine this achieved opacity. Lastly, the question of the impact that transparency has had on the level of opacity has not been addressed by those promoting transparency-geared legislation. I have argued that the latter must be understood in light of three sets of factors and circumstances: the conventions and recommendations by various European-wide bodies that stress individuals' right of access; the reform of the national crime policy; and the devaluation of the protective state that had proven itself capable of misusing its powers. The outcome of the whole process was then formal data protection that in practice resulted in less protection for previously convicted individuals, compared to what had been accomplished by earlier legislation. A partial solution to the problem has been proposed in the 2009 expert commission report, which combines a new vocabulary that stresses the rights of an individual with a revived version of the rehabilitative vocabulary. The proposal aims at preserving the opacity between individuals, although no longer through limitation of access to criminal history records, but rather by restricting who may actually *see* the information on them. This way, it is hoped, it will be possible to maintain the transparency between the citizens and the state, thus ensuring that individuals can still access their personal information and have it rectified where so called for.

Going back to De Hert and Gutwirth's work on opacity and transparency, we might then conclude that also regulation through transparency tools represents a normative practice. In the present case, for example, when subject access rights were incorporated into Swedish law in the 1980s as one way to legitimize data processing, the change had implications for the effectiveness of opacity tools. Two decades later, it was then claimed that opacity could no longer be constructed by limiting transparency; other tools that do not affect transparency would have to be used instead. In this fashion, transparency came to trump opacity. Above, I have outlined no more than some possible factors and circumstances behind this outcome, which, in the case in concern, remains linked to the transformation of the social-democratic welfare state and the growing influence of international institutions as well. More work is therefore needed to fully understand how data protection and transparency interests came to be so dominant and what the consequences of their ascendancy might be, both for the individuals' ability to exercise their right to privacy and in terms of how data processing can be legitimized.

References

Andersson, R.. *Kriminalpolitikens väsen* [The Nature of Swedish Crime Policy]. Stockholm: Stockholm University, Department of Criminology, 2002.

Andersson, Robert, and Roddy Nilsson. *Svensk kriminalpolitik* [Criminal Policy in Sweden]. Malmö: Liber, 2009.

Bennett, C.J. *The Privacy Advocates*. Massachusetts: MIT Press, 2008.

Burke, Kenneth. *Permanence and Change*. Berkeley: University of California Press, 1984 [1935].

Bygrave, L.A. "The Place of Privacy in Data Protection." *University of New South Wales Law Journal* 24, no.1 (2001): 277–283.

Commission Report. *Förslag till förordning angående straffregister* [Proposal for an Ordinance Concerning the Criminal Register]. Stockholm: Kungliga Boktryckeriet and P. A. Norstedt & söner, 1892.

Connerly, M.L., R. D. Arvey, and C. J. Bernardy. "Criminal Background Checks for Prospective and Current Employees: Current Practices among Municipal Agencies." *Public Personnel Management* 30, no. 2 (2001): 173–183.

De Hert, P., and S. Gutwirth. "Privacy, Data Protection Law and Law Enforcement: Opacity of the Individual and Transparency of Power." In *Privacy and the Criminal Law*, edited by E. Claes, A. Duff, and S. Gutwirth, 61–104. Antwerpen and Oxford: Intersentia, 2006.

Demker, M., and G. Duus-Otterström. "Realigning Criminal Policy: Offender and Victim in the Swedish Party System over Time." *International Review of Sociology* 19, no. 2 (2009): 273–296.

Department of Justice. "Promemoria 5 of February." Record Number Dnr 301–380. 1980.

Esping-Andersen, Gøsta. *The Three Worlds of Welfare Capitalism*. Cambridge, UK: Polity Press, 1990.

Flaherty, D.H. *Protecting Privacy in Surveillance Societies*. Chapel Hill: University of North Carolina Press, 1989.

Garland, D. *The Culture of Control*. Chicago: The University of Chicago Press, 2001.

Grier, A., and T. Thomas. "The Employment of Ex-Offenders and the UK's New Criminal Record Bureau." *European Journal on Criminal Policy and Research* 9, no. 4 (2001): 459–469.

Harris, Patricia M., and K.S. Keller. "Ex-Offenders Need Not Apply: The Criminal Background Check in Hiring Decisions." *Journal of Contemporary Criminal Justice* 21, no. 1 (2005): 6–30.

Holzer, H.J., Raphael, S., and M.A. Stoll. "Will Employers Hire Former Offenders?: Employer Preferences, Background Checks, and Their Determinants." In *Imprisoning America: Employer Preferences, Background Checks, and Their Determinants*, edited by M. Pattillo, D. F. Weiman, and B. Western, 205–43. New York: Russell Sage Foundation, 2004.

Holzer, H.J., Raphael, S., and M.A. Stoll. "The Effect of an Applicant's Criminal History on Employer Hiring Decisions and Screening Practises: Evidence from Los Angeles." In *Barriers to Reentry? The Labor Market for Released Prisoners in Post-Industrial America*, edited by S. Bushway, M. A. Stoll, and D. F. Weiman, 117–50. New York, NY: Russell Sage Foundation, 2007.

Häthén, C. *Stat och straff. Rättshistoriska perspektiv* [The State and Punishment: Legal-Historical Perspectives]. Lund: Studentlitteratur, 2004.

Häthén, C. *Straffrättsvetenskap och kriminalpolitik. De europeiska straffteorierna och deras betydelse för svensk strafflagstiftning 1906–1931* [Criminal Jurisprudence and Criminal Policy: European Theories of Punishment and Their Significance for Swedish Criminal Legislation, 1906–1931). Lund: Lund University Press, 1990.

Ilshammar, L. *Offentlighetens nya rum – Teknik och politik* [The New Public Sphere: Technology and Politcs in Sweden, 1969–1999]. Örebro: Universitetsbiblioteket, 2002.

Lahti, R. "Towards a Rational and Humane Criminal Policy: Trends in Scandinavian Penal Thinking." *Journal of Scandinavian Studies in Criminology and Crime Prevention* 1, no. 2 (2000): 141–155.

Lam, A., and B.H. Kleiner. "Criminal Background Checks of Prospective Employees: Why and How Should It Be Done?" *Managerial Law* 43, no. 1/2 (2001): 132–137.

Lam, H., and M. Harcourt. "The Use of Criminal Record in Employment Decisions: The Rights of Ex-Offenders, Employers and the Public." *Journal of Business Ethics* 47, no. 3 (2003): 237–252.

Larsson, B. "Auditor Regulation and Economic Crime Policy in Sweden, 1965–2000." *Accounting, Organizations and Society* 30, no. 2 (2005): 127–144.

Lexbro, L. "Konflikt eller konsensus? Kriminalpolitiken och riksdagen 1946–1965" [Conflict or Consensus? Criminal Policy and the Swedish Parliament, 1946–1965]. *Nordisk Tidskrift for Kriminalvedenskab* 87, no. 1 (2000): 48–58.

Lext, G. *Studier i svensk kyrkobokföring 1600–1946* [Studies on the Keeping of Parish Registers in Sweden, 1600–1946]. Gothenburg: Department of Economic History, School of Business, Economics and Law, University of Gothenburg, 1984.

Louks, N., O. Lyner, and T. Sullivan. "The Employment of People with Criminal Records in the European Union." *European Journal on Criminal Policy and Research* 6, no. 2 (1998): 195–210.

Magnusson Sjöberg, C. "Constitutional Rights and New Technologies in Sweden." In *Constitutional Rights and New Technologies: A Comparative Study*, edited by Ronald; Leenes, B.-J. Koops, P. D. Hert, and S. W. Brenner, 199–224. The Hague: T.M.C. Asser Press, 2008.

Markgren, S. *Datainspektionen och skyddet av den personliga identiteten* [The Data Inspection Board and the Protection of Personal Privacy]. Lund: Studentlitteratur, 1984.

Mills, C.W. "Situated Actions and Vocabularies of Motive." *American Sociological Review* 5, no. 6 (1940): 904–913.

Ministry Publication Series. "Lag om brottsregister m.m." [Criminal Records Act, etc.]. Stockholm: Department of Justice, Ds Ju 1985: 8.

Ministry Publications Series. "De registrerades rätt till insyn i kriminal- och polisregistren" [The Data Subject's Right of Access to Her/His Data Contained in Criminal and Police Records]. Stockholm: Department of Justice, Ds Ju 1981: 6.

Naylor, B., M. Paterson, and M. Pittard. "In the Shadow of a Criminal Record: Proposing a Just Model of Criminal Record Employment Checks." *Melbourne University Law Review* 32, no. 1 (2008): 171–198.

Pratt, J. "Scandinavian Exceptionalism in an Era of Penal Excess: Part I: The Nature and Roots of Scandinavian Exceptionalism." *British Journal of Criminology* 48, no. 2 (2008): 119–137.

Pratt, J. "Scandinavian Exceptionalism in an Era of Penal Excess: Part II: Does Scandinavian Exceptionalism Have a Future?" *British Journal of Criminology* 48, no. 3 (2008): 275–292.

Pratt, J., and A. Eriksson. "Den skandinaviska exceptionalismen i kriminalpolitiken" [Scandinavian Exceptionalism in Criminal Policy]. *Nordisk Tidsskrift for Kriminalvidenskab* 96, no. 2 (2009): 135–151.

Preliminary Report of the Parliament Proceedings. "Onsdagen den 19 maj, kl 09.00–16.37" [Wednesday, May 19, at 09:00–16:37 hrs.], 2009/10:121.

Proposal 1899 No. 18. "Kongl. Maj:ts nådiga proposition till riksdagen med förslag till lag om straffregister; given Stockholms slott den 22 december 1899" [Gracious Proposal by His Royal Majesty for an Act on Criminal Registry, Submitted to the Parliament of Sweden on December 22, 1899].

Proposal 1963:39. "Kungl. Maj:ts proposition till riksdagen med förslag till lag om allmänt kriminalregister m.m." [Proposal by His Royal Majesty for an Act on General Criminal Registry, etc., Submitted to the Parliament of Sweden].

Proposal 1979/1980:2. "Förslag till ny sekretesslag" [Proposal for a New Secrecy Act].

Proposal 1987/1988:122. "Förslag till lag om ändring i lagen (1963:197) om allmänt kriminalregister" [Proposal for an Act to Amend the Act 1963:197 on General Criminal Registry].

Proposal 1997/1998:97. "Polisens register" [Police Databanks].

Proposal 2009/2010:191. "Gallring ur belastningsregistret av uppgifter om unga lagöverträdare" [Expungement of Information on Juvenile Offenders from the Criminal Records Register].

Richardson, L. "Narrative and Sociology." *Journal of Contemporary Ethnography* 1 (1990): 116–135.
Swedish Government Offical Reports. *Enhetligt frihetsstraff.* [Uniform Imprisonment]. Stockholm: Department of Justice, SOU 1953: 17.
Swedish Government Offical Reports. *Den allmänna brottsregistreringen* [The General Criminal Registry]. Stockholm: Department of Justice, SOU 1961: 11.
Swedish Government Offical Reports. *Offentlighet och sekretess.* [Public Access and Secrecy]. Stockholm: Department of Justice, SOU 1966: 60.
Swedish Government Offical Reports. *Data och integritet* [Computers and Privacy]. Stockholm: Department of Justice, SOU 1972: 47.
Swedish Government Offical Reports. *Polisens register. Slutbetänkande.* [The Police Registers: Final Report]. Stockholm: Fritze, SOU 1997: 65.
Swedish Government Offical Reports. *Integritetsskydd i arbetslivet* [Personal Privacy in the Working Life]. Stockholm: Fritzes, SOU 2009: 44.
Söderlind, Å. *Personlig integritet som informationspolitik.* [Privacy as information policy – debate and discussion concerning the first Swedish data protection law, Datalag (1973:289)] Borås and Göteborg: Valfrid, 2009.
Terms of Reference. *Personlig integritet i arbetslivet* [Personal Privacy in Working Life]. Dir. 2006: 55.
Tham, H. "From Treatment to Just Deserts in a Changing Welfare State." In *Beware of Punishment: On the Utility and Futility of Criminal Law*, edited by A. Snare. Oslo: Pax Forlag, 89–122. 1995.
The National Police Board. "Framställning om en översyn av den enskildes rätt att ta del av uppgifter om sig själv i belastningsregistret" [missive to the Government on reforming the way individuals' access to their criminal registry information is handled]. The National Police Board, Legal Secretariat, Record Number RÄS 442-3960/04, 2004.
Thomas, T. *Criminal Records: A Database for the Criminal Justice System and Beyond.* Basingstoke: Palgrave Macmillan, 2007.
Wang, J.-M., and B.H. Kleiner. "Effective Employment Screening Practices." *Management Research News* 23, no. 5/6 (2000): 73–81.

Chapter 7
Ubiquitous Computing, Privacy and Data Protection: Options and Limitations to Reconcile the Unprecedented Contradictions

Johann Čas

7.1 Introduction

The near future promises to turn a new paradigm of information technologies from rather futuristic visions into potential realities. Rapid progress in information and communication technologies increasingly allows the transformation of the visions[1] of ubiquitous computing[2] from the brains of scientists, technology developers and ICT stakeholders into real world applications and services. On one hand, the new

J. Čas (✉)
Institute of Technology Assessment, Austrian Academy of Sciences, A-1030 Vienna, Austria
e-mail: jcas@oeaw.ac.at

This paper was first published in Spanish in the winter-spring 2010 edition of the "Revista española de Protección de Datos", issued by Thomson-Civitas in collaboration with the Agencia de Protección de Datos de la Comunidad de Madrid.

[1] As an evolving concept it would be futile to attempt to provide an exact definition. However, the following description from ISTAG can serve as an approximation of the understanding of ubiquitous computing in this analysis: "According to the ISTAG vision statement, humans will, in an Ambient Intelligent Environment, be surrounded by intelligent interfaces supported by computing and networking technology that is embedded in everyday objects such as furniture, clothes, vehicles, roads and smart materials – even particles of decorative substances like paint. AmI implies a seamless environment of computing, advanced networking technology and specific interfaces. This environment should be aware of the specific characteristics of human presence and personalities; adapt to the needs of users; be capable of responding intelligently to spoken or gestured indications of desire; and even result in systems that are capable of engaging in intelligent dialogue. should also be unobtrusive – interaction should be relaxing and enjoyable for the citizen, and not involve a steep learning curve." IST Advisory Group, "Ambient Intelligence: From Vision to Reality. For Participation in Society & Business." (2003), 8.

[2] Other frequently used terms for ubiquitous computing are pervasive or calm computing, ambient intelligence, wearable computing and also the Internet of things. They are often used synonymously, although they focus on different aspects. In this paper ubiquitous computing is used as a generic term introduced by the originator of this vision (Mark Weiser, "The Computer for the 21st Century," Sci. *Amer.* 265, 3 (1991).). Particular aspects of the different terms (see for instance J Bizer et al., "Technikfolgenabschätzung Ubiquitäres Computing Und Informationelle Selbstbestimmung," In *Studie im Auftrag des Bundesministeriums fur Bildung und Forschung. Online: https://www.datenschutzzentrum.de/taucis/ita_taucis.pdf* (2006), 11f.) will be addressed only insofar as they are of relevance for the assessment of privacy impacts.

S. Gutwirth et al. (eds.), *Computers, Privacy and Data Protection: an Element of Choice*, DOI 10.1007/978-94-007-0641-5_7,

paradigm promises to overcome many insufficiencies and inconveniences of current information systems; on the other, it brings about tremendous threats to individual rights and societal values. Ubiquitous computing technologies have the potential to provide previously inconceivable levels of support for human activities in different spheres of life by systems working unobtrusively in the background, based on technology invisibly embedded in everyday environments and artefacts. Keyboards or other artificial input devices are replaced by natural language interfaces that observe the users and interpret spoken words, gestures or mimes as potential commands. Biometric procedures replace the need to remember passwords or to actively prove any authorization. The envisioned range of services, the aimed unobtrusiveness of their provision and convenience of use require also previously inconceivable levels of knowledge about the inhabitants of ambient intelligence environments, created by corresponding technical capacities of surveillance and dataveillance[3] as well as of merging and processing these data in an unrestricted manner.

This new paradigm also brings about a permanently extending inescapability from pervasive surveillance. Whereas in the past the release of data was mostly tied to activities the data subjects were conscious of and which were therefore, in principle, under the individuals' control, the new paradigm deprives them of the freedom to make such decisions. Although the freedom of choice may in practice not exist for many persons, or be coupled to unacceptable losses in participation in economic or private life, the inevitability of exposing oneself nevertheless creates totally different circumstances for the protection of privacy.

Consequently, ubiquitous computing presents unprecedented challenges not only to privacy, but to the many constituents of democratic and liberal societies in which privacy plays a functional role.[4] These challenges to privacy are widely acknowledged within the R&D community engaged in ubiquitous computing. Considerable research efforts have been devoted to developing privacy with respect to ubiquitous computing environments.

What is missing, however, are convincing concepts for the design of ubiquitous computing systems which could guarantee acceptable levels of personal privacy in the future. Most of the currently discussed solutions may render future technologies, to some extent, less invasive of privacy. However, they are insufficient to overcome their inherently privacy destructive potential; and sometimes even contain new threats, e.g. by the compulsive identification of all data subjects involved. Also in the case of pseudonymous data capture, protective measures are hardly conceivable that could resist the re-personalisation of pseudonymous data by advanced data analysis or mining technologies or by subsequent biometric identification procedures as long as not all data and traces are completely destroyed.

---

[3]This concept was introduced by Roger Clarke: Dataveillance is the systematic use of personal data systems in the investigation or monitoring of the actions or communications of one or more persons. Roger Clarke. Information Technology and Dataveillance. http://www.rogerclarke.com/DV/CACM88.html.

[4]Antoinette Rouvroy. "Privacy, Data Protection, and the Unprecedented Challenges of Ambient Intelligence." *Studies in Ethics, Law, and Technology* 2, 1 (2008).

The main reason for the limits to conceiving and designing privacy friendly ubiquitous computing systems is the fact that this technology vision is in fundamental contradiction with some of the most important principles on which current privacy protection is based. Regardless of whether non-binding international recommendations, EU regulations, or national laws are concerned, data or privacy protection is based on a limited number of similar principles. The most fundamental rules violated by ubiquitous information systems are the Collection Limitation Principle, the Purpose Specification Principle and the Use Limitation Principle. The first of these principles is in apparent contradiction to the uncountable, invisibly embedded processors and sensors forming pervasive computing systems, the last two would render the envisioned functioning of, and service provision by, ubiquitous computing systems practically impossible. On the one hand, they are unachievable for practical reasons as any restriction in the purposes for which the data may be used would imply corresponding restrictions in the usefulness of the systems; on the other hand, the requirement of an informed consent, which is currently sufficient to override the use limitation principle, would also entail ubiquitous and permanent requests for permissions to collect, transfer or use the data. Furthermore, because the elements of ubiquitous information systems are invisible, only specially equipped and trained teams would be able to detect violations of data protection regulations.

In the next chapter the main threats of ubiquitous computing to privacy will be outlined. The following section will analyze contradictions between the principles of privacy protection and ubiquitous computing environments. In the successive chapter selected measures proposed to make pervasive information technologies more privacy compliant will be discussed. In the concluding reflections, the consequences of ubiquitous computing for societal sustainability will be briefly addressed and open questions for further research and public debate will be identified.

7.2 Challenges

The immense increase in the gap between the quantity and quality of data needed and generated by ubiquitous computing systems and the simultaneous elimination of the means available to control the collection, storage, transfer and use of these data constitutes the core challenge inherent to this vision. Several individual factors are responsible for this widening, each of them, taken alone, would be sufficient to raise serious concerns for the protection of privacy; the new and unprecedented dimension of threat is, however, linked to impacts resulting from the equally ubiquitous erosion of the fundaments of current data protection, inherent to the vision of ubiquitous computing itself and therefore also to attempts at its implementation and transfer into ICT systems and services. This section will focus on the data generation issues and discuss briefly the main mechanisms contributing to the multiplication of personal data in terms of quantity and quality accompanying the new paradigm as well as debate the negative impacts on the possibilities of applying standard data protection.

The diffusion of networked information technologies into everyday environments, personal belongings and – in extreme cases – eventually into the bodies of humans inhabiting ambient intelligence surroundings, provides sufficient technical means for establishing a surveillance infrastructure embracing so far untouched spheres of life. The linking and merging of data originating from different sources, and the enrichment by sensors listening, watching and observing human beings without the filtering instances of technical interfaces, adds new qualitative dimensions to the data gathered. Ever increasing capacities and capabilities to store and analyze the huge amounts of data do not only extend surveillance into the history of the concerned persons, they also provide means and incentives to generate predictions about future behaviour and needs of the data subjects. The multitude and abundance of available and accessible data turns all data into personal data and additionally limits the applicability and efficacy of Privacy Enhancing Technologies (PETs). As a consequence, the only remaining rational assumption for individuals inhabiting a world with ubiquitous computing will be to live in a panoptic society, with deep subsequent consequences for individual behaviour, civil liberties, democratic and societal sustainability.[5] Of course, the actual extent to which surveillance is exercised will depend on the degree of materialization of these technologies, the combination of increasing technical capabilities and the central ambitions of ubiquitous computing which will signify that normally no one will be able to exclude the possibility of being observed. It is this potential surveillance which characterizes panoptic societies.

7.2.1 Ubiquitous Surveillance

The increase in the number of sensors, to which data subjects living in ubiquitous computing will be exposed, regardless of whether they are invisibly embedded into environments or into personal devices or belongings carried with oneself, is one factor responsible for a corresponding increase in the generated data. More critical than the sheer number and the resulting growth in the quantity of data that can be captured are the qualitative changes linked to it. A first and crucial change concerns the extension of exposure to potentially all spheres of private or professional activities.

Currently the generation of digital traces is, with a very few exceptions, restricted to active use of information or communication technologies. A major exemption relates to the location information generated by switched-on mobile phones as knowledge of their approximate location is required for routing calls to the nearest base stations. With ubiquitous computing the regular situation is reversed; invisible sensors observe the users and their surroundings permanently to provide services or to adjust the environment according to expressed orders or perceived needs,

[5] Johann Čas. "Privacy in Pervasive Computing Environments - a Contradiction in Terms?." *IEEE Technology and Society Magazine* 24, 1(2004).

interpreting the actual context and relating its users' preferences, gained from past experiences and condensed into constantly refined profiles.

The growing outreach into previously untouched spheres of life is accompanied by a factual impossibility of excluding oneself from being observed. Already today full participation in economic and social life of modern societies is in many cases inseparably linked to the use of communication technologies and Internet services. A renunciation of using such services in order to preserve privacy is hence in a number of situations rather a theoretical concept than a viable alternative, contesting the requirement of "free", informed consent. Nevertheless the use, non-use or shift to more privacy respecting providers or technologies remains basically under the discretion of the individual user. If and to what extent this right can be preserved depends on the concrete architecture and design of the system, e.g., in a world of ambient intelligence this choice will be practically non-existent.[6] Potentially effective solutions to mitigate or eliminate the privacy challenges will necessarily imply refraining from the implementation of fully fledged versions of the new paradigm.

7.2.2 Increases in Data Quality

Here, increase in data quality describes rather more the increase in the dimensions of information and in the conclusions that can be derived and less the completeness, accuracy and up-to-date nature in relation to the purpose of the collected data as referred to in data protection regulations.[7] In these legal dimensions it appears to be impossible to provide general statements about the impact of ubiquitous computing. Whereas the increase in quantity presumably also increases completeness, accuracy may quite as well deteriorate considerably, e.g. due to attribution to the wrong data subjects, which cannot be completely circumvented when using biometric identification methods, or due to wrong interpretations of actions or the contexts in which they take place. In the first definition of data quality – the range of dimensions covered by captured data – profound changes must be expected, both because of the nature of the individual data collected by ubiquitous computing systems and because of the indirect effects of the immense quantitative growth in data generation and the resulting possibilities to merge and analyze these data pools.

The enrichment of information content is a direct consequence of the replacement of artificial interfaces like keyboards, touch screens and mouse devices by natural language interfaces or video observations and interpretations of movement, gestures or mimes. The change from text based information to multimedia data will result in

[6] Under the commercial provision of ubiquitous computing services, the storage and processing of data will probably be restricted to paying clients. However, the exclusion from service provision does not necessarily also imply an exclusion from personal data captured in the first place.

[7] See for instance Art. 6 (d) of the Data Protection Directive (Directive 95/46/Ec of the European Parliament and of the Council of 24 October 1995 on the Protection of Individuals with Regard to the Processing of Personal Data and on the Free Movement of Such Data.).

better quality.[8] In addition to the content of the message, text-only protocols disclose, for instance information about literacy and language skills. Audio recordings furthermore provide information about verbal cues like accent or rhetoric skills, but also about the emotional state, while videos disclose visual cues like dress, physical condition or body language and personal characteristics like age, sex or ethnicity. Sensors with capabilities exceeding those of the human senses, e.g. infrared cameras able to detect the smallest changes in blood circulation far from causing any visible reddening, will result in entirely new qualities of information. Changes in blood circulation or the slightest trembling of the voice could be attributed to nervousness, and such sensors could thus also serve as sophisticated and invisible lie detectors.

The aim to provide useful, non-trivial and non-annoying services implies that they must be profile based and take into account the actual context of service delivery. As a precondition to make new suggestions of potentially interesting offers to a specific customer, a service provider needs for instance to make use of group profiles which are based on experiences of users with similar preferences. Profiling and data mining create the additional challenge that even data that, if taken alone, would appear to be inconsequential from a privacy perspective may become sensitive. "Sets of correlated data that could be considered insignificant or even trivial can provide intimate knowledge about, e.g., life style or health risk, if data mining is applied."[9]

7.2.3 Persistent Data Storage

Technical progress in storage technologies allows the provision of ever increasing capacities at rapidly declining costs; as a consequence costs barriers to long term storage of the exponentially growing amounts of collected data are also quickly loosing their economic relevance. A similar development prevails for data analysis and processing capacities, allowing the application of sophisticated data mining procedures to huge data collections, previously inaccessible to analysis at reasonable costs within tolerable processing time frames. These developments are not specific to ubiquitous computing per se, but they constitute a prerequisite for turning the vision into reality and to extend both, the range and quantity of data and their retention period. These technical advancements will also permit the offering of memory amplifiers, recording any action or expression of ourselves enabling us to browse through our past at a later date.[10]

---

[8]Anne Adams and Martina Angela Sasse. "Privacy in Multimedia Communications: Protecting Users, Not Just Data." In *People and Computers XV - Interaction without Frontiers. Joint Proceedings of HCI2001,* edited by A. Blandford and J. Vanderdonkt (2001).

[9]Hildebrandt, M. "Profiling and the identity of the European citizen." In *Profiling the European Citizen: Cross-Disciplinary Perspectives*, edited by M. Hildebrandt and S. Gutwirth. Dordrecht: Springer, 2008, 304.

[10]Marc Langheinrich. "Privacy by Design - Principles of Privacy-Aware Ubiquitous Systems" (paper presented at the Ubicomp 2001, Atlanta, September 30 – October 2, 2001), 7.

The related privacy threat of lifetime recordings is not bound to sophisticated ubiquitous computing infrastructures, simple personal electronic gadgets are sufficient. Although the prediction that devices capable of storing non-stop lifetime audio recordings would be available at prices of $200 in 2008, those for lifetime compressed video recordings in the same price range in 2010[11] was overestimating the speed of technological progress, a first glance at the potential future impacts on privacy and dignity is already cognizable today; for instance, recordings taken by mobile phones, distributed or uploaded, without the knowledge and consent of the persons concerned. They represent only a small proportion of the privacy threats deriving from the opportunity for practically everybody to make lifetime recordings, which again embodies only a small part of the capacities of full ubiquitous computing systems. Personal recording devices still lack the full surveillance capability, which will be embodied through innumerable sensors and processors. They still miss the full ability of spontaneous networking and access to data stored anywhere, and they do not possess all analytical capacities, like dataveillance, to explore the past, or profiling to generate statements and predictions about the present and the future. Nevertheless, these simple devices already perfectly illustrate the perils; any communication, any personal contacts, any exchange of information supposed to take place in private could be captured and reproduced at any time.

7.2.4 Re-personalization of Data

In ubiquitous computing environments, the existing possibilities of anonymous or pseudonymous use of services, or merely the possibility of being physically present in an unnoticed manner in such systems, will largely disappear. One reason is pertinent to the constituting objective of personalization, a feature that, in itself, would not necessarily require knowing the real identity but could as well be provided with the assistance of identity management systems protecting the real user behind pseudonyms. However, the replacement of artificial interfaces by sensors capturing natural modes of expression eliminates or renders these opportunities practically non-effective.

The abolition of means of pseudonymization as a consequence of the new paradigm does not imply that traditional ways of access provide sufficient levels of protection. Under the currently prevailing paradigm, a single PC or device provides access to information systems or services for a single user. Here, the term "Personal Computer" already indicates that, in general, the widespread assumption of anonymity granted when using the Internet is not justified, but depends to a large extent on the carefulness or reluctance of the individual user to provide identifying data, on the one hand, and, on the other, on the ability to access and to link different

[11] John Alexander Halderman. "Digital Privacy-Rights Management for Ubiquitous Recording Princeton University." 2003.

pieces of data to each other, or on the legal powers of the parties interested in uncovering the real identity of a user. However, the artefact placed between the user and the information system can be used to provide anonymous or pseudonymous access, e.g. to make use of public terminals or to apply identity management technologies. With pervasive computing, and numerous invisibly embedded computers or devices, it is the user himself who initiates the collection and processing of data or the delivery of services. Some form of identification of the person who consciously requests, or unconsciously launches, a process is therefore inevitable.

Furthermore, the possibility of using pseudo-identities is limited because of the process of data capture and the kind of data necessarily involved in this process. The interpretation of natural language commands requires the recognition of the presence of a person; before mimes can be observed, a face must be recognized; faces and persons must be distinguished from each other for any meaningful result. Speech recognition, or the interpretation of gestures or mimes, require the capture of audio or video data, which can easily be (re-)used for biometric identification. The precondition for context-based services is, of course, the analysis of context information; with location data as an important part of context, sufficient data for personal identification are also available in a majority of cases, especially if location data are retained for longer periods. In general, ubiquitous computing makes sense only if the systems are able to learn from the past, to enrich and correct personal profiles permanently and to adapt services accordingly; which in turn requires that the systems are allowed to "remember", i.e., to store personal data. The normal state of ubiquitous computing is that there will always be ways to re-establish the personal identity of individuals once captured. Even if all of the above parameters could be excluded, an unnoticed or forgotten RFDI chip in a personal belonging could still reveal the personal identity. Rather than links missing in the establishment of evidence, the default state will be that the creation of pseudo-identities will be disrupted. The multitude of linking data from different sources, sensors and times implies that data collected by pervasive information systems are, in principle, personal data.[12] The pervasiveness of data collection will also blur the distinction between sensitive and non-sensitive data. First, persistent and pervasive data capture will necessarily also comprise items of a sensitive nature; second, the linking and mining of data pools, consisting of rather uncritical individual pieces of information, can reveal very sensitive knowledge about the data subject concerned.

7.2.5 Increasing Information Asymmetry

Also, without ubiquitous computing systems it was practically impossible to know, in full detail, who collected which data, to whom the data were transferred and for

[12]It is completely sufficient that they can with reasonable efforts be retransformed into personal data (Article 29 Data Protection Working Party. Opinion 4/2007 on the Concept of Personal Data. (2007)).

which purposes they were used. Yet the collection process itself was, by and large, confined to processes where users deliberately filled in offline or online forms, or data were – at least potentially consciously – generated by using communication and information technologies. And the citizens of countries with advanced privacy regulations had the right, although in reality often not or not easily enforceable, to be informed about the collection of data on them and the purpose of such collection as well as to have incorrect data modified, to withdraw formerly given consent or to request the deletion of data that were stored and processed without proper legal basis. Pervasive computing environments are going to worsen the situation dramatically. The desire to provide ambient intelligence in an unobtrusive manner requires a framework in which users are permanently observed and their behaviour and actions autonomously interpreted, taking into account location and other contextual information. The results are then fed into a continuous learning process, which will form the basis for autonomous decisions by the system on how and when to use, or to pass on, the collected information. Ubiquitous computing implies more data about the data subject, at the same time less transparency for and less control by the users; it necessarily enlarges the already existing asymmetry in information and power between the data subjects and the data collectors. The seriousness of this impact depends also on the extent to which the generated data is kept under the control of the user. However, solutions where the data would be stored and maintained at a single place or device would hardly qualify as ubiquitous computing systems as they miss central components of this paradigm, e.g. the seamless integration of users, devices and environments into networking technologies and the additional benefits and dynamic provision of new services which are bound to the exchange of data between different domains. And for a shift in the power relations between citizens and organizations controlling the ubiquitous computing systems the potential use of such information is sufficient. "The crucial issue is not the abuse but rather the fact that we have no effective means of knowing whether and when profiles are used or abused."[13]

7.2.6 Panoptic Society

The only realistic attitude of human beings living in ubiquitous computing environments would be to assume that any activity or inactivity is being monitored, analyzed, transferred, stored, and may be used in any context in the future. This attitude will be justified, notwithstanding the definite limitations in the reach of such environments; it will of course take some time before ubiquitous computing will have been widely diffused, and of course surveillance-free spaces will continue to exist. However, nobody can be sure anywhere that his or her actions are not being observed and conversations not being recorded, or that the presence at

---

[13]Hildebrandt, M. "Profiling and the identity of the European citizen." In *Profiling the European Citizen: Cross-Disciplinary Perspectives*, edited by M. Hildebrandt and S. Gutwirth. Dordrecht: Springer, 2008, 318.

any location is not being stored in some registers, be it by sophisticated ubiquitous computing systems or by simple individual devices. In this way, the "Panopticon" or Inspection-House – a concept developed by Jeremy Bentham[14] in the late eighteenth century primarily for prisons or madhouses – will be extended to public and private spaces in general. As had been intended by Bentham and further discussed by Foucault[15], the possibility of being permanently observed is regarded as sufficient to create strict discipline and uniformity within societies. To what extent the last statement is fully transferable to modern societies under the new information technology paradigm remains an open issue for scientific discussion and research. Nevertheless, we should be aware that the Panopticon is still a very imperfect illustration of the potential disciplining power of pervasive computing systems. Persons living in future ubiquitous computing environments can, with almost 100% certainty, assume that they are being observed – in contrast to the classical Panopticon where nobody could be certain whether he or she was actually monitored. In addition, classical surveillance was restricted in place and time, while data captured by ubiquitous computing will persist across space and time.

7.3 Contradictions to the Current Fundaments of Privacy

Human rights are, at the same time, a consequence of the establishment and a prerequisite for the continued existence and further development of democratic societies. Privacy constitutes a central component of human rights and is consequently established in central international agreements and law like the United Nations' Universal Declaration of Human Rights (Article 12)[16] and in the European Union's Charter of Fundamental Rights (Articles 7 and 8).[17] The right to privacy, ... "given the

[14]Jeremy Bentham. "Panopticon: Or, the Inspection-House : Containing the Idea of a New Principle of Construction Applicable To ... Penitentiary-Houses, Prisons, Houses of Industry, Work-Houses, Poor-Houses, Manufactories, Mad-Houses, Hospitals, and Schools. With a Plan of Management Adapted to the Principle," in *a series of letters, written ... 1787, from Crecheff ... to a friend in England* (Dublin: Thomas Byrne, 1791).

[15]Michel Foucault. *Discipline and Punish: The Birth of the Prison*. London: Penguin, 1977.

[16]"Article 12. No one shall be subjected to arbitrary interference with his privacy, family, home or correspondence, nor to attacks upon his honour and reputation. Everyone has the right to the protection of the law against such interference or attacks." *Universal Declaration of Human Rights*, General Assembly resolution 217 A (III) (10.12.1948).

[17]"Article 7 Respect for private and family life

Everyone has the right to respect for his or her private and family life, home and communications.

Article 8 Protection of personal data

1. Everyone has the right to the protection of personal data concerning him or her.
2. Such data must be processed fairly for specified purposes and on the basis of the consent of the person concerned or some other legitimate basis laid down by law. Everyone has the right of access to data which has been collected concerning him or her, and the right to have it rectified." European Union, "Charter of Fundamental Rights of the European Union," (Official Journal of the European Communities, 2000).
3. Compliance with these rules shall be subject to control by an independent authority.

crucial role it plays in enabling the autonomic capabilities of the individual legal subject, is a precondition to any meaningful exercise of all other rights and freedoms acknowledged by the Council of Europe."[18]

Human rights charters and declarations state these rights in a rather general manner, which does not allow for a detailed examination of possible incompatibilities and contradictions between ubiquitous computing and fundamental rights. For this purpose, concrete recommendations or legal norms developed in the context of information technologies and data processing are better suited. In the following paragraphs, the most important provisions as listed in the OECD Privacy Guidelines[19] and the EU Data Protection Directive 95/46/EC[20] will be compared with some inherent features of pervasive computing systems. Although legally non-binding, the OECD Guidelines – which remained unchanged for almost three decades – have stretched out into many voluntary agreements and regulations enforced by law, including the Directive 95/46/EC. The OECD Guidelines comprise eight principles for the protection of privacy: the Collection Limitation Principle, the Data Quality Principle, the Purpose Specification Principle, the Use Limitation Principle, the Security Safeguards Principle, the Openness Principle, the Individual Participation Principle, and the Accountability Principle. They are also referred to as Fair Information Principles, together with numerous other and similar sets of privacy protecting rules. The latter term originated from a less comprehensive set of rules developed in the early 1970s in the USA. The EU Directive in turn has influenced real world privacy protection regimes in very significant ways. On the one hand, it had to be transposed and implemented into the national laws of the EU member states; on the other, it prohibits the transfer of personal data to countries outside the EU which do not possess an appropriate and comparable level of protection of personal data. The need to ensure an adequate level of protection caused many countries outside the EU to adopt "voluntarily" similar privacy regulations.

Conflicts or contradictions between the ubiquitous computing vision and the OECD Guidelines can be identified for all of the eight principles listed in the guidelines.[21] The first four of the eight principles listed above contain the essence of privacy protection, whereas the last four describe procedural aspects. The first four are particularly important in the sense of forming indispensable pillars of all current

[18]Antoinette Rouvroy and Yves Poullet. "The Right to Informational Self-Determination: Reassessing the Importance of Privacy for Democracy," in *Reinventing Data Protection?*, edited by. S. Gutwirth, et al. Springer, Netherlands, 76. 2009, .

[19]OECD. Guidelines on the Protection of Privacy and Transborder Flows of Personal Data. (1980).

[20]Directive 95/46/Ec of the European Parliament and of the Council of 24 October 1995 on the Protection of Individuals with Regard to the Processing of Personal Data and on the Free Movement of Such Data.

[21]Johann Čas. "UC - Ubiquitous Computing oder Ubiquitous Control?." In *Der Mensch Im Netz - Ubiquitous Computing,* ed. Bernd Britzelmaier, Stephan Geberl, and Siegfried Weinmann, *Reihe Wirtschaftsinformatik* Stuttgart, Teubner, 2002.

privacy protection schemes, they can be found in European as well as in Canadian and US privacy regulations.[22]

7.3.1 Collection Limitation Principle[23]

Already the basic idea of pervasive computing infrastructures totally contradicts the provisions contained in this principle. The first part of this principle refers to a general limitation on the collection of personal data, without specifying the details of such a limitation. Ubiquitous computing is based on the removal of such limitations. Data on persons and objects within the reach of ubiquitous computing systems are actively, pervasively and continuously collected. Even if only part of this huge amount of information will be stored or further processed, the principle of limitation of data collection is fully turned into the reverse. The last part of the principle refers to the awareness and informed consent of the person, whose data are being collected. While a basic awareness is still achievable, e.g. through clearly visible warning tags indicating that ubiquitous computing is in use, detailed knowledge about which objects capture which kinds of data at what time is hardly conceivable, both for practical reasons and for its incompatibility with the inherent goal of unobtrusiveness.

The consent issue is not specified in detail in the OECD Guidelines, therefore we refer to the EU Data Protection Directive for the discussion of conflicts with this requirement. The Directive defines that "'the data subject's consent' shall mean any freely given specific and informed indication of his wishes by which the data subject signifies his agreement to personal data relating to him being processed."[24] One of the requirements of Article 7 of this Directive becomes completely unfeasible, namely the obligation to base the processing of personal data on the unambiguous consent of the affected persons. Also today, the precondition that "the data subject has unambiguously given his consent" is neither accomplishable nor desirable

---

[22]Giovanni Iachello. "Protecting Personal Data: Can IT Security Management Standards Help?." (paper presented at the 19th Annual Computer Security Applications Conference, Las Vegas, Dec. 2003).

[23]"There should be limits to the collection of personal data and any such data should be obtained by lawful and fair means and, where appropriate, with the knowledge or consent of the data subject." Paragraph 7 OECD, "Guidelines on the Protection of Privacy and Transborder Flows of Personal Data."

[24]Article 2(h) Directive 95/46/Ec of the European Parliament and of the Council of 24 October 1995 on the Protection of Individuals with Regard to the Processing of Personal Data and on the Free Movement of Such Data.

in all cases and at any time. Therefore, a set of exceptions from the obligation to seek consent exists, e.g. with respect to the processing of personal data in order to fulfil a contract or to protect vital interests of the person concerned. For the category of "public interest" the exceptions also comprise questions of public security or the efficiency of legal actions, raising well-known basic problems regarding the mutual appraisal of fundamental rights and the often conflicting relation between civil liberties and security. Already today the right to privacy is permanently threatened by security policies and technologies that focus on a one-sided perception of security, neglecting the central role of privacy for individual security, e.g. as a safeguard against state arbitrariness or economic or social discrimination. These threats will, however, dramatically increase under pervasive computing as this technology will immensely enhance the quantitative and qualitative possibilities of monitoring and extend them to areas which are currently out of the reach of permanent and unobtrusive surveillance. Current experiences tell us, that the legal capacities of law enforcement authorities must be expected to be continuously adjusted to the increasing technical possibilities.

The objective of unobtrusiveness is completely incompatible with acquiring individual consent for each data collection activity; a sequence of permanent observations would entail equally permanent requests for consent. Contract based forms of consent provision appear to be a more realistic alternative from a procedural perspective, however, without offering solutions to the fundamental problems related to consent issues. On the contrary, the imbalance in the relation between the feasibility to obtain consent of data subjects, on the one hand, and the observation capacities of UC-systems, on the other, promises to be further increased in dramatic ways. Certainly, parts of the population will not object to total supervision in an UC-world, particularly if, in addition to gains in convenience, arguments of enhanced security are brought into play. Probably, many people will willingly sign pertinent contract clauses, explicitly consenting to the collection, processing and transfer of any data, if providers of UC-services demand it as a precondition for entering into service contracts. Even in the case of voluntary contracts, one could of course discuss whether such agreements would still fall under valid legal or societal norms, or whether they should be considered as illegal or immoral, similar to selling one's own soul. Another much more serious concern is connected with the fact that it is not possible to escape the surveillance infrastructure for those parts of the population who do not want to be permanently observed. If a person decides not to sign a contract he or she will be excluded from the service in question, however, in an ideal ubiquitous computing environment there is no way to escape the ubiquitous surveillance. Therefore, the pervasiveness of ubiquitous computing raises the difficult legal question whether or not a "consent free of doubts" to something that is unavoidable can count as a valid part of individual or collective agreements at all.

The consent issue within ubiquitous computing is further complicated by the comprehensiveness of the captured data in terms of embracing non-sensitive and special categories of data[25] as defined in the Data Protection Directive. Their processing is generally prohibited, with a limited set of listed exceptions for which this general prohibition does not apply.[26] This comprehensiveness implies, that on the one hand, that sensitive data are directly captured – ethnic origin can in many cases be deducted from video recordings, and persisting audio recordings will of course automatically also reveal indications about political opinions, religious or philosophical beliefs. On the other hand, the linking and analysis of data, which would, taken alone, not raise privacy concerns, may expose wide-ranging impressions of the person concerned, including very sensitive personal data.

7.3.2 Data Quality Principle[27]

This principle features two dimensions; firstly, the relevance of the data for the intended purpose, which stands in close relation to two further principles discussed below, secondly, the exactness, completeness and topicality of the data. In general, one may expect that pervasive computing will result in better fulfilling demands within the second dimension. However, only exact knowledge of the particular system in use and empirical data from pilot installations will allow the making of valid statements on these aspects of data quality. If, for instance, the user is identified by means of biometric methods, a certain rate of false allocations of data to persons, and thus of incorrectness and inaccuracy, is unavoidable as a decrease of the FAR (False Acceptance Rate) implies a rise of the FRR (False Rejection Rate) and vice versa. In general, more data do not necessarily lead to better data. In order to get more accurate data, there must be regular controls and corrections as well. This requirement involves another trade-off: without central or coordinated storage in one form or another quality improving procedures are hardly conceivable, while centralized data collections again entail huge incentives for and corresponding high risk of abuse. The purpose related dimension of the data quality principle is discussed under the next two principles which specify it in more detail.

[25]Special data comprise "personal data revealing racial or ethnic origin, political opinions, religious or philosophical beliefs, trade-union membership, and the processing of data concerning health or sex life." Article 8 (1) Ibid.

[26]For this category of data even the possibility of explicit consent by the data subject may be restricted by national law of the Member States: "(a) the data subject has given his explicit consent to the processing of those data, except where the laws of the Member State provide that the prohibition referred to in paragraph 1 may not be lifted by the data subject's giving his consent;" Article 8 (2) Ibid.

[27]"Personal data should be relevant to the purposes for which they are to be used, and, to the extent necessary for those purposes, should be accurate, complete and kept up-to-date." Paragraph 8, "Guidelines on the Protection of Privacy and Transborder Flows of Personal Data."

7.3.3 Purpose Specification Principle[28]

At the core of this principle is the demand that, at least at the time of data acquisition, the purposes must be known and identifiable. In the Data Protection Directive it is further specified that personal data must be "... collected for specified, explicit and legitimate purposes ..."[29] Later changes in the purposes are only allowed if they are compatible with the original intention; in addition, they must be properly indicated.

The aim of ubiquitous information technologies, however, is not to serve single, pre-definable purposes, but to support users in a variety of more or less foreseeable situations. A fundamental problem of ubiquitous computing is the fact that the above principle is simply turned upside down, eliminating one of the central foundations and anchors of contemporary data protection frameworks. The purpose of data collection lies entirely in the accumulation of as much data as processable to generate as much information as possible about individual behaviour patterns and preferences; the contents of and the context in which this knowledge is going to be applied remains necessarily unclear at the time of collecting the data. Ubiquitous computing aiming at the assistance of any arbitrary human activity creates a double dilemma for the data protection framework. The absence of a specific purpose eliminates also an essential criterion for the evaluation of the lawfulness of data collection by such systems; to soften the requirement of specificity in a way sufficient to include ubiquitous computing systems would reduce the applicability and effectiveness of this obligation to an almost certainly unacceptable extent. The lack of specific purposes exterminates the basis for determination whether personal data are "adequate, relevant and not excessive in relation to the purposes for which they are collected and/or further processed;"[30] a central requirement of the Data Protection Directive. Without, or with too general a specified purpose, the principle of data minimization and the principle of proportionality – implicitly included in the cited requirement – also lose the central assessment criteria and hence applicability. The principle of proportionality is rarely directly mentioned in privacy regulations; nevertheless it plays an important role as an element, indirectly included in other provisions or as a generally valid fundamental legal principle of all data protection law.[31]

---

[28] "The purposes for which personal data are collected should be specified not later than at the time of data collection and the subsequent use limited to the fulfilment of those purposes or such others as are not incompatible with those purposes and as are specified on each occasion of change of purpose." Paragraph 9 Ibid.

[29] Article 6 (b) Directive 95/46/Ec of the European Parliament and of the Council of 24 October 1995 on the Protection of Individuals with Regard to the Processing of Personal Data and on the Free Movement of Such Data.

[30] Article 6 (c) Ibid.

[31] Christopher Kuner. "Proportionality in European Data Protection Law and Its Importance for Data Processing by Companies." *Privacy & Security Law Report* Vol. 07, 44 (2008).

7.3.4 Use Limitation Principle[32]

Supplementing and extending the Purpose Specification Principle to subsequent uses, this principle states that data may not be disclosed, transferred or used if this disclosure, transfer or use does not correspond to the purpose specified at the time of the collection of the data. Exceptions to this principle are possible with the consent of the data subject or if the utilization takes place in the framework of the authority of law.

The lack of a specified initial purpose impedes also the use limitation principle and renders it impossible to impose any limits on secondary uses. In addition, the spontaneous linking of innumerable and invisible computers and the exchange of data between them represents a central and indispensable component of ubiquitous computing infrastructures; hence a further fundamental and obvious contradiction exists between the principles of use limitation and purpose specification – and the visions of ubiquitous computing systems. Apart from numerous technical problems a limitation of the transfer and use of data would entail, every attempt to enforce parts of this principle implies also curtailing the potential benefits and the usability of ubiquitous computing infrastructures. The benefits are limited because an invariable assignment of data to applications limits the adaptability and learning abilities of the system; the usability gets restricted because permanent inquiries about consent to or dissent on requests for the transfer of data would contradict the intention to create unobtrusive computing environments and certainly wear out the nerves of any user within a short time.

7.3.5 Procedural Principles

The last four principles mainly describe the technical and procedural aspects and policies, necessary to enforce and safeguard compliance with the first four principles. They also provide for transparency and establish the rights of individual data subjects to be informed about and to challenge data relating to them. They constitute indispensable elements of current data protection regulations and will remain of vital importance for future privacy protection frameworks. Several contradictions between the ubiquitous computing concept and these principles restrict or eliminate their applicability under the new technology paradigm. The Security Safeguards Principle states that "Personal data should be protected by reasonable security safeguards against such risks as loss or unauthorized access, destruction, use, modification or disclosure of data".[33] Obviously these requirements will gain

[32]"Personal data should not be disclosed, made available or otherwise used for purposes other than those specified in accordance with Paragraph 9 [Purpose Specification Principle] except:
(a) with the consent of the data subject; or
(b) by the authority of law." Paragraph 10 OECD, "Guidelines on the Protection of Privacy and Transborder Flows of Personal Data."

[33]Paragraph 11 Ibid.

tremendously in importance in view of the sensitiveness of the personal data processed in ubiquitous computing systems. Their realization, however, will probably encounter insurmountable barriers. It will be practically impossible to provide sufficient levels of security against unauthorized access or disclosure of data when the spontaneous networking of numerous wireless components with limited processing and encryption capacities is a core element of such systems. The general objective of the Openness Principle[34] can be taken into account by creating awareness about the presence of ubiquitous computing technologies, specific requirements like information about the nature of personal data or their main purpose cannot be met because of the dynamic data collection and lack of in advance specified purposes. The huge amount of data involved will seriously complicate the practical execution of the Individual Participation Principle,[35] describing the rights of the data subject. Depending on the concrete implementation of ubiquitous computing systems, it might in addition be difficult or impractical to identify the accountable data controller(s), as required by the last item of the OECD Guidelines, the Accountability Principle.[36]

7.3.6 Automated Individual Decisions

The Data Protection Directive contains another provision obviously in contradiction to envisioned decision making and service provision mechanisms based on the profiling of ubiquitous computing by entitling " . . . the right to every person not to be subject to a decision which produces legal effects concerning him or significantly affects him and which is based solely on automated processing of data intended to evaluate certain personal aspects relating to him, such as his performance at

---

[34]"There should be a general policy of openness about developments, practices and policies with respect to personal data. Means should be readily available of establishing the existence and nature of personal data, and the main purposes of their use, as well as the identity and usual residence of the data controller." Paragraph 12 Ibid.

[35]"An individual should have the right:

(a) to obtain from a data controller, or otherwise, confirmation of whether or not the data controller has data relating to him;
(b) to have communicated to him, data relating to him
  - within a reasonable time;
  - at a charge, if any, that is not excessive;
  - in a reasonable manner; and
  - in a form that is readily intelligible to him;
(c) to be given reasons if a request made under subparagraphs(a) and (b) is denied, and to be able to challenge such denial; and
(d) to challenge data relating to him and, if the challenge is successful to have the data erased, rectified, completed or amended." Paragraph 13 Ibid.

[36]"A data controller should be accountable for complying with measures which give effect to the principles stated above." Paragraph 14 Ibid.

work, creditworthiness, reliability, conduct, etc."[37] "This prohibition seems equally at odds with the logic of adaptive autonomic profiling [...], since most decisions will be taken by machines in a process of machine-to-machine communication."[38]

7.4 Proposals to Overcome the Contradictions

The inherent and obvious threats to privacy from technical progress in information and communication technologies, specifically if embodied in ubiquitous computing did not only provoke the announcement of the end of privacy; they also created an intellectual challenge from the very beginning of development of this vision.[39] As part of an answer, researchers and technology developers concerned with privacy generated numerous concepts to reconcile what appears to be irreconcilable. The main focus of such attempts appears to shift slowly from mainly technical solutions – the transfer or integration of PETs into ubiquitous computing systems – to questioning the adequacy of current data protection regulations to preserve the fundamental right to privacy. In the following sections we will briefly outline some of these attempts and concepts and assess their capabilities and their limitations to mitigate or eliminate inherent threats to privacy. In the discussed examples PETs are applied to limit the generation of personal data, to support the consenting process and to prevent unintended use or transfer of data.

7.4.1 Privacy Enhancing RFID Technologies

RFID (Radio Frequency Identification) technologies are usually seen as one of the central building blocks of future ubiquitous computing systems. They will certainly play an important role, simply by allowing the identification and localization of artefacts present in ambient intelligence environments; including also the identification and localization of persons with whom certain artefacts can be associated. The main critical features from the privacy perspective of RFID are that they provide a unique identification, that they can be read without evidence and that both, RFID readers and tags can be embedded invisibly into the environment or in artefacts. In this way they also allow for unnoticeable identification and tracking of persons

[37] Article 15 Directive 95/46/Ec of the European Parliament and of the Council of 24 October 1995 on the Protection of Individuals with Regard to the Processing of Personal Data and on the Free Movement of Such Data.

[38] Mireille Hildebrandt and Bert-Jaap Koops, editors., *FIDIS Deliverable D7.9: A Vision of Ambient Law* (2007), 43.

[39] See for instance Victoria Bellotti and Abigail Sellen. "Design for Privacy in Ubiquitous Computing Environments." In *Proc. Of the European Conference on Computer-Supported Cooperative Work* (1993).

carrying RFID tags with their personal belongings or within their body. The diffusion of RFID technology brings about severe data protection issues,[40] regardless of the context in which they are applied. Accordingly, attention has been paid to the privacy enhancement of this technology. In a recent overview, more than two hundred research papers dealing with privacy challenges of RFID systems were analyzed and the proposed PETs categorized.[41] The suggested solutions reach from deactivation of the tag – software initiated or by physical destruction – to different schemes that control the access to data stored on the RFID tags. On-tag schemes, the most frequently proposed PETs approach in the analyzed scientific papers, are based on automatic authorization procedures for granting access for the readers to the tags. The tags need to be able to perform complex cryptographic functions and, in addition, depending on the concrete cryptographic technology applied, a back-end database or a public key management infrastructure. This approach can provide technical protection against unauthorized reading of RFID tags, but does not offer possibilities of control or notice for the users. Agent schemes aim to open these possibilities to the users, or owners of tagged items, by delegating privacy management functions to an agent. These agents can perform different tasks, from simply providing information about reading processes and the complete blocking of RFID communication to complex, context dependent, management of individual privacy preferences and correspondingly selective permission for or jamming of RFID communication. The latter functionality could be performed by personal digital assistants described in the next section. The user scheme proposes to lock RFID tags when the owners leave the stores and to put the unlocking procedure under direct control of the owner, e.g. in a simple implementation the owner would select a password that is used for locking and for subsequent access granting. The user scheme would thus involve notice and active consent by the user.

The number of available options for privacy enhancing RFID technologies is drastically reduced if their application is transferred to ubiquitous computing environments. The kill options, the most effective prevention of privacy impacts, would also permanently prevent any desired use of RFID and so consequently do not make sense in the context of ubiquitous computing. The user scheme approach is incompatible with the objective of unobtrusiveness and infeasible for practical reasons. On-tag schemes would probably increase the hardware requirements as well as the system complexity to a non-practicable extent without contributing to users' privacy as regards notice and consent. Agent based concepts appear to be the only approach capable of mitigating some of the privacy threats of ubiquitous computing, as well as protecting RFID communication. However, this concept also bears considerable weaknesses and new risks. The administration of privacy preferences

[40] Article 29 - Data Protection Working Party. Working Document on Data Protection Issues Related to RFID Technology. (The European Commission, 2005).

[41] Sarah Spiekermann and Sergei Evdokimov. "Critical RFID in Privacy-Enhancing Technologies." *IEEE Security & Privacy* 7, 2 (2009).

by agents reduces or eliminates direct cognitive control by the users, hence misinterpretations of preferences or compromised agents can remain undetected, reducing the effectiveness of and trust in such technologies.

7.4.2 Identity Management

Of course, the principles developed for the protection of personal data pertain only to data for which a direct or indirect relation to a person exists or can be established. An obvious approach to avoid privacy problems is therefore to remove this link and to anonymize or pseudonymize the data. In the context of traditional information systems and of the Internet, a number of technical and organizational methods were developed to this end. In principle PETs can also be employed in ubiquitous computing environments. However, deployment of these technologies implies numerous and far reaching restrictions for the formation and functionality of ubiquitous computing systems. Anonymity only makes sense in the context of traditional forms of service provision, for instance, for ubiquitous access to information services, where the user must actively initiate inquiries. And even in these cases they may not feasible in ubiquitous computing environments due to the technical complexity involved in the provision of anonymity services.[42] A constituting feature of ubiquitous computing – personalized services autonomously adapted to individual needs and context – require at least pseudo-identities to which user profiles can be attached. Pseudo-identities could be generated, administered and, if wished, discarded in a user friendly way with the assistance of identity management technologies. This approach offers several advantages when applied in the frame of traditional information systems. The vocational sphere can be separated from private life by using different pseudo-identities, or the linking of data during long periods can be prevented or at least made more difficult by regular creation of new identities.

However, even in current information systems the effectiveness of this approach is quite restricted; increasingly powerful and efficient tools for the linking and analysis of large amounts of data facilitate the disclosure of pseudonyms and limit the protection offered by pseudo-identities. In order to offer any protection in ubiquitous computing environments at all, pseudo-identities must remain the exclusive link in interaction, a demand with far reaching restrictions for the technologies that can be deployed. It would for instance imply that biometric identification methods cannot be applied in parallel; audio or video information must not be stored, as these data could permit a later biometric identification. In addition, no location data may be collected since they allow, firstly, a concatenation in the case of changing pseudo-identities, and secondly, they render obsolete any attempt of anonymization or pseudonymization in case of adequate levels of precision or persistency of observation. Anonymity or pseudonymity can be realized within ubiquitous computing

[42]Langheinrich. "Privacy by Design - Principles of Privacy-Aware Ubiquitous Systems."

systems only if they are so radically restricted in their sensor abilities as well as in their potential of utilization that they would hardly conform to current visions of ubiquitous computing.

Identity management cannot offer sufficient levels of protection against the personalization of data; however, a comparable approach could contribute to mitigate the problem that individual consent in each single act of registration of data is practically impossible in a ubiquitous computing world. Personal digital assistants could store individual privacy preferences and act as agents for their owners, taking over the unrealistic task of permanently consenting or dissenting to data capture activities. Following the Platform for Privacy Preferences specification (P3P) of the World Wide Web Consortium (W3C)[43] as an example of the technical implementation, possible consent and the extent of data exchange would depend on the conformity of the published privacy policy of the respective ubiquitous computing subsystem with the preferences of the user. Although this concept is in principle transferable to ubiquitous computing applications, the failure of this approach in the Internet sphere in terms of the lack of interest and subsequent suspension of further development efforts on this specification by W3C, also raises doubts about the potential for success in ubiquitous computing environments, at least if is not supported by regulatory incentives or requirements to make use of PETs.

Independent of enforcement issues, such schemes should allow the users to keep full control over the profiles and offer a zero-option as an indispensable requirement to provide effective protection. The zero-option means that it must be possible to prevent any ubiquitous computing interactions; in a privacy compatible ubiquitous computing environment, a switched off or missing assistant is to be interpreted as a refusal of any data collecting activity. If several persons are present at the same place, the limitations imposed by the most restrictive participant must apply. This implies, for example, that recordings for the support of the own memory are only possible if all persons present agree to the recording. In addition, again restrictions on biometric identification or retention of data that could be analyzed by biometric methods would apply as well.

7.4.3 *Privacy Respecting Ubiquitous Recording*

Memory enhancement by ubiquitous recording, as an example, represents a rather simple application. The simplicity, compared to full ubiquitous computing systems, results from the single function of audio or video recording without any additional interpretation or learning, and from the fact that the devices or software are under the control of the participating persons.

---

[43] http://www.w3.org/P3P/

The briefly presented example of a cryptographic solution[44] for the elimination of threats to privacy related to ubiquitous recording is based on two principles, consent and confidentiality of policy. The first principle states that no recordings may be made without the consent of all persons present, and that no recordings will be released without consent of all persons involved. The second principle demands that a person's decision to grant or disallow the release of recordings should not be revealed to anyone else. The developed protocols and mechanisms perfectly fulfil the requirements in theoretical terms. For practical applications, they are, however, of little relevance and would bring about new threats to privacy.

Even in the primarily intended application area – privacy enhanced instant messaging software – the consent to use this software cannot prevent cheating and involves new risks. The cheating problem of itself is not a fact sufficient for a refusal of the implementation of privacy enhancing procedures, as, in most cases, they will not provide complete protection against abusive behaviour or covert surveillance. But of course the efforts for privacy protection need to be within a reasonable range compared to the ease of circumvention. More relevant are, however, the negative side effects, e.g. the need of collecting consent from all participating persons in order to be able to access the stored data excludes the anonymous use of the Internet. Additional problems arise when this model is extended to ubiquitous recording in everyday situations. Giving consent would consequently also mean agreeing to be permanently accessible to answer requests for the release of recordings. Rational decisions about granting or disallowing permission for the release may be impossible without prior access to the contents concerned. In other words, the intended service – a perfect memory – is a precondition for using this service in a privacy respecting manner, e.g. to decide whether to release or to block a conversation that happened at a certain date in a certain place.

The lack of practical usefulness of this approach to render ubiquitous recording less privacy invasive is even more obvious if this service is to be offered in business settings. Any participant who fears being disadvantaged by a release could block it and the death of an involved person or the (unfriendly) separation from a participating employee would turn the recordings into inaccessible items and render them worthless. Without back doors or master keys able to overrule blockings in certain circumstances, such systems would hardly find any commercial application. Implementing back doors would, however, sacrifice most of the privacy protecting features of this approach.

This model cannot secure privacy as there is no reasonable protection against covert recordings by non-compliant devices. On the contrary, it may convey an unjustified sense of security; it destroys privacy by limiting the possibilities of anonymous participation in discussions or conversations; and in exchange for a rather doubtful contribution to privacy, the suggested approach considerably

[44] John Alexander Halderman, Brent Waters, and Edward W. Felten. "Privacy Management for Ubiquitous Recording." http://www.cs.princeton.edu/~felten/privman.pdf

increases systems' complexity and makes ubiquitous recording technologies of little use for most of the intended purposes.

7.4.4 Digital Rights Management

Another class of attempts of creating privacy respecting pervasive computing systems is related to the application of Digital Rights Management (DRM) principles and technologies to personal data captured by ubiquitous computing systems. The basic idea behind such approaches is to encrypt the data and make the decryption of, and hence the access to these data dependent on the conformity with (pre-)defined conditions. In contrast to the case of ubiquitous recording discussed above these solutions are intended to be applicable to complex pervasive computing systems, too.

The advantages of the extension of DRM technologies to data captured by pervasive computing systems is that they – in theory – open up the design of more privacy friendly systems, e.g., by integrating new dimensions such as "proximity" or "location". As an example for proximity, memory aiding devices could operate or provide access to stored information only if their owner is present; locality, for instance, could mean that a conference table provides stored information on past discussions to persons present in the conference room, while information requests launched from remote locations are refused. Privacy tagging,[45] using meta-data to identify the kind of permissions assigned to these data could allow the free flow of the encrypted data and still safeguard that the data do not leave the information spaces to which they belong, nor that they transgress the usage boundaries assigned to them. DRM technologies could mandate that a particular privacy policy sticks to the data, travels with it, and decides who may use the data and how they may be used.[46]

In a strict sense, these examples demonstrate the lack of feasibility rather than the possibility to protect privacy in a world with ubiquitous computing systems as they cannot mitigate the primary problem, the permanent and unnoticeable recording itself, taking place without the knowledge and consent of the persons involved. Attempts to restrict the use of these data are either doomed to failure or bear new risks at a socially unacceptable level.

One reason for this pessimistic assessment pertains to the technical problems of encryption in ubiquitous computing systems. The strength of the encryptions

---

[45] Jeremy Goecks and Elizabeth Mynatt. "Enabling Privacy Management in Ubiquitous Computing Environments through Trust and Reputation Systems" (paper presented at the CSCW 2002 workshop Privacy in Digital Environments: Empowering Users, New Orleans, 16 November 2002).

[46] Xiaodong Jiang. "Safeguard Privacy in Ubiquitous Computing with Decentralized Information Spaces: Bridging the Technical and the Social" (paper presented at the Ubicomp 2002 Workshop on Socially-informed Design of Privacy-enhancing Solutions in Ubiquitous Computing, Göteborg, 29 September 2002).

applied will be limited by restrictions in processing capacities and power consumption of the super-miniaturized components. Even if the achieved levels may prevent brute force attacks at the time of capture, increases in processing power may render the protection ineffective within the anticipated periods of data retention. Furthermore, encryption is a double-edged sword; it facilitates both the hiding of content and the verification of identities.[47] A widespread diffusion of encryption technologies and infrastructures, accompanying attempts to make ubiquitous computing privacy compliant, would also create incentives to enforce identification for services previously anonymously accessible. A second reason for the pessimistic judgement on attempts to make ubiquitous computing privacy friendly, concerns new risks emerging from the application of DRM technologies in the context of ubiquitous computing systems. In order to be practically effective, all devices must necessarily be equipped with DRM capabilities implemented into the hardware, otherwise any protection could easily be overcome by using DRM-free devices for the invisible capture, copying, or re-recording, and the successive free distribution of personal data. In such a case, possible violations of privacy as a consequence of allowing DRM-free devices are certainly preferable compared to complete protection by DRM. The latter scenario would imply the creation of a perfect infrastructure for the control of the circulation of any information, which could be abused for censorship much more effectively than any attempt to dominate and influence public opinion in the past.

7.4.5 Legal Proposals

In contrast to the many, very detailed and elaborated technical attempts to make ubiquitous computing privacy compliant, the proposals for legal and regulatory reform usually address this issue on a more general level. Several factors contribute probably to this relative vagueness. One is related to a general observation, that the speed of technical progress is too high to be reflected immediately in the legislation as this process involves a chain of complex activities, ranging from the detection of and awareness building of unresolved problems, discussions of possible causes and the identification of feasible legal actions, and finally, negotiations for balancing conflicting political interest and the time required for decision making in proper democratic processes. A second factor might be regarded as an intended vagueness, or in positive words, general applicability of laws. This built-in flexibility is a precondition that specific regulations can be applied in different situations, and thus also can take into account societal change and technological progress without the need of continuous reformulation. This vagueness does, however, also imply that a time-consuming process of case law decisions is required before concrete interpretations and implications can be provided. In the case of ubiquitous computing,

[47] Lawrence Lessig. "The Architecture of Privacy" (paper presented at the Taiwan Net '98, Taipei, March 1998).

the magnitude of challenges resulting from this new paradigm is certainly a further factor for the lack of concrete and convincing proposals for effective regulations. In addition, the gradual diffusion of ubiquitous computing environments may also erode elements of the legal practice of privacy protection. The European Court of Human Rights has, for instance, introduced the term "reasonable expectation of privacy"[48] in case law decisions. What is reasonable depends largely on the technical capacities of monitoring and the interest and will to make use of these possibilities. The increasing dominance of concerns for inner security in the political debate of the early twenty-first century, largely neglecting the indispensable role of privacy for individual security, is a further central factor contributing to changes in this respect, turning the reasonable expectation of privacy into an "expectation of being monitored".[49] Obviously a more stable replacement for the notion of "reasonable expectation of privacy", invariant to technical progress or to volatile political priorities, will be required for adequate protection of the fundamental right to privacy. The Charter of Fundamental Rights of the European Union[50] refers in Article 8 to established principles of data protection, such as purpose specification or consent, and provides a more detailed basis for future decisions. This charter will probably also considerably reframe the legal discussion as it addresses basic contradictions between these provisions and the ubiquitous computing paradigm on the level of a fundamental right.

Technical progress in information and communication technologies, new forms of service provision or the emergence of innovative forms of using technical platforms, e.g. use of the Internet for peer-to-peer data sharing or for the establishment of social networks, frequently create new problems, requiring new regulations or new or extended interpretations of existing ones; a process that is standard in legislation, e.g. the issuing of new directives or the elaboration of opinions by the Article 29 Working Party at the EU-level.

The fundamental contradictions between ubiquitous computing and principles of data protection raise correspondingly essential doubts, as to the extent to which a regulatory framework, originating from a mainframe computer paradigm,[51] can be adapted and reformed to cope with the new challenges. The conclusion that the usual evolutionary approach of legal reform cannot be followed anymore and that

[48]This term originates from case law on the Fourth Amendment to the United States Constitution. In Katz v. United States, 389 U.S. 347 (1967) Justice Harlan issued a concurring opinion articulating the two-part test later adopted by the U.S. Supreme Court as the test for determining whether a police or government search is subject to the limitations of the Fourth Amendment: (1) governmental action must contravene an individual's actual, subjective expectation of privacy; (2) and that expectation of privacy must be reasonable, in the sense that society in general would recognize it as such. See http://en.wikipedia.org/wiki/Expectation_of_privacy for more details.

[49]Paul De Hert et al. "Legal Safeguards for Privacy and Data Protection in Ambient Intelligence." *Personal and Ubiquitous Computing* 13 (2008).

[50]European Union. "Charter of Fundamental Rights of the European Union."

[51]Bizer et al. "Technikfolgenabschätzung Ubiquitäres Computing und Informationelle Selbstbestimmung." 212.

"... the unprecedented character of some of the issues awaiting regulation in a world of ubiquitous computing and ambient intelligence might well render former anchors irrelevant"[52] opens a wide range of conceivable reactions. We exclude the, in principle, possible, although unrealistic and undesirable options to prohibit ubiquitous computing completely, on the one hand, or to abandon the right to privacy, on the other, as both extremes would be incompatible with democratic societies; notwithstanding that, the protection of democracy may require that substantial restrictions be imposed on the implementation of ubiquitous computing systems.

Manifold safeguards[53] are conceivable which can contribute to the mitigation of negative impacts of ubiquitous computing. As with proposals related to technology, they can either provide only fragmentary protection or are incompatible with the basic principles of ubiquitous computing; and they again contain new risks. Contradictions to the collection limitation and purpose specification principle belong to the most serious and urgent weaknesses of current privacy protection frameworks when applied to the new technological paradigm.

Obviously, the minimization of the generation and collection of personal data is already today increasingly difficult to enforce; in ubiquitous computing environments this endeavour will become a hope – and a meaningless concept. Consequently regulatory attention could be shifted to the use, particularly to the prevention of abuse, of personal data or the knowledge gained from them. This approach would require the reconceptualization of privacy in terms of access to knowledge instead of data and protection against unfair use of that knowledge.[54] Although offering additional and compensating protection, apparently it might prove quite difficult first, to establish clear criteria for the differentiation between permitted use and abuse, and second, to detect and to provide sufficient evidence for actual violations. The enforceability and effectiveness of such regulatory reform will critically depend on the ability to increase the transparency of data processing, analysis and transfers; there is little reason for optimism, taking into account the increasing information asymmetry and the complexity of ubiquitous computing technologies. "How to assign responsibilities in computer-controlled environments where it becomes impossible to locate and isolate the cause of potential damages resulting from combined agencies originating from computer hardware and software, networks, and human beings?"[55] is a general enforcement problem, already existing today, that will be considerably aggravated by ubiquitous computing. Also,

---

[52]Rouvroy. "Privacy, Data Protection, and the Unprecedented Challenges of Ambient Intelligence." 20.

[53]See David Wright et al., editors. *Safeguards in a World of Ambient Intelligence*, Vol. 1 The International Library of Ethics, Law and Technology Springer, New York Inc, 2008. for a comprehensive overview of legal as well as technological and socio-economic safeguards.

[54]Hildebrandt, M. "Profiling and the identity of the European citizen." In *Profiling the European Citizen: Cross-Disciplinary Perspectives*, edited by M. Hildebrandt and S. Gutwirth. Dordrecht: Springer, 2008, 305.

[55]Rouvroy. "Privacy, Data Protection, and the Unprecedented Challenges of Ambient Intelligence." 18.

attempts to embody the protection against abuse of knowledge technologically in the form of sticky policies depend on sufficient levels of transparency. In addition they would share the general limitations and risks of DRM approaches, discussed in the previous section.[56]

A shift in the focus of attention from data to knowledge necessarily emerges from profiling technologies. They can generate highly sensitive information "out of seemingly trivial and/or even anonymous data.[57]" Group profiles, created on the basis of anonymous data and attributed to specific persons on behavioural or biometric data, without any need for any link to identity or identifying data of this person, could become a central part of differentiated service provision or discrimination in ubiquitous computing environments, without the involvement of any of the data protecting provisions as they only apply to personal or personally identifiable data. The only provision applicable in this context, Article 15 on automated individual decisions of the Data Protection Directive,[58] does not provide sufficient protection for citizens against this practice. This lack of protection is occasionally already virulent today, e.g. influencing credit conditions or insurance contracts; it will increase in importance and impact with the expansion of ubiquitous computing and the diffusion of profiling into everyday activities and environments.

An increase in the transparency of data processing, and the use of information generated from this processing, is a prerequisite for conceivable benefits resulting from the suggested shift in the focus of regulation and limitation from data collection to knowledge use. Transparency tools[59] are accordingly regarded as a key element of future legal frameworks capable of restraining the privacy threats of ubiquitous computing. Whereas improved transparency is certainly an important element of data protection regulations, specifically in the support of their enforcement, it is rather questionable to what extent transparency can fulfil the attributed key role under the new paradigm of information technologies. More transparency would – especially in ubiquitous computing environments – contradict the aim of unobtrusiveness and undoubtedly overburden the attention of any user within a short time. The transfer of handling this information and subsequent decision taking to personal digital assistants or agents is coupled with a corresponding transfer of individual autonomy; in addition, the privacy preferences required for the operation of these

---

[56]Digital Rights Management, if implemented voluntarily and incompletely adds considerably to system complexity and little to the protection of privacy, if (enforced to be) embedded generally, Digital Rights Management contains unacceptable risks for the fundamental right of freedom of expression and information.

[57]Mireille Hildebrandt. "Who Is Profiling Who? Invisible Visibility," in *Reinventing Data Protection?*, edited by S. Gutwirth, et al. Springer Netherlands, 2009), 240.

[58]Directive 95/46/Ec of the European Parliament and of the Council of 24 October 1995 on the Protection of Individuals with Regard to the Processing of Personal Data and on the Free Movement of Such Data.

[59]Transparency tools comprise transparency rights and transparency enhancing technologies to support these rights. (See Hildebrandt and Koops, eds., *FIDIS Deliverable D7.9: A Vision of Ambient Law*. or De Hert et al., "Legal Safeguards for Privacy and Data Protection in Ambient Intelligence." for a brief summary of this concept.)

agents themselves constitute personal data of a potentially very sensitive nature. Furthermore, transparency is not sufficient to create freedom of choice in the use that can be made of the data and knowledge gained for the data subjects in general; it can improve information symmetry but it cannot remove asymmetries of power, based on institutional relations like employee-employer dependencies or discrepancies in legal powers between citizens and state, or that are simply based on the knowledge created by ubiquitous computing systems themselves.

A careful consideration of potential impacts on the remaining elements of the regulatory framework is indispensable, particularly when taking into account the limitations of this approach in providing sufficient protection of the right to privacy. It appears to be obvious that the existing regulations need to be supplemented by provisions limiting the use and protecting against the abuse of generated knowledge if the principles of minimization and purpose specification or the requirement of consent related to the collection of data are no longer applicable under the new technological paradigm. The inherent danger of this approach is, however, that it does not only complement the existing protection framework but it might also replace the more direct and effective principles even if they were still applicable. An exemption of ubiquitous computing service providers from the main restrictions of the Data Protection Directive contains immense incentives to declare all information services as belonging to the new domain or to enrich them with ubiquitous computing elements just to avoid being subjected to the basic limitations on data collection. It remains an open question how layered approaches could be designed, implemented and enforced to protect the right to privacy in the future. The Charter of Fundamental Rights of the European Union clearly limits the scope of legal reforms as it reconfirms the central role of purpose specification and of consent of the concerned person in the protection of personal data.[60]

7.5 Concluding Reflections

Ubiquitous computing will erode all central pillars of current privacy protection and we must be aware that "... the world we are entering is about to change these architectures of privacy more completely and more extensively than any such change that we have seen to date".[61] Technical options offering effective protection of privacy are available in principle; their integration into ubiquitous computing systems, however, requires far reaching restrictions to the functionality of these systems and the abandonment of ideas at the core of the ubiquitous computing paradigm. Privacy friendly design, which does not conflict with the framework of ubiquitous computing, is, on the other hand, not able to promise more than marginal improvements for the protection of privacy. Imposing regulatory restrictions on the processing and use

[60] Article 8, European Union. Charter of Fundamental Rights of the European Union.

[61] Lessig. "The Architecture of Privacy." 7.

of the data generated by ubiquitous computing infrastructures is, in principle, necessary and meaningful, however, it may prove very difficult to control and accomplish such restrictions due to the ubiquitous though invisible nature of this technology. "The unavoidable cost of entering in an AmI [Ambient Intelligence] world, and the very condition of possibility of such a world, appears to be the loss of control over personal information: the constitutive ideas of AmI, such as pervasiveness, invisibility of information systems, constant and automatic recording of events etc. render highly implausible that the user will retain control over what and how information is processed."[62] To sacrifice the right to privacy also implies the sacrifice of a central human right and the fundament of democratic societies, with enormous negative consequences for individuals and society. It implies giving up a central precondition of individual and political autonomy, to endanger the very basis of liberal societies and democracy. It will also jeopardize societal sustainability, in terms of the long term potential for individual, societal and democratic development, the creation of and adaptation to new needs and opportunities.

In principle, the ability to adapt would also include the option to give up fundamental rights; privacy could accordingly be abandoned in favour of the opportunities offered by the new paradigm of information systems. However, in the case of privacy the relationship is much more complex. On the one hand, ubiquitous surveillance creates enormous pressure to behave in a "normal" way and not to leave the standardized paths of widely accepted social behaviour. On the other hand, social innovation requires deviations by members of society, both in order to invent new forms of social interactions and to distribute innovative mechanisms throughout society. Ubiquitous computing and the consequently emerging surveillance society might permanently destroy the fundaments for societal renewal.

These fundamental inconsistencies between the visions of ubiquitous computing and the foundations of a central human right in democratic societies implies that maintaining the right to privacy, and the many indispensable values depending on this right, is a challenging and complex task involving several levels of activities. It will of course, be necessary to respect privacy already in the design or to add PETs where possible and to create new regulatory fundaments of privacy where the old ones are becoming inadequate. These efforts may not be sufficient. In addition, it will probably also be necessary to set limits to the application and the implementation of such systems. Specifically, one needs to take into account the lessons learned from environmental pollution: end of pipe approaches are more costly and less efficient than avoiding emissions at the source, and some damage is irreversible. More research and broad political and public debate will be needed to make technology development serves humans rather than force humans to become the servants of technology and to avoid the omnipresence of ubiquitous computing ending in their omnipotence.

[62]Rouvroy. Privacy, Data Protection, and the Unprecedented Challenges of Ambient Intelligence. 6.

References

Adams, A. and M.A. Sasse. "Privacy in Multimedia Communications: Protecting Users, Not Just Data." In *People and Computers XV – Interaction Without Frontiers. Joint Proceedings of HCI2001*, edited by A. Blandford and J. Vanderdonkt 2001. Springer Verlag, London.

Article 29 – Data Protection Working Party. "Working Document on Data Protection Issues Related to RFID Technology." (The European Commission, 2005).

Victoria Bellotti and Abigail Sellen. "Design for Privacy in Ubiquitous Computing Environments." in *Proc. Of the European Conference on Computer-Supported Cooperative Work* (1993).

Bizer, J. et al. "Technikfolgenabschätzung Ubiquitäres Computing und Informationelle Selbstbestimmung." 212. (2006).

Bentham, J. "Panopticon: Or, the Inspection-House : Containing the Idea of a New Principle of Construction Applicable To ... Penitentiary-Houses, Prisons, Houses of Industry, Work-Houses, Poor-Houses, Manufactories, Mad-Houses, Hospitals, and Schools. With a Plan of Management Adapted to the Principle." in *a series of letters, written ... 1787, from Crecheff ... to a friend in England* (Dublin: Thomas Byrne, 1791).

Bizer, J. et al. "Technikfolgenabschätzung Ubiquitäres Computing Und Informationelle Selbstbestimmung." in Studie im Auftrag des Bundesministeriums fur Bildung und Forschung. Online: https://www.datenschutzzentrum.de/taucis/ita_taucis.pdf 2006.

Čas, J. "Privacy in Pervasive Computing Environments – a Contradiction in Terms?" *IEEE Technology and Society Magazine 24*, no. 1 (2004).

Čas, J. "UC – Ubiquitous Computing oder Ubiquitous Control?" In *Der Mensch Im Netz – Ubiquitous Computing*, edited by B. Britzelmaier, S. Geberl and S. Weinmann. *Reihe Wirtschaftsinformatik* Stuttgart, Teubner. 2002.

Clarke, R. "Information Technology and Dataveillance." http://www.rogerclarke.com/DV/CACM88.html

De Hert, P. et al. "Legal Safeguards for Privacy and Data Protection in Ambient Intelligence." *Personal and Ubiquitous Computing* 13 (2008).

Directive 95/46/Ec of the European Parliament and of the Council of 24 October 1995 on the Protection of Individuals with Regard to the Processing of Personal Data and on the Free Movement of Such Data.

European Union. "Charter of Fundamental Rights of the European Union."

Foucault, M. *Discipline and Punish: The Birth of the Prison*. London, Penguin. 1977.

Goecks, J., and E. Mynatt. "Enabling Privacy Management in Ubiquitous Computing Environments through Trust and Reputation Systems." (paper presented at the CSCW 2002 workshop Privacy in Digital Environments: Empowering Users, New Orleans, 16 November 2002).

Halderman, J.A. *Digital Privacy-Rights Management for Ubiquitous Recording*. Princeton University. 2003.

Halderman, J.A. Brent Waters, and Edward W. Felten. "Privacy Management for Ubiquitous Recording." http://www.cs.princeton.edu/~felten/privman.pdf.

Hildebrandt, M. Profiling and the identity of the European citizen. In *Profiling the European Citizen: Cross-Disciplinary Perspectives*, edited by M. Hildebrandt and S. Gutwirth. Dordrecht: Springer, 2008, pp. 303–344.

Hildebrandt, M. "Who Is Profiling Who? Invisible Visibility." In *Reinventing Data Protection?*, edited by S. Gutwirth, et al. Springer, Netherlands, 240. (2009).

Hildebrandt, M., and B.-J. Koops, editors, *FIDIS Deliverable D7.9: A Vision of Ambient Law* (2007).

Iachello, G. "Protecting Personal Data: Can It Security Management Standards Help?" (paper presented at the 19th Annual Computer Security Applications Conference, Las Vegas, Dec 2003).

IST Advisory Group. "Ambient Intelligence: From Vision to Reality. For Participation in Society & Business." (2003), 8.

Kuner, C. "Proportionality in European Data Protection Law and Its Importance for Data Processing by Companies." *Privacy & Security Law Report* 07, no. 44 (2008).

Langherinrich, M. "Privacy by Design – Principles of Privacy-Aware Ubiquitous Systems". In *Proceedings of the 3rd International Conference on Ubiquitous Computing*, GA. London: Springer (2001): 273–291.

Lessig, L. "The Architecture of Privacy" (paper presented at the Taiwan Net '98, Taipei, March 1998).

Rouvroy, A. "Privacy, Data Protection, and the Unprecedented Challenges of Ambient Intelligence." *Studies in Ethics, Law, and Technology* 2, no. 1 (2008).

Rouvroy, A., and Y. Poullet. "The Right to Informational Self-Determination: Reassessing the Importance of Privacy for Democracy." In *Reinventing Data Protection?*, edited by S. Gutwirth, et al. Springer, Netherlands, 76. (2009).

Spiekermann, S., and S. Evdokimov. "Critical RFID in Privacy-Enhancing Technologies." *IEEE Security & Privacy* 7, no. 2 (2009).

Weiser, M. "The Computer for the 21st Century." *Scientific American* 265, no. 3 (1991).

Wright, D. et al. editors, *Safeguards in a World of Ambient Intelligence*, Vol. 1 The International Library of Ethics, Law and Technology Springer, New York Inc). (2008).

Jiang, X. "Safeguard Privacy in Ubiquitous Computing with Decentralized Information Spaces: Bridging the Technical and the Social." (paper presented at the Ubicomp 2002 Workshop on Socially-informed Design of Privacy-enhancing Solutions in Ubiquitous Computing, Göteborg, 29 September 2002).

Chapter 8
EU PNR: European Flight Passengers Under General Suspicion – The Envisaged European Model of Analyzing Flight Passenger Data

Franziska Boehm

8.1 Introduction

One can still remember the outcry of the European Union caused by the American decision to create a passenger name record (PNR) control system to analyze flight passenger information. Policy- and lawmakers all over Europe turned away in shame and declared – not without a certain form of righteous indignation – that such a system of flight passenger screening would never be feasible in an EU that is devoted to the protection of fundamental rights of its citizens.

Now, the EU is on the verge of adopting its own PNR system (EU-PNR proposal), imitating the existing PNR systems of the USA, Canada, and Australia.

At a time when the EU, at an international level, is usually considered as being equipped with a strong fundamental rights framework, it risks losing its credibility if it approves the EU-PNR proposal while simultaneously criticizing the US data protection system.

Questions arise whether the processing of data of thousands of individuals due to a general suspicion, and its application in a wide range of cases, meet the essential data protection requirements laid down in the Lisbon Treaty and in the case law of the European Court of Human Rights (ECtHR).

This contribution seeks to find answers to the questions the EU-PNR proposal raises by looking at the legal background of the European PNR system (Section 8.2) and its legal compliance with basic European data protection rules, particularly with the requirements articulated in the ECtHR's case law (Section 8.3). The objective is to compare the EU-PNR proposal with the criteria the ECtHR developed during the recent years in the context of security, privacy and data protection cases. Towards

F. Boehm (✉)
University of Luxembourg, Luxembourg
e-mail: franziska.boehm@uni.lu

Franziska Boehm is working as a research assistant at the University of Luxembourg where she is also preparing her PhD thesis. E-mail: franziska.boehm@uni.lu; this contribution is based on a presentation held at the CPDP Conference 2010 in Brussels.

S. Gutwirth et al. (eds.), *Computers, Privacy and Data Protection: an Element of Choice*, DOI 10.1007/978-94-007-0641-5_8, © Springer Science+Business Media B.V. 2011

this aim, based on the criteria the ECtHR specified in its respective judgments, the contribution studies the provisions of the EU-PNR proposal in detail.

The analysis shows that under the cloak of law enforcement and anti-terrorism measures, the processing and transmission of personal data in the new PNR system will lead to complex legal structures that have serious consequences on the data protection regime in the EU and will, if adopted, influence the daily life of many EU citizens.

In addition to the fundamental rights analysis, the EU-PNR proposal is briefly compared to other instruments, such as the Data Retention Directive 2006/24 and the EU-US PNR Agreement, which also obligate private actors to participate in law enforcement related activities (Section 8.4).

Although the idea of the European PNR system is quite far advanced, this contribution proposes to considerably rework the EU-PNR proposal and offers some suggestions for improvement (Section 8.5).

8.2 Legal Background and Similarity Between the EU-PNR Proposal and the US-PNR System

Four months after the conclusion of the EU-US Agreement on the exchange of PNR[1], the Commission presented its own EU-PNR proposal (COM (2007) 654).[2] Meanwhile, the Council issued a revised version slightly differing from the Commission's initial proposal.[3] Both versions contain questionable legal provisions whose main statements are summarized and analyzed in the following.[4]

Based on comparable security policy ambitions as the American example in 2007, the provisions of the EU-PNR proposal parallel the wording of the US predecessor in important passages.[5]

---

[1]Agreement between the European Union and the United States of America on the processing and transfer of Passenger Name Record (PNR) data by air carriers to the Unites States Department of Homeland Security (DHS), OJ 2007, L-204/18. In the following: EU-US PNR agreement. A similar agreement exists with Canada: Agreement between the European Community and the Government of Canada on the processing of Advanced Passenger Information and Passenger Name Record data, OJ 2006, L-82/15; on the background of this Agreement and its predecessors: Vagelis Papakonstantinou and Paul De Hert, "The PNR Agreement and Transatlantic anti-terrorism Cooperation: No firm human rights framework on either side of the Atlantic," *Common Market Law Review* 46, 3 (2009): 885–919; Mario Mendez, "Passenger Name Record Agreement, European Court of Justice," *European Constitutional Law Review* 3 (2007): 127–147.

[2]Proposal for a Council Framework Decision on the use of Passenger Name Record (PNR) for law enforcement purposes, COM (2007) 654 from 6 November 2007.

[3]EU-PNR proposal in its latest version, proposal for a Council Framework Decision on the use of Passenger Name Record (PNR) for law enforcement purposes, Council doc. 5618/2/09, interinstitutional file 2007/0237 (CNS) from 29 June 2009.

[4]In each case, reference is made to the specific version the text is referring to.

[5]Patrick Pawlak, "Made in USA? The influence of the US on the EU's data protection regime," *Centre of European Policy Studies, Liberty and Security in Europe, Justice and Home Affairs*

It is worth mentioning that at the time of conclusion, the EU-US PNR agreement was heavily disputed and criticized by many scholars.[6] Criticism focused in particular on the legal basis, the lack of data protection guarantees by the US, and the large volume of transmitted data.[7]

Basically, the EU-US PNR Agreement of 2007 regulates the transmission of 19 data categories which may each entail a variety of different data elements accumulated under one wide-ranging term.[8] In this way, up to 69 different data elements, all containing information related to a plane trip to the USA, are transmitted.[9] They refer to, for instance, "all available contact information", "all available payment/billing information" or "travel agency" and may include, amongst others, the passenger's credit card number, the address, telephone number or information about check-ins, travel status, accompanying persons as well as the name of the travel agency.[10]

---

section, pp. 4-9 (2009), http://www.ceps.be/book/made-usa-influence-us-eu%E2%80%99s-data-protection-regime.

[6]For a profound analysis with further references see: Vagelis Papakonstantinou, and Paul De Hert, "The PNR Agreement and Transatlantic anti-terrorism Cooperation: No firm human rights framework on either side of the Atlantic," *Common Market Law Review* 46 (3) (2009): 885–919; Mario Mendez, "Passenger Name Record Agreement, European Court of Justice," *European Constitutional Law Review* 3 (2007): 127–147.

[7]Vagelis Papakonstantinou, and Paul De Hert, "The PNR Agreement and Transatlantic anti-terrorism Cooperation: No firm human rights framework on either side of the Atlantic," *Common Market Law Review* 46 (3) (2009): 885–919; Mario Mendez, "Passenger Name Record Agreement, European Court of Justice," *European Constitutional Law Review* 3 (2007): 127–147.

[8]Compare for instance category 17 of the agreement between the European Union and the United States of America on the processing and transfer of Passenger Name Record (PNR) data by air carriers to the Unites States Department of Homeland Security (DHS), OJ 2007, L-204/18, "general remarks including OSI, SSI and SSR information", which means "other information", "sensitive security information" and "special service requests" or category 18 referring to "any collected APIS information" which refers to the Advanced Passenger Information System information which includes further personal information such as passport information, country and city of residence as well as first address in the USA.

[9]Examples: date of reservation/issue of ticket, date(s) of intended travel, name(s), available frequent flier and benefit information (i.e. free tickets, upgrades, etc.), other names on PNR, including number of travelers on PNR, all available contact information (including originator information meaning address and telephone number at the final destination), all available payment/billing information linked to the travel transaction, travel itinerary for specific PNR, travel agency/travel agent, code share information, split/divided information, travel status of passenger (including confirmations and check-in status), ticketing information, including ticket number, one-way tickets and Automated Ticket Fare Quote, all baggage information, seat information, including seat number. See also: Edward Hasbrouck, comment on "What's in a passenger name record (PNR)?," http://www.hasbrouck.org/articles/PNR.html (accessed February 05, 2010).

[10]See data categories 7, 8 and 10 of the agreement between the European Union and the United States of America on the processing and transfer of Passenger Name Record (PNR) data by air carriers to the Unites States Department of Homeland Security (DHS), OJ 2007, L-204/18.

Regrettably, the EU-PNR proposal refers to almost the same 19 wide-ranging data categories as the EU-US PNR agreement.[11] As *Pawlak* concluded in this context, the American security policy after 2001 has had a "spill-over effect" in Europe.[12] The EU, and in particular the Commission, reacted and responded to the security related US legislation with no ideas of its own and without taking into account the European (data protection) legislation. The similarity between the EU-PNR proposal and the EU-US PNR Agreement is striking. Not only the amount of data elements to be stored is nearly identical, but also the provisions relating to the purpose, to the data retention period as well as the access conditions are very similar.[13]

Originally, in contrast to the American example, the Commission proposed to limit the scope of the EU-PNR proposal to the prevention of terrorism and organised crime.[14] However, after the recent modifications by the Council, the European PNR are – just as in the EU-US PNR Agreement – intended to be additionally used to investigate other crimes such as illegal immigration.[15]

To achieve these objectives so called Passenger Information Units (PIU) are to be established in each Member State in order to analyze, evaluate and transfer the collected data to the Member States' law enforcement authorities.[16] The information

---

[11] Compare the list of data categories entailed in the annex of the proposal for a Council Framework Decision on the use of Passenger Name Record (PNR) for law enforcement purposes, COM (2007) 654 from 6 November 2007 with the list of data categories of the agreement between the European Union and the United States of America on the processing and transfer of Passenger Name Record (PNR) data by air carriers to the Unites States Department of Homeland Security (DHS), OJ 2007, L-204/18.

[12] Patrick Pawlak, "Made in USA? The influence of the US on the EU's data protection regime," *Centre of European Policy Studies, Liberty and Security in Europe, Justice and Home Affairs section*, p. 9 (2009), http://www.ceps.be/book/made-usa-influence-us-eu%E2%80%99s-data-protection-regime.

[13] Patrick Pawlak, "Made in USA? The influence of the US on the EU's data protection regime," *Centre of European Policy Studies, Liberty and Security in Europe, Justice and Home Affairs section*, pp. 6–7 (2009), http://www.ceps.be/book/made-usa-influence-us-eu%E2%80%99s-data-protection-regime, compare in this context articles 1, 9, 11 as well as the list of stored items in the annex of the Proposal for a Council Framework Decision on the use of Passenger Name Record (PNR) for law enforcement purposes, COM (2007) 654 from 6 November 2007 with points I, III, IV and VII Agreement between the European Union and the United States of America on the processing and transfer of Passenger Name Record (PNR) data by air carriers to the Unites States Department of Homeland Security (DHS), OJ 2007, L-204/18.

[14] Article 1 of the Proposal for a Council Framework Decision on the use of Passenger Name Record (PNR) for law enforcement purposes, COM (2007) 654 from 6 November 2007.

[15] Compare in particular the modification of article 1 of the Proposal for a Council Framework Decision on the use of Passenger Name Record (PNR) for law enforcement purposes, COM (2007) 654 from 6 November 2007 with the EU-PNR proposal in its latest Council version, Council doc. 5618/2/09, interinstitutional file 2007/0237 (CNS) from 29 June 2009.

[16] Article 3 EU-PNR proposal in its latest Council version, Council doc. 5618/2/09, interinstitutional file 2007/0237 (CNS) from 29 June 2009.

that is gathered shall be stored for 10 years; after 3 years the data shall be transferred from an active database into a database with special access conditions.[17]

The EU-PNR proposal will affect all passengers who land in the territory of one Member State originating from a third country or who depart from a Member State's territory to a non-EU country, including any transfer or transit flights.[18] The extension of the scope to intra-Community flights is also under discussion.[19]

With regard to the amount of stored data, the consequences of the adoption of the EU-PNR proposal would go far beyond existing possibilities of data processing and evaluation. The EU-PNR proposal risks to lead to an unprecedented data exchange in the EU between air carriers, the PIUs and national law enforcement agencies. The data of over 10 million passengers would be transferred, stored and analyzed each year.[20] Enormous databases and huge amounts of data analyses would be the consequence. Hence, it is highly questionable whether the EU-PNR proposal complies with European fundamental rights, in particular with data protection and privacy rights.

8.3 Compliance of the EU-PNR Proposal with European Data Protection Rules

The European Data Protection Supervisor, the EU Fundamental Rights Agency, and the European Parliament issued opinions on this project, profoundly criticizing the envisaged measures.[21] The European Data Protection Supervisor assumes that the EU-PNR proposal constitutes a "further step in a movement towards a routine collection of data of individuals who are in principle not suspected of any crime".[22] The main arguments of the three mentioned actors and the compliance of the EU-PNR proposal with European data protection and privacy principles are analyzed hereinafter.

---

[17] Article 9 EU-PNR proposal in its latest Council version, Council doc. 5618/2/09, interinstitutional file 2007/0237 (CNS) from 29 June 2009.

[18] Article 2 (b) EU-PNR proposal in its latest version, Council doc. 5618/2/09, interinstitutional file 2007/0237 (CNS) from 29 June 2009.

[19] See Recital (7) and article 17 of the EU-PNR proposal in its latest version, Council doc. 5618/2/09, interinstitutional file 2007/0237 (CNS) from 29 June 2009.

[20] See statistics: Eurocontrol: http://www.eurocontrol.int/corporate/public/faq/about_us_faq.html#qa12 (August, 25, 2010).

[21] European Data Protection Supervisor, Opinion on the draft proposal for a Council Framework Decision on the use of Passenger Name records (PNR) for law enforcement purposes, OJ 2008, C-110/01, in the following EDPS opinion; European Parliament, Resolution of 20 November 2008 on the proposal for a Council Framework Decision on the use of Passenger Name records (PNR) for law enforcement purposes, B6-0615/2008, in the following EP resolution; Opinion of the European Union Agency for Fundamental Rights on the Proposal for a Council Framework Decision on the use of Passenger Name Record (PNR) data for law enforcement purposes, 28 October 2008, in the following FRA's opinion.

[22] EDPS opinion, point 8.

In addition to fundamental rights problems, a second problem worth mentioning concerns the procedural consequences of the data exchange between private actors (air carriers) and public actors (the PIU's and law enforcement authorities). When the data are transferred from the air carriers to the PIU's, the applicable law changes from private to public law. This problem will be discussed further below in Section 8.4.

8.3.1 Reference Instruments and European Data Protection and Privacy Rules

Prior to evaluating the details of the EU-PNR proposal, the standards, against which the EU-PNR proposal is judged, are briefly illustrated. Considering the intended accession of the EU to the ECHR, particular attention is thereby paid to the ECtHR's interpretation of article 8 European Convention of Human Rights (ECHR).

Due to the former pillar structure, data processing in third pillar security related matters was not included in the relatively comprehensive data protection framework of the first pillar. While, since 1995, the Data Protection Directive 95/46[23] accompanied by sector specific first pillar instruments[24] has established a wide-ranging data and privacy protection for individuals in an economic related first pillar context, data processing for security purposes carried out by governmental law enforcement agencies was excluded from the scope of Directive 95/46.[25]

For a long time, data protection in the framework of former third pillar matters was therefore covered by public international law instruments instead of EU law, most notably by the instruments of the Council of Europe.[26] The ECHR and its interpretation by the Strasbourg Court as well as Convention No. 108 for the

[23]Directive 95/46/EC of the European Parliament and of the Council of 24 October 1995 on the protection of individuals with regard to the processing of personal data and on the free movement of such data, OJ 1995, L-281/31.

[24]For instance: Directive 97/66/EC of the European Parliament and of the Council of 15 December 1997 concerning the processing of personal data and the protection of privacy in the telecommunications sector, OJ 1998, L-24/1.

[25]Article 3 (2) Directive 95/46/EC of the European Parliament and of the Council of 24 October 1995 on the protection of individuals with regard to the processing of personal data and on the free movement of such data, OJ 1995, L-281/31. This statement was clarified by the ECJ in the famous PNR case: joined cases C-317/04 and C-318/04, *Parliament v. Council*, [2006], ECR I-4721.

[26]Mainly by article 8 ECHR, the Convention No. 108 of the Council of Europe for the protection of individuals with regard to automatic processing of personal data from 28 January 1981, the additional protocol to the Convention for the protection of individuals with regard to automatic processing of personal data regarding supervisory authorities and transborder data flows from 2004 and Recommendation R (87) 15 of the Committee of Ministers to the Member States regulating the use of personal data in the police sector, adopted 17 September 1987; compare for a profound analysis: Siemen, Birte (2006). *Datenschutz als europäisches Grundrecht*. Berlin: Duncker & Humblot.

protection of individuals with regard to automatic processing of personal data[27], its respective additional protocols[28] and Recommendation (87) 15 regulating the use of personal data in the police sector[29] built the reference instruments for security-related data processing in the EU. This significant historical background is reflected in the EU-PNR proposal referring to both Council of Europe instruments, to the Convention No. 108 as well as to Recommendation (87) 15.[30]

However, since the adoption of the Framework Decision "on the protection of personal data in the framework of police and judicial cooperation in criminal matters"[31] (DPFD) in 2008, certain minimum requirements also apply in the field of security-related data processing at the EU level.[32] The adoption of the Lisbon Treaty 1 year later additionally strengthened the protection of personal data in this area[33] in two ways: first, its article 16 (TFEU) guarantees the right to the protection of personal data to "everyone" and second, article 6 (3) TEU stipulates that the Charter of Fundamental Rights, which shall have the same legal value as the EU treaties, is additionally applicable when it comes to fundamental rights protection in the EU.[34]

As a starting point, article 8 of the Charter of Fundamental Rights provides for the minimum data protection requirements which encompass the basic guarantees of European data protection principles. Article 8 of the Charter of Fundamental Rights reads as follows:

1. Everyone has the right to the protection of personal data concerning him or her.
2. Such data must be processed fairly for specified purposes and on the basis of the consent of the person concerned or some other legitimate basis laid down by law. Everyone has the right of access to data which has been collected concerning him or her, and the right to have it rectified.
3. Compliance with these rules shall be subject to control by an independent authority.

---

[27] Convention No. 108 of the Council of Europe for the protection of individuals with regard to automatic processing of personal data from 28 January 1981.

[28] In particular the additional protocol to Convention for the protection of individuals with regard to automatic processing of personal data regarding supervisory authorities and transborder data flows, which entered into force in 2004.

[29] Recommendation R (87) 15 of the Committee of Ministers to the Member States regulating the use of personal data in the police sector, adopted 17 September 1987.

[30] Recital (10b) and article 11 (1a) EU-PNR proposal in its latest version, Council doc. 5618/2/09, interinstitutional file 2007/0237 (CNS) from 29 June 2009.

[31] Council Framework Decision 2008/977/JHA of 27 November 2008 on the protection of personal data in the framework of police and judicial cooperation in criminal matters, OJ 2008, L-350/60.

[32] In addition to the Council of Europe instruments, the EU-PNR proposal also refers to the DPFD, compare Recital (10b) and article 11 (1a) EU-PNR proposal in its latest version, Council doc. 5618/2/09, interinstitutional file 2007/0237 (CNS) from 29 June 2009.

[33] Article 16 TFEU clarifies that "everyone has the right to the protection of personal data concerning them".

[34] Article 6 (3) TFEU.

It is important to take into consideration that article 6 TEU not only refers to the Charter, but also to the ECHR by specifying two important principles: the fundamental rights of the ECHR constitute general principles of EU law and the EU shall accede to the ECHR.[35] This transitions clause between EU law and the public international law of the Council of Europe paves the way for a comparison of the ECtHR's interpretation of the right to data protection, enshrined in the right to private life in article 8 ECHR, with the provisions of the EU-PNR proposal. The existing EU instruments applying in the context of security-related data processing (DPFD, Lisbon Treaty and Charter of Fundamental Rights) are relatively new and stipulate important but quite broad principles. While these rules have not yet been subject to the jurisdiction of the European Union Courts, the long established case law of the ECtHR with regard to article 8 ECHR has created concrete and detailed principles in this field over the last decades.[36]

All in all, considering article 6 TEU and the intended accession of the EU to the ECHR, legal proposals should not only comply with the fundamental rights of the Charter of Fundamental Rights, but also with the principles developed by the ECtHR during the recent years.

8.3.2 General Principles of the ECtHR with Regard to Security-Related Data Processing

The case law of the ECtHR provides helpful guidance by concretizing the aforementioned general data protection standard.

The Strasbourg Court refers to the right to private life of article 8 ECHR when data protection infringements are at stake.[37] Even though personal data are not expressly protected by this article, the ECtHR insists that "the protection of personal data" is of "fundamental importance" to a person's enjoyment of his or her right to respect for private and family life.[38]

---

[35] Article 6 (2) and (3) TEU.

[36] Compare for a profound analysis: Siemen, Birte (2006). *Datenschutz als europäisches Grundrecht*. Berlin: Duncker & Humblot.

[37] Compare for instance: ECtHR, *Leander v. Sweden*, Application no. 9248/81 from 26 March 1987; ECtHR, *Amann v. Switzerland*, Application no. 27798/95 from 16 February 2000; ECtHR, *Rotaru against Romania*, Application no. 28341/95 from 4 May 2000; ECtHR, *Panteleyenko v. Ukraine*, Application no. 11901/02 from 29 June 2006; ECtHR, *S. and Marper v the United Kingdom*, Application nos. 30562/04 and 30566/04 from 4 December 2008; ECtHR *Weber and Saravia v. Germany*, Application no. 54934/00 Admissibility Decision from 29 June 2006; ECtHR, *C.G. and others v. Bulgaria*, Application no. 1365/07 from 24 April 2008; ECtHR, *Association for European Integration and Human Rights and Ekimdzhiev v. Bulgaria*, Application no. 62540/00 from 28 June 2007; ECtHR, *Malone v. the United Kingdom*, Application no. 8691/79 from 2 August 1984; ECtHR, *Valenzuela v. Spain*, Application no. 27671/95 from 30 July 1998.

[38] ECtHR, *Z. v Finland*, Application no. 22009/93, from 25 February 1997, para 95; ECtHR, *Peck v. United Kingdom*, Application no. 44647/98 from 28 January 2003, para 78; ECtHR, *L.L. v France* Application no. 7508/02 from 10 October 2006, para 43; ECtHR, *Biriuk v Lithuania*, Application

The jurisprudence of the ECtHR clearly illustrates that governmental data collection and retention interferes with the right to private life as protected by article 8 ECHR.[39] Every transmission of personal data from one authority to another, including the subsequent use of such data, constitutes another separate interference with individual rights under article 8 ECHR. The transmission enlarges the group of individuals with knowledge of the personal data and can therefore lead to investigations being instituted against the persons concerned.[40] The indented Europe-wide PNR collection as contemplated by the EU-PNR proposal therefore undoubtedly interferes with article 8 ECHR.

After the interference has been established, the ECtHR examines whether the measure in question may be justified. In this context, one has to consider three conditions: the act in question must be "in accordance with the law", pursue one of the legitimate aims listed in article 8 (2) ECHR and must additionally be necessary in a democratic society, which means principally that the interfering law must be proportionate to the aim pursued. Whereby in general the ECtHR admits a wide margin of discretion to the Member States when national security is at stake, the interests of the parties, however, have to be reasonably balanced. Moreover, to be in accordance with the law, the measure in question must be "*foreseeable*", which means formulated with sufficient precision to enable an individual to regulate his conduct and to predict the consequences a given action might entail.[41]

To be more precise, in judgments related to governmental data collection and the implementation of surveillance measures in the framework of article 8 ECHR, certain criteria must be fulfilled to guarantee proportionality and in this way the balance of powers between the interests at stake. These criteria include the *limitation on the categories of individuals against whom surveillance measures may be taken* as well as the *clear definition of the circumstances and limits of the storing and the use of the information before the processing*.[42] *Time limits* for storing are essential and

---

no. 23373/03 from 25 November 2008, para 39; ECtHR, *I v Finland* Application no. 20511/03 from 17 July 2008, para 38; ECtHR, *S. and Marper v the United Kingdom*, Application nos. 30562/04 and 30566/04 from 4 December 2008, para 103; ECtHR, *C.C. v. Spain*, Application no. 1425/06 from 6 October 2009, para 31.

[39] ECtHR, *Amann v. Switzerland*, Application no. 27798/95 from 16 February 2000, paras 65–67.

[40] ECtHR, *Weber and Saravia v. Germany*, Application no. 54934/00 Admissibility Decision from 29 June 2006, para 79.

[41] ECtHR, *Sunday Times v. the United Kingdom*, Application no. 6538/74, para 49 from 26 April 1979; ECtHR, *Liberty and others v. the United Kingdom*, Application no. 58234/00 from 1 July 2008, para 68; ECtHR *Silver v. the United Kingdom*, Application no. 5947/72 and others from 25 March 1983, paras 85–88.

[42] ECtHR, *Segerstedt-Wiberg and others v. Sweden*, Application no. 62332/00 from 6 June 2006, paras 88–92; ECtHR, *Liberty and others v. the United Kingdom*, Application no. 58234/00 from 1 July 2008, para 68; ECtHR, *Rotaru v. Romania*, Application no. 28341/954 from 4 May 2000, para 57; ECtHR, *Weber and Saravia v. Germany*, Application no. 54934/00 Admissibility Decision from 29 June 2006, paras 116 and 127.

the *age of the person concerned* must be taken into account to *avoid indiscriminate storing* of personal data in governmental databases.[43]

Prior to surveillance measures and the collection of data in security-related data processing, it is crucial to determine which *kind of data are to be stored and for which purposes the data should be used afterwards* (purpose limitation principle).[44] *Independent review and adequate and effective safeguards* against abuse, including effective remedies, must exist to assure compliance with the rule of law.[45]

With regard to the *subsequent notification of individuals subjected to surveillance measures*, the ECtHR emphasizes that this question is closely linked to the effectiveness of remedies before the courts and therefore to the existence of effective safeguards against the abuse of monitoring powers.[46] In the case *Weber and Saravia v. Germany*, the Strasbourg Court adds: "As soon as notification can be carried out without jeopardizing the purpose of the restriction after the termination of the surveillance measure, [...], information should be provided to the persons concerned".[47]

Against this relatively detailed background, the following section examines the compliance of the EU-PNR proposal with the ECHR standard in accordance with the structure of analysis the ECtHR usually applies.

8.3.3 In Accordance with the Law and Foreseeability

As previously addressed, in order to be in accordance with the law, the EU-PNR proposal should be "accessible" and "foreseeable" to the person concerned, meaning that an individual affected "must be able to foresee its consequences for him".[48] The provisions of the proposal should indicate in what circumstances and on what

[43]ECtHR, *S. and Marper v. the United Kingdom*, Application nos. 30562/04 and 30566/04 from 4 December 2008, para 119; ECtHR, *Segerstedt-Wiberg and others v. Sweden,* Application no. 62332/00 from 6 June 2006, paras 89–92.

[44]ECtHR, *Weber and Saravia v. Germany*, Application no. 54934/00 Admissibility Decision, para 116 from 29 June 2006; ECtHR, *Rotaru v. Romania*, Application no. 28341/954, para 57 from 4 May 2000; see also: ECtHR, *Association for European Integration and Human Rights and Ekimdzhiev v. Bulgaria*, Application no. 62540/00 from 28 June 2007.

[45]ECtHR, *Rotaru against Romania*, Application no. 28341/95 from 4 May 2000, paras 55–63; ECtHR, *Segerstedt-Wilberg and others v. Sweden,* Application no. 62332/00 from 6 June 2006, para 121.

[46]ECtHR, *Weber and Saravia v. Germany*, Application no. 54934/00 Admissibility Decision from 29 June 2006, para 135: "*since there is in principle little scope for recourse to the courts by the individual concerned unless the latter is advised of the measures taken without his or her knowledge and thus able to challenge their legality retrospectively*".

[47]ECtHR, *Weber and Saravia v. Germany*, Application no. 54934/00 Admissibility Decision, para 135 from 29 June 2006.

[48]ECtHR, *Valenzuela v. Spain*, Application no. 27671/95 from 30 July 1998, para 46 et seq.

terms public authorities are empowered to store and use the PNR.[49] The aim and the objectives of the proposal should be formulated in a clear and precise way.

The objectives of the EU-PNR proposal are drafted in an extremely wide ranging manner, embracing in general "the prevention, detection, investigation and prosecution of terrorist offences or serious crimes".[50] These objectives may comply with the aims mentioned in article 8 (2) ECHR (national security, prevention of crime), but they also have to meet the aforementioned foreseeability criterion.

Regrettably, the terms "prevention, detection, investigation and prosecution of terrorist offences or serious crimes" are neither further explained nor defined in the EU-PNR proposal itself.

Against this background, the ECtHR case *Kennedy v. the United Kingdom* is worth remembering because it entails an interesting argument of the applicant which also matters in the EU-PNR context: The applicant *Kennedy* claimed that the term "serious crime", used in a British act to justify restrictive measures, in this case telephone tapping, is not sufficiently clear and therefore blurs the boundaries of what is foreseeable in terms of the ECHR.[51] In view of the Strasbourg Court, the reference to serious crime seems to comply with the foreseeability requirement, although only under the condition that the term is *further explained in the interpretative provisions of the contested act* as well as *in the act itself*.[52] The ECtHR rules: "[...] the reference to serious crime, *together with the interpretative clarifications in the Act*[53], gives the citizens an adequate indication as to the circumstances in which and the conditions on which public authorities are empowered to resort to secret surveillance measures".[54]

The linking of the use of "serious crime" to the presence of additional clarifications provided for within the act, seems to indicate that the term "serious crime" alone would probably not meet the terms of the foreseeability criterion of the ECHR. This question remains regrettably unanswered in the end. However, the wording used by the ECtHR supports the conclusion that supplementary explanations are necessary to be in compliance with the foreseeability standard of the ECHR.

Turning to the provisions of the EU-PNR proposal, its article 2 (h) and (i) as well as its Recital (3) refer to further explanations of the terms used in the EU-PNR proposal.[55] The term "terrorism" should correspond to the terrorism definition of articles 1–4 of the Council Framework Decision on Combating Terrorism from 2002 (which lists a comprehensive catalogue of crimes), and the "serious crime"

---

[49] See a case with regard to governmental data mining: ECtHR, *Rotaru against Romania*, Application no. 28341/95 from 4 May 2000, para 50.

[50] Compare article 1 EU-PNR proposal in its latest version, Council doc. 5618/2/09, interinstitutional file 2007/0237 (CNS) from 29 June 2009.

[51] ECtHR, *Kennedy v. the United Kingdom*, Application no. 26839/05 from 18 May 2010, para 159.

[52] ECtHR, *Kennedy v. the United Kingdom*, Application no. 26839/05 from 18 May 2010, para 159.

[53] Emphasis added.

[54] ECtHR, *Kennedy v. the United Kingdom*, Application no. 26839/05 from 18 May 2010, para 159.

[55] Article 2 (h) and (i) and Recital (3) EU-PNR proposal in its latest version, Council doc. 5618/2/09, interinstitutional file 2007/0237 (CNS) from 29 June 2009.

definition should be equivalent to that of article 2 (2) of the European Arrest Warrant Framework Decision as well as to article 2 of the Framework Decision on the Fight Against Organised Crime.[56] All instruments together, however, regulate over 50 different criminal offences which are broadly formulated and which have to be implemented in 27 different Member States. Article 2 of the European Arrest Warrant Framework Decision for instance refers to 32 different categories, containing also criminal offences related to trafficking in human beings and child pornography, illicit trafficking in endangered animal or plant species and cultural goods as well as illicit trafficking in hormonal substances and other growth promoters.

At first glance various offences are therefore not always obviously connected with the overarching objective of preventing terrorism and serious crime. It is doubtful whether the citizens are provided with "an adequate indication as to the circumstances in which and the conditions on which public authorities"[57] are empowered to use their PNR.

Unless persons concerned study three different Council Decisions, they will not be aware of the offences that fall under the terms "terrorism" and "serious crime". Formulations, including words such as "terrorism" and "serious crime", appear very abstract for the targeted flight passengers and should be therefore explained directly in the EU-PNR proposal itself and not in further reference instruments, which are difficult to consult when booking a flight.

It follows from the foregoing that persons concerned can face difficulties when assessing whether their conduct falls within the scope of the envisaged measures. A general suspicion jeopardizes the "naturalness" of behavior and leads to uncertainty: individuals do not know which conduct complies with "normal" behavior patterns and which conduct might be suspicious.[58] Having in mind that the instrument in question regulates data processing of all European flight passengers, the scope and its definitions must be formulated in unambiguous terms in order to comply with the "foreseeability" requirement of the ECtHR. In its current version, the EU-PNR proposal violates this ECHR obligation.

8.3.4 Necessary in a Democratic Society

As mentioned above, the ECtHR leaves a wide margin of discretion when national security is at stake, although the formerly stipulated minimum conditions have to be met. There are some strong arguments to be made here that the interference with

[56] Articles 1–4 Council Framework Decision on Combating Terrorism, OJ 2002 L-164/03, article 2 (2) of the Council Framework Decision on the European Arrest Warrant and the surrender procedures between Member States, OJ 2002, L-190/1 and article 2 Council Framework Decision of 24 October 2008 on the Fight Against Organised Crime, OJ 2008, L-300/42.

[57] ECtHR, *Kennedy v. the United Kingdom*, Application no. 26839/05 from 18 May 2010, para 159.

[58] See arguments of the German Constitutional Court delivered in a judgment in context of a governmental profiling case: Bundesverfassungsgericht, 1 BvR 518/02, para 117.

fundamental rights is not necessary in a democratic society. The most crucial points are analyzed hereinafter.

8.3.4.1 Purpose Limitation

In light of the foregoing considerations, the main concerns relate to the compliance with the purpose limitation principle, specifically with the central rule that data originally collected and used for one purpose (to make a flight reservation) are not allowed to be later used for another (security-related) purpose. Usually, the purpose allowing the subsequent collection and processing of personal data has to be determined *before* starting to gather and/or to process data and is *not allowed to be changed* during the retention or use of the data.[59] Derogations from this principle may only take place in few restricted cases and only insofar as they are proportionate, indispensable and foreseeable and can outweigh the serious infringement caused.[60]

Taking into account the imprecise and vague formulations analyzed above (prevention, detection, investigation and prosecution of terrorist offences or serious crimes), additional evidence should be given to justify such a serious infringement of the purpose limitation principle. The analysis of the European Data Protection Supervisor however points out that precise information and concrete results relating to the achievements of other PNR systems in third states are missing altogether.[61] No case so far has been cited as evidence to prove the effectiveness of a PNR analysis system.[62]

This should be taken into account when assessing the necessity of the planned measure. If other less intrusive methods exist to obtain the indented aim, these methods have to be used in the first place.[63] In this context, the suitability of existing law enforcement databases and systems monitoring individuals in Europe, including the Schengen-, Visa- and Customs-Information Systems, as well as the API Directive

---

[59] Compare for instance: ECtHR, *Weber and Saravia v. Germany*, Application no. 54934/00 Admissibility Decision from 29 June 2006, paras 121–122.

[60] ECtHR, *Weber and Saravia v. Germany*, Application no. 54934/00 Admissibility Decision from 29 June 2006, paras 80 et seq.

[61] EDPS opinion, points 27-29.

[62] Vagelis Papakonstantinou and Paul De Hert, "The PNR Agreement and Transatlantic anti-terrorism Cooperation: No firm human rights framework on either side of the Atlantic," *Common Market Law Review* 46, 3 (2009): 917.

[63] Compare for instance: ECtHR, *K.U. v. Finland*, Application no. 2871/02 from 2 December 2008, para 26; ECtHR, *Copland v. the United Kingdom*, Application no. 62617/00 from 3 April 2007, para 38.

2004/82/EC[64] should have been assessed before implementing such far reaching measures.

8.3.4.2 Clear Definition of the Circumstances and the Limits of Processing

Closely connected to the limitation of the purpose is the requirement to clearly define *the circumstances and limits of the storing and the use of the information before the processing.*[65]

In addition to the mentioned objectives defined in article 1 of the EU-PNR proposal (prevention, detection, investigation and prosecution of terrorist offences or serious crimes), Recital (9) of the EU-PNR proposal makes the attempt to clarify the proposal's objectives by stipulating that the collected data should be kept "for a sufficiently long period for carrying out trend analysis and for using in investigations".[66] The exact meaning of "trend analysis" or "using in investigations" is not further elucidated, although an accompanying document to the EU-PNR proposal issued by the Commission gives some indication about the content of these vague formulations.[67]

The PNR are considered to be very important for making associations between known and unknown people. More specifically, "once a known terrorist or criminal is identified, the PNR can be used to identify another passenger who is connected

---

[64] In addition to the EU-US PNR Agreement, the EU-PNR proposal is closely related to Directive 2004/82/EC at EU level. While the Directive regulates the transfer of advanced passenger information (API) from air carriers to border control authorities of the Member States, the EU-PNR proposal attempts to harmonize the legal provisions of the Member States regarding the duties of air carriers to transfer their PNR to law enforcement authorities of the Member States for crime prevention purposes. Whereas the API Directive obliges air carriers to transmit on prior request of border control authorities information relating to the passengers they will carry, the EU-PNR proposal provides for a general obligation for air carriers to transfer their passenger data to law enforcement authorities without the request requirement. Moreover, in contrast to API, which principally contains passport information, PNR information includes more data categories, mainly based on the information the passenger provides him- or herself during an airline ticket reservation. For an overview on the similarity and the provisions of API Directive 2004/82 see: Evelien Brouwer, "Towards a European PNR system? Questions on the added value and the protection of fundamental rights," study requested by the European Parliament's Committee on civil liberties, justice and home affairs (LIBE) (2009), pp. 2–3.

[65] ECtHR, *Segerstedt-Wiberg and others v. Sweden,* Application no. 62332/00 from 6 June 2006, paras 88–92; ECtHR, *Liberty and others v. the United Kingdom*, Application no. 58234/00 from 1 July 2008, para 68; ECtHR, *Rotaru v. Romania*, Application no. 28341/954 from 4 May 2000, para 57; ECtHR, *Weber and Saravia v. Germany*, Application no. 54934/00 Admissibility Decision from 29 June 2006, paras 116 and 127.

[66] Recital (9) EU-PNR proposal in its latest version, Council doc. 5618/2/09, interinstitutional file 2007/0237 (CNS) from 29 June 2009.

[67] Accompanying document to the proposal for a Council Framework Decision on the use of Passenger Name Record (PNR) for law enforcement purposes from 12 November 2007 – summary of the impact assessment, 2007/0237 (CNS), 14922/07, p. 3, para 2.

to the known terrorist/criminal".[68] Evidence should be obtained by making associations to data of other flight passengers while at the same time the data should identify previously unknown passengers. Travel patterns of passengers are intended to be established.[69] Additionally data should be run "against a combination of characteristics and behavioral patterns, aimed at creating risk-assessment".[70] "When a passenger fits within a certain risk assessment, he could be identified as a high-risk passenger".[71]

These formulations make clear what is to be understood under the euphemistic terms "trend analysis" and "for using in investigations". The objective of the EU-PNR proposal is not only limited to the detection and identification of terrorists and criminals, it also consists of risk assessment and the systematic creation of travel patterns leading to the development of abstract profiles which distinguish between normal (i.e. not dangerous) and suspicious flight passengers.[72] The risk assessment may thereby base on pre-determined characteristics and behavior patterns.[73]

As follows from the foregoing, the circumstances and limits of the storing and the use of the PNR before the processing, risks to be very unclear. An individual disclosing his personal data to make a flight reservation can not necessarily foresee that his data are used for law enforcement purposes or to develop profiles to track terrorists.

8.3.4.3 Limitation of the Individuals Subject to Surveillance

The proposed measures apply to all European flight passengers, whether law enforcement authorities conduct investigations concerning them or not. The planned data processing can be carried out without having *any initial suspicion*. It is simply based on the assumption that on the 28,000 daily flights which are handled across Europe there could be terrorists or criminals.[74] There is no limitation at all as regards the individuals subject to the indented analyses and investigations.[75]

---

[68] Accompanying document to the proposal for a Council Framework Decision on the use of Passenger Name Record (PNR) for law enforcement purposes from 12 November 2007 – summary of the impact assessment, 2007/0237 (CNS), 14922/07, p. 3, para 2.

[69] Article 3 (3) (c) EU-PNR proposal in its latest version, Council doc. 5618/2/09, interinstitutional file 2007/0237 (CNS) from 29 June 2009.

[70] Accompanying document to the proposal for a Council Framework Decision on the use of Passenger Name Record (PNR) for law enforcement purposes from 12 November 2007 – summary of the impact assessment, 2007/0237 (CNS), 14922/07, p. 3, para 2.

[71] Accompanying document to the proposal for a Council Framework Decision on the use of Passenger Name Record (PNR) for law enforcement purposes from 12 November 2007 – summary of the impact assessment, 2007/0237 (CNS), 14922/07, p. 3, para 2.

[72] Compare: EDPS opinion, point 18 and FRA's opinion point 12.

[73] EDPS opinion, points 18–25.

[74] Annual report of the European Organisation for the Safety of Air Navigation, Eurocontrol, http://www.eurocontrol.int/epr/public/standard_page/AnnualReport.html, p. 27.

[75] Compare article 1 EU-PNR which refers to the scope of the EU-PNR proposal and article 5 (1) EU-PNR proposal referring to the obligations of air carriers which are obliged to "make available

This clearly contradicts ECtHR jurisdiction. In *Weber and Saravia v. Germany*, the ECtHR clarified that the Bundesverfassungsgericht (German Constitutional Court) only adequately counterbalanced an interference, provoked by the collection and transmission of security-related personal data to another authority, by strictly limiting the types of offences on behalf of which data transmission was permitted.[76] The restriction referred to the order of the Bundesverfassungsgericht that the law in question could only be applied and data could only be transmitted, if *specific facts* – as opposed to mere factual indications – aroused the suspicion that someone had committed one of the *limited offences* listed in a special article of the challenged act.[77]

The Bundesverfassungsgericht repeatedly points in this context to the risks of misuse and errors arising out of the storage of a large amount of data. Intimidation effects which could have a negative impact on the exercise of fundamental rights and the "feeling of living in a surveillance state" can be further consequences of a mass profiling.[78]

While the German solution might constitute only one possibility to limit the circle of persons concerned by surveillance measures and governmental data mining, the interpretation of the ECtHR's jurisdiction suggests that the collection of personal data regardless of any suspicion in a wide range of cases would most likely contradict the guarantees of article 8 ECHR.[79] Having the enormous amount of persons concerned in mind, strict criteria to limit the circle of targeted flight passengers must be developed to comply with the ECHR.

8.3.4.4 Time Limit

No agreement has been reached so far on the questions of the exact time limit of the PNR storing.[80] The data should be kept "for a sufficiently long period".[81] As mentioned above, article 9 EU-PNR proposal indeed foresees a (provisional) 3 years retention period of PNR, with an additional 7 years period in a dormant database

the PNR data of all passengers of the flight" to the PIUs of the Member States, EU-PNR proposal in its latest version, Council doc. 5618/2/09, interinstitutional file 2007/0237 (CNS) from 29 June 2009.

[76] ECtHR, *Weber and Saravia v. Germany*, Application no. 54934/00 Admissibility Decision from 29 June 2006, para 129.

[77] ECtHR, *Weber and Saravia*, Application no. 54934/00 Admissibility Decision from 29 June 2006, para 127.

[78] Judgment on governmental profiling of the German Constitutional Court, Bundesverfassungsgericht, 1 BvR 518/02, para 117.

[79] ECtHR, *Weber and Saravia*, Application no. 54934/00 Admissibility Decision from 29 June 2006, paras 125–129.

[80] Compare commentary in footnote 56 of EU-PNR proposal in its latest version, Council doc. 5618/2/09, interinstitutional file 2007/0237 (CNS) from 29 June 2009 referring to the different positions regarding the data retention period provided for in article 9 of the EU-PNR proposal.

[81] Recital (9) EU-PNR proposal in its latest version, Council doc. 5618/2/09, interinstitutional file 2007/0237 (CNS) from 29 June 2009.

underlying special access conditions[82]. However, the Council considers it "highly unlikely that a consensus can be reached among 27 delegations on an exact and obligatory retention period".[83] It clearly follows from ECtHR's case law that a time limit is an indispensable requirement for any data storing.[84]

Regardless of this discussion, even the provided 10 years retention period *per se* seems to fail to strike the right balance between the rights of, in principle, unsuspected individuals and the Member States' crime prevention interests, in particular with regard to the risk of the possible stigmatising effect the long-term data storage might have. Whereby the total duration of 10 years might be necessary if the data are relevant to *specific cases* or for the duration of ongoing investigations, the storage of *all* PNR for 10 years *in absence of any suspicion* seems to be disproportionate.

Another aspect concerning the deletion of the data additionally deserves attention. The EU-PNR proposal foresees the deletion "from all databases" of the PIU after the expiry of 10 years.[85] While a deletion requirement is generally to be welcomed, no regulation regarding the whereabouts or the deletion of the data retrieved during the 10 years period seems to apply. Member States could easily circumvent the initial storage period by transferring the desired data to their national databases. Recital (9a) EU-PNR proposal refers to this possibility by clarifying that "the retention periods for PNR data set by this Framework Decision are without prejudice to different, possibly longer periods during which PNR data which are being processed by police or judicial authorities in the context of a criminal investigation or prosecution, may be retained".[86]

A provision restricting the time of the use of the data after the retrieval from the PIU would avoid an unlimited exploitation of the PNR.

8.3.4.5 Risk of Stigmatization and Discrimination

The relevance of the rights of unsuspected individuals in the field of security related data storing and processing has recently been underlined in *S. and Marper v. the United Kingdom.*[87] The ECtHR found a strong violation of article 8 ECHR in the case concerning the storage of DNA and fingerprint information of suspected, but

---

[82]Article 9 EU-PNR proposal in its latest version, Council doc. 5618/2/09, interinstitutional file 2007/0237 (CNS) from 29 June 2009.

[83]Compare commentary in footnote 56 of EU-PNR proposal in its latest version, Council doc. 5618/2/09, interinstitutional file 2007/0237 (CNS) from 29 June 2009 referring to the different positions regarding the data retention period provided for in article 9 of the EU-PNR proposal.

[84]*S. and Marper v. the United Kingdom*, Application nos. 30562/04 and 30566/04 from 4 December 2008, para 119; *Segerstedt-Wiberg and others v. Sweden* Application no. 62332/00 from 6 June 2006, paras 89–92.

[85]Article 9 (3) EU-PNR proposal in its latest version, Council doc. 5618/2/09, interinstitutional file 2007/0237 (CNS) from 29 June 2009.

[86]Recital (9a) EU-PNR proposal in its latest version, Council doc. 5618/2/09, interinstitutional file 2007/0237 (CNS) from 29 June 2009.

[87]ECtHR, *S. and Marper v. the United Kingdom*, Application nos. 30562/04 and 30566/04 from 4 December 2008.

not convicted individuals. The challenged data storage took place "irrespective of the nature or gravity of the offence with which the individual was originally suspected or of the age of the suspected offender".[88] The ECtHR clearly opposed data retention carried out "indefinitely whatever the nature or seriousness of the offence of which the person was suspected".[89]

When comparing this case with the EU-PNR proposal, two things are worth pointing out: first, the data to be stored for 10 years exclusively concern persons not suspected of any crimes and second, the EU-PNR proposal completely disregards the age of the person concerned.

While the ECtHR acknowledges that DNA data might not be directly compared with other categories of personal data retained, it also emphasized that every indiscriminate data retention regime calls "for careful scrutiny regardless of these differences".[90]

The European Data Protection Supervisor and the Agency of Fundamental Rights additionally highlight the discriminatory effect on certain ethnic or religious groups that the proactive investigation methods, based on "pre-determined risk criteria"[91], may have.[92]

In particular decisions taken about one individual which result from analyzing patterns derived from other individuals, raise fundamental rights problems and risk to have a high error rate.[93] An individual, whose data have been linked to data of another, possibly suspicious person, might be treated himself with more suspicion than before.

The classification of individuals without suspicion and the separation between "high risk"[94] and "normal" passengers is strongly reminiscent of methods of governmental profiling which were also declared void in the aforementioned judgment of the German Constitutional Court in 2006.[95] Already in 1983, the same Court

---

[88]ECtHR, *S. and Marper v. the United Kingdom*, Application nos. 30562/04 and 30566/04 from 4 December 2008, para 119.

[89]ECtHR, *S. and Marper v. the United Kingdom*, Application nos. 30562/04 and 30566/04 from 4 December 2008, para 119.

[90]ECtHR, *S. and Marper v. the United Kingdom*, Application nos. 30562/04 and 30566/04 from 4 December 2008, para 120.

[91]Article 3 (a) EU-PNR proposal in its latest version, Council doc. 5618/2/09, interinstitutional file 2007/0237 (CNS) from 29 June 2009.

[92]EDPS opinion, point 19 and FRA's opinion point 12. To the general concerns raised by data mining including further references, see: Paul De Hert and Rocco Bellanova, "Data protection from a transatlantic perspective: The EU and US move towards an international data protection agreement?," study requested by the European Parliament's Committee on civil liberties, justice and home affairs (LIBE) (2008), pp. 25–26 and 37–38.

[93]EDPS opinion, point 22.

[94]Compare wording in article 18 (2) EU-PNR proposal in its latest version, Council doc. 5618/2/09, interinstitutional file 2007/0237 (CNS) from 29 June 2009.

[95]Judgment on governmental profiling of the German Constitutional Court, Bundesverfassungsgericht, 1 BvR 518/02.

stated in its famous census decision that a law regulating the comprehensive registration and the "cataloguing" of the personality through the connection of personal data for the purpose of creating profiles and patterns is not permitted.[96] This applies also in the context of anonymous statistical surveys.

Clear and unambiguous criteria for the data processing as well as the risk-assessment including clear limits on the use of "pre-determined risk criteria"[97], would therefore avoid stigmatization effects.

8.3.4.6 Independent Control and Notification

In light of the ECtHR requirements concerning the independent review and the existence of adequate and effective safeguards against abuse[98], special attention has to be paid to the control of the data processing of the PIUs. The EU-PNR proposal shifts the responsibility for the protection of the PNR data completely to the Member States' data protection authorities (DPAs).[99] In absence of a special control mechanism monitoring the PIUs, only the national DPAs should carry out this extremely exhaustive task. As a consequence, in practice, 27 different legal regimes would apply to the exercise of the rights of the passengers concerned. Against this background, the question arises whether the national DPAs dispose of the necessary financial or personal resources to monitor effectively and independently the PNR processing.[100] Further research to clarify this question must be undertaken before adopting the proposal.

In addition to this shortcoming, although required by the ECtHR[101], notification of passengers, when their data were used in investigations or for "trend analysis", is not intended. A clause containing a notification duty, as soon as this notification can be carried out without jeopardizing the purpose of the measure taken, should be introduced to satisfy the criteria of the Strasbourg Court.[102]

---

[96] Census decision of the German Constitutional Court, Bundesverfassungsgericht, Volkszählungsurteil, BVerfGE 65, 1, para 177.

[97] Article 3 (a) EU-PNR proposal in its latest version, Council doc. 5618/2/09, interinstitutional file 2007/0237 (CNS) from 29 June 2009.

[98] ECtHR, *Rotaru against Romania*, Application no. 28341/95 from 4 May 2000, paras 55–63; ECtHR, *Segerstedt-Wiberg and others v. Sweden,* Application no. 62332/00from 6 June 2006, para 121.

[99] Article 11 et seq. EU-PNR proposal in its latest version, Council doc. 5618/2/09, interinstitutional file 2007/0237 (CNS) from 29 June 2009.

[100] Evelien Brouwer, "Towards a European PNR system? Questions on the added value and the protection of fundamental rights," study requested by the European Parliament's Committee on civil liberties, justice and home affairs (LIBE) (2009), p. 26.

[101] *Weber and Saravia v. Germany*, Application no. 54934/00 Admissibility Decision, para 135 from 29 June 2006.

[102] *Weber and Saravia v. Germany*, Application no. 54934/00 Admissibility Decision, para 135 from 29 June 2006.

8.3.4.7 Interim Findings

When looking at the foregoing arguments, the necessity and the proportionality of the EU-PNR proposal are extremely questionable. Data protection rules as well as rights against discrimination have to be considerably improved and it has to be examined whether existing law enforcement databases can function as an alternative solution.

8.4 Applicable Law: From Private to Public Law

A particular legal problem which is also linked to the practical enforcement of data protection rights is mentioned above and arises out of the fact that the applicable law changes from private to public law at the moment when the data are transferred to public databases. This shift has serious consequences, which implicitly lead to a change in the applicable data protection rights and their connected procedural guarantees such as access, appeal and correction rights. A very similar problem exists with regard to the EU-US PNR Agreement as well as with regard to other EU legal instruments, such as the Data Retention Directive[103] or the API Directive.[104]

8.4.1 No Coherent Solution by the European Court of Justice

What is the legal significance of a case, where data which are regulated by private law and which were collected and stored for an economic purpose (such as booking a flight) are subsequently used for law enforcement purposes? On two occasions, the European Court of Justice (ECJ) has faced this problem, but regrettably the Court missed the opportunity to go into the substance or the fundamental rights implications of this question.[105] While both cases involved the choice of the legal basis (first or third pillar) for measures obliging private actors to hold their data available for law enforcement agencies, the ECJ reached two different conclusions.

---

[103] Spiros Simitis, "Der EuGH und die Vorratsdatenspeicherung oder die verfehlte Kehrtwende bei der Kompetenzregelung," *Neue Juristische Wochenzeitschrift* 25 (2009): 1782–1786.

[104] Council Directive 2004/82/EC of 29 April 2004 on the obligation of air carriers to communicate passenger data, OJ 2004, L-261/24, compare footnote 64.

[105] Joined cases C-317/04 and C-318/04, *Parliament v. Council*, [2006], ECR I-4721 and case C-301/06 *Ireland v. Parliament and Council* [2009] ECR I-593. To the similiarity of the cases, see: Spiros Simitis, "Der EuGH und die Vorratsdatenspeicherung oder die verfehlte Kehrtwende bei der Kompetenzregelung," *Neue Juristische Wochenzeitschrift* 25 (2009): 1782–1786.

8.4.1.1 The Annulment of the Legal Basis of the First EU-US PNR Agreement

The first case concerned the legal basis for the first EU-US PNR Agreement of 2004[106]: the PNR data transfers to the US and their processing were initially treated as economic related first pillar data processing, because the PNR were originally collected by the airlines.[107]

The ECJ however, ruled that the *use* and the *purpose of processing* of the data, and not the purpose initially justifying their collection, should decide the legal basis of the EU-US PNR Agreement. The Court came to the conclusion that "the transfer of PNR data [. . .] constitutes processing operations concerning public security and the activities of the State in areas of criminal law".[108] Therefore, the PNR data transfers were regarded as security-related third pillar data processing. First pillar decisions of the Council and of the Commission leading to the conclusion of the agreement were annulled.[109]

Although being challenged by the Parliament as well as by the European Data Protection Supervisor, intervening in support of the Parliament[110], the ECJ limited its further findings to the discussion of the legal basis of the PNR processing and avoided the question of the data protection implications of the EU-US PNR Agreement on the rights of individuals.[111]

8.4.1.2 The Legal Basis of Data Retention

The second case brought before the ECJ involved the choice of the legal basis of the Data Retention Directive 2006/24.[112]

---

[106] Joined cases C-317/04 and C-318/04, *Parliament v. Council*, [2006], ECR I-4721.

[107] For a more detailed analysis and the consequences of this case, compare: Vagelis Papakonstantinou, and Paul De Hert, "The PNR Agreement and Transatlantic anti-terrorism Cooperation: No firm human rights framework on either side of the Atlantic," *Common Market Law Review* 46, 3 (2009): 885–919 and Mario Mendez, "Passenger Name Record Agreement, European Court of Justice," *European Constitutional Law Review* 3 (2007): 127–147.

[108] Joined cases C-317/04 and C-318/04, *Parliament v. Council*, [2006], ECR I-4721, para 56.

[109] The Court added: "While the view may rightly be taken that PNR data are initially collected by airlines in the course of an activity which falls within the scope of Community law, namely sale of an aeroplane ticket which provides entitlement to a supply of services, the data processing which is taken into account [. . .] is, however, quite different in nature", as a result, the PNR transfers did not concern "data processing necessary for a supply of services, but data processing regarded as necessary for safeguarding public security and for law-enforcement purposes", see: Joined cases C-317/04 and C-318/04, *Parliament v. Council*, [2006], ECR I-4721, para 57.

[110] Joined cases C-317/04 and C-318/04, *Parliament v. Council*, [2006], ECR I-4721, paras 33–50.

[111] Spiros Simitis, "Der EuGH und die Vorratsdatenspeicherung oder die verfehlte Kehrtwende bei der Kompetenzregelung," *Neue Juristische Wochenzeitschrift* 25 (2009): 1782–1786, in particular 1782.

[112] Case C-301/06 *Ireland v. Parliament and Council* [2009] ECR I-593. Directive 2006/24/EC of the European Parliament and of the Council of 15 March 2006 on the retention of data generated or processed in connection with the provision of publicly available electronic communications services or of public communications networks and amending Directive 2002/58/EC, OJ 2006, L-105/54.

Pursuant to its article 1, Directive 2006/24 harmonizes the Member States' provisions concerning the obligation of electronic communication service providers to store the "traffic and location data on both legal entities and natural persons" and "the related data necessary to identify the subscriber or registered userclient data" processed by them, "*in order to ensure that the data are available for the purpose of the investigation, detection and prosecution of serious crime*".[113]

Taking into account the EU-US PNR Agreement case, Ireland, supported by Slovakia, challenged the first pillar legal basis (article 95 EC Treaty/now article 114 TFEU) and asked the central question of whether Directive 2006/24 should not have been based on a third pillar legal basis, as it regulates the data retention for law enforcement purposes, or whether the Parliament and the Council were correct in choosing article 95 EC Treaty as the legal basis. Article 95 EC Treaty can be invoked "when disparities exist between national rules which are such as to obstruct the fundamental freedoms or to create distortions of competition and thus have a direct effect on the functioning of the internal market".[114]

Similar to the EU-PNR Agreement case, one of the underlying questions however, from a fundamental rights point of view, concerned the limits of the use of personal data originally stored for an economic purpose (electronic communication services) and later used for law enforcement purposes.[115] Disappointingly, as in the EU-PNR Agreement case, the ECJ totally sidestepped this problem and completely focussed on the choice of the legal basis.

Regardless of the clear wording of article 1 of Directive 2006/24 cited above, the Court ruled that Directive 2006/24 regulates operations which "are independent of the implementation of any police and judicial cooperation in criminal matters"[116] and exclusively relate to the harmonization of the activities of service providers in the relevant sector of the internal market.[117] The Court distinguished between retention and storing of the data and its subsequent use and the access to them.[118] Consequently, the ECJ approved the first pillar choice of article 95 EC Treaty as the correct legal basis for the directive.

---

[113] Article 1 (1) and (2) Directive 2006/24/EC of the European Parliament and of the Council of 15 March 2006 on the retention of data generated or processed in connection with the provision of publicly available electronic communications services or of public communications networks and amending Directive 2002/58/EC, OJ 2006, L-105/54.

[114] Case C-301/06 *Ireland v. Parliament and Council* [2009] ECR I-593, para 63.

[115] Spiros Simitis, "Der EuGH und die Vorratsdatenspeicherung oder die verfehlte Kehrtwende bei der Kompetenzregelung," *Neue Juristische Wochenzeitschrift* 25 (2009): 1782–1786, in particular 1783.

[116] Case C-301/06 *Ireland v. Parliament and Council* [2009] ECR I-593, para 83.

[117] Case C-301/06 *Ireland v. Parliament and Council* [2009] ECR I-593, para 84.

[118] Case C-301/06 *Ireland v. Parliament and Council* [2009] ECR I-593, para 84.

8.4.1.3 Two Cases, Two Different Solutions

With the ruling in the data retention case, the ECJ contradicts its own jurisprudence in the EU-PNR Agreement case, which evidently focused on the *use* and the *access* to the data as well as their *purpose of processing* as being the decisive factor in search of a legal basis.[119]

As a result of the rulings, both measures have completely different legal bases, despite the fact that both cases concern the interest of law enforcement agencies in the personal data stored by private actors.

While the reasons for the ECJ's turnaround in the data retention case might be well-intended, it again disregards the fundamental rights dimension.[120] By disconnecting the storage of the data from its subsequent *use* and the *access* to these data as well as their *purpose of processing*, the ECJ creates an artificial distinction, which fundamentally challenges the purpose limitation principle. Only the *connection of the purpose of processing* with *the reason for the storage* assures that the data are not disproportionally used for other purposes.

Meanwhile, several national Courts, such as the German Bundesverfassungsgericht as well as the Romanian Constitutional Court, annulled the respective national acts implementing Directive 2006/24 on grounds of non-compliance with their constitutions, notably with the proportionality test and the presumption of innocence.[121] Although these judgments did not touch upon the question of the lawfulness of the provisions of Directive 2006/24 itself, they clearly demonstrate the fundamental rights implications inherent to them.

8.4.2 *Consequences for the EU-PNR Proposal*

Applying the outcomes of the two ECJ decisions to the EU-PNR proposal, it becomes evident that the solution found in the data retention case can not be transposed to the EU-PNR proposal. The artificial distinction between the storage and the retention of data on the one hand, and the access, use and processing on the other, can not be upheld in respect of the EU-PNR proposal, because it combines all these measures in one single act.

---

[119] Spiros Simitis, "Der EuGH und die Vorratsdatenspeicherung oder die verfehlte Kehrtwende bei der Kompetenzregelung," *Neue Juristische Wochenzeitschrift* 25 (2009): 1782–1786, in particular 1784.

[120] A third pillar basis would have for instance hindered a subsequent judicial control in front of the European Courts.

[121] Bundesverfassungsgericht, 2 March 2010, 1 BvR 256/08 and Curtea Constitutionala, 8 October 2009 number 1258, Romanian Official Monitor no. 789 of 23 November 2009. For more details, see: Katja de Vries, Rocco Bellanova and Paul de Hert, "Proportionality overrides Unlimited Surveillance, The German Constitutional Court Judgment on Data Retention," *Centre of European Policy Studies, Liberty and Security in Europe*, (2010), http://www.ceps.eu/book/proportionality-overrides-unlimited-surveillance and Bogdan Manolea, "Implementation of EU Data Retention Directive Unconstitutional," *Computer Law Review International* 2 (2010): 49–51.

Decisions such as the data retention and the PNR ruling put the emphasis on the need to go beyond the formal debate on the legal basis by discussing the fundamental problem inherent to the change in the purpose of processing.

The practical results for the EU-PNR proposal are twofold:

First: a European instrument based on articles regulating police cooperation[122] obliges private actors, to which characteristically European criminal law does not apply, to systematically contribute to national law enforcement measures. Consequently, private actors become indirectly responsible for the enforcement of criminal law which traditionally is – and rightly so should stay – a proper governmental task.

Second: different data protection instruments apply to the same data elements. At the outset, the PNR are protected by the provisions of the Data Protection Directive 95/46 applicable in the framework of the collection and the initial processing through the air carriers.[123] The transfer of the PNR to the PIUs is covered by the provisions of the EU-PNR proposal as well as by the DPFD.[124] Finally, the transfer from the PIUs to national law enforcement authorities is regulated by national data protection laws and the DPFD.[125]

The different legal regimes would in practice have a major impact on the applicable data protection rights.[126] When judicial review, access rights and responsible data processors/controllers vary with every transfer of the PNR, the individuals concerned face difficulties when looking for the applicable rights as well as the actors responsible for the processing of their data. Differences in the legal regimes make it complicated to know which rules are applicable in cases of misuse or incorrectly entered information. Regrettably, the EU-PNR proposal does not provide for responsible data protection authorities or appeal committees in the specific context of EU-PNR processing.[127] A *single clear and precise rule* regulating *one* data protection standard *for the collection, as well as the processing of data* would be a significant improvement compared to the current state of affairs. This common

[122] Former articles 29, 30 (1) (b) and 34 (2) (b) TEU; now 67, 87 (2) (a) TFEU. Article 34 (2) (b) TEU was repealed, compare EU-PNR proposal in its latest version, Council doc. 5618/2/09, interinstitutional file 2007/0237 (CNS) from 29 June 2009.

[123] Recital (6) EU-PNR proposal in its latest version, Council doc. 5618/2/09, interinstitutional file 2007/0237 (CNS) from 29 June 2009.

[124] Recitals (10a) and (10b) and article 11 (1) EU-PNR proposal in its latest version, Council doc. 5618/2/09, interinstitutional file 2007/0237 (CNS) from 29 June 2009.

[125] Ibid.

[126] Compare remarks of the EDPS in this context, EDPS opinion, points 54–66.

[127] Compare articles 11d and 11e EU-PNR proposal in its latest version, Council doc. 5618/2/09, interinstitutional file 2007/0237 (CNS) from 29 June 2009, which state that Member States should decide whether the rights of the individuals should be exercised via the national DPA's or directly asserted against the PIU.

standard should cover the data processing by air carriers, the PIUs, as well as the competent national authorities.[128]

8.5 Conclusion and Improvement Suggestions

There is substantial uncertainty related to the lawfulness as well as to the applicable data protection rights of the EU-PNR proposal.

The creation of passenger patterns and the enormous amount of provided risk analyses lead to data processing to an extraordinary extent that has never been reached in Europe before. In search of potentially dangerous individuals, millions of pieces of data could be compared based on pre-determined characteristics. They could be put in another context when connecting them to data of other flight passengers. Thereby, the analysis of such a huge amount of data can easily lead to the entry of irregular or incorrect data which are then hard to correct. Effective judicial review in such an anonymous system seems impossible. Therefore, the EU-PNR proposal, in its current version, is not in accordance with the European data protection standard.

This might be partly explained by the fact that the EU-PNR proposal in its current state of affairs is strongly influenced by instruments of the American and Canadian security policy. These measures are to a certain extent of transnational nature and coercive, but this does not explain why the Commission and the Council ignored basic European data protection mechanisms when preparing the European PNR system.

The entry into force of the Lisbon Treaty can now correct this failure by providing for an obligatory participation of the European Parliament in the upcoming legislative process. Future measures in the field of police and judicial cooperation in criminal matters are now subject to the usual effects of EU law (direct effect and supremacy) and legislative acts will have the form of directives and regulations which are adopted in accordance with the so called "ordinary legislative procedure" where Council and European Parliament decide together and which are subject to the jurisdiction of the Court of Justice.[129] Data protection scholars are putting their hopes on the European Parliament which now has the power to oppose the Council's and the Commission's EU-PNR proposal. The following suggestions, resulting from the foregoing analysis, should be taken into consideration in future negotiations on the EU-PNR proposal:

[128] This solution was strongly opposed by Denmark and France, see footnote 61 relating to article 11 EU-PNR proposal in its latest version, Council doc. 5618/2/09, interinstitutional file 2007/0237 (CNS) from 29 June 2009.

[129] Steve Peers, comment on "The Third Pillar acquis after the Treaty of Lisbon enters into force," comment posted on December 1, 2009, http://www.statewatch.org/analyses/86-third-pillar-acquis-post-lisbon.pdf.

(i) To be in accordance with the law, the scope of the EU-PNR proposal should be defined more precisely in order to comply with the "foreseeability" requirement of the ECtHR.

(ii) To justify the necessity of the EU-PNR proposal in a democratic society, several points need to be clarified:

- Before seriously violating the purpose limitation principle, the effectiveness of a PNR analysis system must at least be undoubtedly proven by showing concrete cases and results of PNR processing. Even then, further analyses considering the interests at stake are necessary.
- If such analyses reveal other, less intrusive methods to obtain the same result, these methods have to be used in the first place. In particular, the efficiency of the existing systems monitoring individuals in Europe, such as the Schengen-, Visa- and Customs-Information Systems, as well as the API Directive 2004/82/EC should be fundamentally assessed.
- The meaning of terms such as "trend analysis" and "for using in investigations" must be specified to clearly define the circumstances and limits of the storing and the use of the PNR.
- The number of targeted flight passengers appears exaggerated and should be limited to comply with ECtHR requirements.
- A time-limit for the use of the data after the retrieval from the PIUs should be introduced to avoid infinite exploitation of the PNR by the Member States.
- The total length of storage (10 years) might be necessary in specific cases or for the duration of ongoing investigations, but the indiscriminate storage of all PNR for 10 years in absence of any suspicion needs to be reconsidered.
- Clear and unambiguous criteria for the data processing, as well as the risk-assessment, including strict limits on the use of pre-determined risk criteria, would avoid stigmatization effects.
- The role of national DPAs in the control of the PIUs should be clarified. An authority specialized exclusively in the monitoring of the PNR exchange could be established.
- A notification duty, as soon as this notification can be carried out without jeopardizing the purpose of the PNR processing, should be introduced to comply with the criteria of the ECtHR.

(iii) The basic question to what extent private actors are allowed to contribute systematically to traditional governmental tasks, such as security-related data processing, should be answered before the adoption of the EU-PNR proposal. This discussion will necessarily lead to a debate about the importance of the purpose limitation principle and its fundamental breach being inherent to systematic data transfers from private to public actors.

References

Accompanying document to the proposal for a Council Framework Decision on the use of Passenger Name Record (PNR) for law enforcement purposes from 12 November 2007 – impact assessment, 2007/0237 (CNS), 14922/07.

Additional protocol to the Convention for the protection of individuals with regard to automatic processing of personal data regarding supervisory authorities and transborder data flows from 2004.

Agreement between the European Community and the Government of Canada on the processing of Advanced Passenger Information and Passenger Name Record data, OJ 2006, L-82/15.

Agreement between the European Union and the United States of America on the processing and transfer of Passenger Name Record (PNR) data by air carriers to the Unites States Department of Homeland Security (DHS), OJ 2007, L-204/18.

Annual report 2009 of the European Organisation for the Safety of Air Navigation, Eurocontrol, http://www.eurocontrol.int/epr/public/standard_page/AnnualReport.html (accessed August 25, 2010).

Brouwer, E. "Towards a European PNR system? Questions on the added value and the protection of fundamental rights," study requested by the European Parliament's Committee on civil liberties, justice and home affairs (LIBE) (2009).

Bundesverfassungsgericht, 1 BvR 518/02, Judgment on governmental profiling of the German Constitutional Court.

Bundesverfassungsgericht, 2 March 2010, 1 BvR 256/08, judgment related to the implementation of the Data Retention Directive.

Bundesverfassungsgericht, Volkszählungsurteil, BVerfGE 65, 1, Census decision of the German Constitutional Court.

Case C-301/06 *Ireland v. Parliament and Council* [2009] ECR I-593.

Convention No. 108 of the Council of Europe for the protection of individuals with regard to automatic processing of personal data from 28 January 1981.

Council Directive 2004/82/EC of 29 April 2004 on the obligation of air carriers to communicate passenger data, OJ 2004, L-261/24.

Council Framework Decision 2008/977/JHA of 27 November 2008 on the protection of personal data in the framework of police and judicial cooperation in criminal matters, OJ 2008, L-350/60.

Council Framework Decision on Combating Terrorism, OJ 2002 L-164/03.

Council Framework Decision on the European Arrest Warrant and the surrender procedures between Member States, OJ 2002, L-190/1.

Curtea Constitutionala, 8 October 2009 number 1258, Romanian Official Monitor no. 789 of 23 November 2009, judgment related to the implementation of the Data Retention Directive.

De Hert, P., and R. Bellanova. "Data protection from a transatlantic perspective: The EU and US move towards an international data protection agreement?" study requested by the European Parliament's Committee on civil liberties, justice and home affairs (LIBE) (2008).

De Vries, K., R. Bellanova, and P. de Hert"Proportionality overrides Unlimited Surveillance, The German Constitutional Court Judgment on Data Retention", *Centre of European Policy Studies, Liberty and Security in Europe*, (2010), http://www.ceps.eu/book/proportionality-overrides-unlimited-surveillance (accessed August 25, 2010).

Directive 2006/24/EC of the European Parliament and of the Council of 15 March 2006 on the retention of data generated or processed in connection with the provision of publicly available electronic communications services or of public communications networks and amending Directive 2002/58/EC, OJ 2006, L-105/54.

Directive 95/46/EC of the European Parliament and of the Council of 24 October 1995 on the protection of individuals with regard to the processing of personal data and on the free movement of such data, OJ 1995, L-281/31.

Directive 97/66/EC of the European Parliament and of the Council of 15 December 1997 concerning the processing of personal data and the protection of privacy in the telecommunications sector, OJ 1998, L-24/1.

ECtHR, *Amann v. Switzerland*, Application no. 27798/95 from 16 February 2000.

ECtHR, *Association for European Integration and Human Rights and Ekimdzhiev v. Bulgaria*, Application no. 62540/00 from 28 June 2007.

ECtHR, *Biriuk v Lithuania*, Application no. 23373/03 from 25 November 2008.

ECtHR, *C.C. v. Spain*, Application no. 1425/06 from 6 October 2009.

ECtHR, *C.G. and others v. Bulgaria*, Application no. 1365/07 from 24 April 2008.

ECtHR, *Copland v. the United Kingdom*, Application no. 62617/00 from 3 April 2007.

ECtHR, *I. v Finland* Application no. 20511/03 from 17 July 2008.

ECtHR, *K.U. v. Finland*, Application no. 2871/02 from 2 December 2008.

ECtHR, *Kennedy v. the United Kingdom*, Application no. 26839/05 from 18 May 2010.

ECtHR, *L.L. v France* Application no. 7508/02 from 10 October 2006.

ECtHR, *Leander v. Sweden*, Application no. 9248/81 from 26 March 1987.

ECtHR, *Liberty and others v. the United Kingdom*, Application no. 58234/00 from 1 July 2008.

ECtHR, *Malone v. the United Kingdom*, Application no. 8691/79 from 2 August 1984.

ECtHR, *Panteleyenko v. Ukraine*, Application no. 11901/02 from 29 June 2006.

ECtHR, *Peck v. United Kingdom*, Application no. 44647/98 from 28 January 2003.

ECtHR, *Rotaru against Romania*, Application no. 28341/95 from 4 May 2000.

ECtHR, *S. and Marper v the United Kingdom*, Application nos. 30562/04 and 30566/04 from 4 December 2008.

ECtHR, *Segerstedt-Wiberg and others v. Sweden,* Application no. 62332/00 from 6 June 2006.

ECtHR, *Silver v. the United Kingdom*, Application no. 5947/72 and others, from 25 March 1983.

ECtHR, *Sunday Times v. the United Kingdom*, Application no. 6538/74.

ECtHR, *Valenzuela v. Spain*, Application no. 27671/95 from 30 July 1998.

ECtHR, *Weber and Saravia v. Germany*, Application no. 54934/00 Admissibility Decision from 29 June 2006.

ECtHR, *Z. v Finland*, Application no. 22009/93, from 25 February 1997.

Edward Hasbrouck Blog, http://www.hasbrouck.org/ (accessed August 25, 2010).

EU-PNR proposal in its latest version, Council doc. 5618/2/09, interinstitutional file 2007/0237 (CNS) from 29 June 2009.

European Data Protection Supervisor, Opinion on the draft proposal for a Council Framework Decision on the use of Passenger Name records (PNR) for law enforcement purposes, OJ 2008, C-110/01.

European Parliament, Resolution of 20 November 2008 on the proposal for a Council Framework Decision on the use of Passenger Name records (PNR) for law enforcement purposes, B6-0615/2008.

Joined cases C-317/04 and C-318/04, Parliament v. Council, [2006], ECR I-4721.

Manolea, B. "Implementation of EU Data Retention Directive Unconstitutional." *Computer Law Review International* 2 (2010): 49–51.

Mendez, M. "Passenger Name Record Agreement, European Court of Justice." *European Constitutional Law Review* 3 (2007): 127–147.

Opinion of the European Union Agency for Fundamental Rights on the Proposal for a Council Framework Decision on the use of Passenger Name Record (PNR) data for law enforcement purposes, 28 October 2008.

Papakonstantinou, V., and P. De Hert. "The PNR Agreement and Transatlantic anti-terrorism Cooperation: No firm human rights framework on either side of the Atlantic." *Common Market Law Review* 46, 3 (2009): 885–919.

Pawlak, P. "Made in USA? The influence of the US on the EU's data protection regime," *Centre of European Policy Studies, Liberty and Security in Europe, Justice and Home Affairs section*, (2009), http://www.ceps.be/book/made-usa-influence-us-eu%E2%80%99s-data-protection-regime (accessed August 25, 2010).

Peers, S. "The Third Pillar acquis after the Treaty of Lisbon enters into force" comment posted on December 1, 2009, http://www.statewatch.org/analyses/86-third-pillar-acquis-post-lisbon.pdf (accessed August 25, 2010).

Proposal for a Council Framework Decision on the use of Passenger Name Record (PNR) for law enforcement purposes from 6 November 2007, COM (2007) 654.

Recommendation R (87) 15 of the Committee of Ministers to the Member States regulating the use of personal data in the police sector, adopted 17 September 1987.

Siemen, B. *Datenschutz als europäisches Grundrecht*. Berlin: Duncker & Humblot, 2006.

Simitis, S. "Der EuGH und die Vorratsdatenspeicherung oder die verfehlte Kehrtwende bei der Kompetenzregelung." *Neue Juristische Wochenzeitschrift* 25 (2009): 1782–1786.

Chapter 9
Options for Securing PCs Against Phishing and Espionage: A Report from the EU-Project "Open Trusted Computing"

Arnd Weber and Dirk Weber*

9.1 Problems

This paper addresses the problem of malicious code from the Internet infecting computers. Malicious code includes viruses as well as so-called "Trojan horses". Trojan horses typically appear to be something useful, e.g., an update of a program or an email attachment worth reading, while in reality they perform an attack, e.g., by collecting passwords or other user data and reporting these data back to the attacker. In particular, we are thinking of Trojan horse attacks that lead to significant costs, e.g., attacks on home banking or economic espionage[1]. Such attacks can have severe consequences for the individual victim, whether this is a private person or a company. However, other types of malicious code, such as viruses, also lead to significant costs, e.g., in terms of the labour needed by the victim to restore systems, the expenses necessary for purchasing the usual means of protection, or for the administrators who are continuously necessary to update systems and clean up the systems after an attack has taken place, e.g., from a new virus not yet known to the existing protection tools. In considering the future use of digital signatures, for instance in eGovernment, faked signatures due to weaknesses in the signing environment (often a PC) will also form a potential threat to organisations, but in particular to the individual signing or relying person.

A. Weber (✉)
Institute for Technology Assessment and Systems Analysis, Karlsruhe Institute of Technology, Hermann-von-Helmholtz-Platz 1, D-76344 Eggenstein-Leopoldshafen, Germany
e-mail: arnd.weber@kit.edu

*Contribution based on work at Institute for Technology Assessment and Systems Analysis, KIT.

[1] See: Chris Dalton. "A Hypervisor Against Ferrying Away Data," Interview by Franco Furger and Arnd Weber. *OpenTC Newsletter,* April 2009. http://www.opentc.net/publications/OpenTC_Newsletter_07.pdf. MI5: *Espionage.*http://www.mi5.gov.uk/output/espionage.html.

S. Gutwirth et al. (eds.), *Computers, Privacy and Data Protection: an Element of Choice*, DOI 10.1007/978-94-007-0641-5_9, © Springer Science+Business Media B.V. 2011

9.2 Approaches

There appear to be various ways to address these problems. Let us quickly review their feasibility:

- Future updates of Microsoft Windows might address this problem. However, completely securing Windows "isn't going to work", as Paul England of Microsoft put it, due to the complexity of the system.[2]
- Other operating systems, such as the Apple Macintosh system or Linux, are similar to Windows and would also be attacked as soon as the user base is large enough for criminals to consider attacks worthwhile.
- Yet another approach would be to redesign computers from scratch. However, in practice such a system would not be very useful as existing user applications and data would not be usable on such a system.
- Another approach would be to use physically separate machines, and only use them for certain security critical actions. While this works for small devices such as smartcard readers with a display and may also work for certain applications such as military ones, these solutions are costly and inconvenient.
- We believe that only one viable solution is left, namely to develop a system which allows work with existing operating systems to be continued, while isolating other applications, be they security-critical or risky ones. In other words, all applications would run in sandboxes which are designed in such a way that malicious code cannot spread from one to the other. Such sandboxes, or compartments, could isolate particularly sensitive data as well as new, secure applications. However, they could also be used to isolate potentially malicious applications. For example, dubious emails could be isolated, arbitrary websites could be visited, or new software, operating systems, or drivers could be tested. If necessary, an infected compartment could be deleted and reinstalled from scratch. In order to have isolated compartments and manage them, a new layer would be needed, running "beneath" the operating systems. Such a layer is called a hypervisor or a virtual machine monitor. In such a system, an operating system would no longer communicate with the hardware, but only with a virtual hardware layer, provided by the hypervisor, hence one speaks of virtualisation.

[2]Paul England. "Practical Techniques for Operating System Attestation". Presentation given at: *Trusted Computing - Challenges and Applications, First International Conference on Trusted Computing and Trust in Information Technologies, Trust 2008,* Villach, Austria, March 11–12, 2008.

9.3 Progress

Figure 9.1 shows the architecture of such a system at the highest level. While the approach has been known in principle for some time[3], it would be desirable to use secure hardware to authenticate its code. Otherwise laypersons might be fooled into installing bad updates. Secure hardware could be used to block fake updates, or to alert to any straying from the secure path.

The authors participated in a research project which aimed at building a prototype of such a system. This was the Open Trusted Computing project which ran from 2005 to 2009.[4] Several prototypes have been built. It has been shown that an open-source hypervisor works with Microsoft Windows and can use the Trusted Platform Module as a hardware security anchor to monitor the system. Openness can be regarded as an essential characteristic, as it allows error detection by the interested public, and also could be used to ensure that there are no backdoors for eavesdropping on the users.

We discovered that Trusted Computing used with virtualization can make a lot of sense for individuals or corporations wishing to check whether their computers are in a known good state. Also, corporations might wish to check whether a computer requesting access to their Intranet has been properly set up (using remote attestation). It has often been assumed that, with Trusted Computing, the hardware vendors

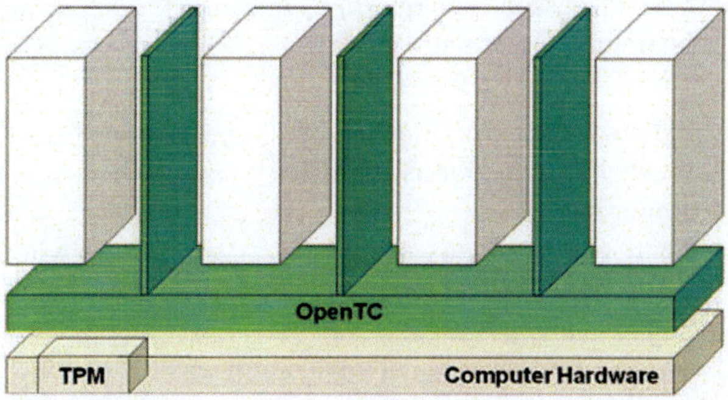

Fig. 9.1 Open trusted computing hypervisor high level architecture. Computer with a hypervisor (*green*), providing isolation to the operating system compartments (*white*), measured using the Trusted Platform Module. The compartments isolate legacy applications from new, secure ones, as well as from potentially risky code

---

[3] See: William Arbaugh, David Farber, and Jonathan Smith. "A Secure and Reliable Bootstrap Architecture," Proceedings of the 1997 IEEE Symposium on Security and Privacy: 65–71. Birgit Pfitzmann, James Riordan, Chris Stüble, Michael Waidner, and Arnd Weber. The PERSEUS System Architecture. IBM Research Report RZ 3335, IBM Research – Zurich, April 2001. http://www.zurich.ibm.com/security/publications/2001.html.

[4] The OpenTC-project was supported by the European Commission (project IST-027635). ITAS was responsible for work on requirements, specifications, and dissemination.

need to provide some guarantee about their computers, that long lists of good values would be needed to keep track of ever-changing hardware components, or that a global public key infrastructure is needed for using Trusted Computing, etc. None of these is necessary if a computer buyer trusts his vendor and sets up a closed system. Still, that buyer can verify all of his or her own machines. Trusted Computing could then evolve. Maintenance of these good values could be outsourced to a computer vendor, and values and keys could be exchanged with business partners, so that ultimately a global system might emerge.

In the framework of the project, the authors developed key aspects of a user interface. Our starting questions were: Is it possible to design a user interface for this kind of system such that it is usable by laypersons? The challenge is based, for instance on the fact that users need to understand that there are programs outside their usual operating system. Furthermore, the task is challenging as part of the screen space would be needed to inform the user about the new layer and the new programs, so this might reduce the usability of legacy applications. Furthermore, questions emerged for the design of such a user interface such as: How can the owner be protected against Trojan horses in the form of, e.g., pop-up windows claiming to represent a software update and asking for passwords or linking into confidential areas? Should a physically separate display provide information about the state of a compartment? Should a new hardware key on the keyboard be used to switch between compartments, which would also allow each compartment to obtain access to the full screen? We discussed such questions in a small in-depth survey of CTOs and leading administrators in Germany in 2006[5]. The answers were, in short:

- The hypervisor should have a simple GUI, e.g., with buttons using left and right mouse clicks. Neither administrators nor users want to spend time learning how to use new computer interfaces.
- Switching between compartments should be as simple as switching between applications is today, e.g., using a mouse click.

Figure 9.2 shows that a taskbar similar to today's taskbars could be used to manage the various sandboxes. This novel taskbar would use a secure part of the screen in order to unambiguously display whether the hypervisor, or certain sandboxes, are trustworthy.

The same approach to secure hardware-based virtualisation can be applied to servers or mobile devices. It can also be used to create compartments with digital rights management, such that an administrator cannot eavesdrop on the data, but delete the whole compartment if needed. The latter approach can in principle also be used to protect data on cloud computing servers.

---

[5]Dirk Weber, Arnd Weber, Stéphane Lo Presti. *Requirements and Design Guidelines for a Trusted Hypervisor User Interface.* Paper presented at: Future of Trust in Computing. Berlin, Germany, 30 June – 2 July, 2008. Proceedings published by Vieweg & Teubner, Wiesbaden 2009

Fig. 9.2 A possible future user interface. The figure shows a novel taskbar, with a sealed image as a means of protection against mimicry by malicious code. The image is only displayed if the hypervisor provided correct measurements. The taskbar also shows buttons for a corporate Windows (running), and buttons for switching to Linux, to a player, or to the hypervisor management interface. Note that the proposed amendment to the user interface requires only little screen space, so that the user interface available for normal work is hardly changed. On wide screens, the novel taskbar could be moved to a narrow side; it could even be hidden if not in uses

In the OpenTC project, several key modules were evaluated and any errors detected were corrected[6]. However, a research prototype is not a final product. The project showed that much more work is needed to build a whole, completely verified system.

9.4 Conclusions

For practitioners of technology assessment or data protection, we want to highlight several points which merit observation.

- The fight against attacks from the Internet may benefit from the use of virtualisation. It can be used to provide an additional layer of protection by isolating

[6]For technical details, see the information available at www.opentc.net. Use, e.g., the project's final report or the newsletter as a guidance.

sensitive data. Hypervisors might become mainstream technology, and in the future more and more applications might run in isolated compartments.

- Industry is not only working on hypervisors, but also working on producing the necessary hardware with "curtaining" features.[7]
- While any usable, unmeasured, unverified solution may help, ultimately a high quality system is needed. It might be advisable to watch out for products of a suitably assured quality, e.g., evaluated by a competent body and certified for use on open networks with malicious code. This applies to the whole system of hard- and software.
- To allay fears of security holes built in for law protection purposes, open source products might be attractive.

For protecting users, the development of such secure systems could be observed and influenced at the political level. Auditing requirements might further promote this approach, similar to the incentives provided by PCI-DSS (Payment Card Industry Data Security Standard) or the Sarbanes-Oxley Act on auditing requirements. Governments could procure open source systems certified for use on open networks.

Acknowledgements We wish to express our thanks to Dirk Kuhlmann, Armand Puccetti and Matthias Schunter and to all OpenTC-partners: Technikon Forschungs- und Planungsgesellschaft mbH (project coordination, AT); Hewlett-Packard Ltd (technical leader, UK); AMD Saxony LLC & Co. KG (DE); Budapest University of Technology and Economics (HU); Commissariat à l'Energie Atomique – LIST (FR); COMNEON GmbH (DE); Forschungszentrum Karlsruhe GmbH – ITAS (DE); Horst Goertz Institute for IT Security, Ruhr-Universitaet Bochum (DE); IBM Research GmbH (CH); Infineon Technologies AG (DE); INTEK Closed Joint Stock Company (RU); ISECOM (ES); Katholieke Universiteit Leuven (BE); Politecnico di Torino (IT); Portakal Teknoloji (TR); Royal Holloway, University of London (UK); SUSE Linux Products GmbH (DE); Technische Universitaet Dresden (DE); Technische Universitaet Graz (AT); Technische Universitaet Muenchen (DE); Technical University of Sofia (BR); TUBITAK – UEKAE (TR); and University of Cambridge (UK).

References

Arbaugh, W., D. Farber, and J. Smith. "A Secure and Reliable Bootstrap Architecture," *Proceedings of the 1997 IEEE Symposium on Security and Privacy*, (1997): 65–71.

Dalton, C. "A Hypervisor Against Ferrying Away Data," Interview by Franco Furger and Arnd Weber. *OpenTC Newsletter*, April 2009. http://www.opentc.net/publications/OpenTC_Newsletter_07.pdf.

England, P. Practical Techniques for Operating System Attestation. Presentation given at: *Trusted Computing - Challenges and Applications, First International Conference on Trusted Computing and Trust in Information Technologies, Trust 2008,* Villach, Austria, March 11–12, 2008.

Grawrock, D. *The Intel Safer Computing Initiative.* Intel Press, Hillsboro, 2006.

[7] David Grawrock. *The Intel Safer Computing Initiative.* Intel Press, 2006.

Kuhlmann, D., and A. Weber. *OpenTC Final Report. The Evolution of the OpenTC Architecture Illustrated via its Proof-of-Concept-Prototypes.* Bristol, Karlsruhe 2009, http://www.opentc.net/.

MI5: *Espionage.*http://www.mi5.gov.uk/output/espionage.html.

OpenTC. *Project website.* http://www.opentc.net/

OpenTC. Project newsletter, available at www.opentc.net.

Pfitzmann, B., J. Riordan, C. Stüble, M. Waidner, and A. Weber. "*The PERSEUS System Architecture.*" IBM Research Report RZ 3335, IBM Research – Zurich, April 2001. http://www.zurich.ibm.com/security/publications/2001.html.

Weber, D., A. Weber, and S. Lo Presti. "*Requirements and Design Guidelines for a Trusted Hypervisor User Interface.*" (Paper presented at: Future of Trust in Computing. Berlin, Germany, 30 June – 2 July, 2008). Proceedings published by Vieweg & Teubner, Wiesbaden 2009.

Part III
Privacy Practices as Vectors of Reflection

Chapter 10
Keeping Up Appearances: Audience Segregation in Social Network Sites

Bibi van den Berg and Ronald Leenes

10.1 Introduction

Millions of users worldwide use the internet to communicate and interact with others and to present themselves to the world via a variety of channels. These include, among others, personal and professional home pages, forums, online communities, blogs, dating sites, and social network sites such as Facebook, LinkedIn and MySpace. In this article we discuss some of the privacy-issues surrounding the presentation of personal content and personal information[1] in social network sites (SNSs). Particularly, we examine users' abilities to control who has access to the personal information and content they post in such communities. We conclude that social network sites lack a common mechanism used by individuals in their everyday interactions to manage the impressions they leave on others and protect their privacy: *audience segregation*. The lack of this mechanism significantly affects the level of users' control over their self-presentation in social network sites. In this article we argue that adding a virtual version of this real-world mechanism would contribute to enhancing privacy-friendliness in social network sites. We show that audience segregation is not only important in real life, but vital, yet currently undervalued and overlooked for the protection of one's self-images and privacy in social network sites.

B. van den Berg (✉)
Tilburg Institute for Law, Technology and Society (TILT), Tilburg University, Tilburg, The Netherlands
e-mail: bibi.vandenberg@uvt.nl

[1] By 'personal content' we mean any content (i.e. text, pictures, sounds, movies etc.) that can be attributed to and/or is experienced as 'personal' by the person posting it. By 'personal information' we mean any attribute (i.e. name, address, work or leisure affiliation, etc.) that can be attributed to and/or is experienced as 'personal' by the person posting it. This definition is broader than the definition of 'personal data' within Directive 95/46/EC and that of 'Personally Identifiable Information' as used in the US.

S. Gutwirth et al. (eds.), *Computers, Privacy and Data Protection: an Element of Choice*, DOI 10.1007/978-94-007-0641-5_10, © Springer Science+Business Media B.V. 2011

At the end of this article we present a privacy-preserving social network site called Clique [2] that we have built to demonstrate the mechanism. We discuss Clique and the three tools we have developed for it: contact-management, setting visibility rights, and managing multiple faces in a single social network environment.

10.2 Privacy Issues in Social Network Sites: Overview and Discussion

One of the fastest growing online fora for self-presentation and social interaction in recent years are "*social network sites*" (SNSs). In June 2008 these sites attracted "*an average of 165 million unique visitors a month*"[3]. Currently, Facebook claims to have over active 500 million users.[4] In these online domains, users can present themselves using a so-called "profile", and they can engage in interactions with a network of "contacts"[5] also active in the same environment. One of the most oft-quoted definitions of social network sites was developed by boyd and Ellison, who write that these are

> web-based services that allow individuals to (1) construct a public or semi-public profile within a bounded system, (2) articulate a list of other users with whom they share a connection, and (3) view and traverse their list of connections and those made by others within the system. The nature and nomenclature of these connections may vary from site to site.[6]

Despite the fact that social network sites are a recent phenomenon, there is quite a bit of variation in the intended goals of individual social network sites – ranging from dating and meeting friends, to connecting with work relations and finding new jobs, to providing recommendations for products, services and information[7]. Moreover, not all social network environments have the same *make-up*. Gross and Acquisti write:

---

[2]See http://clique.primelife.eu. Clique was built using Elgg [see http://elgg.com], an open source social networking engine.

[3]Kirsti Ala-Mutka, et al., The impact of social computing on the EU information society and economy. (Seville: IPTS/JRC, 2009), 16

[4]http://www.facebook.com/press/info.php?statistics, last accessed on 11 January 2011.

[5]Confusingly, in many current-day social network sites a person's contacts are called 'friends', regardless of the actual relation (friend, relative, colleague, acquaintance, and so on) the person has to these others. This issue will be discussed in more detail below. Following James Grimmelmann, we prefer to use the term 'contacts' for the collection of connections that a person gathers in a social network site, since "*. . .it's more neutral about the nature of the relationship than the terms used by many sites, such as 'friend'* [. . .] *. . .'friends' include not just people we'd call 'friends' offline but also those we'd call 'acquaintances'* [. . .] *Contact links are a mixture of what sociologists would call 'strong ties' and 'weak ties.'*" James Grimmelmann, "Facebook and the social dynamics of privacy [draft version]," (2008), http://works.bepress.com/james_grimmelmann/20/, 5 and 28.

[6]danah boyd and Nicole B. Ellison, "Social network sites: Definition, history, and scholarship." *Journal of Computer-Mediated Communication* 13 (2007): 211.

[7]Ralph Gross and Alessandro Acquisti, Information revelation and privacy in online social networks, (paper presented at WPES'05, Alexandria, Virginia, USA, 2005), 71

> The most common model is based on the presentation of the participant's profile and the visualization of her network of relations to others – such is the case of Friendster. This model can stretch towards different directions. In matchmaking sites, like Match.com or Nerve and Salon Personals, the profile is critical and the network of relations is absent. In diary/online journal sites like LiveJournal, profiles become secondary, networks may or may not be visible, while participants' online journal entries take a central role. Online social networking thus can morph into online classified in one direction and blogging in another.[8]

Sharing personal content and personal information is one of the key elements of social network sites. Individuals join these networks to present information about themselves, for instance through text (blogs, descriptions of their current activities etc.), through pictures, movies and sound clips, and through listing their "favorites" – a broad category of pre-defined and user-generated labels to help categorize oneself, ranging from clothing and other commercial brands, to music and movies, to locations and activities. Thus, an image of each individual user emerges. Most, though not all, information is added to the profile by users themselves. Other users can also add information to one's profile, thereby further refining the image created.

One of the most fascinating aspects of this emerging field of self-presentation is the fact that users put so much and such personal information about themselves in their profiles[9]. It is not surprising, therefore, that much of the research revolving around social network sites has focused on the *privacy* and *security issues* involved in individuals' self-presentations and the sharing of personal content and personal details. Acquisti and Gross write: "*...one cannot help but marvel at the nature, amount, and detail of the personal information some users provide, and ponder how informed this information sharing is*"[10]. In an article on the privacy risks for individuals using Facebook Grimmelmann dryly points out:

> Facebook knows an immense amount about its users. A fully filled-out Facebook profile contains about 40 pieces of recognizably personal information, including name; birthday; political and religious views; online and offline contact information; sex, sexual preference and relationship status; favorite books, movies, and so on; educational and employment history; and, of course, picture. [...] Facebook then offers multiple tools for users to search out and add potential contacts. [...] By the time you're done, Facebook has a reasonably comprehensive snapshot both of who you are and of who you know.[11]

---

[8]Ralph Gross and Alessandro Acquisti, Information revelation and privacy in online social networks, (paper presented at WPES'05, Alexandria, Virginia, USA, 2005), 72

[9]See for example: Zeynep Tufekci, "Can you see me now? Audience and disclosure regulation in online social network sites," *Bulletin of Science, Technology and Society* 28 (2008), and Alyson L. Young and Anabel Quan-Haase, Information revelation and internet privacy concerns on social network sites: A case study of Facebook, (paper presented at C&T '09, University Park, Pennsylvania, USA, 25–27 June, 2009)

[10]Alessandro Acquisti and Ralph Gross, "Imagined communities: Awareness, information sharing, and privacy on the Facebook," (paper presented at 6th Workshop on Privacy Enhancing Technologies, Cambridge, UK, 2006), 2

[11]James Grimmelmann, "Facebook and the social dynamics of privacy [draft version]," (2008), http://works.bepress.com/james_grimmelmann/20/, 9

So what makes people behave this way, given that there obviously are security and privacy issues? Why do they provide such detailed, and true[12], personal information on their social network site profile? Many explanations can be given, but we restrict ourselves to some of the most familiar. Acquisti and Gross say: "*Changing cultural trends, familiarity and confidence in digital technologies, lack of exposure or memory of egregious misuses of personal data by others may all play a role in this unprecedented phenomenon of information revelation*"[13]. Grimmelmann argues that the reason is actually much more straightforward: people misunderstand the risks involved in presenting detailed and personal information online. This misunderstanding takes a number of forms. For one thing, users are often unaware of who has access to their personal profile and to the content they place online, because the architecture and design of social network sites is such that it provides individuals with a false sense of security and privacy. These sites "*systematically* [deliver] *them signals suggesting an intimate, confidential, and safe setting*"[14], an environment that is private, "*closed to unwanted outsiders.*"[15]. Second, users falsely believe that there is safety in numbers, in two senses of the expression. They believe that when everyone else around them massively starts using social network sites, these sites must be safe to use, because otherwise others would avoid them (a line of reasoning

[12]There are some interesting differences between the level of truthfulness in self-presentations across different social network sites. Research has shown, for instance, that while the overwhelming majority of members use their real name on their Facebook profile (a staggering 94,9% according to Tufekci (Zeynep Tufekci, "Can you see me now? Audience and disclosure regulation in online social network sites," *Bulletin of Science, Technology and Society* 28 (2008)). An even higher number, 99,35%, was found in a 2009 study by Young and Quan-Haase (Alyson L. Young and Anabel Quan-Haase, "Information revelation and internet privacy concerns on social network sites: A case study of Facebook," (paper presented at C&T '09, University Park, Pennsylvania, USA, 25-27 June, 2009)). In the above-cited article Tufekci shows that, by contrast, in MySpace a substantial amount of users (38,2%) provide a nickname on their profiles. There are many explanations for such differences. One of the most straightforward ones is the fact that Facebook actively, and quite strictly, discourages the use of fake names, as was made clear by a tell-tale example presented by Grimmelmann: "*Facebook applies* [its] *policy* [regarding the ban on the use of fake names] *rigorously almost to the point of absurdity. It refused to let the writer R.U. Sirius sign up under that name, even though he'd written six books and hundreds of articles under it and he uses it in everyday life.*" (James Grimmelmann, "Facebook and the social dynamics of privacy [draft version]," (2008), http://works.bepress.com/james_grimmelmann/20/, 6). Another explanation could be that users want to avoid the fact that their friends cannot find them online. As boyd writes: "*While teens are trying to make parental access more difficult, their choice to obfuscate key identifying information also makes them invisible to their peers. This is not ideal because teens are going online in order to see and be seen by those who might be able to provide validation.*" (danah boyd, "Why youth (heart) social network sites: The role of networked publics in teenage social life," In *MacArthur Foundation Series on Digital Learning – Youth, Identity, and Digital Media Volume*, edited by David Buckingham. (Cambridge, MA: MIT Press, 2008b), 131-132)

[13]Alessandro Acquisti and Ralph Gross, "Imagined communities: Awareness, information sharing, and privacy on the Facebook," (paper presented at 6th Workshop on Privacy Enhancing Technologies, Cambridge, UK, 2006), 2

[14]James Grimmelmann, "Facebook and the social dynamics of privacy [draft version]," (2008), http://works.bepress.com/james_grimmelmann/20/, 17

[15]James Grimmelmann, "Facebook and the social dynamics of privacy [draft version]," (2008), http://works.bepress.com/james_grimmelmann/20/, 18

that runs the obvious risk of being flawed if everyone follows it), and they believe the risks they run are very limited since there are so many members in social network sites that chances are in fact really small that something will befall them as individuals (Grimmelmann, 2008: 17–18).

Or, as boyd argues,

> [m]ost people believe that security through obscurity will serve as a functional barrier online. For the most part, this is a reasonable assumption. Unless someone is of particular note or interest, why would anyone search for them? Unfortunately for teens, there are two groups who have a great deal of interest in them: those who hold power over them – parents, teachers, local government officials, etc. – and those who wish to prey on them – marketers and predators.[16]

Taking things to a more general level one can argue that there are four fundamental issues surrounding privacy and (unintended) information disclosure in relation to online worlds[17]. These can be summarised as follows:

- It is difficult or even impossible for users to know what the composition or the reach of the *audience* is for whom they are presenting their personal information and content;
- Since information on the internet can easily be recorded, copied and stored, it gets a degree of *persistence* that most information in the real world lacks. This means that information may (intentionally) reach audiences in the (far) future;
- Information shared in one internet environment may easily be *transported* (copied, linked) to other contexts. Thus, information that had one meaning in the original context may gain a different meaning in another context, possibly reflecting back on the individual in unintended and unforeseen ways;
- Our online self-presentations are the result of content and information posted by both ourselves and others, and made up of an amalgam of images ranging from deliberate and explicit self-presentations to more implicit "traces of self" of which users are not especially aware. *Controlling* these self-presentations and the possible deductions others may make on the basis of them is difficult, if not wholly impossible, for the individual.

These four issues are highly relevant to social network sites as well. For one, when posting content or personal information in a profile, individuals do not know (exactly) who will be able to access this information. The audience, to phrase it differently, is in-transparent. Now, while some social network sites allow users some level of control over the visibility of the information placed in profiles (e.g., changing personal information to "visible to friends only"), the default privacy settings

---

[16]danah boyd, "Why youth (heart) social network sites: The role of networked publics in teenage social life," In *MacArthur Foundation Series on Digital Learning – Youth, Identity, and Digital Media Volume*, edited by David Buckingham. (Cambridge, MA: MIT Press, 2008b), 133

[17]See for example: Leysia Palen and Paul Dourish, "Unpacking 'privacy' for a networked world," (paper presented at Computer-Human Interaction (CHI) Conference 2003, Ft. Lauderdale, Florida, USA, 5-10 April, 2003), and Daniel J. Solove. *The future of reputation: Gossip, rumor, and privacy on the Internet.* (New Haven, CT: Yale University Press, 2007)

are usually set to "public", which means that individuals' profiles and the information contained therein can be viewed by anyone accessing the social network site. This means, Acquisti and Gross conclude, "*that the network is effectively an open community, and its data effectively public.*"[18]

Second, since information can be copied, saved and stored easily and infinitely, information placed online at any particular moment may come back to haunt the individual years down the line. This means that the audience is unlimited both in terms of its size and makeup (in contrast to audiences in the physical world), but also in terms of temporality. In the words of Tufekci, the temporal boundaries shift in such a way that "*the audience can now exist* ***in the future****.* [. . .] *Not only are we deprived of audience management because of spatial boundaries, we also can no longer depend on simultaneity and temporal limits to manage our audiences.*"[19]

Third, as we will discuss more extensively below, when presenting disparate identities in various online domains, there is a risk of information from one of these domains, for instance personal or professional home pages, seeping into another, such as someone's social network site profile. Since different behavioural rules guide these various domains mixing and merging information about the person behind all of these various roles can lead to serious problems. Tufekci gives a very simple, yet illuminating example:

> For example, a person may act in a way that is appropriate at a friend's birthday party, but the photograph taken by someone with a cell phone camera and uploaded to MySpace is not appropriate for a job interview, nor is it necessarily representative of that person. Yet that picture and that job interview may now intersect.[20]

Last, and this is related to the previous point, in social network sites who we are is expressed by an online representation of ourselves, which may be composed, for instance, of a profile with personal details, stories and pictures. Now, while we have some level of control over the type and content of information we put online, our control only goes so far. Other users can add or change information in a person's personal profile, put pictures or information about the person on their own or other people's profiles, and tag pictures to reveal the identities of those portrayed in them. Tufekci's example in the previous paragraph is a case in point: placing a picture of another person online affects the image of that person to the audience viewing it, and hence may have an effect on the (current and future) self-presentations and impressions of that individual.

---

[18]Alessandro Acquisti and Ralph Gross, "Imagined communities: Awareness, information sharing, and privacy on the Facebook," (paper presented at 6th Workshop on Privacy Enhancing Technologies, Cambridge, UK, 2006), 3

[19]Zeynep Tufekci, "Can you see me now? Audience and disclosure regulation in online social network sites," *Bulletin of Science, Technology and Society* 28 (2008), 22, emphasis in the original

[20]Zeynep Tufekci, "Can you see me now? Audience and disclosure regulation in online social network sites," *Bulletin of Science, Technology and Society* 28 (2008), 22

The central question we posed ourselves in our own research on privacy issues in social network sites was how we could contribute to solving some of the issues outlined in this section. We will turn to a description of some of our ideas now.

10.3 Privacy-Preserving Social Networking: Audience Segregation

In our view, there are two central issues to be addressed in providing users with more privacy-respecting or -preserving social network environments:

- User *awareness* of the privacy issues discussed in the previous section should be raised, i.e., users ought to become more aware of the fact that, and the ways in which, personal information and personal content may "leak" to unintended audiences and places on the internet;
- Users should be provided with *tools* to help them manage their personal information and content in a more privacy-friendly manner.

To maximise awareness and usability, these tools ought to be easily recognisable for users. This is why we have taken a social mechanism that individuals use in everyday life contexts to control the image others have of them and the information they disclose about themselves: *audience segregation*. Mirroring or mimicking this real-life strategy in a virtual environment, we have developed a social network site, Clique, that implements it.

10.3.1 Audience Segregation

The concept of "*audience segregation*" was coined by Erving Goffman[21] as part of a perspective on the ways in which identities are constructed and expressed in interactions between human beings in everyday contexts. According to Goffman, whenever individuals engage in interactions with others they *perform roles*, the goal of which is to present an image of themselves which is favourable, not only to the personal goals they are attempting to achieve within the context in which they find themselves (strategic interaction), but at the same time also meets with the approval of those with which they engage in the interaction ("public validation"[22]). To Goffman, then, *impression management* is key in such self-presentations.

Individuals performs a wide variety of roles in their everyday lives, relating to both the places they visit, and the other people present there[23]. For instance, when

---

[21] Erving Goffman. *The presentation of self in everyday life*. (Garden City, NY: Doubleday, 1959)

[22] Ann Branaman, "Goffman's social theory," In *The Goffman reader*, edited by Charles C. Lemert and Ann Branaman. (Cambridge, MA: Blackwell Publishers, 1997), xlvi

[23] See for example: Joshua Meyrowitz. *No sense of place: The impact of electronic media on social behavior*. (New York, NY: Oxford University Press, 1985), and Bibi Van den Berg. The situated self: Identity in a world of Ambient Intelligence. (Nijmegen: Wolf Legal Publishers, 2010)

at work, individuals will display different images of themselves than when they are at home, or when they buy groceries at a local store, or when they visit a movie theatre. However, the *location* a person finds himself in is not the only relevant parameter; so is the *presence* (or absence) of *specific other people* in that location. Individuals will show different sides of themselves when they are at home with their family than when they are hosting a party for their colleagues in that same home. The presentation of selves, then, is *situated* or *contextual* – it relates to *where* one is, and *who else is there* [24].

One of the key elements of Goffman's perspective on identity its the fact that individuals attempt to present self-images that are both *consistent* and *coherent*. To accomplish this, performers engage in what Goffman calls "audience segregation", "*...so that the individuals who witness him in one of his roles will not be the individuals who witness him in another of his roles*"[25]. With segregated audiences for the presentation of specific roles, people can "maintain face" before each of these audiences. Their image will not be contaminated by information from other roles performed in other situations before other audiences, particularly not by information that may *discredit* a convincing performance in the current situation[26]. For example, a person whose professional role consists of displaying a role of authority, such as a political leader or a judge, may try to shield aspects of his private life from the public, such as the fact that in his relationship his partner is the one in charge and he is not an authoritative person at all when at home. He shields this information from those he may encounter in his professional life to prevent his professional authority being undermined by their knowing about this aspect of his personal life.

While Goffman's idea of audience segregation didn't originally relate directly to privacy, it is easy to see that audience segregation and privacy are, in fact, closely linked. Helen Nissenbaum has famously argued that privacy revolves around "*contextual integrity*", which means that individuals' personal integrity ought to be maintained across and between the various contexts they engage in each day[27]. Nissenbaum starts from the following observation:

> Observing the texture of people's lives, we find them [...] moving about, into, and out of a plurality of distinct realms. They are at home with families, they go to work, they seek medical care, visit friends, consult with psychiatrists, talk with lawyers, go to the bank,

---

[24] Bibi Van den Berg, "Self, script, and situation: Identity in a world of ICTs," in The future of identity in the information society: Proceedings of the third IFIP WG 9.2, 9.6/11.6, 11.7/FIDIS International Summer School on the Future of Identity in the Information Society, ed. Simone Fischer-Hübner, Penny Duquenoy, Albin Zuccato and Leonardo Martucci. (New York, NY: Springer, 2008), and Bibi Van den Berg. The situated self: Identity in a world of Ambient Intelligence. (Nijmegen: Wolf Legal Publishers, 2010)

[25] Erving Goffman. *The presentation of self in everyday life*. (Garden City, NY: Doubleday, 1959), 137

[26] Erving Goffman. *The presentation of self in everyday life*. (Garden City, NY: Doubleday, 1959), 137

[27] Helen Nissenbaum, "Privacy as contextual integrity," *Washington Law Review* 79 (2004), also see Kieron O'Hara and Nigel Shadbolt. *The spy in the coffee machine*. (Oxford: Oneworld Publications, 2008), 77 ff.

> attend religious services, vote, shop, and more. Each of these spheres, realms, or contexts involves, indeed may even be defined by, a distinct set of norms, which governs its various aspects such as roles, expectations, actions, and practices.[28]

Following Michael Walzer[29], Nissenbaum argues that what privacy is, is the fact that we respect the contextual boundedness of the (personal) information individuals share in each of these distinct realms. Phrased differently, according to this view privacy revolves around a person's ability to keep audiences separate and to compartmentalise his or her (social) life.

10.3.2 Audience Segregation in Social Network Sites: Why?

Above we have argued that in current social network sites users lack mechanisms to *separate and manage the various audiences for whom they perform.* Many social network sites only provide their users the option to collect one list of contacts, called "friends". Given the fact that Facebook users, for instance, on average have 130 "friends"[30], this necessarily conflates different contexts. Providing users with mechanisms to control access over the information they present in such online communities would improve the quality of interactions and self-presentations for three reasons. First of all, it would mimic real life interaction patterns to a larger degree, and align more closely with the ways in which individuals tend to engage with others in everyday settings. As we have seen, audience segregation is a common feature of self-presentations in everyday life, and even a necessary *requirement* for optimal impression management and role performance. Second, enabling access control and audience segregation in social network sites could be a first step in countering some of the privacy and security risks we have discussed above and, therefore, make social network sites more privacy-friendly. Considering the numbers of people active on social network sites today it seems that this is a worthwhile goal to strive for indeed. Third, enabling users to compartmentalise the audiences for whom they perform in social network sites provides them with an opportunity to present different sides of themselves to different audiences, thereby allowing each (partial!) self-presentation to be textured and full of depth. Audience segregation enables users to avoid what danah boyd calls "*social convergence*"[31]. If individuals do not have enough facilities to properly manage impressions in front of various separate audiences, they need to present one single "face" that works for all of these audiences. While these conflated self-presentations might be acceptable for a wide range of audiences and a wide assortment of social contexts, they will at the same time lack the depth,

---

[28] Helen Nissenbaum, "Privacy as contextual integrity," *Washington Law Review* 79 (2004): 137.

[29] Michael Walzer. *Spheres of justice: A defense of pluralism and equality*. (New York, NY: Basic Books, 1983)

[30] See http://www.facebook.com/press/info.php?statistics, last visited 11 January 2011.

[31] danah boyd, "Facebook's privacy trainwreck," Convergence: The International Journal of Research into New Media Technologies 14 (2008a)

breadth, variety and uniqueness of socially constricted, contextual ones. Moreover, with multiple audiences to keep into account, it becomes very difficult to decide what "face" to show. The result, says boyd, is social convergence:

> Social convergence occurs when disparate social contexts are collapsed into one. Even in public settings, people are accustomed to maintaining discrete social contexts separated by space. How one behaves is typically dependent on the norms in a given social context. How one behaves in a pub differs from how one behaves in a family park, even though both are ostensibly public. Social convergence requires people to handle disparate audiences simultaneously without a social script. While social convergence allows information to be spread more efficiently, this is not always what people desire. As with other forms of convergence, control is lost with social convergence.[32]

Therefore, audience segregation offers users the opportunity to be "round characters" in each role, rather than merely "flat ones", to borrow some terminology from literature studies.

Now, not all social network sites have the same intended *goals*. Some cater specific needs, such as providing opportunities for finding a date or meeting new friends, while others cater to specific groups, such as professionals, or provide opportunities for finding specific products, services and information. When social network sites cater individuals' specific needs or revolve around particular groups, it is easy to see that audience segregation is both relevant and desirable. A person presenting himself in a profile on a dating network may feel uncomfortable if the information displayed there "spills over" into other domains and networks, for instance into their work-related network. Alternatively, a person presenting himself in a network providing professional connections will want to avoid information regarding his (all too) personal sphere or background from seeping in.

However, audience segregation does not merely apply to the spill-over of information from one online environment into another, but is also an issue *within* one and the same environment. We envision that users would find it convenient and worthwhile to be able to control their various kinds of online profiles using a single dashboard. This would entail that, for instance, a person's work profile, his personal profile and the profile for his avatar in an online role-playing game such as Second Life would be combined within a single social network site. Moreover, a person's profile information from collaborative workspaces such as wikis and forums could be stored in the same place as well. Facebook and Friendster already cater to the more "general" goal of connecting individuals without a particular shared interest or aspect of self, and hence it seems likely that social network sites such as these will most easily grow into the "central identity management platforms" that we envisage.

In these multipurpose social network sites individuals connect with both friends, family members, distant relatives, colleagues, acquaintances, old schoolmates, members of their local community, etc. – some of whom are intimately known to them, while others are distant, loose, or even unknown connections. It is easy to see why individuals using such sites might want to make distinctions between the *types*

---

[32]danah boyd, "Facebook's privacy trainwreck," Convergence: The International Journal of Research into New Media Technologies 14 (2008a), 18

of information they want to make available to each of these different categories of connections, and give different connections access to different *content*. For instance, an individual might want to share his holiday pictures with close friends, family members and other relatives, but not with his colleagues or old schoolmates. Or, more specifically, he might want to share his holiday pictures with his close friends and family members – but *not* with Mom and Aunt So-and-so. Alternatively, an individual might want to share work-related documents or postings with his colleagues, but not with his friends, *except* for Friend So-and-so, and so on and so forth.

Currently, most social network sites provide limited options for making one's profile or its content (in)visible for specific others or specific collections of others. Generally, users can choose from: "visible to everyone" (i.e. all members of the social network site), "visible only to friends" (i.e. all of a user's contacts!), "visible only to friends and friends of friends", and in some cases "invisible to everyone"[33]. In some social network sites, the user can specify the (in)visibility settings of specific *types* of information, e.g. they can make their basic information (name, home town etc.) available to all members of the platform, while keeping their pictures only for their contacts. Assigning different "collections" within one's own network of contacts has recently been added as an option to Facebook, but at the moment none of the other major social network sites (e.g. Friendster, LinkedIn, MySpace) have it, let alone assigning different access rights to different individuals and for different kinds of content within one's own network of contacts.

10.4 A Note on Terminology

Before turning to a presentation of the way in which we've translated the conceptual ideas of audience segregation into a working demonstrator, we address an issue concerning terminology. The language used to discuss online communities, the users participating in them, and the connections between these users is often quite fuzzy and imprecise. This is why we pause to define each of these concepts.

1. The terms "*platform*" and "social network site" (which we've defined in the introduction to this article) will be used interchangeably;
2. On the platform a person can create a "*face*", a profile page on which he displays particular information about himself. The totality of all the faces a person manages within a platform makes up his identity. While users currently tend to have only one face in social network sites catering specific needs (e.g. dating or professional self-presentation), those catering to several needs, or those catering no specific need at all, might invoke users to create *multiple* faces within the same domain. In such social network sites, then, the personal information making up various identities may be brought together for each individual user;

[33]This applies, for instance, to one's e-mail address.

3. "*Contacts*" are all the individuals with whom a users is connected within the platform;
4. "*Collections*" are sets of contacts selected and assigned by the individual from the totality of his network. Collections can consist of anywhere between zero and an unlimited amount of contacts. The individual can assign a name to each collection to identify them as a collection (e.g., "colleagues" or "old schoolmates" or "boring people"). Collections have labels that have meaning for their creator. The labels are not visible to the members of a particular collection. They need not know that they are grouped into a cluster "distant friends". The distant friends may know or realise that they don't belong to someone's inner circle, but usually this is not made explicit in real life interactions.
   Each time content is added to the profile, it can be made available for specific collections, or even for specific members of each collection, based on the user's own preferences (more on this below). The management of collections and the content available to them should be dynamic, transparent and open to change at all times.
5. A "*context*" is each instance in which a *particular face* and a *particular collection* come together. For instance, a "work context" is one in which a user presents his "work identity" (face) to his "collea-gues" (collection). Similarly, a "reminiscence context" arises when a user presents information (pictures, documents, text in chat relays) (face) regarding his younger years to his "old school friends" (collection). A third example is that of a person making his holiday pictures available, i.e. information that is often regarded as quite personal (face) to all of his family members (collection) and some individuals from his friends (collection).

Figure 10.1 presents a graphic depiction of the structures and concepts we distinguish in relation to social network sites and collaborative workspaces.

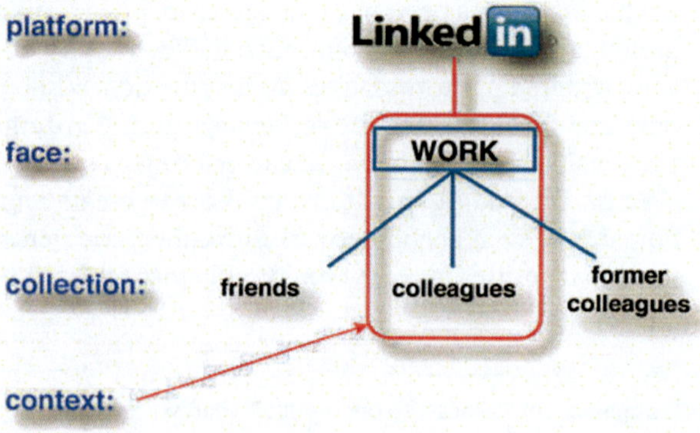

Fig. 10.1 Terminology

10.5 Transforming the Conceptual Framework into Practical Tools

In the remainder of this article we will present our proposals for realising audience segregation within a social network site. We have implemented this mechanism into three tools: a tool for contact-management, one for setting access control policies, and one for managing multiple faces.

10.5.1 Contact-Management: Collections

Our starting point for realising audience segregation in social network sites is the introduction of *nuance* in connections[34]. By this we mean: enabling users to create their own labels for "collections" in which they may cluster one or more of their contacts. As we have seen above, in most current-day social network sites all contacts in a user's network are lumped together in one category. No distinction is made between the different social networks a person may participate in, as all of us do in our everyday lives. This means that a) it is impossible for users to hide parts of their network of contacts from other contacts (e.g., a person does not want his colleagues to see his friends, or he does not want his mother to see his colleagues); and b) that it is impossible to show particular information to one portion of one's network, while hiding it from others. All information displayed on one's profile is there for all to see, at least for one's entire network of contacts.

By allowing users to create collections within their list of contacts, they can cluster social relations according to their own preferences, and thereby mimic the actual practice of building and maintaining separate social spheres in real life in the process. It is important that users are free in labelling their own set of collections, since they themselves know best what the fabric of their own social lives consists of and how it could be divided into relevant and meaningful categories.

James Grimmelmann has argued that offering what he calls "technical controls" to manage the (in)visibility of a person's profile in social network sites is not a workable solution. He claims that if the provider of the social network site offers the possibility to place contacts in clusters (such as "family" or "friends") then these clusters are never going to be an adequate representation of the complexity of social relationships in real life. He writes:

> Consider the RELATIONSHIP project, which aims to provide a "vocabulary for describing relationships between people" using thirty-three terms such as "apprenticeTo," "antagonistOf," "knowsByReputation," "lostContactWith," and "wouldLikeToKnow."[...] Clay Shirky shows what's wrong with the entire enterprise by pointing out that RELATIONSHIP's authors left out "closePersonalFriendOf," "usedToSleepWith," "friendYouDontLike," and every other phrase we could use to describe our real, lived

[34] J. Donath and danah boyd, "Public displays of connection," *BT Technology Journal* 22 (2004): 72.

> relationships. [. . .] We shouldn't expect Facebook's formal descriptors to be precise approximations to the social phenomena they represent.[35]

Grimmelmann is absolutely right, of course, in claiming that the social network site *provider* can never manage to capture the complexity of individuals' many social spheres and connections. However, we argue that the *individuals themselves* are fully capable of doing so, and this is why it is important to place access control mechanisms into their hands. Users can then choose which labels to use for which collections and also how granulated they want their own set of collections to be. This solves the problem signalled by Grimmelmann above. Having said that, with regard to user-friendliness a number of standard options might be included as labels for collections (e.g., "family", "relatives", "friends", "colleagues", "acquaintances", etc.).

In Clique, the creation and management of collections was one of the first functionalities introduced. Users in Clique can cluster contacts into self-assigned and self-labelled sets. After inviting contacts, they can assign them to one or more collections, and change or delete these ascriptions at any time. Figure 10.2 shows what collection management in Clique looks like. Notice that the collection "colleagues" is marked as Ronald's primary audience (marked as default).

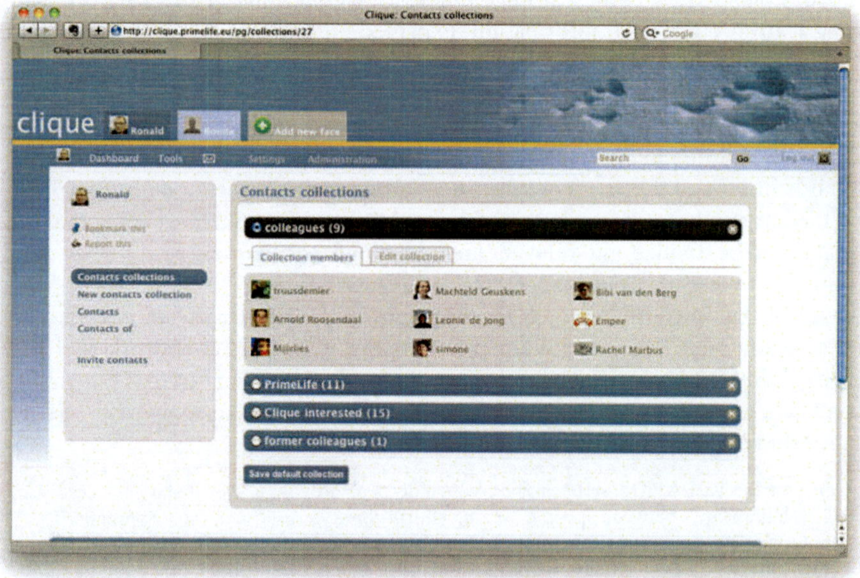

Fig. 10.2 Managing collections in clique

[35]James Grimmelmann, "Facebook and the social dynamics of privacy [draft version]," (2008), http://works.bepress.com/james_grimmelmann/20/, 27

10.5.2 Setting Visibility Rights

The second principle in realising audience segregation in social network sites is *contextualising* the user's profile and all the information gathered there[36]. This means that a person's specific "face" is combined with information made public for a specific collection. Such contextualisation mimics the maintenance of different social spheres as we have them in real life. In most social network sites the user builds one single profile, in which all of his information is stored. All of his contacts see the same information. However, as we have argued in this article it is important to allow users to diversify the information and content they present to various audiences. Moreover, many people now maintain different profiles in different social network sites, which is cumbersome and time-intensive. As we have argued above it seems reasonable to suspect that users would prefer gathering all of the various profile pages in one single social network site. Obviously this development makes it all the more important that users can contextualise the content and information they share in each face.

We have developed two tools for contextualising content and information in Clique. The first is the use of *visibility rights,* which enables users to assign access rights to different collections and individuals. Each time users post items of information (personal information in a profile, pictures, text, documents, etc.) within a context, they can choose for which contacts (both collections and individuals) this item will be visible. For example, a user may decide to make his holiday pictures invisible to his colleagues but visible to his relatives and some members of his collection of friends, or he may decide to prevent acquaintances from reading his diary entries, but leave them visible to everyone else in his contacts list.

In Clique we provide individual users as much control over the visibility settings of each individual item of information as possible for two reasons. First, individuals use social network sites to present personal information and personal content with different goals and purposes in mind. Some may use them, for instance, only to stay in touch with people they know intimately in the real world, whereas others may want to use them especially to present (aspects of) themselves before an audience of strangers. Obviously, the needs of these people, in terms of the visibility of their information, varies. Therefore, it would be patronising and limiting if the social network provider or the software designer would decide for users which information to share and for which (limited or unlimited) audience.

Second, users' ideas of which kinds of information are deemed "private" vary. As O'Hara and Shadbolt write:

> Different people have different views of what should be private. [. . .] People must be able to reach their own decisions about what should be private, and what gains they would hope to make by releasing information about themselves.[37]

---

[36] J. Donath and danah boyd, "Public displays of connection," *BT Technology Journal* 22 (2004): 72.

[37] Kieron O'Hara and Nigel Shadbolt. *The spy in the coffee machine.* (Oxford: Oneworld Publications, 2008), 74

Now, one of the most obvious objections to this choice would be the idea that users do not *want* to have this much control over their personal information and personal content in social network sites. In fact, in the past researchers regularly argued that users wouldn't be interested in having possibilities for more fine-grained control over the display of personal data, for instance because making the profile invisible makes it harder for other people to find them[38], or because they would simply find it too much hassle. However, recent research has shown that, when given the opportunity, many people do in fact want to shield some of their information[39], especially since a number of negative examples regarding information spill and privacy issues with respect to social network sites have been published in the press in many Western countries.[40]

We have built a fine-grained architecture for setting access control policies, in which each consecutive element of the profile can be made visible for either collections, or individuals, or a mixture of both. This means, for instance, that a user can make his name and date of birth visible to everyone while keeping his address invisible for anyone, and allowing only some of his contacts, of his own choosing, to see his mobile phone number. The picture below shows the user profile page in Clique. With each entry there is an icon, which displays who can access that particular datum. Figure 10.3 displays these visibility settings in Clique.

Users can choose between the following access control options for the content published on their profile: "only visible to me", "contacts/collections", "all contacts", and "public".

When users publish information they are presented with an access control dialogue as shown in Fig. 10.4 below. In this dialogue window we "nudge"[41/42] the user to act in a privacy savvy manner without undermining sociality. By default, the user's primary audience (default collection, see Fig. 10.2) is selected as having access to the content to be published. The user can drag collections and individual contacts to the red and green boxes to grow or shrink the audience. Note that in this case, Ronald's colleagues have access to the content to be published, with the

[38]See for example: danah boyd, "Why youth (heart) social network sites: The role of networked publics in teenage social life," In *MacArthur Foundation Series on Digital Learning – Youth, Identity, and Digital Media Volume*, edited by David Buckingham. (Cambridge, MA: MIT Press, 2008b)

[39]See for example: Zeynep Tufekci, "Can you see me now? Audience and disclosure regulation in online social network sites," *Bulletin of Science, Technology and Society* 28 (2008)

[40]On 21 November 2009, for instance, the Canadian Broadcasting Corporation presented a story of a Canadian woman who was on long-term sick leave due to depression. This woman's health benefits were allegedly terminated after the health insurance company discovered pictures of the woman tanning on a beach and having a good time at a party with strippers on her Facebook page. See http://www.cbc.ca/canada/montreal/story/2009/11/19/quebec-facebook-sick-leave-benefits.html [last accessed 25 November 2009].

[41]The Nudge 'methodology' consists of: provide iNcentives, Understand mappings, Defaults, Give feedback, Expect error, Structure complex choices

[42]Richard H. Thaler and Cass R. Sunstein. *Nudge: Improving decisions about health, wealth, and happiness*. (New Haven, CT: Yale University Press, 2008)

Fig. 10.3 Visibility cues in clique

exception of Arnold Roosendaal, and four other individuals. While enabling access to a collection, thus, the user can still choose to make information unavailable for particular individuals.

The icon associated to the published content reveals the audience when hovering over (see Fig. 10.5).

10.5.3 Managing Multiple Faces in One Social Network Site: Tabs

The second tool we have developed to contextualise information is the introduction of *tabs* to represent the different faces a user may want to combine within the same social network environment. Each tab functions as a separate social sphere, representing one aspect of the user's identity. For instance, users may create a tab for their private face and for their professional face. Each of these faces contains a network of contacts, who can be assigned to the various collections within each tab. Access rights can be defined for collections and contacts with regard to all personal information and content presented in a context (i.e. using a specific face in front of a specific collection). Contacts only get access to the information that is made visible for them. This means that a) contacts who only know the individual professionally, for instance, are prevented from acquainting themselves with his digital representation from a leisurely profile; and b) within each face, contacts can only access the information that is made available for them through the use of visibility rights.

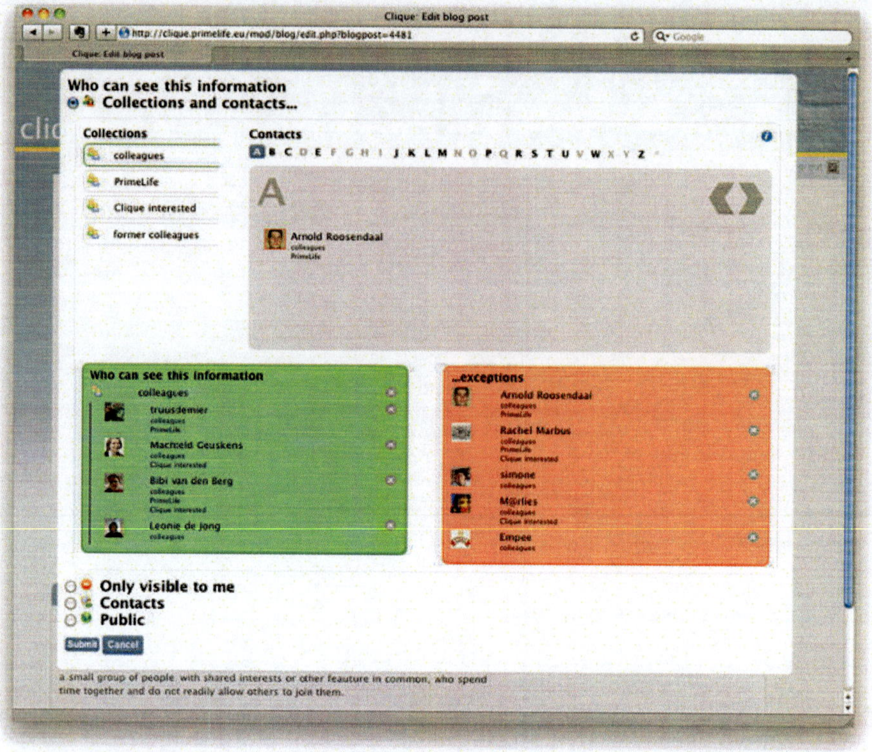

Fig. 10.4 Extended access control dialogue in clique

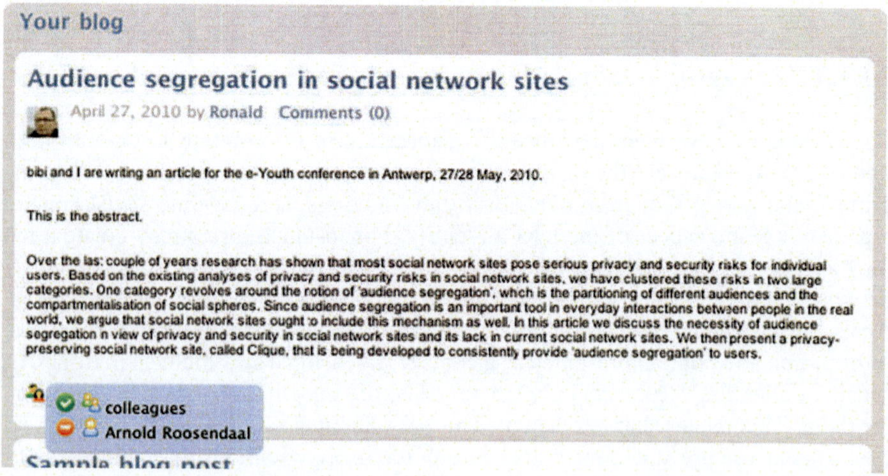

Fig. 10.5 Audience indicator in clique

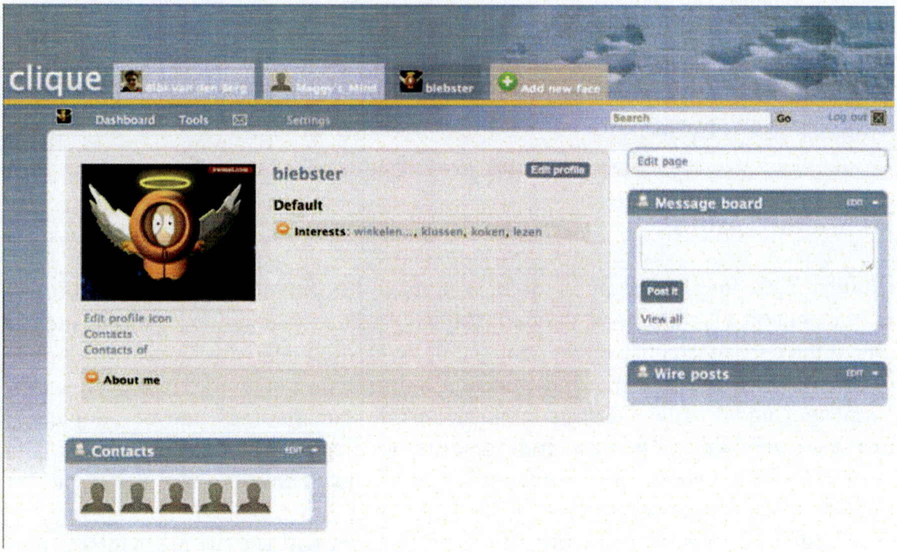

Fig. 10.6 Managing multiple faces in clique

Using tabs to distinguish between different contexts is a visually appealing and easy way for the individual to manage his or her own profile and the various faces contained therein. Information added to one of the faces (e.g. the "Biebster" tab in Fig. 10.6 above) is invisible in all other tabs, and hence it is easy for the user to manage who sees what. Clique can therefore be seen as a dashboard for multiple social contexts. By simply clicking through the different tabs a user can see what information is accessible there, and by hovering over the icons attached to each item of information, he or she can easily keep track of what information is made available to whom. Figure 10.6 displays multiple tabs, each representing a different face, for a single user.

Creating new faces is a bit cumbersome, since it means that users need to build a new profile, set the security and privacy settings, and add contacts and content for each individual face. This means users need to invest energy and time in setting up a new profile. Particularly when users create multiple faces for which the contact list shows a significant overlap we may wonder whether users are willing to make this investment, and whether they may see (enough of) the benefits and advantages of creating separate faces. However, this objection can be remedied by allowing users to import existing profile pages and contact lists, for instance from LinkedIn or Facebook, into separate tabs in Clique. Moreover, once the face has been created it is instantly clear what the advantages of this system are, and that they outweigh the initial energy to be invested. The visual separation of different social spheres and the division of content between these spheres, entails that users can effortlessly see which contact sees which information, both in terms of the profile and the content he or she has posted on his page. Managing audience segregation has thus been reduced to an intuitive, easy-to-manage and basic element of the social network site. This means that the user can engage in interactions with his contacts in a safer and more "natural" way, without having to manage his information with a high level of vigilance and privacy-awareness.

10.6 Conclusion

Context is a central concept in the disclosure of information. What is appropriate in one context may not be in another. We have argued that audience segregation is one of the core mechanisms that people employ in their everyday life to accomplish contextual integrity and that most current online social network sites have a very simplistic model of social structures. In our view, technology can be adopted to help users maintain different partial identities en control who can access their data even in social networks. We have taken the first steps in developing a prototype that implements audience segregation.

Whether or not social network site users can, and will use the mechanisms provided remains to be seen. To test whether they do, we have set up an experimental site consisting of the Clique prototype (http://clique.primelife.eu). The reader is invited to participate in this experiment.

Part of the research leading to these results has received funding from the European Community's Seventh Framework Programme (FP7/2007-2013) under grant agreement No. 216483. The authors want to thank Joeri de Ruiter who did a tremendous job of implementing the ideas of the authors into the Clique application.

References

Acquisti, A., and R. Gross, "Imagined communities: Awareness, information sharing, and privacy on the Facebook." (paper presented at 6th Workshop on Privacy Enhancing Technologies, Cambridge, UK, 2006): 36–58.

Ala-Mutka, K., D. Broster, R. Cachia, C. Centeno, C. Feijóo, A. Haché, S. Kluzer, S. Lindmark, W. Lusoli, G. Misuraca, Y. Punie, and J.A. Valverde, *The impact of social computing on the EU information society and economy*. Seville: IPTS/JRC, 2009.

boyd, d., "Facebook's privacy trainwreck." *Convergence: The International Journal of Research into New Media Technologies* 14 (2008a): 13–20.

boyd, d., "Why youth (heart) social network sites: The role of networked publics in teenage social life." In *MacArthur Foundation Series on Digital Learning – Youth, Identity, and Digital Media Volume*, edited by D. Buckingham, 119–142. Cambridge, MA: MIT Press, 2008b.

boyd, d., and N.B. Ellison, "Social network sites: Definition, history, and scholarship." *Journal of Computer-Mediated Communication* 13 (2007): 210–230.

Branaman, A., "Goffman's social theory." In *The Goffman reader*, edited by C.C. Lemert and A. Branaman, xlv-lxxxii. Cambridge, MA: Blackwell Publishers, 1997.

Donath, J., and d. boyd, "Public displays of connection." *BT Technology Journal* 22 (2004): 71–83.

Goffman, E., *The presentation of self in everyday life*. Garden City, NY: Doubleday, 1959.

Grimmelmann, J., "Facebook and the social dynamics of privacy [draft version]," (2008), http://works.bepress.com/james_grimmelmann/20/ (last accessed on July 6, 2009).

Gross, R., and A. Acquisti, "Information revelation and privacy in online social networks." (paper presented at WPES'05, Alexandria, Virginia, USA, 2005): 71–81.

Meyrowitz, J. *No sense of place: The impact of electronic media on social behavior*. New York, NY: Oxford University Press, 1985.

Nissenbaum, H. "Privacy as contextual integrity." *Washington Law Review* 79 (2004): 119–159.

O'Hara, K., and N. Shadbolt. *The spy in the coffee machine*. Oxford: Oneworld Publications, 2008.

Palen, L., and P. Dourish. "Unpacking 'privacy' for a networked world" (paper presented at Computer-Human Interaction (CHI) Conference 2003, Ft. Lauderdale, Florida, USA, 5–10 April, 2003): 129–137.

Solove, D.J. *The future of reputation: Gossip, rumor, and privacy on the Internet*. New Haven, CT: Yale University Press, 2007.

Thaler, R.H., and C.R. Sunstein. *Nudge: Improving decisions about health, wealth, and happiness*. New Haven, CT: Yale University Press, 2008.

Tufekci, Z. "Can you see me now? Audience and disclosure regulation in online social network sites." *Bulletin of Science, Technology and Society* 28 (2008): 20–36.

Van den Berg, B. "Self, script, and situation: Identity in a world of ICTs." In *The future of identity in the information society: Proceedings of the third IFIP WG 9.2, 9.6/11.6, 11.7/FIDIS International Summer School on the Future of Identity in the Information Society*, edited by S. Fischer-Hübner, P. Duquenoy, A. Zuccato and L. Martucci, 63–77. New York, NY: Springer, 2008.

Van den Berg, B. *The situated self: Identity in a world of Ambient Intelligence*. Nijmegen: Wolf Legal Publishers, 2010.

Walzer, M. *Spheres of justice: A defense of pluralism and equality*. New York, NY: Basic Books, 1983.

Young, A.L., and A. Quan-Haase, "Information revelation and internet privacy concerns on social network sites: A case study of Facebook." (paper presented at C&T '09, University Park, Pennsylvania, USA, 25–27 June, 2009): 265–274.

Chapter 11
Avatars Out of Control: Gazira Babeli, Pose Balls and "Rape" in Second Life

Katja de Vries

Computer-based simulated environments, often called "virtual worlds", have only been around for a little bit more than 3 decades. As such they are a relative novelty. Moreover, the developments since the emergence of the first simple textual *Multi-User Dungeons* (MUD's) have been enormous: looking at today's graphically advanced virtual worlds like *World of Warcraft,*[1] *Sims Online*[2] or *Second Life*[3] it is almost hard to believe that it has only been a little bit more than 30 years. In this paper I present some theoretical explorations with regard to the question whether there are certain ways of relating to these novel environments which are more *felicitous*, and in particular more *enlightened*, than others. I also argue that the way in which inhabitants of virtual worlds relate to their environment affects how these digital realms should be regulated legally.

11.1 How to Relate to the Novel?

What happens when one encounters, discovers, or fabricates something which is completely novel: for instance, a new species, a new land, a new person? Every confrontation with the new poses the question how to *relate* to this novelty. To relate means both: "to act with" or "to act upon", but also "to name". *Acting* and *naming* are of course not independent of each other: in calling the being in front of us "human", "fiend" or "beast" we guide our actions in certain directions, and in pulling our sword, planting a flag or stretching out a hand, certain kind of names will be more likely to roll of our tongue than others. Of the many ways of relating

K. de Vries (✉)
Center for Law, Science, Technology and Society Studies (LSTS), Vrije Universiteit Brussel, Pleinlaan 2, 1050, Brussels, Belgium
e-mail: edevries@vub.ac.be

[1] A massively multiplayer online role-playing game (MMORPG). See <http://www.worldofwarcraft.com> (last visited June 3, 2010).

[2] This massively multiplayer online (MMO) management game does no longer exist. *Electronic Arts*, the publisher of Sims Online, closed the game down in 2008.

[3] A MMO social game. See <http://secondlife.com> (last visited June 3, 2010).

S. Gutwirth et al. (eds.), *Computers, Privacy and Data Protection: an Element of Choice*, DOI 10.1007/978-94-007-0641-5_11, © Springer Science+Business Media B.V. 2011

to a new phenomenon, some ways might be more felicitous than others. Although putting a pair of glasses on your nose is clearly not the only way to use them – they can, for example, also come very handy when you try to create a fire in the middle of nowhere – some ways of usage are more felicitous. In a famous Russian fable an elderly and weak-sighted monkey hears that spectacles might improve its vision, but lacking the knowledge how to use this tool it ends up disappointed:

> So it gets half-a-dozen pairs of spectacles, turns them now this way and now that, puts them on the top of its head, applies them to its tail, smells them, licks them; still the spectacles have no effect at all on its sight.[4]

Krylov concludes the fable by stating its moral:

> However useful a thing may be, an ignorant man, who knows nothing about its value, is sure to speak ill of it, and, if he possesses any influence, he persecutes it too.[5]

What could the monkey have done to improve its odds to find a more felicitous way of relating to the spectacles? Or, returning to the subject of this paper, which ways of relating to so-called virtual[6] or digital worlds should be considered felicitous? Translating Krylov's fable to the context of digital worlds is complicated by the fact that in the latter case at least two kinds of "monkeys" are involved: firstly, the users who inhabit these worlds, and secondly the law enforcement which is responsible for the regulation of behaviour within these digital worlds.

In this paper I will argue that both direct users, as well as those who enter the stage to evaluate behaviour from a legal perspective, might benefit from looking for the *affordances* and *constraints* governing the world they inhabit or attempt to regulate.

---

[4]Ivan Andreevich Krylov, "The Monkey and the Spectacles," In *The Russian Fabulist Krilof and His Fables*, ed. William Ralston Shedden Ralston (London: Strahan and co., 1869), 121.

[5]Ibid., 122.

[6]As one of the anonymous reviewers of this paper kindly pointed out, the term "virtual" is rather confusing in this context. Even though it is very common to use the expression "virtual worlds" in order to oppose it to the "real world", all the interactions, interactions and events which are performed, entertained or happening in this so-called "virtual" world are no less real than their counterparts in the "real" world. Thus, the philosopher Deleuze famously argued that the "real" and the "virtual" are no opposing concepts – only the "actual" and the "virtual" are. See e.g. Gilles Deleuze, "L'actuel Et Le Virtuel," In *Dialogues*, ed. Gilles Deleuze and Claire Parnet (Paris: Flammarion, 1996), Gilles Deleuze, *Bergsonism* (New York: Zone, 1991). Nevertheless, *both* the actual and the virtual belong to the realm of the real. In order to avoid too many conceptual complications I try to stick as much as possible to the word "digital", instead of the more equivocal "virtual". However, the adjective virtual has become so commonplace with regard to certain phenomena that in certain instances (e.g. "virtual goods" or "virtual rape") it was impossible to circumvent it.

11.2 Affordances and Constraints

> The things we call "technologies" are ways of building order in our world. Many technical devices and systems important in everyday life contain possibilities for many different ways of ordering human activity. [. . .] In that sense technological innovations are similar to legislative acts or political foundings that establish a framework for public order that will endure over many generations. For that reason the same careful attention one would give to the rules, roles, and relationships of politics must also be given to such things as the building of highways, the creation of television networks, and the tailoring of seemingly insignificant features on new machines.[7]

Since Winner's seminal paper "*Do Artifacts Have Politics?*" many authors have engaged with the question how the set-up of devices (tools, artifacts, machines, objects, mechanisms, etc.) – especially if they are construed in a "user friendly"[8] way! – both allows and forbids certain behavior. The understanding of *affordances* and *constraints* as described in this paper is particularly inspired by the writings of Donald Norman and Bruno Latour. In his book "*The Design of Everyday Things*"[9] the former famously appropriated and extended the term "affordances" (originally coined and used by the psychologist J. J. Gibson in his research of human perception[10]) to be applicable in the process of designing objects.

> A rock can be moved, rolled, kicked, thrown, and sat upon-not all rocks, just those that are the right size for moving, rolling, kicking, throwing, or sitting upon. The set of possible actions is called the affordances of the object.[11]

Norman also underlines that this "set of possible actions" is not simply a property which is built into a device but that it has to be understood as "a relationship that holds between the object and the organism that is acting on the object" and that the "same object might have different affordances for different individuals".[12]

> A rock that affords throwing for me does not for a baby. My chair affords support for me, but not for a giant. My desk is not throwable by me, but might be by someone else.[13]

Another important distinction introduced by Norman is the one between "real" and "perceived" affordances:

---

[7]Langdon Winner, "Do Artifacts Have Politics?," In *The Whale and the Reactor: A Search for Limits in an Age of High Technology*, ed. Langdon Winner (Chicago, IL: University of Chicago Press, 1986), 28–29. The chapter is a reprint from: Langdon Winner, "Do Artifacts Have Politics?," *Daedelus* 109, 1 (1980).

[8]Bruno Latour, "Where Are the Missing Masses? The Sociology of a Few Mundane Artifacts," In *Shaping Technology/Building Society: Studies in Sociotechnical Change*, ed. Wiebe E. Bijker and John Law (Cambridge, MA: MIT Press, 1992), 232.

[9]Donald A. Norman, *The Design of Everyday Things* (New York, NY: Doubleday/Currency, 1990).

[10]Donald A. Norman, *The Invisible Computer: Why Good Products Can Fail, the Personal Computer Is So Complex, and Information Appliances Are the Solution* (Cambridge, MA: MIT Press, 1998), 123.

[11]Ibid.

[12]Ibid.

[13]Ibid.

> Perceived affordances are often more about conventions than about reality. [...] Perceived affordances are not necessarily the same as actual ones. If I saw a realistic painting of a door on a wall, the perceived affordance would be that I could open the door and walk out of the room, and I might even try to do so, but the painting would not really afford those actions. Similarly, if a cupboard door has no perceivable handle, it may be impossible to figure out how to open it, even if the cupboard affords opening.[14]

Recently Sicart has shown the importance of Norman's writings with regard to ethical game design:

> Games as objects can condition what the ethical practices and values of the players will be through their affordances and constraints.[15]

Around the same period when Norman showed how the concepts "affordance" and "constraint" could guide the design process, the philosopher-sociologist Latour pointed to ways in which social scientists and empirical philosophers could bring these embedded and often hardly noticed "prescriptions" to the surface:

> How can the prescriptions encoded in the mechanisms be brought out in words? By replacing them by strings of sentences (often the imperative) that are uttered (silently and continuously) by the mechanisms for the benefit of those who are mechanized: do this, do that, behave this way, don't go that way, you may do so, be allowed to go there.[16]

Though Latour's concept of "prescription" is aimed at a different audience (sociologists, philosophers) than Norman's "affordance" (designers), I think it is possible to use these terms as synonyms:

> *Prescription; proscription; affordances; allowances:* What a device allows or forbids from the actors – humans and nonhuman – that it anticipates; it is the morality of a setting both negative (what it prescribes) and positive (what it permits).[17]

My own use of the notions "affordance" and "constraint" has an aim that differs slightly from the paths followed by Latour and Norman, in that its specific aim is to present a philosophical exploration of the ways in which inhabitants of digital worlds, as well as those who have to enforce the law in such environments, could find more felicitous ways of relating to these virtual universes. In the following sections I will argue that the *felicity* of a certain way of inhabiting a digital world is better understood through the notion of *enlightenment* than through *freedom* – as the latter can easily become an endless source of misconceptions about SL.

---

[14] Ibid., 124.

[15] Miguel Sicart, *The Ethics of Computer Games* (Cambridge, MA; London: MIT, 2009), 102.

[16] Latour, "Where Are the Missing Masses? The Sociology of a Few Mundane Artifacts," 232.

[17] Madeleine Akrich and Bruno Latour, "A Summary of a Convenient Vocabulary for the Semiotics of Human an Nonhuman Assemblies," In *Shaping Technology/Building Society: Studies in Sociotechnical Change*, ed. Wiebe E. Bijker and John Law (Cambridge, MA: MIT Press, 1992), 261.

11.3 An Imitation of Real Life *Without* Constraints or Simply with *Different* Constraints?

Why do some people spend large parts of their life in digital worlds like *World of Warcraft*, *Sims Online* or *Second Life*? Even though there might be many different personal reasons for doing so, it is likely that one of the most important motives will often be the *freedom* that such environments grant their inhabitants in creating their digital lives; it is the *freedom* from bodily, social, geographical or financial constraints. Here the poor can be rich, the unattractive can be staggering beauties and we all can defy the laws of gravity by simply taking of and flying wherever we want. Thus the slogan of Second Life (SL) seems rather befitting: "*Your World. Your Imagination*". However this does not imply that participating in digital world amounts to mere escapism. In the introduction to a Cooper's beautiful book *Alter Ego*, which juxtaposes avatars and their creators, Dibbell[18] remarks:

> On the one hand, for instance, the abundance of powerful, beautiful avatars posed to glamour-challenged, suburban nobodies seems to argue the proposition that we fly to virtual worlds as a departure from quotidian reality; yet just as striking is the number of avatars shaped to look precisely like the people who play them, suggesting just as forcefully that virtual worlds are better understood as an extension of reality and no escape from it at all

The appropriateness of this observation is particularly striking in SL. If we would take the SL slogan – *Your World. Your Imagination* – seriously and assume that you can create whatever world you imagine then it is actually surprising that, if you look at what people actually *do* in Second Life, it all looks very much like real life (RL). The Dutch writer Ilja Pfeijffer, who worked for a while as an undercover reporter in SL, said in a interview:

> In Second Life you have architects realizing dreamlike designs, but most users are simply themselves, while they have the freedom to become whatever they want. They build an *All-American Dallas-house*, with a swimming pool, although they could have built pyramids floating upside down in thin air as well.[19] [translation mine, *KdV*]

Apart from some striking exceptions confirming the rule (such as, for example, the work of Babeli, discussed below, in Section 11.7 of this paper), navigating through SL one notes that this world is predominantly used as a stage for a rather slavish reproduction of RL – a house, a garden, a car, a flirtation, buying clothes, taking a holiday, having a friendly chat in a bar, shopping, intercourse, building a pool, creating a personal look, taking a stroll through town – only without the constraints of the latter: everything is larger, richer and faster. The flying, the big breasts and the shiny Ferrari are not so much a radical alternative to RL but an extension. One of the SL encounters Pfeijffer has is with an girl who is sitting in a wheelchair – just

---

[18] Julian Dibbell, "Introduction," In *Alter Ego. Avatars and Their Creators.*, edited by Robbie Cooper (London: Chris Boot, 2007).

[19] Maartje Bakker and Martine Peters van Ton, "Ilja Pfeijffer: Tussen Kunst En Kritiek," *ANS-Online (Website of the "Algemeen Nijmeegs Studentenblad")* (2007, May), www.ans-online.nl.

like its owner in RL – because creating an avatar with legs "didn't quite feel like herself"[20]. The avatar explains:

> This wheelchair is so much better than my real wheelchair. I even can climb stairs with it.[21] [translation mine, *KdV*]

Yet the fact that certain burdening constraints from RL are lacking, or at least less stringent (e.g., currently the exchange rate for Linden dollars, the internal currency of SL, is on average L$265 per US $1, which makes your buying power in SL significantly higher than in RL) does not imply that SL is a world *without* constraints but merely that those constraints are *different* from those we experience within RL. Although SL lacks such constraints as DNA or gravity, its particular architecture which is, for example, embedded in the fact that it is based on underlying computer code and that you rely on the capacities of a graphic card, creates its own affordances and limitations. As Velleman[22] has pointed out, this is what makes a digital world like SL so very different from the *pretend play* or *make-believe* that children like to engage in:

> What is true in a make-believe world includes only what the players have stipulated or enacted, plus what follows from those overt contributions; what is true in a virtual world is usually far more determinate than the players know or can infer. When the children begin playing at pirates, the objects in their environment have no determinate roles in the fictional world of the game, and their characters have no determinate histories. If the children do not assign a fictional role to the coffee table, either explicitly or implicitly, then there is no fact of the matter as to what it stands for in the fiction. [...] By contrast, a virtual world has determinate features that outrun what is known to any one of the players. Each player has to explore this world in order to learn what it is like, and he will then encounter others whose knowledge of the world is largely disjoint from his. [...] He sees only from the avatar's perspective, and he cannot see around corners unless the avatar turns to look.

With Velleman's dinstinction in mind the SL slogan suddenly sounds far less convincing. After all our universe and the digital worlds of the so-called "metaverse" are not that different: depending on one's abilities and skills to play with the constraints and affordances both worlds have the potential to be either a stage for *Your Imagination* and the creation of *Your World*, or a dungeon of determinations in which we are helplessly chained. Both universe and metaverse are governed by their own "natural" laws. Therefore I will argue in the remainder of this paper that to take SL seriously would imply the acknowledgement that it is a place with its *own* constraints and its *own* affordances.

---

[20] I would like to thank one of the anonymous reviewers for the helpful remark that, psychologically speaking, it is not surprising that people try to have their avatars look in a way that is similar to the way they represent themselves in the physical world: after all, *recognition* will be more gratifying if it resonates with the personality and characteristics that persons believe are theirs. However, a deeper investigation into the motives for modeling one's avatar in accordance to one's appearance in the physical world, falls beyond the scope of this paper.

[21] Ilja Leonard Pfeijffer, *Second Life: Verhalen En Reportages Uit Een Tweede Leven [Second Life: Stories and Reportages from a Second Life]* (Amsterdam: De Arbeiderspers, 2007), 44.

[22] J. D. Velleman, "Bodies, Selves," *American Imago* 65, 3 (2008): 408.

11.4 Lost in Translation Between RL and SL

In SL many things are imitations of RL. These SL imitations mimic RL objects from the native grounds of their creators and are often utterly useless within their new SL environment. Magritte's famous saying "*Ceci n'est pas une pipe*" applies perfectly: though it looks like a pipe, and is held like a pipe, an SL-pipe can never be filled with tobacco and enjoyed in the same way as in the real world.[23] In SL a pipe is not a pipe, a stair is not a stair, and a kiss is not a kiss. Thus it appears as if the functions these objects and actions perform in RL are completely superfluous within the internal logic of SL. Who needs a stair if you can fly? Yet, the fact that things are not what they appear to be does *not* mean that they are *nothing* at all – it only means that they have their own constraints and affordances. Of course this also implies that objects and actions, such as a stair or a kiss, which have been simply imported from RL and only function through their semiotic value, will not be very well adapted to their new environment. However, when one is faced with phenomena that have emerged *within* the digital worlds another problem arises, that is: how to name or define such a native phenomenon that has no equivalent within RL? This becomes especially clear when we look at so-called virtual or digital crime (a category of actions which comprises a wide range of wrongdoings which take place in digital environments). Lacking proper names we speak, for instance, of "cyber bullying", "virtual stalking", "virtual murder", "virtual rape", "virtual trespassing" or "virtual theft". However, because the way we *name* something also influences the way we *relate* to a phenomenon (see above, Section 11.1), the transferal of such RL categories to SL phenomena is not without controversy. A quick glance at the endless discussions[24] at forums devoted to these topics shows that often victims of RL crimes feel that their suffering is belittled if they are placed in one category with the victims of so-called "virtual" crimes. For example a victim of real rape might feel offended when put under the same denominator as a victim of virtual rape. On the other hand victims of digital crimes feel that the harm that they suffered is belittled by the adjective "virtual" which seems to suggest that nothing real or serious took place. Finding the right name for digital crimes thus becomes of pivotal importance. Once legal authorities are involved, the question how SL objects and actions should be named or characterized is no longer academic tittle-tattle but becomes of crucial importance.[25] A nice example hereof is a recent Dutch criminal case in which the

[23]Nevertheless one could argue that the mimicry of SL provokes a "ghostly" form of enjoyment, a phenomenon related to what Deleuze called the "technological production of ghosts". Gilles Deleuze, "Course of 10–23 February 1982 - Two Transcriptions by Carine Baudry," (1982), http://www.univ-paris8.fr/deleuze/article.php3?id_article=136. I would like to thank one of the anonymous reviewers for this useful reference.

[24]See e.g. the fierce discussion following the post "Future and virtual rapists" (December 15, 2006) at http://feministing.com/archives/006218.html

[25]Susan W. Brenner, "Fantasy Crime: The Role of Criminal Law in Virtual Worlds," *Vanderbilt Journal of Entertainment and Technology Law* 11, no. 1 (2008); Orin S. Kerr, "Criminal Law in Virtual Worlds," *University of Chicago Legal Forum* (2008).

question was raised whether "virtual goods" are goods in the legal sense.[26] In this case the court had to decide whether the charge with violent theft – the suspect had used RL threats and violence to force the victim to transfer a virtual amulet and mask from the game *RuneScape*[27] – was applicable given the fact that no "real" goods had been removed. The Court of Leeuwarden answered this question affirmative:

> The idea that the requirement of materiality has to be fulfilled in order for a good to fall within the reach of the aforementioned article has been abandoned already since the Electricity case of 1921 [a Dutch case in which it was decided that electricity could be the object of theft, *KdV*]. In order to answer the question whether electricity was a good that could be prone to theft the High Court [of the Netherlands] deemed it to be of greater importance that electricity had – and has – to be considered as an asset with utility value than that it has an immaterial nature. Moreover in [Dutch] case law the notion of economic value has become gradually more and more relative and subjective. In particular it is relevant whether the good has value for its possessor. In the current case it is obvious that the possession of virtual goods and the points which could be earned were extremely desirable for the plaintiff, the suspect and the co-suspect [translation mine, *KdV*]

The *RuneScape* case showed how important it is to ask whether a phenomenon from a virtual world can be understood as equivalent to a real world phenomenon. However, within the functional and pragmatic approach of the court of Leeuwarden ("If it looks like a duck, swims like a duck, and quacks like a duck, then it probably is a duck"[28]) the distinction between virtual and real goods was shown to be superfluous: according to the court something should be considered a good when – independent of its materiality of immateriality! – it *functions* as a good (represents value, is an desirable asset, etc.). In this way the court managed to avoid getting mixed up in complicated translations between different worlds. To underline this it added:

> The act of theft was committed outside the context of the game. Therefore this is not about virtual acts in a virtual world, but about factual acts which influenced the virtual world

Yet, the possibility to avoid the question of translation is quite exceptional. Only because of the fact that part of the crime (the violence committed in RL) took place outside the virtual world and the strictly functional approach of the court, it became possibly to establish equivalence between real and virtual goods (and given

---

[26] *RuneScape*, Gerechtshof Leeuwarden 10 november 2009, LJN: BK2773

[27] An adventurous MMORPG set in the Middle Ages. See <http://www.runescape.com/> (last visited June 3, 2010).

[28] Such a functional approach is not undisputed. For example, De Hert and Gutwirth argue that the notion of "virtual theft" overlooks the particularities of information which make it ill-suited to be the object of theft. Paul De Hert and Serge Gutwirth, "Informatie: Wel Beschermd, Doch Niet Vatbaar Voor Diefstal. Denkoefeningen over Het Juridisch Statuut Van Informatie Vanop Het Grensvlak Tussen Strafrecht En De Intellectuele Rechten ["Information: Protected, but Not Susceptible to Be the Object of Theft. Reflections from the Border between Criminal an Ip Law on the Legal Statute of Information"]," In *Tendenzen in Het Economisch Recht ["Tendencies in Economical Law"]*, edited by K. Byttebier, E. De Batselier, and R Feltkamp (Antwerpen: Kluwer, 2006).

that these goods can be considered the same the question of translation becomes superfluous).

In contrast to the Dutch judgment, Kerr[29] leaves no doubt about the fact that there is an unbridgeable difference between virtual and real crimes:

> Existing law will not recognize virtual murder, virtual threats, or virtual theft. While these "offenses" may appear to users as the cyber-version of traditional crimes, existing law requires proof of physical elements rather than virtual analogies.

Although the Dutch case of virtual theft showed that it is impossible to exclude virtual offenses from the realm of criminal law based on their lack of physicality, it is also clear that in a case of virtual rape or murder the immateriality of the acts seem to make it impossible to consider them equivalent to their real life counterparts. Nevertheless in 2008 a 43-year old female piano teacher was arrested in Japan for virtual murder in *Maple Story*.[30] The woman became so outraged when the virtual husband of her avatar, whom she had never met in RL, decided to divorce her that she hacked his account and deleted his user-profile which put an abrupt end to his existence.[31] The public prosecutor took the case very seriously and had the woman arrested. All over the world the story got covered under the heading of "virtual murder", even though this category does not exist legally, and the public prosecutor was planning to charge her instead with data manipulation and illegal access to an internet account (for which the maximum punishment is 5 years of imprisonment and a fine of up to $5,000). The Japanese prosecutor was not the first one to take a crime in a virtual world this seriously. One year earlier, in 2007, a reported case of virtual rape[32] led the Brussels public prosecutor to ask members of the Federal Computer Crime Unit to patrol and investigate the case in SL. Even though both the virtual murder and the virtual rape case never got any further coverage, which most likely means that in both cases the charges were dropped, it nevertheless raises many questions. Are virtual murder and rape indeed such serious offenses that they deserve the attention of legal authorities? Especially the notion of virtual rape[33] seems to be an easy target for RL ridicule – after all how can there be rape without a RL body?

---

[29] Kerr, "Criminal Law in Virtual Worlds," 416.

[30] A South-Korean MMORPG. See <http://www.maplestory.com/> (last visited June 3, 2010).

[31] Martin Kölling, "Virtuelle Gewalt Vor Gericht," *Technology Review* (2008, 30 October), http://www.heise.de/tr/blog/artikel/Virtuelle-Gewalt-vor-Gericht-272014.html.

[32] Benjamin Duranske, "Reader Roundtable: "Virtual Rape" Claim Brings Belgian Police to Second Life," *Virtually Blind: Virtual Law & Legal Issues that Impact Virtual Worlds* (2007, 24 April), http://virtuallyblind.com/2007/04/24/open-roundtable-allegations-of-virtual-rape-bring-belgian-police-to-second-life/.

[33] Michael Bugeja, "Avatar Rape," *Inside Higher Ed* (2010, February 25), http://www.insidehighered.com/views/2010/02/25/bugeja.

11.5 A Rough Wake-Up Call from the Illusion That the Metaverse Is a Place of Pure Freedom

Notwithstanding the seeming paradox of a virtual rape,[34] it is one of the earliest crimes to be reported about in a digital world.[35] The first described virtual rape took place in 1992 in a so-called Multi-User Dungeon (MUD), i.e., a non-graphical, purely text-based predecessor of today's massively multiplayer online (MMO) games. Avatars, locations and objects only existed as textual descriptions appearing on a screen. One evening a group of avatars is gathered in a virtual room, enjoying themselves peacefully, when another avatar – a certain Mr Bungle – suddenly launches a malicious computer subprogram or a so-called "voodoo doll" that seems to take control over all their actions. The users have to read helplessly how their beloved avatars start to participate in a series of sadistic, sexist and self-destructive events:

> They say that by manipulating the doll he forced them to have sex with him, and with each other, and to do horrible, brutal things to their own bodies.[36]

The inhabitants of this MUD were as shocked as if there real bodies would suddenly have been taken over by an alien force. Even though this digital world looked pretty different from RL, users expected that they would have the same amount of control over their avatar as over their real life body. One victim reported that posttraumatic tears were streaming from her eyes after the event.[37]

In this most archetypical case of digital or virtual rape there is not even a graphical depiction of the bodies that are assaulted, let alone a real body. The bodies that were harmed consisted merely out of words. As this incident clearly shows, a digital rape is not "real" in the classical sense – no bodily fluids are involved – but it is an act which betrays the user's trust in the slogan that this is *Your World. Your Imagination*. It is a betrayal of the *belief* that you are as much in control over your avatar as you are over your body in real life. This betrayal can occur in many different ways: the classical case of virtual rape is a malicious script which takes control of an avatar and forces it to submit to sexual acts, but also other ways of being

---

[34]Some prefer to speak of "sexual griefing" instead of virtual rape Kimban, Dandellion [pseudonym], "Many Ways to Rape," *Living in the Metaverse. Gonzo phenomenology of virtual worlds* (2007), http://metaverse.acidzen.org/2007/many-ways-to-rape.

[35]Julian Dibbell, "A Rape in Cyberspace. How an Evil Clown, a Haitian Trickster Spirit, Two Wizards, and a Cast of Dozens Turned a Database into a Society," *The Village Voice*, no. (23 December 1993) (1993), http://www.villagevoice.com/2005-10-18/specials/a-rape-in-cyberspace/.

[36]Julian Dibbell, *My Tiny Life: Crime and Passion in a Virtual World* (London: Fourth Estate, 1999), 11, Dibbell, "A Rape in Cyberspace. How an Evil Clown, a Haitian Trickster Spirit, Two Wizards, and a Cast of Dozens Turned a Database into a Society."

[37]Dibbell, "A Rape in Cyberspace. How an Evil Clown, a Haitian Trickster Spirit, Two Wizards, and a Cast of Dozens Turned a Database into a Society."

tricked into virtual intercourse[38] or simply being exposed to sexual harassment[39] are categorized as such. Lynn[40] describes vividly why a virtual rape is much more traumatic than the mere humping of a Ken and Barbie in a children's pretend-play:

> There is no question that forced online sexual activity – whether through text, animation, malicious scripts or other means – is real; and is a traumatic experience that can have a profound and unpleasant aftermath, shaking your faith in yourself, in the community, in the platform, even in sex itself. [...] A virtual rape is by definition sudden, explicit and often devastating. If you've never immersed yourself in online life, you might not realize the emotional availability it takes to be a regular member of an internet community. The psychological aspects of relating are magnified because the physical aspects are (mostly) removed. [...] Some suggest that the best way to deal with a virtual rape is to ignore it, or simply log off and come back as another user. But in a game, you don't want to lose the long-term investment you've made in your character. And these days, your real world income or professional reputation can depend on your online self.

The pain that is inflicted in a virtual rape will always partly consist in the sobering realization that the digital world is not a simple "real-life-only-better-because-lacking-any-constraints" but that it has simply different dangers and limitations.

---

[38] A post (20 October 2009) reacting at Duranske ("Reader Roundtable: "Virtual Rape" Claim Brings Belgian Police to Second Life") recalls how one can be tricked into having virtual sex with a double: "It may seem impossible. . .and outlandish, but I actually experiences being raped in SL a few weeks ago by an avatar pretending to be an alt of someone I am very friendly with. I fell for the deception (stupidly). This individual posted videos some real some fake on two SL websites to hurt our reputation and Linden Labs is not able to do anything about these sites. I am still traumatized and upset by the whole incident and it has had terrible ripple effects into both of our real lives and our community in sl. If one looks as Second Life as just game of pixels with cartoon character who have sex sometimes or shoot each other. . .then rape does not exist. Some people see it as is just a game and that is their prerogative, that is the freedom of SL. They put in their profile sl = sl rl = rl for example. For others, their avatars in SL are surrogates for and extensions of their real life selves. It is these people who are vulnerable to all the joys and sorrow and even rape (unfortunately) of the "real world". Many come to Second Life thinking that it is just a game and then discover as they grow "older" that there are real people behind the avis".

[39] A post (29 October 2008) reacting at the entry "Rape again" on the blog *Living in the Metaverse. Gonzo phenomenology of virtual worlds, by Dandellion Kimban* describes the sexual harassment that can occur to new and naïve inhabitants of SL: "[. . .] 2 h after I became a new resident of SL I went to see [. . .] a huge desert setting of an ancient alien civilization. No one else was there. About 20 min after being there I suddenly had two [avatars] standing one foot away from me, a guy and a girl. The guy had voice chat and was commenting on my butt, saying how good in bed he was. . .it really freaked me out. I walked away as fast as I could and they followed. The girl (if it was a girl) said "oh how fun, follow the leader". I finally realized I could teleport out (my second teleport ever). It shook me. I didn't even know about the abuse button. It was attempted mental rape. They were getting their jollies from sexually harassing someone who was clearly new to SL, and the fact there were two of them made it even more scary. I will never forget it. I am not damaged by it because I am old enough to chalk it up to an experience, but it was unforgivable behavior. [. . .] Some wouldn't be able to handle it mentally as well as I did. In some circumstances I think it should be a crime." Available at: http://metaverse.acidzen.org/2007/rape-again Victims of any kind of harassment in SL can report this (http://secondlife.com/policy/security/harassment.php) but how complaints are handled is not completely clear.

[40] Regina Lynn, "Virtual Rape Is Traumatic, but Is It a Crime?," *Wired*, no. (5 April 2007) (2007), http://www.wired.com/culture/lifestyle/commentary/sexdrive/2007/05/sexdrive_0504.

While the occurrence of virtual rapes cannot be fully avoided, it might be possible that the preventive confrontation with this sobering truth about virtual worlds can turn virtual rapes into mere sexual nuisances. Kerr[41] argues that such a mental vaccination should be possible:

> Is it not better to say, "You can't rape me. I don't have a body" than "I believe rape is an assault upon the mind, and so, even though I don't have a body, you can rape me anyway"? Although we cannot currently break the lifegiving mind-body link [. . .], we can build protective walls and make the mind impervious from virtual assaults. The interface which generates the persona and the virtual reality should be that barrier to harm, and in the lapse of technology, we must rely upon social construction to ensure it.

11.6 The Art of Scripted Objects: Pose Balls and Virtual Role-Play Rape

Let's assume you are in SL: you meet an attractive avatar and the feelings are mutual. How to proceed if you want to get more intimate than having a mere chat? What you need then is a pose ball – blue for a boy and pink for a girl – that will animate both of you in a passionate tango or, more likely, sexual intercourse. Such a pose ball is a scripted object, whose most common graphic depiction is that of a floating small pink or blue ball, and by clicking on it you submit voluntary and consensually to a scripted animation. However, even though clicking on a pose ball is in most cases a free decision, it still implies an act of *submission*. This realization has opened up a whole range of possibilities for inhabitants of SL who have a fetish for sadomasochistic practices (BDSM). What happens in such BDSM practices in SL is that a person submits his or her beloved avatar to a scripted animation and gets a kick out of this submission which plays on the affordances created by the fact that this is a coded and scripted world: submitting to code is a very specific, and in certain respects far-reaching, kind of submission that has no equivalent in RL. Whereas most of SL comprises of simple mimicking of RL, here we have found a practice which plays on the specific affordances and constraints of SL. It is one of the still rare examples of a practice which has emerged *within* the internal logic of SL: not an imported but a truly native practice. Or, to put it in Norman's terms (see above, Section 11.2): he BDSM practice does not merely play on the *perceived* affordances, but also on the digitally encoded yet *real* affordances.

Thus, next to the categories of RL rape and virtual rape, a third category emerges: that of consensual or role-play rape. In fact, most so-called virtual rapes are fully consensual. In 2008 Gawker Media[42] reported in the entry "Second Life: Rape for Sale" that SL inhabitants can use the well-known pink and blue pose balls to "indulge in rape fantasies (options: "Rape victim," "Get raped," or "Hold victim") for a trifling 220 Linden dollar things. Nice that the purchase takes place in

---

[41] Kerr, "Criminal Law in Virtual Worlds."

[42] http://gawker.com/222099/second-life-rape-for-sale

an evocative back alley, with the actual rape set in some kind of red cobblestone gimp-dungeon".

Should we take a stern feminist stance towards this practice, arguing that it reinforces the false idea that women *ask* to be raped, or is this practice a promising fulfillment of the Foucauldian imperative to create new relational possibilities?

> No! Let's escape as much as possible from the type of relations which society proposes for us and try to create, in the empty space where we are, new relational possibilities.[43]

To answer the question which stance we should take would go beyond the scope of this paper, but leaving aside which normative value should be assigned to consensual rape it is at least clear that the way in which this practice plays on the fact that SL is a coded and scripted world is of a refreshing creativity. Moreover, one can of course also be very creative in the *design* of scripted objects, such as for instance scripted SM collars or leashes! However, the members of the BDSM community in SL are not the only ones who play around with the possibilities of scripted objects.

11.7 Gazira Babeli: An SL Artist Who Is Truly Native

The possibilities of scripted objects are endless – possibilities that are not only interesting for the BDSM community within SL but which also allows for thought provoking artworks made by the artist and "code performer"[44] Gazira Babeli.[45]

One of the many things that makes Babeli unique as an artist is that she seems to live truly within SL: nobody knows who created her in RL but because of her autonomy in SL the answer to this question does not appear to be very significant either.

> . . .for Gazira, the subject – be it a man or a woman – that created her, is not her "real" alter ego, but simply the stupid deity that manipulates the interface she lives in, the mysterious being that governs her actions from on high. In this way, Second Life becomes her real plane of action, and it is from this perspective that her radical identification between social life and manipulation of code acquires meaning. Living in any world means acting with an awareness of the rules that govern that world. But the social conventions that rule the virtual world of Second Life, just like the linguistic conventions that support its interface, only work on the surface: the world that Gazira has chosen for herself is based on other laws, those written in programming code. This is why her performances are not based on acting – like any normal avatar – on the Second Life platform, but on manipulating and activating its

[43]Michel Foucault, "Le Triomphe Social Du Plaisir Sexuel: Une Conversation Avec Michel Foucault ["the Social Triumph of the Sexual Will: A Conversation with Michel Foucault", Interview Held in 1981 and Originally Published In "Christopher Street", 6, 4, May 1982, p. 36–41]" In *Dits Et Écrits Ii, 1976–1988*, ed. D. Defert and F. Ewald (Paris: Gallimard, 2001), 1150.

[44]Domenico Quaranta, "Gaz', Queen of the Desert," in *Catalogue Text for the Exhibition Gazira Babeli – [Collateral Damage], Exhibita, Odyssey, Second Life, April 16 / June, 2007, Curated by Sugar Seville and Beavis Palowakski* (2007).

[45]Information can be found at http://gazirababeli.com/ but of course it is better to experience Gazira's works in SL in her gallery in the Locusolus region.

> code. She is not a performer, but a "code performer". She does not pretend, like everyone else, to be in a world made of objects and atoms, but is aware of inhabiting a world made of code, and being made of code herself. Performance art is always a critique of the norms the surrounding world is based on. And Gazira operates precisely in this way, which is why she appears like some kind of bizarre shaman to those who see her. [. . .] Gazira runs scripts as if they were magic spells, unleashing earthquakes, natural disasters and invasions of pop icons like plagues of locusts.[46]

From the 16th of April till the 30th of May 2007 Babeli held an exhibition named "Collateral Damage" in *ExhibitA Gallery*, *Odyssey* in SL. When one enters the exhibition, which is now archived in *Locusolus* and which can still be visited, one suddenly experiences what SL truly is and could be: a world full of scripted objects with effects that are at least as bewildering and unsettling as the bottle saying "Drink me" and the cake labeled "Eat me" in Alice's Wonderland. Entering the gallery (in front of which it says: "enter at your own risk") is like jumping into the endless rabbit hole of Carroll's imagination.

An exemplary work of art is a triptych named "Avatar on Canvas" (2007) which is strongly reminiscent of the series painted by Francis Bacon.[47] However, contrary to the paintings by Bacon this triptych does not depict any tormented human figures, but instead in front of each of Babeli's paintings a three dimensional chair is floating, implicitly alluring us to sit down. Once you do, the true nature of the work of art becomes clear: the chairs turn out to be scripted objects and your avatar becomes as wildly deformed as the characters depicted on Bacon's canvases. Even if you get up from the chair the effect lingers on for quite a while, causing an acute sense of panic ("Will my avatar, to whom I feel as attached as to my own body, stay deformed forever?). Though every sexual connotation is lacking, the sensation must be close to an experience of virtual rape: a sensation of betrayal of the link between me and my virtual body.

> . . . The avatars sitting on these chairs were thus suddenly wildly deformed, and their terror and embarrassment betrayed the – entirely irrational – attachment that residents have to their virtual bodies, held to be sacred and inviolable exactly like our physical bodies. The logic behind the "fake Bacon" [. . .]: like a deforming mirror, *Avatar on Canvas* cuts the illusory link that forms between the subject and his or her "second" image, the Second Life avatar. I am not my avatar and I can't see myself in it any longer.[48]

In the performance-sculpture "Come together" (2007) Babeli shows that she is aware of the similarities between her work and that of the pose ball-fuelled vibrant sex industry of SL. The sculpture which is placed in the middle of the gallery consists out of a pedestal surrounded by blue and pink pose balls. By clicking on one of those balls your avatar is placed on the pedestal where it starts to dance. However, the effect is particularly impressive if several avatars participate simultaneously in the performance-sculpture:

---

[46] Quaranta, "Gaz', Queen of the Desert."

[47] Domenico Quaranta, *Gazira Babeli* (Brescia: Fabio Paris Editions, 2008), 38–43.

[48] Ibid., 39.

> If you click on one of the *pose balls* you will start to dance or make other movements up on the pedestal. Hopefully someone will join you and you will get very intimate, even luckier you may experience a treesome, fivesome. No, of course this is not about sex, it's about sculpture. You have to make the sculpture yourself, like in the *Avatar on Canvas*, you are becoming art or a part of Gaziras art work. [...] It does not hurt at all to be intersected by other avatars.[49]

In the act "Come to Heaven" (2006) Babeli shows another fascinating SL constraint that is often overlooked and forgotten: our dependence on the characteristics of our particular graphic card. The act consists of the visitor of the exhibition being invited to fall down at a speed of 900 km/h – a challenging task that will put your graphic card to the test and scramble your face according to the its specific characteristics:

> In some cases the polygons shatter and the result no longer resembles anything human, while in others the body appears to have gone through a turbine: limbs multiplied and blown apart, and the body a confused mess of flesh and hair.[50]

The three artworks which are described above, "Avatar on Canvas", "Come Together" and "Come to Heaven", all point into possible directions how to acknowledge and act upon the fact that SL is a place with its *own* constraints and its *own* affordances. As I have argued before this awareness is important because it does not only bring along a sense of freedom to be able to play around with the laws that govern one's environment – a sensation which clearly differs from the freedom experienced when SL is understood as a place without constraints! – but it also creates a more enlightened and mature way of living and relating. Gazira's artworks

> ...[r]emove an avatar from its self-imposed state of immaturity, by showing it that the consensual hallucination it inhabits is not real, or a poor imitation of a mistaken idea of reality, but an imperfect mishmash of code, textures and polygons, in which Gaz too lives and works.[51]

Of course, there is nothing bad in mimicry of RL within SL – this can be a psychologically deeply satisfying experience. But when this mimicry is accompanied by the ability to play on the affordances and constraints governing the digitally simulated phenomena, one's way of relating to the digital environment can become more enlightened.

11.8 Naming the Offenses of the New World: Will We Make a Law for Enlightened Adults or for Minors?

During the last decades mankind has slowly colonized the virtual worlds of the metaverse. In a similar way as Adam names the animals of the newly created world ("And Adam gave names to all cattle, and to the fowl of the air, and to every beast

---

[49] Plurabelle Posthorn, "Gazira Babeli at Exhibit A," *Virtual Artist Alliance. The official blog for the Virtual Artists Alliance group in You-Know-Where* (2007, April 18), http://virtualartistsalliance.blogspot.com/2007/04/gazira-babeli-at-exhibit.html.

[50] Quaranta, *Gazira Babeli*, 11.

[51] Ibid., 67.

of the field"[52]) the settlers of the newly created virtual worlds will have to find the right words to give to the objects and actions which populated those spheres. To suggest that many inhabitants, as well as many representatives of the law trying to regulate behavior in digital worlds, relate to this environment in an *immature* way, might sound as a revolting and, above all, as a politically incorrect proposition. The reason why this proposition is easily perceived as revolting probably results from the fact that our understanding of enlightenment and maturity, largely shaped by Kant's seminal essay "*What is Enlightenment?*" (1784), is that it is perceived as something which comes naturally: lack of enlightenment results from a lack of courage, or simple laziness.

> *Enlightenment is man's emergence from his self-incurred immaturity. Immaturity* is the inability to use one's own understanding without the guidance of another. This immaturity is *self-incurred* if its cause is not lack of understanding, but of resolution and courage to use it without the guidance of another. [...] Laziness and cowardice are the reasons why such a large proportion of the population of men, even when nature has long emancipated them from alien guidance (*naturaliter maiorennes*), nevertheless gladly remain immature for life.[53]

However, as Foucault[54] and Stiegler[55] have convincingly argued, enlightenment is not a natural tendency but something which has to be construed – Foucault stresses the importance of so-called "technologies of the self", Stiegler underlines the importance of a good education[56] – often at great pains and costs. As Stiegler[57] argues the distinction made in all Western legal systems between minors – who cannot be blamed for their actions, nor possess the right to vote – and responsible adults, depends on the capacity of society to construct this difference.[58] enlightened maturity will not emerge by itself.

---

[52]Genesis 2:20, *King James Bible.*

[53]Immanuel Kant. "An Answer to the Question: 'What Is Enlightenment?'" in *Political Writings (Cambridge Texts in the History of Political Thought)*, ed. H.S. Reiss (Cambridge: Cambridge University Press, 1991), 54

[54]Michel Foucault, "What Is Enlightenment?," in *The Foucault Reader*, ed. Paul Rabinow (New York, NY: Pantheon, 1984).

[55]Bernard Stiegler, *Prendre Soin De La Jeunesse Et Des Générations* (Paris: Flammarion, 2008). Translated as: Bernard Stiegler, *Taking Care of Youth and the Generations*, trans. Stephen Barker (Stanford, CA: Stanford University Press, 2010).

[56]Stiegler blames Foucault for only stressing the *disciplining* effect of educational systems (putting it at same level as prisons or mental clinics), while ignoring its enlightening effect. Stiegler, *Prendre Soin De La Jeunesse Et Des Générations*, 208 ff.

[57]Ibid., 11 ff.

[58]It is important to underline that I do not try to diminish the often astonishing capacities and creativity of children, or to deny the fact that minors (contrary to, for instance Krylov's weak-sighted monkey) are citizens as much as adults. The distinction to which I refer, that is between responsible adults and minors, is in itself a legal, educational and societal invention and a rather recent one as well – only established after the French revolution. I thank one of the anonymous reviewers for pointing out that in attaching importance to this distinction I could easily be misunderstood as suffering from a pathological hatred of children in general (*misopedia*).

This raises the question which techniques have to be put into place to create a more enlightened ways of relating to digital environments. Realizing that simple translations and transferals from RL will not do ("*Ceci n'est pas une pipe*") would be a first step in the right direction. Education, possibly making use of the work of code performers like Babeli, would play an important role in this process. The second step would be to find alternative names and definitions. Especially in those instances where the law is involved the right definitions are of pivotal importance. In order to know how to deal with acts that we now clumsily define as "virtual theft", "virtual murder" or "virtual rape" we will need the right words to describe them. Replacing the word "virtual" by "digital" could be a first step in the right direction. However, the description of our experience of such acts will also depend on how we live with these new phenomena: do we live like children of the pre-enlightenment era, inhabiting the surface of digital appearances, believing in the in a world of myths, badly fitting explanations and crooked analogies, or do we live like enlightened adults who know the world and the natural laws which govern it? The answer is not given: after all, enlightenment comes often at the cost of a sobering disenchantment. As Coleridge famously wrote: *A sadder and a wiser man / He rose the morrow morn*. Is it not better to live in the illusion that we finally discovered a world without constraints, of pure freedom – even if it comes at the risk of a brusque disillusion when confronted with constraints and affordances? Whatever we decide, one thing is for sure: only once we have decided who the addressees of our virtual laws are – minors or enlightened citizens – we can begin to write those laws down.

References

Akrich, M., and B. Latour. "A Summary of a Convenient Vocabulary for the Semiotics of Human an Nonhuman Assemblies." In *Shaping Technology/Building Society: Studies in Sociotechnical Change*, edited by Wiebe E. Bijker and John Law, 259–64. Cambridge, MA: MIT Press, 1992.

Bakker, M., and M. Peters van Ton. "Ilja Pfeijffer: Tussen Kunst En Kritiek." *ANS-Online (Website of the "Algemeen Nijmeegs Studentenblad")* www.ans-online.nl. (May, 2007)

Brenner, S. W. "Fantasy Crime: The Role of Criminal Law in Virtual Worlds." *Vanderbilt Journal of Entertainment and Technology Law* 11, 1 (2008): 1–97.

Bugeja, M. "Avatar Rape." *Inside Higher Ed*, http://www.insidehighered.com/views/2010/02/25/bugeja. (February 25, 2010)

De Hert, P., and S. Gutwirth. "Informatie: Wel Beschermd, Doch Niet Vatbaar Voor Diefstal. Denkoefeningen over Het Juridisch Statuut Van Informatie Vanop Het Grensvlak Tussen Strafrecht En De Intellectuele Rechten ["Information: Protected, but Not Susceptible to Be the Object of Theft. Reflections from the Border between Criminal an Ip Law on the Legal Statute of Information"]." In *Tendenzen in Het Economisch Recht ["Tendencies in Economical Law"]*, edited by K. Byttebier, E. De Batselier and R Feltkamp. Antwerpen: Kluwer, 2006.

Deleuze, G. *Bergsonism*. New York, NY: Zone, 1991.

Deleuze, G. "Course of 10–23 February 1982 – Two Transcriptions by Carine Baudry." http://www.univ-paris8.fr/deleuze/article.php3?id_article=136 (1982).

Deleuze, G. "L'actuel Et Le Virtuel." In *Dialogues*, edited by Gilles Deleuze and Claire Parnet, 177–185. Paris: Flammarion, 1996.

Dibbell, J. "Introduction." In *Alter Ego. Avatars and Their Creators.*, edited by Robbie Cooper. London: Chris Boot, 2007.

Dibbell, J. *My Tiny Life: Crime and Passion in a Virtual World*. London: Fourth Estate, 1999.

Dibbell, J. "A Rape in Cyberspace. How an Evil Clown, a Haitian Trickster Spirit, Two Wizards, and a Cast of Dozens Turned a Database into a Society." *The Village Voice*. http://www.villagevoice.com/2005-10-18/specials/a-rape-in-cyberspace/ (23 December, 1993).

Duranske, B. "Reader Roundtable: "Virtual Rape" Claim Brings Belgian Police to Second Life." *Virtually Blind: Virtual Law & Legal Issues that Impact Virtual Worlds* http://virtuallyblind.com/2007/04/24/open-roundtable-allegations-of-virtual-rape-bring-belgian-police-to-second-life/ (24 April, 2007).

Foucault, M. "Le Triomphe Social Du Plaisir Sexual: Une Conversation Avec Michel Foucault ["the Social Triumph of the Sexual Will: A Conversation with Michel Foucault", Interview Held in 1981 and Originally Published In "Christopher Street" 6, 4, May 1982, pp. 36–41]" In *Dits Et Écrits II, 1976–1988*, edited by D. Defert and F. Ewald, 1127–1133. Paris: Gallimard, 2001.

Foucault, M. "What Is Enlightenment?" In *The Foucault Reader*, edited by Paul Rabinow, 32–50. New York, NY: Pantheon, 1984.

Kant, I. "An Answer to the Question: 'What Is Enlightenment?'" In *Political Writings (Cambridge Texts in the History of Political Thought)*, edited by H.S. Reiss, 54–60. Cambridge: Cambridge University Press, 1991.

Kerr, O. S. "Criminal Law in Virtual Worlds." *University of Chicago Legal Forum* (2008): 415–429.

Kimban, D [pseudonym]. "Many Ways to Rape." *Living in the Metaverse. Gonzo phenomenology of virtual worlds*. http://metaverse.acidzen.org/2007/many-ways-to-rape. (2007)

Kölling, M. "Virtuelle Gewalt Vor Gericht." *Technology Review*. http://www.heise.de/tr/blog/artikel/Virtuelle-Gewalt-vor-Gericht-272014.html. (30 October, 2008)

Krylov, I. A. "The Monkey and the Spectacles." In *The Russian Fabulist Krilof and His Fables*, edited by William Ralston Shedden Ralston, 121–122. London: Strahan and co., 1869.

Latour, B. "Where Are the Missing Masses? The Sociology of a Few Mundane Artifacts." In *Shaping Technology/Building Society: Studies in Sociotechnical Change*, edited by Wiebe E. Bijker and John Law, 225–258. Cambridge, MA: MIT Press, 1992.

Lynn, R. "Virtual Rape Is Traumatic, but Is It a Crime?" *Wired*, no. http://www.wired.com/culture/lifestyle/commentary/sexdrive/2007/05/sexdrive_0504. (5 April, 2007)

Norman, D. A. *The Design of Everyday Things*. New York, NY: Doubleday/Currency, 1990.

Norman, D. A. *The Invisible Computer: Why Good Products Can Fail, the Personal Computer Is So Complex, and Information Appliances Are the Solution*. Cambridge, MA: MIT Press, 1998.

Pfeijffer, I. L. *Second Life: Verhalen En Reportages Uit Een Tweede Leven [Second Life: Stories and Reportages from a Second Life]*. Amsterdam: De Arbeiderspers, 2007.

Posthorn, P. "Gazira Babeli at Exhibit A." *Virtual Artist Alliance. The official blog for the Virtual Artists Alliance group in You-Know-Where*. http://virtualartistsalliance.blogspot.com/2007/04/gazira-babeli-at-exhibit.html. (April 18, 2007)

Quaranta, D. "Gaz', Queen of the Desert." In *Catalogue Text for the Exhibition Gazira Babeli – [Collateral Damage], Exhibita, Odyssey, Second Life, April 16 / June, 2007, Curated by Sugar Seville and Beavis Palowakski*, 2007.

Quaranta, D. *Gazira Babeli*. Brescia: Fabio Paris Editions, 2008.

Sicart, M. *The Ethics of Computer Games*. Cambridge, MA; London: MIT, 2009.

Stiegler, B. *Prendre Soin De La Jeunesse Et Des Générations*. Paris: Flammarion, 2008.

Stiegler, B. *Taking Care of Youth and the Generations*. Translated by Stephen Barker. Stanford, CA: Stanford University Press, 2010.

Velleman, J. D. "Bodies, Selves." *American Imago* 65, 3 (2008): 405–426.

Winner, L. "Do Artifacts Have Politics?" *Daedelus* 109, 1 (1980): 121–136.

Winner, L. "Do Artifacts Have Politics?" In *The Whale and the Reactor: A Search for Limits in an Age of High Technology*, edited by Langdon Winner, 19–39. Chicago, IL: University of Chicago Press, 1986.

Chapter 12
Privacy as a Practice: Exploring the Relational and Spatial Dynamics of HIV-Related Information Seeking

Fadhila Mazanderani and Ian Brown

12.1 Introduction

This paper is an exploration of the relational and spatial dynamics of privacy practices in the context of HIV-related information seeking. It is a study of how a specific group of people, African women living with HIV in London, go about "doing privacy" while seeking health information in relation to living with HIV. Based on material (primarily qualitative interviews, but including focus groups and observations) collected as part of a broader research project on Internet use by women living with HIV, the paper explores alternative methods for researching privacy as an embedded and contingent practice. More specifically, it does so in the context of an increasing interest in the use of the Internet as a source of health information for people living with a stigmatised illness.[1] While the literature on privacy has long stressed that what is considered private information is highly contextual,[2] until relatively recently there has been little in-depth empirical work that attempts to unwrap what constitutes that context across different domains. However, if we are to develop socially sensitive privacy policies and protection mechanisms this is a necessary first step.

During the course of this research when participants spoke of seeking information and support in relation to HIV they supplemented what they were doing and how they were doing it with where and in relation to whom they did it. Therefore, HIV-related information seeking was articulated as strongly relationally and spatially contingent. Building on this we focus on the relational and spatial parameters

F. Mazanderani (✉)
Oxford Internet Institute, The University of Oxford, Oxford, UK
e-mail: mazanderani@gmail.com

[1]Magdalena Berger et al. Internet use and stigmatized illness. *Social Science & Medicine* 61 (2005): 1821–1827.

[2]Alan Westin, *Privacy and Freedom* (New York: Atheneum, 1970); Irwin Altman. *The Environment and Social Behaviour* (Belmont, California: Wadswirth Publishing Company, 1975); Ferdinand Schoeman, *Privacy and Social Freedom* (Cambridge: Cambridge University Press, 1992); Helen Nissenbaum, *Privacy in Context* (Stanford, California: Stanford University Press, 2009).

S. Gutwirth et al. (eds.), *Computers, Privacy and Data Protection: an Element of Choice*, DOI 10.1007/978-94-007-0641-5_12, © Springer Science+Business Media B.V. 2011

of privacy practices. Of particular interest in this is how these parameters were not reducible to or superimposable on traditional demarcations of public versus private spheres. Rather than equating the private with the domestic, participants spoke of how they worked towards making things private across different spheres. From this perspective our focus on privacy is not on privacy as a state – for example, privacy as intimacy – but rather privacy as an accomplishment – for example, privacy for intimacy.[3]

The paper is structured as follows. The following section gives an overview of the literature on privacy that informed the analysis, combined with a brief background to the empirical case under consideration and an outline of the methods used. Next the substantive portion of the paper is broken into two sections. The first is an exploration of the tension participants expressed around seeking information and support in relation to HIV while simultaneously keeping information about an HIV positive diagnosis private. The second section describes privacy practices in relation to information seeking on the Internet in more detail, and specifically how a sense of place continues to persist and shape online interactions. Given the widely accepted emphasis on the embedded "everyday" nature of Internet use generally[4] and with regards to health more specifically,[5] this latter discussion is situated within the broader understanding of the landscape of relationships and spaces developed in the preceding section.

12.2 Background and Context

When faced with health concerns people seek out information and support from a range of sources, such as family and friends, allopathic and alternative healthcare practitioners, books and magazines, and more recently the Internet. These informal and formal help seeking practices, or "lay" referral systems, have long been recognised as important for people's health.[6] Numerous factors, such as culture, age, gender, type of illness, formal healthcare infrastructure, income and education have

[3]Ferdinand Schoeman. *Privacy and Social Freedom* (Cambridge: Cambridge University Press, 1992).

[4]Barry Wellman and Caroline Haythornthwaite. *The Internet in Everyday Life* (Oxford: Blackwell, 2002); Leslie Haddon, *Information and Communication Technologies in Everyday Life: A Concise Introduction and Research Guide* (Oxford: Berg, 2004); Maria Bakardjieva, *Internet Society: The Internet in Everyday Life* (London: Sage, 2005).

[5]Sarah Nettleton et al. "The mundane realities of the everyday lay use of the internet for health, and their consequences for media convergence." *Sociology of Health & Illness* 27 (2005): 7; Joelle Kivits. "Researching the 'Informed Patient': The case of online health information seekers." *Information, Communication & Society* 7 (2004): 4; Elizabeth Sillence et al. How do patients evaluate and make use of online health information? *Social Science & Medicine* 64 (2007): 1853–1862.

[6]Eliot Freidson, "Client Control and Medical Practice." *The American Journal of Sociology* 65 (1960): 4; David Mechanic. *Medical Sociology* (New York: Free Press, 1968); John B. McKinlay, "Social Networks, Lay Consultation and Help-Seeking Behavior." *Social Forces* 51 (1972): 3;

all been suggested as playing a role in how, why and where people search for health information. However, one aspect of information seeking that has received relatively little attention is the tension between that which is sought and that which is withheld. This tension is particularly relevant in the case of stigmatised illness where peoples' unwillingness to disclose information has been shown to have a negative impact on their ability to seek out help.[7] It has been suggested that the Internet, in enabling people to search for information with relative anonymity, has the potential to be a privileged source of information about stigmatised illness.[8] In this paper we explore this tension, and use of the Internet in relation to it, in more depth in the context of HIV-related information seeking by African women living with HIV in London.

In 2009 there were an estimated 89,531[9] people living with HIV/AIDS in the UK, with people identifying as "black African" constituting the second largest group.[10] In the last UK census there were over 480,000[11] people recorded as living in England who self-identified as "black African",[12] (0.97% of the population), over three-quarters of whom were living in Greater London.[13] Considering these regional discrepancies, it is unsurprising that 53% of the HIV positive diagnoses amongst Africans in the UK have been in London.[14] Moreover, clinics in east London (where most of the interviews and observations took place) treat a large proportion of the female African HIV positive patients in public sector HIV specialist centers, where the majority of HIV care takes place in the UK.[15]

Annette Scambler et al. "Kinship and Friendship Networks and Women's Demands for Primary Care." *Journal of the Royal College of General Practitioners* 26 (1981): 746–750.

[7]Shayna D. Cunningham et al. "Attitudes about sexual disclosure and perceptions of stigma and shame." *Sexually Transmitted Infections* 78 (2002): 5; J. Dennis Fortenberry et al. "Relationships of stigma and shame to gonorrhea and HIV screening." *American Journal of Public Health* 92 (2002): 3.

[8]Magdalena Berger et al. "Internet use and stigmatized illness." *Social Science & Medicine* 61 (2005): 1821–1827.

[9]Health Protection Agency. "United Kingdom HIV New Diagnoses to end of June 2009," page 7, New HIV Diagnoses National tables: Table 1, published in June 2009 and available online at: http://www.hpa.org.uk/web/HPAwebFile/HPAweb_C/1237970242135. Last accessed 23 January 2010.

[10]Audrey Prost et al. "Social, Behavioural, and Intervention Research among People of Sub-Saharan African Origin Living with HIV in the UK and Europe: Literature Review and Recommendations for Intervention." *AIDS and Behavior* 12 (2008): 2.

[11]UK 2001 Census data, http://www.ons.gov.uk/census/index.html. Last accessed 12 March 2010.

[12]For the sake of brevity from now on we will use the term African when we mean "black Africans."

[13]Department of Health. *HIV and AIDS in African communities: A framework for better prevention and care*. London, 2004.

[14]Gill Green and Richard Smith. "The psychosocial and health care needs of HIV-positive people in the United Kingdom." *HIV Medicine* 5 (2004): 1.

[15]Jonathan Elford et al. "HIV in East London: ethnicity, gender and risk. Design and methods." *BMC Public Health* 6 (2006): 150.

Issues around information management have been highlighted as a crucial challenge for people living with HIV in the UK generally[16] and among African women specifically.[17] This was reflected in our research where issues of disclosure, confidentiality and privacy were raised as barriers for accessing health information by the women who took part in the study. The reason usually given for this difficulty was stigma, and while we do not go into the details of this here it is important to note that some of the women who took part in the research had experienced situations – ranging from losing a job to physical abuse – that were deeply upsetting. This is in keeping with an earlier study of a similar group of women where approximately a third of participants reported direct experience of HIV-related stigmatisation,[18] often with profound negative effects on their mental wellbeing. Although we do not elaborate on stigma here, except when it was raised directly in relation to privacy and information seeking, it formed an omnipresent backdrop to the research and has been discussed in relation to HIV in more detail elsewhere.[19] However, one aspect of how participants expressed their experiences of HIV-related stigma that emerged as particularly relevant in relation to health information seeking and Internet use was how they felt an HIV positive diagnosis affected their most intimate relationships.

Like with other stigmatised illnesses, decisions to disclose an HIV positive status occur in the context of specific relationships.[20] However, because HIV is a sexually communicable virus issues around disclosure can be particularly complex and difficult within the most personal of these.[21] In a similar vein participants often expressed how they felt it was crucial for them to maintain privacy regarding an HIV positive status in the areas of life typically associated with the "private" domain: home and family life. Instead of privacy being equated with the domestic sphere where it has traditionally been placed, both in philosophy and legal scholarship, it was distributed across a range of different spaces, most notably, in relation to HIV,

[16]Gill Green and R. Smith. "The psychosocial and health care needs of HIV-positive people in the United Kingdom." *HIV Medicine* 5 (2004): 1; Leslie Doyal and Jane Anderson, "My fear is to fall in love again" How HIV-positive African women survive in London. *Social Science & Medicine* 60 (2005): 1729–1738.

[17]Leslie Doyal and Jane Anderson. "'My fear is to fall in love again' How HIV-positive African women survive in London." *Social Science & Medicine* 60 (2005): 1729–1738.

[18]Ibid.

[19]Angelo A. Alonzo and Nancy R. Reynolds. "Stigma, HIV and AIDS: An exploration and elaboration of a stigma trajectory." *Social Science & Medicine* 41 (1995): 3; Paul Flowers et al. "Diagnosis and stigma and identity amongst HIV positive black Africans living in the UK." *Psychology & Health* 21 (2006): 1; J. Dennis Fortenberry et al. "Relationships of stigma and shame to gonorrhea and HIV screening." *American Journal of Public Health* 92 (2002): 3.

[20]Valerian J. Derlega et al. "Perceived HIV-related stigma and HIV disclosure to relationship partners after finding out about the seropositive diagnosis." *Journal of Health Psychology* 7 (2002): 4.

[21]Kathryn Greene et al. *Privacy and Disclosure of HIV in Interpersonal Relationships* (New Jersey: Lawrence Erlbaum Associates, 2003).

the clinic and the community support group. Thus, the "intuitive" boundaries of what constitute the private and the public shifted in relation to HIV in this research.

Critiquing the dichotomy between the private and the public has been central to feminist writing and political struggle.[22] Indeed, some feminists have explicitly criticised privacy as being dangerous for women as it can be used as a means and justification for covering up domestic violence and subjugation.[23] Others have suggested that although notions of privacy can be subject to abuse this is not necessarily the case.[24] In our research the need for rethinking the public/ private divide emerged as being salient in the sense that privacy should not be considered as necessarily associated with the domestic domain, but rather that it needs to be "exploded"[25] across numerous domains, people, practices and objects. However, this exploding does not, as in Catharine MacKinnon's original suggestion,[26] amount to the dissolution of privacy. Instead privacy shifts from being associated with a specific place towards being associated with practices that enact different places and spaces of relative privacy. What we mean by this will be discussed in more detail in the substantive discussion below, but first it is necessary to outline our approach to privacy.

Since Warren and Brandeis's early definition of privacy as the "right to be let alone"[27] researchers have been grappling with how to delineate this elusive concept. While this research has extended our understanding of privacy there is still no unified definition of what privacy is. In this paper we draw on and extend two broad conceptualisations of privacy:[28] a dialectic one, as proposed by Irwin Altman[29] and that of privacy as contextual integrity, proposed by Helen Nissenbaum.[30] In the former privacy is seen as a process of selective control of access to the self, an idea that has been elaborated on further in work on information disclosure.[31] However, our focus within this discussion is not on disclosure per se, an area that has been

[22]Carole Pateman. *The Disorder of Women* (Stanford, California: Stanford University Press, 1989), 118.

[23]Catharine MacKinnon. *Toward a Feminist Theory of the State* (Cambridge, Mass: Harvard University Press, 1989).

[24]Anita Allen. *Uneasy Access: Privacy for Women in a Free Society* (Totowa, New Jersey: Rowman and Littlefield, 1988).

[25]Catharine MacKinnon. *Toward a Feminist Theory of the State* (Cambridge, Mass: Harvard University Press, 1989).

[26]Ibid.

[27]Samuel Warren and Louis Brandeis. "Right to Privacy." *Harvard Law Review* 193 (1890).

[28]There are many different versions of what privacy is, of which these are only two. In selecting these we am not suggesting that this is all that privacy entails, but simply that these versions of privacy emerged as particularly appropriate in the context of this research.

[29]Irwin Altman. *The Environment and Social Behaviour* (Belmont, California: Wadswirth Publishing Company, 1975).

[30]Helen Nissenbaum. *Privacy in Context* (Stanford, California: Stanford University Press, 2009).

[31]Sandra S. Petronio. *Boundaries of Privacy* (Albany, New York: State University of New York Press, 2002); Kathryn Greene et al. *Privacy and Disclosure of HIV in Interpersonal Relationships* (New Jersey: Lawrence Erlbaum Associates, 2003).

covered extensively in relation to HIV/AIDS,[32] but on the privacy practices participants adopted while seeking HIV-related information and support, practices which sometimes, although not always, facilitated situations participants felt were conducive for the disclosure of an HIV positive status. In the latter, Helen Nissenbaum argues that questions of privacy are tied to specific contextual norms that are not dichotomised by a public versus private distinction, but are distributed across a plurality of different spheres of life.[33] Building on this in our particular case of HIV-related information seeking we examine how relational and spatial parameters form part of these contextual norms.

12.3 On Method

Not only is privacy notoriously hard to define it is also difficult to study. One of the methods traditionally employed, as exemplified by the work of Westin,[34] has been the survey. There is, however, an increasing awareness that this only captures a specific, and often narrow, vision of what privacy entails. As a consequence of this limitation, different approaches, such as grounded theory,[35] experiments,[36] and diary methods,[37] have been adopted to study privacy. Despite these methodological developments, concerns persist with regards to how we should best study privacy in relation to Internet use. These include how to conduct research on privacy sensitive individuals,[38] how to avoid prompting an increased privacy sensitive response by the

---

[32] Sandra S. Petronio. *Boundaries of Privacy* (Albany, New York: State University of New York Press, 2002); Jonathan Elford et al. "Disclosure of HIV status: the role of ethnicity among people living with HIV in London." *Journal of Acquired Immune Deficiency Syndromes* 47 (2008): 4; Rosalie Corona et al. "Do Children Know Their Parent's HIV Status? Parental Reports of Child Awareness in a Nationally Representative Sample." *Ambulatory Pediatrics* 6 (2006): 3; Martha B. Lee and Mary Jane Rotheram-Borus. "Parents' disclosure of HIV to their children." *AIDS* 16 (2002): 16; Valerian J. Derlega et al. "Perceived HIV-related stigma and HIV disclosure to relationship partners after finding out about the seropositive diagnosis." *Journal of Health Psychology* 7 (2002): 4.; Kathryn Greene et al. *Privacy and Disclosure of HIV in Interpersonal Relationships* (New Jersey: Lawrence Erlbaum Associates, 2003).

[33] Helen Nissenbaum. *Privacy in Context* (Stanford, California: Stanford University Press, 2009).

[34] Alan Westin. "Social and Political Dimensions of Privacy." *Journal of Social Issues* 59 (2003): 2

[35] Anne Adams and Martina Angela Sasse. "Privacy in Multimedia Communications: Protecting Users, Not Just Data." In *People and Computers XV – Interaction without frontiers*, edited by Ann Blandford, Jean Vanderdonckt and Philip D. Gray (Lille: Springer, 2001), 49–64.

[36] Alessandro Acquisti and Jens Grossklags. "Losses, gains, and hyperbolic discounting: An experimental approach to information security attitudes and behavior." 2nd Annual Workshop on Economics and Information Security- WEIS'03 (2003); Kai-Lung Hui et al. The value of privacy assurance: A field experiment. *MIS Quarterly* 31 (2007): 1.

[37] Denise Anthony et al. "Privacy in Location-Aware Computing Environments." *IEEE Pervasive Computing* 6 (2007): 4

[38] Carina Paine et al. "Internet users' perceptions of privacy concerns and privacy actions." *International Journal of Human-Computer Studies* 65 (2007): 526–536.

inclusion of explicit privacy references[39] and the persistence of a "privacy paradox" in which a discrepancy is noted between reported and actual behaviour.[40] Although we do not seek to resolve these issues, building on a belief that "robust intuitions about privacy norms" are rooted in the texture of peoples' lives[41] our approach here is to situate questions of privacy within broader narratives of information seeking, narratives which are in turn situated within wider experiences of being diagnosed and living with HIV.

Although narrative interviews have a respected pedigree in medical sociology and anthropology,[42] our aim was not to collect and analyse privacy narratives per se. Instead our goal was to contextualise questions of privacy in relation to health information seeking within broader narratives of living with HIV. As part of a wider study on Internet use and HIV, 41 women from a range of sub-Saharan African countries including Angola, Burundi, Gambia, Ghana, Kenya, Nigeria, Sierra Leone, Somalia, South Africa, Uganda, Zambia and Zimbabwe participated in the research. They were recruited at one of three HIV specialist centres in east London where they were receiving treatment and care. Rather than asking participants about privacy directly, they were prompted to talk about their experiences of living with HIV. The first part of the interview (which lasted between 30 minutes to 3 hours, with an average of 45 minutes) was unstructured with participants talking retrospectively about their experiences of looking for information and help, online and off, in relation to their health from the point they were diagnosed to the present. In the second half, specific questions were asked about their history of and current use of the Internet. Participants' levels of Internet access and use differed greatly, ranging from daily broadband access via personal laptops to sporadic access on terminals at Internet cafes.

The majority of the participants were interviewed in a private hospital room, although some, depending on their preferences were interviewed at their homes or in community support groups. In some cases more than one interview was conducted. Most of the interviews were audio recorded, then transcribed and analysed, but in the cases where research participants were not comfortable with the interviews being recorded simultaneous notes were taken instead. The interview transcripts and field notes were analysed thematically throughout the course of the research and issues raised by participants were included in subsequent interviews. Initial coding was used to highlight areas of pertinence to privacy. Although the words

---

[39] Adam Joinson et al. "Measuring self-disclosure online." *Computers in Human Behavior* 24 (2006): 5.

[40] Carina Paine et al. "Internet users' perceptions of privacy concerns and privacy actions." *International Journal of Human-Computer Studies* 65 (2007): 526–536.

[41] Helen Nissenbaum. "Privacy as Contextual Integrity." *Washington Law Review* 79 (2004): 119–157.

[42] Arthur Kleinman. *The illness narratives: Suffering, healing, and the human condition* (New York: Basic Books, 1988); Arthur Frank. *The wounded storyteller: Body, Illness and Ethics* (Chicago: University of Chicago Press, 1995); Michael Bury. "Illness narratives: fact or fiction." *Sociology of Health and Illness* 23 (2001): 3.

"private" or "privacy" were brought up relatively infrequently by the interviewees associated words, such as "confidentially", "disclosure", "safe", "free", "comfortable", were regularly used, and issues of stigma, disclosure and confidentiality were omnipresent. More detailed thematic coding was then carried out on this "privacy aware" data in order to generate an in-depth understanding of how participants spoke about "doing" privacy while searching for health information.

12.4 Exploring the Relational and Spatial Dynamics of Privacy

12.4.1 Practices of Demarcating HIV and Non-HIV Places

There are few places within which space and time are as overtly marked out as the settings of contemporary medical care,[43] and an HIV clinic is no exception. There are different wards, different waiting rooms, whole areas marked with large staff only signs, rooms which require security passes to access and others that require codes. Who gets access, when and where, is not static. Patients and staff move between these spaces, and this movement changes over time. Where patients and staff have access, where they go and do not go, is neither trivial nor arbitrary and the HIV patients who took part in this research expressed this strongly when they spoke about seeking and receiving HIV treatment and care. Building on this, our aim in this section is to describe how rather than speaking about public and private places in general, participants distinguished between HIV and non-HIV places in particular. This distinction was enacted and maintained in various ways as was made clear while carrying out the interviews, as illustrated below in the case of Frances.[44]

Frances is a patient who was introduced to the first author by a nurse as she had expressed a willingness to take part in the research. However, after reading the information sheet she said, "I am happy to talk to you, but I am not HIV positive." Frances had been approached to take part in the research explicitly because of her HIV positive status and she had accepted this status in front of the nurse. Yet barely 5 minutes had passed and a short corridor traversed before her status had changed, before she was no longer HIV positive to the researcher. Not to the doctors and nurses, but to the researcher. We were still technically in the same location as when we were introduced, but when alone Frances made it clear that outside the hospital she was no longer an HIV patient. Three weeks later, when Frances was interviewed again, she appended not being HIV positive with a spatial specification: "when I leave this place I am no longer HIV positive." What this draws attention to is the way in which privacy was articulated by participants as spatially specific, and yet

[43]For a historical description of how this has changed within British general practice see David Armstrong. "Space and Time in British General Practice." *Social Science & Medicine* 20 (1985): 7.

[44]All names have been changed as have any details that could indicate the identity of the women who took part in the research.

the salience of this spatiality was not necessarily bounded within a given location. Rather, participants demarcated these different places through the management of physical geography, but also interpersonal relations.

The reason Frances had chosen this particular hospital was its distance from where she lived: "I just wanted to get out of my borough. You know where not anyone is knowing me from my borough. So I decided to come here. I could have gone to Barking,[45] but I preferred this hospital." For Frances, who has only disclosed her HIV status to her partner and the healthcare staff directly involved in her care, the physical distance between her house and the clinic was one of the ways in which she protected her privacy. This practice of physical distancing was not unique to Frances. A number of other research participants travelled long distances to seek out treatment and care in order to maintain a separation between places in which they were HIV positive and ones where they felt they simply could not be. Interestingly, in medical geography, an area where issues of space and health are of central importance, access to healthcare services is normally measured based on proximity to people's homes,[46] yet here, for privacy reasons, a certain distance between healthcare services and other relevant locations such as home and work was seen as desirable rather than detrimental for access. Of course, not all participants selected hospitals far from where they lived, but what they did do was employ a range of demarcation practices in order to keep different spheres of life separate. These practices became particularly visible when, as above, they emerged as apparent contradictions or paradoxes. They could also be seen in the case where participants resisted changes to these demarcations, as discussed below.

12.4.2 *The Difficulty of Moving Between HIV Places and Non-HIV Places*

> By that time I am here (department of sexual health), but before coming here I was in big fight. I said I am not coming here. I said to them anything they want to do to me should be in the private place because I don't want to see my country people because they will pick up phone and tell my family and friends back home and my daughter will not be able to be in peace. So I said I am not coming to this building, this particular building I will never ever come here. They talk to me and they take me to one room in this building.

In the excerpt above, Wendy one of the research participants, speaks about her difficulty in receiving treatment in the department of sexual health after being diagnosed during pregnancy. As pregnant women in the UK are routinely offered HIV tests it was unsurprising that a number of the women interviewed for this research had moved from the antenatal clinic to the HIV clinic, a move they often described as extremely traumatic. In the former they were focussed on being an expectant mother, where in the latter they shifted to being an HIV patient. One of the key things Wendy

---

[45]Name of hospital referred to changed.

[46]Robin A. Kearns and Alun E. Joseph. "Space in its place." *Social Science & Medicine* 37 (1993): 6.

objected to in this transition was being seen at the department of sexual health by people from her country. In Frances's case the demarcation between HIV and non-HIV places was centred on physical distancing, while in Wendy's the nested rather than Euclidian nature of these demarcations is more apparent. Although Wendy resisted moving between the antenatal clinic and the department of sexual health in a hospital in east London, the implications of this resistance, for her, stretched in a widening arc that included Africa.

This particular nesting of locations occurred frequently. Participants distinguished between Africa and the UK, within the UK between home and hospital, within the hospital between the different departments, and within the departments between different people. Although these distinctions did shift over time, once they had been made a great deal of work was required for the old demarcations to be replaced by new ones. In Wendy's case it took time, persuasion and care on the part of the healthcare practitioners for her to become happy to attend the HIV clinic. This resistance related to physical movement between departments as well as the distribution of her information. Wendy did not want her information to be distributed between departments unless it was absolutely necessary. She even objected to it being visible when she was receiving in-patient care at the hospital. Underpinning this local resistance to the distribution and display of information was a diffuse and constant fear of people "back home in Africa" finding out about an HIV positive status. This separation between HIV in London, where treatment and care was possible, and HIV in Africa, where people were dying, highlights how decisions around what and where to keep things private were not only embedded in specific locations, but nested within broader socioeconomic factors.

The idea that different privacy norms and expectations are applied in different "spheres of life" is not a new one.[47] In this research these spheres were not only defined in relation to specific places – the HIV clinic, the community group, the home, the Internet cafe, the GP surgery – but also nested in much broader socio-political networks. It is in the creation of these networks, and the spaces of possibility they engender, where the relevance of the feminist injunction that the personal is the political can be seen most clearly. While Catharine MacKinnon claims that "to see the personal as political means to see the private as public" and that "the very place (home, body), relations (sexual), activities (intercourse and reproduction), and feelings (intimacy, selfhood) that feminism finds central to women's subjection form the core of privacy doctrine",[48] this is based on the equation of privacy as intimacy, as domesticity, rather than for intimacy, for domesticity, and for multiple other worlds. "Exploding" the private does not have to necessitate its destruction, but rather a splintering into multiple interrelated private spheres.[49] And

[47] Ferdinand Schoeman. *Privacy and Social Freedom* (Cambridge: Cambridge University Press, 1992).

[48] Catharine MacKinnon. "Feminism, Marxism, Method, and the State." *Signs* 8 (1983): 656–657.

[49] Catharine MacKinnon. *Toward a Feminist Theory of the State* (Cambridge, Mass: Harvard University Press, 1989), 190.

it is in understanding how these spheres came into being and are maintained, that we can develop a richer understanding of privacy as a relational concept.

As one might expect these demarcations, and the manner in which they were made, changed over time. In a similar way to the unfolding of a stigma trajectory,[50] changes in participants' health and life experiences resulted in changes in what they considered HIV or non-HIV places. As such, privacy practices not only took place within these demarcations, but also were performative of new ones. In the excerpt below, Wendy describes how this happened in relation to the department of sexual health becoming a place in which she was comfortable receiving treatment:

> Before I met Dr. Thompson[51] and Angela [another doctor] it was a big trouble. I said I don't want to see any of them. They said why? I said no I don't want to see them because it is better I die because I am nothing now. Because this is not going to heal up, so its better I die, so I don't want to see anybody. Lucy [midwife who initially told Wendy about her HIV positive diagnosis] was very very nice. She would drive, go to my house, talk with me, bring all sorts of things, to talk to me, go to this home treatment people, bring the head of the home treatment people to my house. She went there about three times on her own. And that was the time she get me to come here and before I come here again, when I came to hospital it was very bad again. We went to the antenatal because I tell them I am not coming here, so we went to the antenatal in one of the doctors room and when I went there Dr. Thompson was there, but before Dr. Thompson came it was a black lady supposed to take back my blood test and as soon as I realised it was a black lady and I said no! I don't want anyone of that colour to look after me. I was shouting. They said why? I said no. Because I don't know where she is from.

As can be seen above it took a great deal of work to make the department of sexual health a safe HIV place for Wendy. Healthcare practitioners provided reassurances of confidentiality and specific interpersonal relations were developed over time. Moreover, Wendy explained how her initial aversion to people from her country treating her gave way after a while once she got to know individuals. Often these processes of her becoming familiar with people involved her checking for explicit privacy indicators such as badges of identification and evidence of professional status.

12.5 Privacy Practices and HIV-Related Internet Use

12.5.1 Putting the Internet in Its Place

Up to this point we have articulated a sense of place as an important parameter for privacy in offline HIV-related information seeking. Additionally, we highlighted that this sense of place did not simply mirror existing physical locations. Instead of being self-evident "safe" HIV places were brought into being through practices

---

[50] Angela A. Alonzo and Nancy R. Reynolds. "Stigma, HIV and AIDS: An exploration and elaboration of a stigma trajectory." *Social Science & Medicine* 41 (1995): 3.

[51] All healthcare practitioners' names have been changed.

such as the use of privacy indicators, reassurances of confidentiality, practices of care, specific interpersonal relations and disclosure. The word place is usually used to indicate an actual physical location.[52] Moreover, it is often a place with very specific associations and meanings, in the sense of "anthropological place".[53] In this section we take this sense of place and look at how it interacts with online information seeking.

While the fact that the Internet is not removed from people's lives, that it did not "fall out of the sky,"[54] is well recognised in the privacy literature, most research on privacy in relation to the Internet focuses on privacy on the Internet. Although this remained important in our research, another aspect to privacy and Internet use emerged, that of privacy when using the Internet. Where participants used the Internet mattered for privacy, as can be seen in the example below:

Interviewer: "Do you have Internet access at home?"

Emma: "I don't use it at home 'cause you know I don't want people to see'. Sometimes, you know, I just go, you know, to the business centres, cyber cafes. Because I don't like to go down because the computer is in the kitchen. We put it in the kitchen. And you know because of the boys in the house, they are in school, they use the computer all the time. So I said this week I have to get a laptop so I can stay at home and I can do anything I want. And I am going to lock it up, you know because they are kids, I can't tell them. So next week I will get a laptop for myself."

Many contemporary conceptualisations of privacy have been developed based on the assumption that the home is, or at least should be, a private place.[55] Yet, in our research whether the home was demarcated as a private HIV place depended on a range of factors influenced by the socioeconomic circumstances of the participants, family arrangements and disclosure status. In the excerpt above Emma, who has not disclosed her HIV positive status to anyone apart from her partner, spoke of how she was not comfortable using her home PC for HIV-related information because it was situated in the kitchen and shared with her children. For her, in relation to HIV, she would rather go to an Internet cafe to find information about HIV. In Emma's case an apparently public place of Internet use became, when compared to her home, relatively private. This apparent inverting of the private and the public was extremely common amongst research participants and often the home was considered the least private place with regards to HIV. In many cases, like Emma's, this

---

[52] Stuart Shapiro. "Places and Spaces: The Historical Interaction of Technology, Home, and Privacy." *Information Society* 14 (1998): 14.

[53] Christine Milligan. "Location or dis-location?" *Social & Cultural Geography* 4 (2003): 4.

[54] Adam Joinson et al. "Measuring self-disclosure online." *Computers in Human Behavior* 24 (2006): 5, 242.

[55] Stuart Shapiro. "Places and Spaces: The Historical Interaction of Technology, Home, and Privacy." *Information Society* 14 (1998): 14.

was primarily because participants did not want family members, friends or housemates finding any evidence of them searching for HIV information online, and so they went elsewhere to use the Internet even when they had access at home:

> I don't use it a lot [the Internet] because I share the computer at home and the bookmarks stay on. I can use the computer at home to find things not related to HIV, but for HIV things I go to the Internet cafe.

However, the relationality of these demarcations needs to be stressed. It was not that the home was not deemed private, but rather it was not deemed private in the context of living with HIV. For the women who took part in this research the Internet of the home was markedly different from the Internet of the cyber cafe, of the library, of the community support group. More importantly, and what is so interesting about these different Internets, is that their relative privacy or publicity were emplaced. The practices of demarcation, discussed earlier, impinged not only on face-to-face information seeking, but also online. However, even when the home was not considered an HIV place participants did sometimes use the Internet in relation to HIV there and when they did so they used a variety of mechanisms to increase the relative privacy of this use. In the following section we discuss two prominent examples of this: deleting and depersonalising.

12.5.2 Practices for Making Internet Use Private

Some participants used the Internet at home even when they did not consider the home an HIV place. In these cases, they spoke of practices they adopted in order to make their use of the Internet in relation to HIV private, the most common example of which was to delete any evidence of their use – to render it invisible.

> The problem with HIV is that you can't really, that is me, well I am talking on my behalf. I find it hard asking people, but I can go like on the website and ask Google, ask whatever I want to ask about it. I am not good with computers but sometimes I try because I have a family and my children are teenagers and they don't know my status. So I have to find a time whereby they are not even in the house so they don't look at what I am searching for. If I am finished I make sure I delete the history, the history of the website I was on so they don't say oh my God mum was, who was looking at this?

For participants, like Grace above, use of the Internet at home for HIV-related information was relatively private as long as they felt comfortable they could delete the evidence of that use. Therefore, use of the Internet was only considered partially private – something that you did covertly and not too prolifically, something that had the potential to be private. Yet, the ability to bring this potentiality into being was highly contingent on your familiarity and comfort with computer technology. In some cases, however, the removal of evidence of Internet use in relation to HIV was not necessary even when participants had not disclosed their status to the people they lived with. In these cases instead of deleting or hiding evidence of HIV-related Internet use participants depersonalised that use.

The most common mechanism for depersonalising HIV-related Internet use was to associate it with something else, most typically work or studies. For example Olivia, who has not told her teenage children about her HIV status, felt comfortable using the Internet in relation to HIV at home because of her position as a nurse: "I feel safe to use the Internet. Being a nurse if anybody was checking I am covered, because I am a nurse and I need to know what goes on". It was not the act of searching for HIV information that was at stake here, but its legitimisation. A number of participants were working as or studying to be healthcare professionals in the UK and they developed a sense of privacy by disassociating their search for HIV information from themselves. What this draws attention to is that the affect of associations on privacy are not always threatening, but can also be a form of protection. So while Wendy resisted going to the department of sexual health because of its association with sexually transmitted infections and HIV, Olivia uses her position as a nurse to associate her interest with HIV with work and hence prevent this interest from being interpreted as personal. Instead of keeping her HIV status private she turns her interest in HIV into something public in order to keep it private. This highlights how in some case people actively leverage the entanglement of private versus public information as a means of privacy enhancement.

These two examples of practices for rendering Internet use private focus on privacy when using the Internet, but what about privacy on the Internet? In order to illustrate this we turn to cases where the home was considered more straightforwardly private in relation to HIV; where the participants lived alone or with people who knew their HIV positive status. In these cases there was typically more overlap between the designation of domesticity with privacy, and this filtered through to the privacy practices participants adopted when using the Internet. It is important to note, however, that this overlap was not given simply as a characteristic of the technology, but dependant on whether they were living alone, whether they had children, whether they had an Internet connection and whether they could use it on their own or had someone they trusted who could help them. And in cases where the Internet was used in homes considered HIV places, other forms of spatial and relational privacy dynamics emerged, which we discuss below using Harriet's particular case as an example.

12.5.3 Places and Spaces of Privacy Online

Harriet lives alone and has a laptop with an Internet connection that she uses regularly in relation to HIV. She does not only search for information about HIV medication, but subscribes to newsletters, reads up on research and has even contacted doctors via e-mail to ask questions and get quotes regarding treatment. It would appear that she is comfortable about using the Internet in relation to living with HIV and is not too concerned with issues of privacy. Yet, through two interviews and an analysis of an information-seeking diary, a more nuanced sense of her perceptions and practices around privacy on the Internet emerged.

Harriet was happy to search for information across a range of websites but had a strong preference for trusted sources of information, such as NAM.[56] Her two most frequent methods of finding information online were to put queries into Google and click through the websites that came up, or to go directly to sites she knew and trusted. She e-mailed doctors and kept in contact with people from her HIV community support group, but when it came to meeting new people online, as shown in the excerpt below, she was more wary. This opens up a range of questions with regards to the relationship between offline HIV places and online spaces.

Interviewer: "You said you go to the community support groups to speak to people who have had the same experiences as you; do you ever do that online?"

Harriet: "No I don't do that online, because online you know I don't want to expose myself to people who don't know me. I don't want to chat to someone I don't know [pause], because you know it's not a secret, HIV is stigmatised, someone will come in like an HIV person and yet they are not. And they will get everything, chat, chat, chat, the next thing, before you know it's on Facebook and everywhere."

Interviewer: "But you feel at the community support group its ok?"

Harriet: "It's ok because these are positive people [pause], why should they be malicious? They are suffering like me. So that community I don't mind exposing myself to them, no problem, they are like me, I am like them."

Even in purely online interaction the importance for a sense of place for privacy persists. While Harriet is happy to e-mail the people she has met at a community support group and chat to them online she does not trust people she has only met online. In addition to trust relations developed offline persisting online, Harriet's description of her Internet use in relation to HIV was broken up into the designation of specific HIV spaces. These online spaces often, but not always, had a strong connection with an offline place (the community support groups for example), and a strong preference for trusted information sources emerged most notably amongst research participants who had been using the Internet for a while in relation to their HIV health concerns and questions. The majority of these were based in the UK: the NHS choices website,[57] NAM, i-Base[58] and Avert,[59] but participants did use sites from other countries such as The Body[60] in the US. However, while they felt this information was useful for them they also felt it was often not relevant, both

[56]National Aids Manual. http://www.aidsmap.com. Last accessed 2 November 2009.

[57]http://www.nhs.uk/Pages/HomePage.aspx. Last accessed 2 November 2009.

[58]http://www.i-base.info/. Last accessed 2 November 2009.

[59]http://www.avert.org/. Last accessed 2 November 2009.

[60]http://www.thebody.com/. Last accessed 26 May 2010.

culturally as it was more US centric, but also medically as the healthcare services and treatments being discussed were sometimes not applicable to the UK context.

12.6 Conclusion

Through the situated case of HIV-related information seeking by women living with HIV in London, this paper explored some of the relational and spatial dynamics of privacy when using the Internet for health in the context of a stigmatised illness. In examining how participants spoke about these practices the persistence of a sense of place in relation to privacy and the Internet emerged. However, this sense of place was not one that was simply geographically bounded but delineated through sets of relations and associations that often challenged traditional notions of Euclidian space. Rather than being reducible and super-imposable on physical locations the salience of place for privacy was in how it was actively performed by participants and those around them.

We discussed these relational and spatial dynamics in more detail in relation to practices of demarcating, deleting, and depersonalising. However, each of these practices deserves to be explored in more detail in relation to privacy, both online and off. Using narrative interviews we demonstrated that it is not sufficient to talk of the Internet as a source of private information without situating it within a broader understanding of different "spheres of life". Privacy practices on the Internet are informed by where the Internet (or often more aptly the computer in question) is placed as well as the specificities of online spaces. These are in turn embedded within broader socioeconomic and political circumstances. This draws attention to how being able to make the Internet "private" involves work that is contingent on where you live and your pre-existing knowledge, opening new avenues for research on privacy in relation to different aspects of people's lives and different technologies.

Acknowledgments The authors gratefully acknowledge support from the Engineering and Physical Sciences Research Council (grant EP/G002606/1). They would also like to thank all the patients whose generosity with their time and willingness to share their experiences made this research possible. Considerable support was provided by staff working at the Homerton, Newham and Barking hospital departments of sexual health whose contribution and commitment were essential for this project; in particular Dr. Jane Anderson. Thanks to Noortje Marres for her valuable input on an earlier draft of this paper. The research was carried out with ethical approval from the Outer North East London Research Ethics Committee (reference 08/H0701/75).

References

Acquisti, A., and Jens G. "Losses, Gains, and Hyperbolic Discounting: An Experimental Approach to Information Security Attitudes and Behavior." In *2nd Annual Workshop on Economics and Information Security – WEIS '03.*, (2003).

Adams, A., and M. A. Sasse. "Privacy in Multimedia Communications: Protecting Users, Not Just Data." In *People and Computers XV – Interaction without frontiers*. Lille: Springer, 49–64. (2001).

Allen, A. L. *Uneasy Access*. Rowman & Littlefield, New Jersey, 1988.
Alonzo, A. A., and N. R. Reynolds, "Stigma HIV and AIDS: An Exploration and Elaboration of a Stigma Trajectory." *Social Science & Medicine* 41, 3 (August 1995): 303–315.
Altman, Irwin. *The Environment and Social Behaviour*. Monterey, CA: Brooks/Cole Publishing Co, 1976.
Anthony, D., H. Tristan, and K. David. "Privacy in Location-Aware Computing Environments." *IEEE Pervasive Computing*, 6, 4 (2007): 64–72.
Bakardjieva, M. *Internet Society: The Internet in Everyday Life*. Thousand Oaks, CA: Sage Publications Ltd, 2005.
Berger, M., T. H. Wagner, and L. C. Baker. "Internet Use and Stigmatized Illness." *Social Science & Medicine* 61, (2005): 1821–1827.
Bury, M. "Illness Narratives: Fact or Fiction." *Sociology of Health & Illness* 23, 3 (2001): 263–285.
Corona, R., M. K. Beckett, B. O. Cowgill, M. N. Elliott, D. A. Murphy, A. J. Zhou and M. A. Schuster. "Do Children Know Their Parent's HIV Status? Parental Reports of Child Awareness in a Nationally Representative Sample." *Ambulatory Pediatrics* 6, 3 (2006): 138–144.
Cunningham, S. D., J. M. Tschann, J. E. Gurvey, J. D. Fortenberry, and J. M. Ellen. "Attitudes About Sexual Disclosure and Perceptions of Stigma and Shame." *Sexually Transmitted Infections* 78, 5 (2002): 334–338.
Department of Health. *HIV and AIDS in African communities: A framework for better prevention and care*. London, 2004.
Derlega, V. J., B. A. Winstead, K. Greene, J. Serovich, and W. N. Elwood. "Perceived HIV-Related Stigma and HIV Disclosure to Relationship Partners After Finding Out About the Seropositive Diagnosis." *Journal of Health Psychology* 7, 4 (2002): 415–432.
Doyal, L., and J. Anderson. "My fear is to fall in love again" How HIV-positive African women survive in London. *Social Science & Medicine* 60 (2005): 1729–1738.
Elford, J., J. Anderson, C. Bukutu, and F. Ibrahim. "HIV in East London: ethnicity, gender and risk. Design and methods." *BMC Public Health* 6, 150 (2006).
Elford, J., F. Ibrahim, C. Bukutu, and J. Anderson. "Disclosure of HIV status: the role of ethnicity among people living with HIV in London." *Journal of Acquired Immundeficiency Syndrome* 47, 4 (2008): 514–521.
Flowers, P., M. Davis, G. Hart, M. Rosengarten, J. Frankis, and J. Imrie. "Diagnosis and stigma and identity amongst HIV positive black Africans living in the UK." *Psychology and Health* 21, 1 (2006): 109–122.
Fortenberry, J. D., M. McFarlane, A. Bleakley, S. Bull, M. Fishbein, D. M. Grimley, C. K. Malotte, and B. P. Stoner. "Relationships of stigma and shame to gonorrhea and HIV screening." *American Journal of Public Health* 92, 3 (2002): 378–381.
Foucault, M., and J. Miskowiec. "Of Other Spaces." *Diacritics* 16, 1 (1986): 22–27.
Frank, A. *The wounded storyteller: Body, Illness and Ethics*. Chicago, IL: University of Chicago Press, 1995.
Freidson, E. "Client Control and Medical Practice." *The American Journal of Sociology* 65, 4 (1960): 374–382.
Green, G. and R. Smith. "The psychosocial and health care needs of HIV-positive people in the United Kingdom." *HIV Medicine* 5, 1 (2004): 5–46.
Greene, K., V. J. Derlega, G. A. Yep, and S. Petronio. *Privacy and Disclosure of HIV in Interpersonal Relationships: A Sourcebook for Researchers and Practitioners*. 1st ed. London: Routledge, 2003.
Haddon, L. *Information and Communication Technologies in Everyday Life: A Concise Introduction and Research Guide*. Oxford: Berg Publishers Ltd, 2004.
Hui, K.-L., H. H. Teo, and S.-Y. T. Lee. "The value of privacy assurance: A field experiment." *MIS Quarterly* 31, 1 (2007): 19–33.
Joinson, A., C. Paine, T. Buchanan, and U.-D. Reips. "Measuring self-disclosure online: Blurring and non-response to sensitive items in web-based surveys." *Computers in Human Behavior* 24, 5 (9, 2008): 2158–2171.

Kearns, R. A. and A. E. Joseph. "Space in its place: Developing the link in medical geography." *Social Science & Medicine* 37, 6 (September 1993): 711–717.

Kivits, J. "Researching the 'Informed Patient': The case of online health information seekers." *Information, Communication & Society* 7, 4 (2004): 510–530.

Kleinman, A. *The illness narratives: Suffering, healing, and the human condition*. New York, NY: Basic Books, 1989.

Lee, M. B., and M. J. Rotheram-Borus. "Parents' disclosure of HIV to their children." *AIDS* 16, 16 (2002): 2201–2207.

Mackinnon, C. "Toward a Feminist Theory of the State." *Women's Rights Law Reporter* 12 (1990): 205.

MacKinnon, C. A. "Feminism, Marxism, Method, and the State: Toward Feminist Jurisprudence." *Signs* 8, 4 (Summer 1983): 635–658.

MacKinnon, C. A. *Toward a Feminist Theory of the State*. 2nd ed. Cambridge, MA: Harvard University Press, 1991.

Margulis, S. T. "On the Status and Contribution of Westin's and Altman's Theories of Privacy." *Journal of Social Issues* 59, 2 (6, 2003): 411–429.

McKinlay, J. B. "Social Networks, Lay Consultation and Help-Seeking Behavior." *Social Forces* 51, 3 (1972): 275–292.

Mechanic, D. *Medical Sociology*. New York, NY: Free Press, 1968.

Milligan, C. "Location or dis-location? Towards a conceptualization of people and place in the care-giving experience." *Social & Cultural Geography* 4, 4 (2003): 455.

Nettleton, S., R. Burrows, and L. O'Malley. "The mundane realities of the everyday lay use of the internet for health, and their consequences for media convergence." *Sociology of Health & Illness* 27, 7 (2005): 972–992.

Nissenbaum, H. F. "Privacy as Contextual Integrity." *Washington Law Review* 79 (2004): 119.

Nissenbaum, H. F. *Privacy in Context: Technology, Policy, and the Integrity of Social Life*. Stanford, CA: Stanford University Press, 2009.

Orgad, S., "The cultural dimensions of online communication: a study of breast cancer patients' internet spaces." *New Media & Society* 8, 6 (2006): 877–899.

Paine, C., U.-D. Reips, S. Stieger, A. Joinson, and T. Buchanan. "Internet users' perceptions of privacy concerns and privacy actions." *International Journal of Human-Computer Studies* 65, 6 (June 2007): 526–536.

Pateman, C. *The Disorder of Women: Democracy, Feminism and Political Theory*. Polity Press, Cambridge, 1990.

Petronio, S. S. *Boundaries of Privacy: Dialectics of Disclosure*. Illustrated edition. Albany, NY: State University of New York Press, 2002.

Prost, A., J. Elford, J. Imrie, M. Petticrew, and G. J. Hart. "Social, Behavioural, and Intervention Research among People of Sub-Saharan African Origin Living with HIV in the UK and Europe: Literature Review and Recommendations for Intervention." *Aids Behaviour* 12 (2008): 170–194.

Scambler, A., G. Scambler, and D. Craig. "Kinship and Friendship Networks and Women's Demands for Primary Care." *Journal of the Royal College of General Practitioners* 26 (1981): 746–750.

Schoeman, F. D. *Privacy and Social Freedom*. 1st ed. Cambridge, MA: Cambridge University Press, 1992.

Shapiro, S. "Places and Spaces: The Historical Interaction of Technology, Home, and Privacy." *The Information Society* 14 (1998): 275–284.

Sillence, E., P. Briggs, P. R. Harris, and L. Fishwick. "How do patients evaluate and make use of online health information?" *Social Science & Medicine* 64 (2007): 1853–1862.

Warren, S. D., and L. D. Brandeis. "Right to Privacy." *Harvard Law Review* 4 (1890): 193.

Wellman, B., and C. Haythornthwaite. *The Internet in Everyday Life*. Oxford: Blackwell, 2002.

Westin, A. F. *Privacy and Freedom*. New York, NY: Atheneum, 1970.

Westin, A. F. "Social and Political Dimensions of Privacy." *Journal of Social Issues* 59, 2 (6, 2003): 431–453.

Chapter 13
Rise and Phall: Lessons from the Phorm Saga

Paul Bernal

The saga of Phorm's Webwise behavioural targeting system has been fraught with problems: legal challenges, technical disputes, serial campaigning, police action, EU action, smear campaigns and propaganda. Until the spring of 2009 it had looked as though Phorm would succeed, with the UK government firmly behind it, three of the biggest ISP's planning to use its service, an endorsement of sorts from noted privacy advocates and guarded approval from the Information Commissioner's Office. Then, however, things began to fall apart, and by the autumn of 2009 Phorm's business model was in tatters.

Phorm's Webwise has to most intents and purposes failed. Seen through the lens of symbiotic regulation,[1] it was through a failure to understand the complexity and nature of the regulatory matrix in which Phorm operated that lay behind many of their mistakes, and in the end led to the failure of their business idea. Ultimately, however, Phorm failed because it did not understand the nature of the symbiosis between businesses and users that exists in the web, where users effectively sacrifice privacy or personal data in exchange for free, cheap or improved services.[2] Symbiosis succeeds when both sides of the symbiosis benefits from the relationship – as happens with models like those of Google, Facebook and others. With Webwise, only Phorm stood to benefit – in effect, they were offering parasitism rather than symbiosis. The parasite was rejected and effectively purged from the system. That purging was a painful process, and has had some serious implications, not all of which are positive. Indeed, the whole of behavioural targeting is under threat as a consequence of the saga – and technologically innovative and potentially beneficial business ideas may be lost, delayed or hamstrung as a result.

P. Bernal (✉)
Information Technology Law, University of East Anglia, Norwich, UK; London School of Economics and Political Science, London, UK
e-mail: p.a.bernal@lse.ac.uk

[1] As set out in A.D. Murray, *The Regulation of Cyberspace: Control in the Online Environment*. (Milton Park, Abingdon, UK, 2006), Chapter 8

[2] The theory behind this symbiosis is set out in P. Bernal. Web 2.5: The Symbiotic Web. *International Review of Law, Computers & Technology* 24 (2010): 25–37

S. Gutwirth et al. (eds.), *Computers, Privacy and Data Protection: an Element of Choice*, DOI 10.1007/978-94-007-0641-5_13, © Springer Science+Business Media B.V. 2011

13.1 Behavioural Targeting

The marketing industry are highly enthusiastic about behavioural targeting, suggesting that not only does it work well for advertisers but that it gives customers what they want, "improving user experiences" and that audiences welcome it. Mark Wilmot, writing in Marketing Daily in 2009, said "Something amazing happens when marketing efforts are actually relevant to people. We see this step as initiating that crucial dialogue. And shoppers, for their part, are replying; essentially giving permission to marketers to learn their habits and respond accordingly".[3] What Wilmot means by "essentially giving permission" reveals a great deal about the way that the advertising industry views the issue of consent – as something that customers do automatically, implicitly, just by participating in their programmes or accepting their services, without debate or discussion.

Privacy advocates take a diametrically opposite view, seeing behavioural targeting as a pernicious and potentially dangerous practice.[4] A 2009 study by a group from the University of Pennsylvania and the Berkeley Center for Law & Technology suggested that the American public is closer to the views of the privacy advocates than those of the marketing industry. According to that study, 66% of adult Americans do not want marketers to tailor advertisements to their interests. More significantly, when informed of three of the most common ways that marketers gather data in order to tailor ads, even higher percentages – between 73 and 86% – say they would not want such advertising.[5] No similar surveys have been done in the UK or Europe to date, but in the absence of evidence to the contrary it would at least be unreasonable to suggest that the opposite – that people in the UK or Europe want to have advertising tailored to their interests – would necessarily be true.

13.2 Webwise

With Webwise, Phorm took behavioural targeting to a new level, not just tracking particular aspects of surfers' web activities, or activities on particular websites or uses of particular web services, but attempting to track their entire web activity – every website visited, every click made, every service used. Achieving this depth of monitoring involved two key things: some very inventive technology, and an alliance with cooperative ISPs. Detailed work on how the technology works has been done, most notably by Richard Clayton of the Computer Laboratory at the University of Cambridge.[6] As Clayton puts it:

---

[3] http://www.mediapost.com/publications/?fa=Articles.showArticle&art_aid=110489.

[4] E.g http://www.privacyinternational.org/article.shtml?cmd%5B347%5D=x-347-559082

[5] J. Turow et al. *Americans Reject Tailored Advertising and Three Activities that Enable It.* (Annenberg: University of Pennsylvania, 2009), particularly p. 3

[6] R. Clayton, *The Phorm "Webwise" System* (2008). http://www.cl.cam.ac.uk/~rnc1/080518-phorm.pdf:

> The basic concept behind the Phorm architecture is that they wish to take a copy of the traffic that passes between an end-user and a website. This enables their systems to inspect what requests were made to the website and to determine what content came back from that website. An understanding of the types of websites visited is used to target adverts at particular users.[7]

This monitoring mechanism involves putting a "false" cookie onto the user's computer by masquerading as the website which the user wishes to visit. That cookie will contain an individual identifier that is then used by Phorm to monitor the activities of the user on the relevant domain. That identifier (known as a UID) is used as the principle way of identifying a user throughout the Webwise process. It is a randomly generated number, with no connection to the individual, and as such maintains (in Phorm's opinion) anonymity and privacy. Phorm uses it not only as what they believe is a way to ensure that they are not covered by data protection law (since the data they hold is not linked to an individual, just to a UID) but as a way of portraying themselves not just as a "privacy friendly" company but as a company in the vanguard of the fight in favour of privacy.

13.3 Practical and Legal Implications

In many ways Webwise is similar to more conventional behavioural targeting systems – Phorm's system of channels, for example, is similar to the profiles used by other systems. The way that advertisements are served to participating websites is effectively identical. There is one very significant difference. Webwise monitors and analyses all an individual's activities, not just those on a particular site or system. Google's behavioural targeting, by comparison, works only on data gathered through searches made using Google's search engine and other Google services – so if a user searches with Yahoo or ASK instead, that data will neither be available nor be used by Google for their analysis.

Webwise does this by working at the ISP level. It is a system that needs to be deployed by an ISP, so that it can be in a position to intercept all the web-surfer's activities. Indeed, the key to the Phorm business model, as it first became apparent, was that Phorm was aiming to work with three of the UK's largest ISPs: BT, TalkTalk and Virgin Media.

Nicholas Bohm, of the Foundation for Internet Policy Research, in his legal analysis of Phorm,[8] suggested that the deployment by an ISP of the Phorm architecture would involve four different forms of illegality, for which the ISP would be primarily liable, and for which Phorm would be liable as an inciter:

---

[7]Ibid. p2

[8]N. Bohm, The Phorm *"Webwise" System – a Legal Analysis."* Foundation for Information Policy Research, 2008

1. Interception of communications, an offence contrary to Section 1 of the Regulation of Investigatory Powers Act 2000 (RIPA). This relates to the monitoring phase, where the Phorm architecture, as managed and operated by the ISP, intercepts the instructions sent by the surfer to a website in order to copy them.[9]
2. Fraud, an offence contrary to Section 1 of the Fraud Act 2006. This relates to the way in which the Phorm server masquerades as the target server, in order to make the surfer's web browser accept the Phorm cookie.[10]
3. The risk of committing civil wrongs actionable at the suit of website owners – Bohm suggests the example of the Bank of England, which like many other websites states categorically in its published privacy policy that it does not "use cookies to collect information about you." When Phorm is in action, it would look to most users that the Bank of England is doing precisely that – though the monitoring cookie would actually have been placed by Phorm, it would look to a user as though it were a Bank of England cookie. The owner of the site might therefore have civil remedies for false implication – or even defamation.[11]
4. Unlawful processing of sensitive personal data, contrary to the Data Protection Act 1998.

These four hint at the deeper issues that lie behind not just Phorm but most behavioural targeting systems – and other forms of data gathering on the Internet. The RIPA issue concerns the privacy of an individual's actions on the net to start with – whether people want or expect their web browsing to be private or not. The second and third issues are issues of good faith – when a surfer visits a website, can they expect that their interactions with that website just to be with that website, and not with another, unconnected third party?

13.4 Data Protection and Sensitive Personal Data

The fourth of these legal issues, the data protection issue, is perhaps the most important. Browsing activities can be some of the most personal, most intimate, and most sensitive of activities – concerning everything from personal tastes to relationships, jobs and finance, plans for the future, even personal peccadilloes. Is there an expectation that this kind of thing is considered private? From a legal perspective, as outlined by Bohm, there are a number of different ways in which the processing of data by Phorm might be considered illegal.

Firstly, if the issues relating to RIPA and the Fraud Act above are accepted, then the purpose for which the data is being gathered and processed cannot be legal, and hence the processing itself cannot be legal. Secondly if all web browsing is

---

[9]Ibid. pp. 3–11

[10]Ibid. pp. 11–12

[11]Ibid. p. 16

intercepted, that web browsing will be likely to include information about the browser that would be classified as "sensitive personal data" according to Section 2 of the Data Protection Act 1998.

This latter point is contentious – can a presumption be made that the data is likely to include sensitive data, or, if not sensitive data, data from which sensitive data can be derived or revealed? According to the rules set out in the DPA, data concerning whether a person suffers from diabetes would be classified as "sensitive personal data". Data about whether the subject is a regular purchaser of sugar-free chocolate, or has ordered books about treatment for diabetes would not. Similarly, data about whether a man was a homosexual would be considered to be "sensitive personal data", but data suggesting they were members of the Barbra Streisand fan club, and or that they spent large sums of money on hairdressing would not. Of course none of these facts specifically indicate that the individuals are diabetics or homosexuals respectively – but if profiling is applied, even automatically, the chances of the individuals being classified within categories that consist almost entirely of diabetics or homosexuals respectively would be high.

Whether the data technically satisfies the requirements for sensitive personal data is, however, is only one aspect of the issue – whether the users would consider their browsing habits to be sensitive and personal is another issue, and one which is perhaps even more important.

13.5 The Rise and Fall of Phorm?

Phorm raises a wide range of issues, from the technological nature of its interception and inspection systems and the various technical legal issues highlighted by Bohm's legal analysis to the deeper and less concrete concerns over people's every activity being monitored and exploited for financial gain. Whether the legal issues put forward by Bohm and others have technical merit (or would actually succeed in court) does not appear, in practice, to have been as important as the part that their existence as challenges has played in what appears to be the ultimate demise of Phorm.

The controversy over Phorm has been played out in public. Hackers, digital rights and privacy groups reacted strongly from the moment the proposed service became known. The Open Rights Group started a "Stop Phorm" campaign,[12] while Professor Ross Anderson said "The message has to be this: if you care about your privacy, do not use BT, Virgin or Talk-Talk as your Internet provider."[13] Tim Berners-Lee told the BBC that he would change his ISP if it introduced a system like Webwise.[14]

[12] http://www.openrightsgroup.org/campaigns/stop-phorm

[13] Quoted in the Evening Standard, 6th March 2008, at http://www.thisislondon.co.uk/standard-home/article-23449601-web-users-angry-at-isps-spyware-tie-up.do;jsessionid=D5AA1541C914 46314EAD7013363AB159

[14] See "Web creator rejects net tracking" at http://news.bbc.co.uk/1/hi/technology/7299875.stm

13.5.1 BT's "Secret" Trials

One of the most contentious issues was the discovery that in 2006 and 2007, prior to the existence of Phorm's Webwise becoming public, BT had carried out "secret" trials of the system, involving tens of thousands of end-users. These trials were carried out without the consent of the end users, and when their existence became public, through a report leaked onto the Internet, there was not just an outcry from privacy groups but the City of London Police met with BT representatives to informally question them about the trials. The City of London Police decided not to pursue a formal investigation, suggesting that there was no criminal intent on behalf of BT, and, crucially, that there was "implied consent" by the end-users.[15] This latter claim is highly contentious, while Bohm suggested that the police claim that there was no criminal intent was simply a misunderstanding of the legal requirements for criminal consent.[16] Nonetheless, no police action followed immediately, though BT may still face civil action from customers who were unknowingly involved in the trials[17] and it was reported by The Register in February 2010 that the CPS was still considering taking legal action.[18] Moreover, though nothing specific has yet materialised from the controversy, the outcry caused BT embarrassment, provided a weapon for anti-Phorm campaigners, and added to the impression that Phorm itself was somehow "underhand", secretive and potentially illegal.

13.5.2 Phorm's Defence

Phorm's defence to these attacks included a PR campaign that included founder Kent Ertugrul talking directly to the media, including being interviewed by the BBC,[19] The Guardian,[20] and The Register,[21] as well as engaging directly with the UK Government, firstly to ask the Information Commissioner's Office to confirm that Phorm's UID anonymity system meant that it was compliant with the Data Protection Act. Phorm believed that data protection did not apply to their system, as the data they gather, process and use does not constitute "personal data", let alone "sensitive personal data". The ICO did, effectively, confirm that this was the case, though they also expressed the view that "opt-in" consent would be required for any trials and for any eventual rollout of the service, and suggested that they would be continuing to monitor the situation very closely.[22]

---

[15] See for example http://www.theregister.co.uk/2008/09/22/bt_phorm_police_drop/

[16] Also see http://www.theregister.co.uk/2008/09/22/bt_phorm_police_drop/

[17] See http://www.theregister.co.uk/2008/03/17/bt_phorm_lies/

[18] See http://www.theregister.co.uk/2010/02/25/bt_cps/

[19] http://www.bbc.co.uk/blogs/ipm/2008/03/phorm_an_interview_with_kent_e.shtml

[20] http://www.guardian.co.uk/technology/blog/2008/mar/06/yourquestionspleaseforkent

[21] http://www.theregister.co.uk/2008/03/07/phorm_interview_burgess_Ertugrul/

[22] http://www.ico.gov.uk/upload/documents/pressreleases/2008/new_phorm_statement_040408.pdf

Phorm also sought advice from the Home Office on the RIPA issue. There were two questions: firstly, do Phorm's actions constitute "interception of communications" or not, and secondly if they do, is it lawful interception. On the second question, interception can only be lawful if both the sender and the intended recipient of the communication have consented to that interception. Phorm relied on the idea that surfers have consented to their service in some form (through the ISP's terms and conditions, or through some kind of direct consent yet to be determined) and on the assumption that if a website consents to be spidered for search engine purposes, then they have consented to have communications to them intercepted for Phorm's purposes. The memo that they received in response has become available on the Internet ultimately suggested that Webwise would be legal if the users gave explicit consent.[23]

The issuing of this advice became the centre of another controversy, as emails between the Home Office and Phorm was released that appeared to show that the company had helped edit this draft legal interpretation of Phorm by the Office, in an attempt to ensure that the service would be seen as appropriately "legal". Baroness Sue Miller, the Liberal Democrat spokeswoman on Home Affairs, accused the Home Office of "collusion", calling the exchange of emails "jaw-dropping", and said that "The fact the Home Office asks the very company they are worried is actually falling outside the laws whether the draft interpretation of the law is correct is completely bizarre."[24] Both the Home Office and Kent Ertugrul vigorously denied this interpretation of the exchange of emails.

13.5.3 European Involvement

In response to "several questions from UK citizens and UK Members of the European Parliament", the European Commission inquired into how the UK government had responded to the complaints about Phorm by users.[25] EU Telecoms Commissioner Viviane Reding sent a letter to the UK Government – this time to the Department for Business, Enterprise and Regulatory Reform ("BERR") – asking for an explanation as to how Phorm's technology conformed with EU data protection and privacy laws. BERR replied, after a delay, providing an explanation whose key points depended on Phorm's UID-based anonymity, together with a confirmation of the requirement in the Home Office memo that explicit consent would be required.[26] Phorm, in BERR's opinion, complied with EU privacy law.

---

[23] See for example http://cryptome.org/ho-phorm.htm.

[24] See http://news.bbc.co.uk/1/hi/technology/8021661.stm

[25] http://europa.eu/rapid/pressReleasesAction.do?reference=IP/09/570&format=HTML&aged=0&language=EN&guiLanguage=en

[26] BERR's reply to Commissioner Reding was not made public, but BERR did disclose to The Register the key points, which were then published on the Internet at http://www.theregister.co.uk/2008/09/16/phorm_eu_berr/

This view was immediately challenged by the Open Rights Group and others, noting specifically the requirement for both sides of a communication to need to consent to an interception of a communication – so not only did web surfers need to consent, but website owners, and stressing the inadequacy of Phorm's UID-based anonymity. As a result of this and other responses, the EC inquiry concluded that if the UK believed that Phorm complied with UK privacy law, then that law must not be a correct implementation of the relevant EU directives. After much communication with the UK Government, in April 2009 the Commission launched an action against the UK government, calling for changes in UK law. In the words of EU Telecoms Commissioner Viviane Reding,

> We have been following the Phorm case for some time and have concluded that there are problems in the way the UK has implemented parts of EU rules on the confidentiality of communications.[27]

This action has yet to be concluded. Both Phorm and the ISBA ("the voice of British advertising") have been trying to dissuade the EC from continuing their action,[28] while the Open Rights Group and others have been actively supporting it, and publicising the existence of the action through the media.

13.5.4 Privacy Friendly?

During all these disputes, Phorm has been portraying itself as a "privacy friendly" company, suggesting that rather than being a threat to privacy, Phorm would be providing something that was positive for privacy. Webwise, according to Phorm, meant that you could have the targeted advertising that people wanted without the need for gathering or holding personal data. Phorm engaged a specialist consultancy service, 80/20 Thinking, to perform a "Privacy Impact Assessment" on the service. That assessment appeared largely positive. As 80/20's Simon Davies puts it to the BBC: "We were impressed with the effort that had been put into minimising the collection of personal information." The Privacy Impact Assessment was subsequently used by Phorm to demonstrate their "privacy-friendly" credentials. However, this was not without issues. Simon Davies, as well as being CEO of 80/20 Thinking, is a noted privacy advocate and one of the founding members of Privacy International – and Kent Ertugrul tried to suggest that 80/20's positive assessment of the Phorm system meant that Privacy International had endorsed Phorm, something that he later had to retract.[29]

The disputes between privacy advocates and Phorm became increasingly rancorous as the affair wore on. A number of "anti-Phorm" websites appeared such

[27] http://europa.eu/rapid/pressReleasesAction.do?reference=IP/09/570&format=HTML&aged=0&language=EN&guiLanguage=en

[28] See for example http://www.isba.org.uk/isba/news/657

[29] In a live webchat on Phorm's own blog. See http://www.webwise.com/how-it-works/transcript_080311.html

as Badphorm[30] and Dephormation[31]. In response to some of the more vociferous of anti-Phorm campaigners, Alexander Hanff and Marcus Williamson, Phorm set up their own campaigning site, Stopphoulplay.com. Phorm were forced to admit to "overzealous" editing of their Wikipedia entry, after having deleted sections critical of Phorm and links to some further stories.[32] In the words of BBC correspondent Darren Waters, "This is a battle with no sign of a ceasefire, with both sides settling down to a war of attrition, and with governments, both in the UK and the EU, drawn into the crossfire."[33]

13.5.5 Rise and Phall?

This, then, was the far from simple background. Legal challenges, technical disputes, serial campaigning, possible police action, EU action against the UK government, smear campaigns and propaganda, while Phorm attempted to get its business into action. The result began to become clear in 2009. Though before that stage it had looked as though Phorm was likely to succeed, with the UK government apparently firmly behind it, three of the biggest ISP's planning to use its service, an endorsement of sorts from noted privacy advocates and a guarded approval from the Information Commissioner's Office. Then, however, business reality began to kick in, as other businesses and other government departments began to respond seriously to the furore generated by the whole affair.

In April 2009, Amazon.com announced that it would not allow Phorm to scan any of its domains.[34] Others followed, including the Nationwide Building Society.[35] Then the hammer blow fell when BT announced that it would not be implementing Phorm – followed immediately by Talk-Talk, and then Virgin Media. Phorm's shares fell 40% on the announcement, and it looked as though Phorm's business model was in danger of total collapse. Then, in August, the Office of Fair Trading announced that it was investigating the use of personal information in Internet advertising, questioning the use of tailored advertising and the possibility of tailored prices based on personal information.[36] Phorm's share price fell once more, this time more than 20%, as a result of the announcement of that investigation.[37] The All Party Parliamentary Communications Group ("apComms") has also undertaken its own inquiry into Internet Traffic, covering amongst other things, behavioural

[30]http://www.badphorm.co.uk

[31]https://www.dephormation.org.uk/index.php

[32]See http://www.theregister.co.uk/2008/04/08/phorm_censors_wikipedia/

[33]http://www.bbc.co.uk/blogs/technology/2009/04/phorm_hoping_to_stop_phoul_pla.html

[34]See http://news.bbc.co.uk/1/hi/technology/7999635.stm

[35]See http://www.guardian.co.uk/business/marketforceslive/2009/jul/21/phorm

[36]See http://www.guardian.co.uk/media/2009/aug/20/internet-targeted-advertising-oft-investigation

[37]See http://www.guardian.co.uk/media/2009/aug/20/advertising-digital-media

advertising – the report was issued in October 2009.[38] This report came out with strong conclusions, including the recommendation that:

> . . . the Government review the existing legislation applying to behavioural advertising, and bring forward new rules as needed, to ensure that these systems are only operated on an explicit, informed, opt-in basis.[39]

In September 2009 potentially the final blow fell with the resignation of Phorm's Chief Technology Officer, Stratis Scelparis[40].

13.6 The Fall Out from Phorm

The Phorm saga has had an impact on more than just the company itself. The aforementioned investigations by the OFT and apComms are just part of the fall out, a fall out that has led to serious contemplation of regulation on both sides of the Atlantic.

In the US, a coalition of privacy and consumer rights groups have written an open letter to the House Committee on Energy and Commerce calling for the regulation of behavioural advertising.[41] A bill to be put before Congress is being drafted by Congressman Rick Boucher, who heads the House Energy and Commerce Subcommittee on Communications Technology and the Internet.[42] In Europe Meglena Kuneva, the consumer affairs Commissioner, told a gathering of ISPs, major websites and advertising firms that they were violating "basic consumer rights in terms of transparency, control and risk", through data collection and behavioural targeting[43] and aims to produce a Green Paper on the subject early in 2010.[44]

Some of that regulation has already come into action. The most dramatic example so far has been the quiet adoption in October 2009 of an EU directive that has such far-reaching implications that Struan Robertson, the editor of OUT-LAW.COM and a respected blogger in the field, has called it "breathtakingly stupid."[45] This

---

[38] All Party Parliamentary Communications Group (apComms), *"Can we keep our hands off the net?"* Report of an Inquiry by the All Party Parliamentary Communications Group, October 2009. http://www.apcomms.org.uk/uploads/apComms_Final_Report.pdf

[39] Ibid. p21

[40] See http://www.telegraph.co.uk/finance/newsbysector/mediatechnologyandtelecoms/media/6209787/Phorm-loses-technology-chief.html

[41] http://www.uspirg.org/uploads/Lh/2Y/Lh2Y_vpDJ2A5maDU214SFw/WaxmanBartonLetterSEPT091.pdf

[42] See http://jetl.wordpress.com/2009/09/13/privacy-on-the-web-congress-set-to-curb-online-behavioral-advertising/

[43] Reported in http://www.theregister.co.uk/2009/03/31/kuneva_behavioural/

[44] Reported in http://www.theregister.co.uk/2009/09/15/brussels_behavioural_targeting/

[45] http://www.out-law.com/page-10510

directive modifies existing European legislation[46] and effectively appears to require that any cookie can only be stored on a user's computer, or accessed from that computer, with that user's explicit, informed consent. This would cover not just such things as advertising but any kind of web analytics – indeed the functioning of most modern websites. The directive has been passed, and must come into force in all 27 member-states of the EU by 26th April 2011. When respected lawyers like those of Pinsent Masons who provide OUTLAW.COM suggest such an approach it gives an indication of how serious the implications of this directive might be. There is, however, significant doubt as to whether the directive really means what it appears to mean – advertising trade bodies have suggested that the law can be satisfied by a user's browsers' settings.[47] The eventual outcome is hard to predict until the member-states start to implement the directive.

Whether the effective failure of Phorm is an individual incident or representative of an overall movement has yet to be seen – but the regulatory crackdown does suggest the latter. Why is this happening and what, if anything should be done about it? Murray's theory of symbiotic regulation[48] can help to provide some of the answers to these questions.

13.7 Phorm: Symbiotic Regulation in Practice?

It can be argued that what is happening to Phorm is happening to a great extent because Phorm has failed to fully understand the complexity of the regulatory matrix. From this perspective, Phorm appears to an excellent example of how symbiotic regulation really works, and why, if a good regulatory result is to be achieved, it needs to be harnessed.

The regulatory matrix in which Phorm operates is complex. As the story related above has shown, many of the different relationships within it have had their impact: Phorm's relationship to their customers, Phorm's relationships with their business allies and with their competitors, all the various different parts of the UK Government's relationships both with Phorm, and with the people, the hackers and the advocacy groups' relationships with people, with other businesses, with the UK government – and with the EU; the EU's relationship with the UK; and, as the culmination of all these things, other businesses' relationships with their customers.

It appears that Phorm took too simplistic a view of the regulatory environment, relying on its ability to lobby and negotiate with government, to form alliances with businesses. They looked to find solutions that could be argued to meet with the letter

---

[46]The directive, labeled PE-CONS 3674/09, modifies Directives 2002/22/EC and 2002/58/EC (the ePrivacy Directive) and Regulation (EC) No 2006/2004. It is available at http://register.consilium.europa.eu/pdf/en/09/st03/st03674.en09.pdf

[47]The Interactive Advertising Bureau (IAB) Europe and publishers' trade body the European Publishers' Council (EPC), quoted in http://www.out-law.com/page-10550

[48]As set out in A.D. Murray, *The Regulation of Cyberspace: Control in the Online Environment.* (Milton Park, Abingdon, UK, 2006), Chapter 8

of the law, for the arguments that Phorm put forward about compliance with data protection law have some substance to them, enough to convince the ICO, the Home Office and BERR to give Phorm their support and approval. Kent Ertugrul made a similar misunderstanding when he tried to make the distinction between legal adware and illegal spyware in his earlier business activities, but then abandoned even adware.[49] Then, as for Phorm, he did not understand sufficiently that what people understood and felt was more important than the letter of the law. The public does not like adware, even if it's legal, and sees very little difference between it and spyware. Effectively, Kent Ertugrul was saying that the public was "wrong" not to distinguish between the two, but that he would have to bow to their "wrong view" in abandoning his system. He did not appear to accept that the public might actually be right – in the sense that they didn't like adware because it interfered with what they considered to be their rights. The letter of the law was not what the public cared about, rather what they thought to be right.

Compliance with the letter of the law is not enough when the community and the market come into play. Phorm underestimated the feelings of the community with regard to privacy[50] and the power of the community to influence other parts of the regulatory matrix. Through the various advocacy groups, through public campaigning, and through the EU, the community managed to get its view across. Phorm became perceived as "anti-privacy" and this perception gathered momentum, regardless of Phorm's efforts to portray itself as a privacy-friendly company.

Whether these perceptions actually lay behind the key events in Phorm's ultimate downfall – BT's withdrawal, or the refusal of websites like Amazon.com to be scanned – is questionable. In Amazon's case, letting Phorm scan their website could have robbed them of some of their competitive advantages, while for BT, it might simply have been a matter of not wanting to throw good money after bad. In Amazon's case, however, the fact that they chose to talk about the privacy issues as a part of their reasoning was very revealing. BT did not mention privacy – they said very little except that they were no longer pursuing Phorm as an option. The adverse publicity and overall image of Phorm, however, cannot have helped the cause of BT's continued participation.

13.7.1 Facebook's Beacon and Google StreetView

The story of Facebook's Beacon is another case in point. Through Beacon, Facebook shared data with an alliance of online retailers, allowing each to use the other's information about individuals in order to better target advertising and services.[51] Beacon was controversial, and just as for Phorm the public outcry was vociferous.

---

[49]The business that became Phorm began as 121 Media, an 'adware' company whose products were labeled by many, including F-Secure, as spyware.

[50]See J. Turow et al. *Americans Reject Tailored Advertising and Three Activities that Enable It.* (Annenberg: University of Pennsylvania, 2009)

[51]For a summary of how Beacon worked from Facebook's perspective, see the launch press release at: http://www.facebook.com/press/releases.php?p=9166

Facebook's initial response was to change in the way Beacon operated – primarily to change it from "opt out" to "opt in" – but ultimately Facebook abandoned the system completely.[52] Just as in the case of Phorm, community reaction was strong enough to bring about the end of a service that went beyond what people thought was right. Furthermore, just as in Phorm's case, it was through the manipulation of all the various relationships in the regulatory matrix – relationships between individuals and Facebook, between individuals and governments, through the use of the law, through working with businesses – that this result was brought about. And just as in the case of Phorm, much of the trouble could have been avoided if Facebook had been more aware of both public opinion and of the ability of the public to bring that opinion to bear.

Some other services – for example Google StreetView – have produced somewhat similar reactions from privacy advocates, and in some ways appear even more intrusive, and yet have not suffered the fate of Phorm and Beacon, at least within the UK.[53] The reasons for this are not simple – but in symbiotic regulation terms, the regulatory matrices in which they operate are different. From the start, Google had both a stronger base position and a better reputation with the public, and, it appears, a better grasp of how to get the community on its side. Moreover, StreetView offers a service that is both useful and attractive to users – a benefit in exchange for the intrusion.[54]

13.7.2 Ramifications for Government and Business

The Phorm affair has caused the UK government considerable problems. It faces a lawsuit from the EU and accusations of collusion with what is perceived to be a "dodgy" business, and being portrayed itself as riding roughshod over people's privacy and rights – and all of this to back a business idea that has ultimately ended in failure. If it had had a better idea of the likely outcome, and a better understanding of what it was that mattered to people – why, in the end, people were sufficiently distressed by Phorm to bring about its downfall – then the government could have avoided the whole farrago.

Finally, it is not just Phorm and the UK government who have found themselves in difficulties, but the whole of the online advertising industry. They're facing a

---

[52]Facebook abandoned Beacon on 21st September 2009. See http://www.wired.com/images_blogs/threatlevel/files/facebook_beacon_complaint0812081.pdf

[53]Action has been taken against Google StreetView in some countries – Switzerland is one example, see http://www.techradar.com/news/internet/swiss-take-legal-action-over-google-street-view-650241, while in Japan there are concerns about the misuse of images – see http://www.searchenginejournal.com/google-street-view-in-japan-faces-various-complaints/13048/ – but Google StreetView appears to have been accepted in most countries.

[54]StreetView has had more problems since – in May 2010, for example, it was the center of a controversy surrounding the way that StreetView cars had been "harvesting data" from people's wifi networks (see http://news.bbc.co.uk/1/hi/technology/10122339.stm) – but those problems have largely been unconnected with the basic principle of the service, and show little sign of halting the success of the service.

regulatory crackdown not only in Europe but potentially in the US as well – a crackdown that could potentially damage their entire business models. That crackdown has yet to fully materialise, but at the very least they are faced with the need for some serious lobbying – and at a time when finances are being stretched to breaking point for many, that is a distraction and a drain on resources that they can little afford. There are many who have most of their eggs in the behavioural targeting basket, and if the eventual result of the Phorm farrago is that this basket is broken, their businesses could break with it.

13.8 Maintaining the Beneficial Symbiosis

There is another way of looking at the reasons that Phorm appears to have failed. Businesses and individuals operate in the Internet in a kind of mutually beneficial symbiosis, individuals effectively exchanging their personal data for better, cheaper and often free services. This symbiosis does not function with Phorm's Webwise. With Webwise, only Phorm benefits, not the users – unlike models like those of Google, who for almost all their services, including the most apparently intrusive ones like StreetView, offer something new or improved in return for information or monitoring. Phorm doesn't improve the services, or offer anything new to the user, but just uses existing services and acts in a way that could be described as parasitic. Phorm takes but gives little in return – and in a world in which the value of data is becoming increasingly understood, not just by businesses but by individuals, this, in the end, cannot work – and hence Phorm failed.

It failed in a painful way for almost all concerned. For the users and privacy advocates who, though they ultimately reached what seems to be a positive outcome, were forced to mount a long and serious campaign to fight against Phorm. For Phorm itself, which struggled mightily to launch what they had hoped would be a technologically innovative and potentially highly lucrative business. For their business partners like BT, who have been highly embarrassed, spent significant amounts money and even now face the possibility of legal action. And for the UK government, who have also been extremely embarrassed and face legal action from the EU and the possibility of having to rewrite a number of key laws. The pain for all concerned could have been reduced – or perhaps even avoided – if the symbiosis had been better understood.

Acknowledgements This paper is based on research for my doctorate funded by the Arts and Humanities Research Council.

References

All Party Parliamentary Communications Group (apComms), "Can We Keep Our Hands off the Net?" Report of an Inquiry by the All Party Parliamentary Communications Group, October 2009. Available at: http://www.apcomms.org.uk/uploads/apComms_Final_Report.pdf

Bernal, P.A. "Web 2.5: The Symbiotic Web." *International Review of Law, Computers & Technology* 24 (2010), 25–37

Bohm, N. "The Phorm "Webwise" System – a Legal Analysis." Foundation for Information Policy Research, (2008). http://www.fipr.org/080423phormlegal.pdf

Clayton, R, "The Phorm "Webwise" System" (2008). http://www.cl.cam.ac.uk/~rnc1/080518-phorm.pdf:

Murray, A. D. *The Regulation of Cyberspace: Control in the Online Environment*, Milton Park, Abingdon, UK, 2006; New York, NY, Routledge-Cavendish.

Turow, J., King, J., Hoofnagle, C. J., Bleakley, A. & Hennessy, M. "Americans Reject Tailored Advertising and Three Activities that Enable It." Annenberg: University of Pennsylvania, 2009. Available at SSRN: http://ssrn.com/abstract=1478214

Note: All web references throughout the article accessed 18th May 2010

Chapter 14
Disclosing or Protecting? Teenagers' Online Self-Disclosure

Michel Walrave and Wannes Heirman

14.1 Introduction

In the past decade there has been an enormous growth in the number of households using information and communication technologies.[1] Also children and teenagers take part in this growing wired population. In Belgium, approximately 96% of 12- to 18-year-olds uses the Internet regularly, most of them accessing the Internet at home (92.8%) or at school (62.6%).[2] Parents often encourage their children to use the Internet, since they believe it assists them in developing valuable skills for education and future position in the labour market. An American study shows that a majority of parents (54%) considers the Internet as having a more positive influence on children than television.[3]

Following the recognition by the end of the nineties that the Internet could be used for commercial purposes, there has been a steep increase in commercially fuelled websites destined for children. Parallel with this growth, also the number of online advertisements targeting young people has exploded. Approximately two out of three of child-targeting websites has advertising as backbone of revenues.[4]

M. Walrave (✉)
Department of Communication Studies, Research Group MIOS, University of Antwerp, Sint Jacobstraat 2, BE 2000, Antwerpen, Belgium
e-mail: michel.walrave@ua.ac.be

[1]Esther Thorson and David W. Shumann, "The Internet waits for No One," in Internet Advertising: Theory and Research, ed. David W. Shumann and Esther Thorson (Mahwah: Lawrence Erlbaum Associates, 2007), 6.

[2]Michel Walrave, Sunna Lenaerts and Sabine Demoor, Cyberteens @ risk? Tieners verknocht aan internet, maar ook waakzaam voor risico's? (Brussels: BELSPO, 2008),16.

[3]Sabrina M. Neeley, "Internet Advertising and Children," in Internet Advertising: Theory and Research, ed. David W. Shumann and Esther Thorson (Mahwah: Lawrence Erlbaum Associates, 2007), 343.

[4]Kathryn C. Montgommery, "Digital kids: The new on-line children's consumer culture," in Handbook of Children and the Media, ed. Dorothy G. Singer and Jerome L. Singer (Thousand Oaks: Sage, 2001), 638.

S. Gutwirth et al. (eds.), *Computers, Privacy and Data Protection: an Element of Choice*, DOI 10.1007/978-94-007-0641-5_14, © Springer Science+Business Media B.V. 2011

Neuborne[5] describes the downfall of non-commercial children oriented websites from 10% in 1999 to 2% in 2001. Moreover, an analysis of 294 websites targeting children and youngsters found that 82% of these sites process personal data.[6]

This evolution can be attributed to both technological and economic forces.[7] Technological innovations of the past decade have laid the foundations for commercial Internet applications and the infrastructure for the Internet to turn into a valid marketplace. At the economical level, marketeers have been inspired by the growing popularity of interactive media among children to adapt their marketing communication strategies. This adaptation was considered necessary to stay in touch with a minor audience, since young people represent profit for marketeers. According to McNeal[8] children and teenagers are important consumers in three ways: as current customers with a considerable autonomy to spend money, as important influencers of household purchases and finally as future prospects in adult life.

As children and teenagers are increasingly approached by interactive marketing techniques, public concerns on minors' online privacy rights have risen. More specifically, the personal data collecting practices towards minors are an important issue in the online privacy debate.[9]

14.2 Policy Framework

In an attempt to respond to these concerns, policy initiatives have been developed in order to protect personal data of minors. In this paragraph a broad outline describes some important policy work at the regulatory and self-/co-regulatory level.

14.2.1 Regulatory Policy Initiatives

US Congress adopted the Children's Online Privacy Protection Act (COPPA) in 1998. This law orders that any person, who operates a child-oriented website and collects personal data of children younger than 13 years old, has to comply with certain rules. The website must specify what information is collected from children and for which purposes these personal data are gathered. Moreover, verifiable parental

---

[5]Ellen Neuborne, "For kids on the Web, it's an ad, ad, ad, ad world," Business Week 3745(2001): 108–109.

[6]Michel Walrave, Cyberkids' e-privacy at stake? Data processing and privacy policies in websites aimed at minors. Privacy Paper N°4 (Antwerp: University of Antwerp, 2005), 24.

[7]Kathryn C. Montgommery, "Digital kids: The new on-line children's consumer culture," in Handbook of Children and the Media, ed. Dorothy G. Singer and Jerome L. Singer (Thousand Oaks: Sage, 2001), 640.

[8]James U McNeal. Children as Consumers (Lexington: Lexington Books, 1987), 40–55.

[9]Seounmi H. Youn, "Teenagers' Perception of Online Privacy and Coping Behaviors: A Risk-Benefit Appraisal Approach," Journal of Broadcasting & Electronic Media 1(2005): 90.

consent is a precondition before starting data collection. Also other requirements are made.[10] This legislation only applies to adolescents younger than 13 years old. Thereby older teenagers are treated as adults by COPPA in this respect. In absence of legal protection, teenagers themselves have to be aware and concerned about data collecting practices to enable them to protect their personal data in the online environment.[11]

In the European Union, the Article 29 Data Protection Working Party, an independent EU advisory body on data protection, issued guidelines on the protection of personal data of minors in general and in the educational field in particular. The Working Party interprets the principles of the Data Protection Directive (95/46/EC) in accordance with the principle of the child's best interest. On the one hand, the Working Party stresses that one has to discern the varying levels of maturity between children as a basis to determine when children can start dealing with their own data. On the other hand, their legal representatives (for instance parents) have the right to represent minors in cases where the personal data disclosure could prejudice the best interests of the child. Parents can therefore be seen as a "legal guardian" in specific cases where their children's best interest could be at stake. But also in the application of individuals' data protection rights, specific arrangements have to be made for children. For example, it is recommended to provide information to children by using layered notices and using specially adapted language that makes the purposes of data processing practices easily understandable. Although the right of access to personal data is normally exercised by the child's legal representative, also the maturity and the specific situation have to be taken into account. When highly personal data and individual rights are concerned (for instance in a medical context or youth welfare services) children could be entitled to exercise their right alone, excluding their legal representatives from access to specific information. In these situations a careful balancing exercise has to be made in deciding if a child's privacy right prevails over the access to this information by the child's legal representative. On the right to object, the Working Party recalls that also children are entitled to object to the processing of their personal data for direct marketing purposes.[12]

In the context of direct marketing, the Belgian Data Protection Authority also recommends parental consent. Especially when the child has not yet reached the "age of discernment" – estimated around 13 or 14 years – parental consent is

---

[10]Federal Trade Commission, "Children's Online Privacy protection Act of 1998," Federal Trade Commission, http://www.ftc.gov/ogc/coppa1.htm.

[11]Deborah M. Moscardelli and Richard Divine, "Adolescents' Concern for Privacy When Using the Internet: An Empirical Analysis of Predictors and Relationships with Privacy-Protecting Behaviours," Family and Consumer Sciences Research Journal 3(2007): 236.

[12]Article 29 Working party, "Working document 1/2008 on the protection of children's personal data. General guidelines and the special case of schools," Article 29 Working Party, http://ec.europa.eu/justice_home/fsj/privacy/workinggroup/wpdocs/2008_en.htm

recommended in data processing situations that are not primarily in the direct interest of the child, like for instance direct marketing.[13,14]

14.2.2 *Self- and Co-regulatory Initiatives*

Also the Federation for European Direct and Interactive Marketing (FEDMA) pays special attention to data collection from minors in its code of conduct. FEDMA has negotiated a co-regulatory code with the national data protection authorities (grouped in the Article 29 Working Party) and the European Commission.[15] This code applies to any company performing direct marketing initiatives in Europe, and is not limited to those companies associated with FEDMA or other national Direct Marketing Associations in the European Union. A specific provision (2.6) states that organisations processing data of young children (under 13 years old) have to make all reasonable efforts to properly inform the child and parent about the purposes of the data processing. In commercial messages directed to children this information should not only be displayed prominently but also formulated understandably for children. Whenever national or European data protection legislation requires consent to personal data processing, organisations should obtain informed and prior consent from the child's parent. Furthermore, a child's participation in an online activity, for instance a game, should not be conditional on the willingness to disclose more personal data than strictly necessary. Also other provisions are formulated.[16]

Several academic researchers have attempted to verify whether marketeers comply with these policy lines. Research outcomes indicate that despite regulatory efforts many companies do not behave accordingly. In an American study 162 child-oriented websites were scrutinised. Although a majority of websites collected personal data, only 4 websites did fully comply with the main components of the law.[17] These outcomes were confirmed in a Belgian study showing that only a minority (39%) of child-oriented websites was equipped with a privacy policy.

---

[13]Commission for the Protection of Privacy, "Advies Nr. 38/2002 van 16 september 2002: Advies uit eigen beweging betreffende de bescherming van de persoonlijke levenssfeer van minderjarigen op Internet," Commission for the Protection of the Privacy, http://www.privacycommission.be/nl/docs/Commission/2002/advies_38_2002.pdf.

[14]Commission for the Protection of Privacy, "Direct Marketing en de Bescherming van Persoonsgegevens," Commission for the Protection of Privacy, http://www.privacycommission.be/nl/static/pdf/direct-marketing/20080805-nota-direct-marketing-nl.pdf.

[15]FEDMA, "European Code of Practice for the Use of Personal Data in Direct Marketing," FEDMA, http://img.custompublish.com/getfile.php/342991.1014.xacscqtseu/FEDMACodeEN.pdf?return=fedma.custompublish.com

[16]FEDMA, "European Code of Practice for the Use of Personal Data in Direct Marketing," FEDMA, http://img.custompublish.com/getfile.php/342991.1014.xacscqtseu/FEDMACodeEN.pdf?return=fedma.custompublish.com.

[17]Xiaomei Cai, Walter Gantz, Nancy Schwartz and Xinje Wang, "Children's Website Adherence to the ftc's Online Privacy Protection Rule," Journal of Applied Communication Research 5(2003): 350.

Furthermore, the privacy statements were often incomplete, difficult to understand and did not use a phrasing adjusted to minors. Only a small minority of websites did involve parents when collecting personal data, by informing them (12%) or by asking (3.5%) their permission.[18]

Despite a considerable amount of marketing communication campaigns targeting children and teens, and the policy initiatives taken so far, academic studies on teenagers' handling with online privacy and data disclosure in commercial websites are scarce. Therefore, the purpose of the present study is to analyse how teenagers react on personal data requests of marketeers and how these data collecting practices are perceived. The choice to focus on older teenagers is inspired by the fact that much policy work (COPPA and others) aims to protect teenagers younger than thirteen, while older teenagers are often not envisioned by policy makers.[19]

14.3 Teenagers' Online Disclosure in Websites: A Literature Review

When reviewing the literature on teenagers' personal data disclosure, several factors are discerned that could predict and explain youngsters' data disclosure. The following variables seem to be relevant: types of data, privacy concern, benefits of data disclosure and parental mediation. Furthermore, some demographic variables (age and gender) and ICT-use have been related to teenagers' privacy concern and behaviour and are therefore included as control variables in this study.

14.3.1 Types of Personal Data

In adult privacy literature, the type of personal information being requested by marketers is an important factor influencing consumer's decision whether or not to disclose information to a specific website.[20,21,22] Results yielded by these earlier studies indicate that adult consumers are more protective of financial data, personal identifiers, and personal data that are perceived as likely to lead to more marketing

[18]Michel Walrave, Cyberkids' e-privacy at stake? Data processing and privacy policies in websites aimed at minors. Privacy Paper N°4 (Antwerp: University of Antwerp, 2005), 26.

[19]Seounmi H. Youn, "Parental Influences and Teens' Attitude Toward Online Privacy Protection," Journal of Consumer Affairs 3(2008): 363.

[20]Daniel R. Horne and David A. Horne, "Domains of Privacy: Toward an Understanding of Underlying Factors," (Paper presented at the Direct Marketing Educators' Conference, San Francisco, CA, October 11, 1998).

[21]Paul Wang and Lisa A. Petrison, "Direct Marketing Activities and Personal Privacy: A Consumer Survey," Journal of Direct Marketing 1(1993): 208–210.

[22]Tiffany B. White, "Consumer Disclosure and Disclosure Avoidance: A Motivational Framework," Journal of Consumer Psychology 1(2004): 41–51.

efforts or spammed mailboxes.[23,24,25,26] The types of data most protected by consumers are those through which they can be contacted. The same studies found that adult consumers were more willing to disclose demographic, lifestyle and other non-identifiable information.[27] Also Castaneda and Montoro found that respondents' sensitivity to providing personal identification data is greater than to giving demographic details.[28]

Only few studies have examined teenagers' willingness to provide different types of personal data to marketeers. Most studies simply assess teenagers' overall willingness to disclose personal information. Therefore, one purpose of the present study is to verify whether teenagers are, like adults, more willing to disclose demographic, lifestyle and other data (i.e. profile data) than data through which they can be contacted (i.e. contact data).

14.3.2 Privacy Concern and Perceived Benefits

Although in an American study the majority of teenagers (79%) admits being concerned about their privacy online, these concerns are not reflected in their online behaviours.[29] This phenomenon is sometimes referred to as the "privacy paradox".[30] As suggested by Youn, this contradiction may be explained by a risk-benefit appraisal made by teenagers when asked for personal data.

In this respect, she observed that 45% of teenagers would disclose personal information and 54% would be prepared to tell advertisers about the favourite shops of their parents in return for an online gift.[31]

---

[23] Joseph Phelps, Glen Nowak and Elizabeth Ferrell, "Privacy concerns and consumers' willingness to provide personal information," Journal of Public Policy & Marketing 1(2000): 36–38.

[24] Daniel R. Horne and David A. Horne, "Domains of Privacy: Toward an Understanding of Underlying Factors," (Paper presented at the Direct Marketing Educators' Conference, San Francisco, CA, October 11, 1998).

[25] Glenn Nowak and Joseph Phelps, "Understanding Privacy Concerns," Journal of Direct Marketing 6(1992): 35–39.

[26] Paul Wang and Lisa A. Petrison, "Direct Marketing Activities and Personal Privacy: A Consumer Survey," Journal of Direct Marketing 1(1993): 193–220.

[27] Joseph Phelps, Glen Nowak and Elizabeth Ferrell, "Privacy concerns and consumers' willingness to provide personal information," Journal of Public Policy & Marketing 1(2000): 27–41.

[28] Alberto J. Castaneda and Francisco J. Montoro, "The effect of Internet general privacy concern on customer behavior," Electronic Commerce Research 7(2007): 129–135.

[29] Joseph Turow and Lilach Nir, The Internet and the Family 2000: The View from Parents, The View from Kids (Philadelphia: The Annenberg Public Policy Center, 2000), 6–7.

[30] William G. Staples, Encyclopedia of Privacy (Westport: Greenwood Publishing Group), 2006, 22.

[31] Seounmi H. Youn, "Teenagers' Perception of Online Privacy and Coping Behaviors: A Risk-Benefit Appraisal Approach," Journal of Broadcasting & Electronic Media 1(2005): 98.

Several studies have therefore investigated whether the level of privacy concern also influenced the amount of data Internet users disclose. Sheehan found that less concerned Internet users are more inclined to register for websites.[32] Moreover, they are more willing to provide complete information. Also Paine et al. found that the majority of adult respondents having privacy concerns take action to protect their privacy when they are online (e.g. being careful about the amount and type of information they disclose online).[33] Furthermore, Casteneda and Montoro[34] observed that customer's Internet privacy concerns related to the transmission of personal information has a strong negative effect on the user's intention to provide data. A similar negative relationship has been found between privacy concern and the willingness to provide personal data to engage in online transactions in several other studies.[35,36]

Although only few studies were conducted on teenagers' disclosure of personal data towards online businesses, results are in line with studies conducted with adult respondents. Moscardelli and Divine found that adolescents' concern for privacy was significantly related to privacy protecting behaviours (e.g. providing inaccurate information and requesting removal from mailing lists). What is more, when comparing the results of the adolescent respondents to those of adults surveyed by Sheehan and Hoy, young respondents seemed to score higher on some protecting and lower on divulging behaviours.[37] Youn also observed that teenagers' susceptibility to the severity of privacy risks related to data disclosure had a negative effect on the willingness to divulge personal information.[38] Also concerning teenagers' self-disclosure in social networking sites, a negative correlation has been found between privacy value and information disclosure.[39]

---

[32] Kim B. Sheehan. and Maria G. Hoy, "Flaming, Complaining, Abstaining: How Online Users Respond to Privacy Concerns," Journal of Advertising 3(1999): 37–51.

[33] Carina Paine, Ulf-Dietrich Reips, Stefan Stieger, Adam Joinson and Tom Buchanan, "Internet users' perceptions of 'privacy concerns' and 'privacy actions'," International Journal of Human Computer Studies 65(2006): 530–532.

[34] Alberto J. Castaneda and Francisco J. Montoro, "The Effect of Internet General Privacy Concern on Customer Behavior," Electronic Commerce Research 7(2007): 134–140.

[35] Tamara Dinev, Paul Hart, "Privacy Concerns and Levels of Information Exchange: An Empirical Investigation of Intended e-services Use," E-Services Journal 3(2006): 48–49.

[36] Tamara Dinev, Paul Hart and Michael R. Mullen, "Internet Privacy Concerns and Beliefs About Government Surveillance – An Empirical Investigation," Journal of Strategic Information Systems 17 (2008): 228–232.

[37] Deborah M. Moscardelli and Richard Divine, "Adolescents' Concern for Privacy When Using the Internet: An Empirical Analysis of Predictors and Relationships with Privacy-Protecting Behaviours," Family and Consumer Sciences Research Journal 3(2007): 232.

[38] Seounmi H. Youn, "Teenagers' Perception of Online Privacy and Coping Behaviors: A Risk-Benefit Appraisal Approach," Journal of Broadcasting & Electronic Media 1(2005): 86–110.

[39] Zaineb De Souza and Geoffrey N. Dick, "Disclosure of Information by Children in Social Networking – Not Just a Case of 'You Show Me Yours and I'll Show You Mine'," International Journal of Information Management 29(2009): 255.

Krasnova found that especially concerns regarding organisational threats, namely fears of data processing by the provider and third parties, have a negative influence on the amount of information individuals disclose in their profile.[40]

Based on these findings, this study hypothesizes that teenagers would be less inclined to provide contact data (H1a) and profile data (H1b), when they are more concerned about their online privacy. Conversely, the more they are convinced of possible benefits, the more they are inclined to disclose personal data to marketers (contact data H2a, profile data H2b).

14.3.3 Parental Mediation

Various studies have examined in which way parents can influence online risk coping behaviour by minors.[41,42,43,44] These mediation strategies do not only encompass monitoring of children's media use in general and Internet use in particular, but also conversations and co-use of media to interpret content, assess risks and discuss coping strategies.[45,46,47,48,49,50]

---

[40] Hanna Krasnova, Oliver Günther, Sarah Spiekermann and Ksenia Koroleva, "Privacy concerns and identity in social networks," Identity in the Information Society 1(2009):39.

[41] Sonia Livingstone and Ellen J. Helsper, "Parental Mediation of Children's Internet Use," Journal of Broadcasting & Electronic Media 4(2008): 581–599.

[42] Sook-Jung Lee and Young-Gil Chae, "Children's Internet Use in a Family Context: Influence on Family Relationships and Parental Mediation," Cyberpsychology and Behavior 5(2007): 640–644.

[43] Albert K. Liau, Angeline Khoo and Peng Hwaang, "Factors Influencing Adolescents Engagement in Risky Internet Behaviour," Cyberpsychology and Behavior 6(2005): 513–520.

[44] Victoria Rideout, Donald F. Roberts and Ulla G. Foehr, Generation M: Media in the Lives of 818 Year Olds (Menlo Park: The Henry J. Kaiser Family Foundation, 2005), 31–35.

[45] Sonia Livingstone and Ellen J. Helsper, "Parental Mediation of Children's Internet Use," Journal of Broadcasting & Electronic Media 4(2008): 581–599.

[46] Margaret Kerr and Hakan Stattin, "What Parents Know, How They Know It, and Several Forms of Adolescent Adjustment: Further Support for a Reinterpretation of Monitoring," Developmental Psychology 3(2000): 366–370.

[47] Amy I. Nathanson, "Identifying and Explaining the Relationship Between Parental Mediation and Children's Aggression," Communication Research 2(1999): 124–130.

[48] Patti M. Valkenburg, Marina Krcmar, Allerd L. Peeters, Nies M. Marseille. "Developing a Scale to Assess Three Styles of Television Mediation: 'Instructive Mediation,' 'Restrictive mediation', and 'Social coviewing',". Journal of Broadcasting & Electronic Media 1(1999): 52.

[49] Erica W. Austin, "Effects of Family Communication on Children's Interpretation of Television," in Television and the American family, ed. Jennings Bryant and Alison J. Bryant (Hillsdale, NJ: Lawrence Erlbaum Associates Inc., 1990), 377–392.

[50] Ron Warren, Phil Gerke, Mary A. Kelly, "Is There Enough Time on the Clock? Parental Involvement and Mediation of Children's Television Viewing," Journal of Broadcasting & Electronic Media 1(2002): 88.

Therefore, the present study made a distinction between three broad mediation strategies: *active cosurfing*, *restriction* and *monitoring*. An active cosurfing mediation strategy involves discussions between children and parents while surfing online. This approach integrates two strategies that have been discerned for parental (television) monitoring, on the one hand coviewing and on the other factual or evaluative mediation, namely discussing about the medium content.[51,52,53] While in TV-viewing, family members can sit in front of the medium and watch it with little or no discussion, it is less likely that parents surfing with their children will do this in silence. Therefore we assume that when parents are cosurfing with their child, they will inform them about websites, possibly warn them for specific risks and hint them on how to cope with online risks.

Parents can also submit explicit rules, such as restricting the duration of online sessions and forbidding specific behaviour like for instance downloading software, participating in online contests or disclosing personal data.[54,55] This type of parental mediation is in line with parental restrictions on specific TV-programs. However, when mediating Internet use, also other strategies can be implemented. Parents can monitor the Internet use of their children by checking the navigation history. In this study, we therefore question whether parental monitoring of children's online behaviour would influence their online data disclosure. Based on previous research, we expect that discussion may lead to a more critical attitude towards marketing and its objectives, and may in the end lead to higher levels of privacy concern. Research confirms that discussion between children and parents fosters privacy concern.[56] Also other research has found a positive relationship between cosurfing and the level of privacy concern, while rulemaking did not reach significance threshold.[57] Moreover, various studies show that heightened levels of privacy concern lead to a higher chance that privacy statements are being read and an elevated

---

[51] Amy I. Nathanson, "Identifying and Explaining the Relationship Between Parental Mediation and Children's Aggression," Communication Research 2(1999): 124–143.

[52] Matthew S. Eastin, Bradley S. Greenberg and Linda Hofshire, "Parenting the Internet," Journal of Communication (2006): 486–490.

[53] Amy I. Nathanson, "Parent and Child Perspectives on the Presence and Meaning of Parental Television Mediation," Journal of Broadcasting & Electronic Media 2(2001): 203–207.

[54] Sonia Livingstone and Ellen J. Helsper, "Parental Mediation of Children's Internet Use," Journal of Broadcasting & Electronic Media 4(2008): 581.

[55] Matthew S. Eastin, Bradley S. Greenberg and Linda Hofshire. "Parenting the Internet," Journal of Communication (2006): 486–490.

[56] Deborah M. Moscardelli and Richard Divine, "Adolescent's Concern for Privacy When Using the Internet: An Empirical Analysis of Predictors and Relationships with Privacy-Protecting Behaviours," Family and Consumer Sciences Research Journal 3(2007): 232–252.

[57] Seounmi H. Youn, "Parental Influences and Teen's' Attitude Toward Online Privacy Protection," Journal of Consumer Affairs 3(2008): 378.

level of teenagers' resistance to communicate personal information online.[58,59,60] Therefore we expect that active cosurfing would incline youngsters to disclose less personal data (contact data H3a, profile data H3b). Also restrictive mediation (contact data H4a & profile data H4b) and monitoring (contact data H5a & profile data H5b) are expected to be negatively related with willingness to disclose personal information.

14.3.4 Other Variables: Gender, Age and ICT-Use

Also other factors, such as ICT-use and sociodemographics (gender and age), may impact Internet users' willingness to disclose of personal information.

Studies on the influence of ICT-use on teenagers' privacy concern level have yielded mixed results. Whereas some studies have found that heavy use of the Internet is related with low levels of perceived online risk,[61, 62] conflicting results were found in other studies: Paine et al.[63] found that Internet experience was the best predictor for privacy protection measures taken by the consumer.

With regard to gender previous studies have pointed out that females are less inclined to divulge personal information for commercial purposes.[64] Moscardelli and Divine[65] found that female teens are significantly more concerned about protecting their privacy than their male counterparts. These findings are in accordance with other research stating that girls have less trust in privacy policies of websites and perceive higher risks associated with online advertising. Boys provide more

---

[58] Robert Larose and Nora J. Rifon, "Promoting of Safety: Effects of Privacy Warnings and privacy Seals on Risk Assessment and Online Privacy Behaviour," Journal of Consumer Affairs 1(2007): 143–149.

[59] George R. Milne and Mary J. Culnan, "Strategies for Reducing Online Privacy Risks: Why Consumers Read (Or Don't Read) Online Privacy Notices," Journal of Public Policy & Marketing (3)2004: 25–29.

[60] Kim B. Sheehan and Mariea G. Hoy, "Flaming, complaining, abstaining: how online users respond to privacy concerns," Journal of Advertising 3(1999): 45–51.

[61] Yehoshua Liebermann and Shmuel Stashevsky, "Perceived Risks as Barrier to Internet and E-Commerce Usage," Qualitative Market Research: An International Journal 4(2002): 297.

[62] Anthony D. Miyazaki and Ana Fernandez, "Consumer Perceptions of Privacy and Security Risks for Online Shopping," *The Journal of Consumer Affairs* 1(2001): 36–38.

[63] Carina Paine, Ulf-Dietrich Reips, Stefan Stieger, Adam Joinson and Tom Buchanan, "Internet Users' Perceptions of 'Privacy Concerns' and 'Privacy Actions'," *International Journal of Human Computer Studies* 65(2006): 530–532.

[64] Seounmi H. Youn, "Teenagers' Perception of Online Privacy and Coping Behaviors: A Risk-Benefit Appraisal Approach," Journal of Broadcasting & Electronic Media 1(2005): 100.

[65] Deborah M. Moscardelli and Richard Divine, "Adolescent's Concern for Privacy When Using the Internet: An Empirical Analysis of Predictors and Relationships with Privacy-Protecting Behaviours," Family and Consumer Sciences Research Journal 3(2007): 236.

personal details in online forms, read more unwanted commercial e-mail and have more negative reactions towards spam.[66]

Also in using social networking sites and disclosing profile data gender differences have been found. Girls are more inclined to post personal pictures online, while boys are more eager to divulge their address, their school's name and their mobile phone number. However, boys more often disclose false data on a profile page.[67] In another study concerning teenagers' activities on social networking sites, girls appear to be more privacy sensitive, whilst boys are more tending to disclose certain data (for example MSN-address and phone number).[68] Moreover, in a study assessing profiles' accessibility, it was observed that girls are significantly more likely to have private profiles than boys.[69] Various studies have found that these gender differences stay valid during adulthood[70,71,72,73] whilst other research did not find significant differences between men and women.[74,75]

Previous research on age differences in the context of data disclosure suggests that young consumers are more likely aware of data collection practices and therefore use relatively more privacy protection strategies than older consumers.[76] In addition, Youn[77] found that among minors young children were more inclined to

---

[66] Joseph Turow and Lilach Nir, The Internet and the Family 2000: The View from Parents, The View from Kids (Philadelphia: The Annenberg Public Policy Center, 2000), 32.

[67] Amanda Lenhart and Mary Madden. Teens, Privacy & Social Networks (Washington: Pew Internet & American Life Project, 2007), 24.

[68] Joshua Fogel and Elham Nehmad, "Internet Social Network Communities: Risk Taking, Trust and Privacy Concerns," Computers in Human Behaviour 1(2009): 157.

[69] Kevin Lewis, Jason Kaufman and Nicholas Christakis, "The Taste for Privacy: An Analysis of College Student Privacy settIngs in An Online Social Network," Journal of Computer-Mediated Communication 14(2008): 89.

[70] Susannah Fox, Trust and privacy online: Why Americans want to rewrite the rules. (Washington: Pew Internet & American Life Project, 2000), 17.

[71] Kim Bartel Sheehan, "An Investigation of Gender Differences in Online Privacy Concerns and Resultant Behaviors," Journal of Interactive Marketing 4(1999): 24.

[72] Colleen M. Kehoe and James E. Pitkow, "Surveying the Territory: GVU's Five WWW User Surveys Surveying the Territory: GVU's Five WWW User Surveys," WWW User Survey Home Page, http://www.cc.gatech.edu/gvu/user_surveys/papers/w3j.html

[73] Timothy R. Graeff and Susan Harmon, "Collecting and Using Personal Data: Consumers' Awareness and Concerns," Journal of Consumer Marketing 4(2002): 310–312.

[74] Joseph Phelps, Glen Nowak and Elizabeth Ferrell, "Privacy Concerns and Consumers' Willingness to Provide Personal Information," Journal of Public Policy & Marketing 1(2000): 35–41.

[75] George R. Milne and Andrew J. Rohm, "Consumer Privacy and Name Removal Across Direct Marketing Channels: Exploring Opt-in and Opt-out Alternatives," Journal of Public Policy & Marketing 2(2000): 238.

[76] Joseph Phelps, Glen Nowak and Elizabeth Ferrell, "Privacy Concerns and Consumers' Willingness to Provide Personal Information," Journal of Public Policy & Marketing 1(2000): 35–41.

[77] Seounmi H. Youn, "Teenagers' Perception of Online Privacy and Coping Behaviors: A Risk-Benefit Appraisal Approach," Journal of Broadcasting & Electronic Media 1(2005): 100.

ask their parents or teachers for advice before disclosing personal information in the online environment. This corresponds with the result that older teenagers (age 15–17) are more inclined to disclose personal data than younger teenagers do (age 12–14).[78] Based on the findings above Internet use, age and gender were entered in the analyses as control variables.

14.4 Survey Among Teenagers

14.4.1 Method

A survey was conducted among 1,318 twelve to eighteen year-old secondary school pupils in Belgium. As education is a regional competency in Belgium, a stratified random sample of twenty-eight schools was drawn in both the Francophone and the Flemish communities. Data were collected through anonymous questionnaires distributed in the classroom. Provisions were made to guarantee the participants' privacy and confidentiality during the administration of the questionnaire.

The survey was first tested for comprehensibility and question clarity in a class of 12- to 13-year-olds. Subsequently, some terminology was briefly defined and certain questions were rephrased. Prior to their answering the question on online privacy concerns, the respondents were asked to provide sociodemographic details (gender, age, education etc). ICT-use was assessed by asking the daily number of hours respondents spent online.

Several scales have been constructed to measure parental mediation styles, privacy concern and the amount of personal data that are disclosed (contact data versus profile data).

To probe the respondents' willingness to disclose personal data, respondents were asked whether they were prepared to provide a specific piece of personal information in a commercial website in exchange for a gift.

A list of fourteen types of personal data was presented and responses were measured using a 4-point Likert scale ranging from 1 to 4 (respectively "would certainly not disclose information" and "would certainly disclose information"). In this way the Likert scale provided more information on the respondents' willingness to disclose each of these fourteen data.

Exploratory factor analysis with varimax rotation showed that two factors could be distinguished with an eigenvalue greater than one.

Tabachnick and Fidell[79] cite 0.32 as loading threshold for an item of a latent construct to be interpreted. Yet, the higher the item loads, the more it is considered as a pure measure of the factor. Comrey and Lee[80] provide us with an interpretation,

---

[78] Amanda Lenhart and Mary Madden, Teens, Privacy & Social Networks (Washington: Pew Internet & American Life Project, 2007), 4.

[79] Tabachnick, B.G., and L.S. Fidell. *Using Multivariate Statistics*. Boston: Allyn and Bacon, 2007.

[80] Comrey, A.L., and H.B. Lee. *A First Course in Factor Analysis*. Hillsdale, NJ: Lawrence Erlbaum Associates, Inc., 1992.

indicating that loadings of 0.32 are rather poor, 0.45 are fair, 0.55 are good, 0.63 very good and 0.71 are excellent. Compared to these levels, the loadings in the study showed that all items but three had very good relationships with the supporting latent construct (namely >0.63) and low cross-loadings on the other factor. Therefore, the three items that did not load strongly on one specified factor were excluded from subsequent analysis. Eleven items were thus retained for the analyses.

The items that loaded highest on the first factor were five contact details: home address, home phone number, mobile number, e-mail address and e-mail address parents. This factor accounted for 25.4% of the variance. It was labelled *contact data*. The second factor accounted for 36.1% of the variance and was named *profile data*. The personal data that loaded highest on the second factor were: forename, age, gender, hobbies, favourite products, favourite shops.

The reliability coefficient (Cronbach's alpha) for the 5-item contact data scale was 0.85, for the 6-item profile data scale 0.92. For further analysis, raw scores were aggregated, with higher values indicating a higher inclination to disclose a specific category of personal data.

To probe the respondents' level of privacy concern six statements were formulated. Responses were measured with a 4-point Likert scale ranging from 1 ("totally disagree") to 4 ("fully agree").

Factor analysis identified one single factor with three items loading highest on the factor that explained 46.3% of the variance ("I am concerned about what websites do with my personal data", "I sometimes question why websites collect personal data", "I look for information about the protection of personal data in a website, before answering to personal data requests"). Following Comrey and Lee,[81] items that loaded poorly on the factor were excluded from subsequent analyses. The reliability coefficient (Cronbach's alpha) for this 3-item scale was 0.71. Raw scores were summed, with higher values indicating higher levels of privacy concern.

With a single item (with a 4-point Likert scale), the attitude towards the benefits of disclosing personal data was measured. Respondents were asked whether they did or did not appreciate that entrusting personal data leads to reception of interesting personalised offers.

The three parental mediation strategies that were discerned in this study, were assessed by asking the teenagers if they discussed with their parents about their Internet use and surfed together to certain websites, suspected their parents to monitor their online activities or if some specific rules were set by their parents.

Active cosurfing was measured by asking the adolescents how frequent they go online with their parents (i.e. cosurfing), how often their parents were present while surfing and how frequent parents and children discussed these online activities. The measures were rated on a 4-point Likert scale ranging from 1 ("never") to 4 ("very often").

---

[81] Comrey, A.L., and H.B. Lee. *A First Course in Factor Analysis*. Hillsdale, NJ: Lawrence Erlbaum Associates, Inc., 1992.

Restrictive mediation was assessed with a total of ten items dealing with parental rules (e.g. "My parents tell me which websites I am permitted to visit", "My parents forbid me to surf when I'm home alone") were rated using a "yes" and "no" dichotomous format, with a summated index adding the items, and higher scores indicating more parental rules.

Parental monitoring of children's Internet use was assessed by a single variable, namely the question whether teenagers suspected their parents of checking their online activities.

From all items dealing with parental restrictive mediation and co-use, two factors were discerned using exploratory factor analysis. The first factor consisted of three items, accounted for 13.2% of the variance and was labeled *active cosurfing*. The second factor accounted for 8.4% of the variance and dealt with specific rules parents established and was called *restrictive mediation*. From a total of ten rules, six loaded higher than 0.45 on the latent construct. A total of four items that did not load well on the factor were excluded from subsequent analyses. The reliability coefficient (Cronbach's alpha) for the active cosurfing and the restrictions' scale was above the 0.60 recommended threshold for exploratory research.[82]

For both mediation styles an index was used in subsequent analyses, with higher scores indicating higher levels of parental mediation.

14.4.2 Results

14.4.2.1 Descriptive Findings

A first important finding is that the majority of teenagers share a critical attitude towards data collecting practices initiated by marketeers. Three out of four teens (72.8%) question why websites are requesting personal data. A similar proportion (69.2%) expresses concerns about the further use of personal data. Moreover, survey results show that a majority of teenagers seek information about the website's data processing policy before disclosing any personal details (73.5%). Our results indicate that six out of ten (60.6%) teenagers confess having ever deliberately provided false personal data in an electronic form.

Although teenagers share a rather sceptical attitude towards data processing by websites, survey data indicate that still a lot of youngsters are prepared to disclose a considerable amount of personal data for marketing purposes, as summarised in the next figure.

Figure 14.1 shows to what degree teenagers are willing to disclose different types of personal information. The results for profile data are grouped at the left side of this figure and contact data on the right. Both sides of the figure show a sharp contrast in respondents' willingness to disclose profile data on the one hand and contact data on the other hand. While first name, surname, age, gender, hobby, favourite shops and favourite brands are surely or potentially ("maybe yes") disclosed by a

[82]Earl Babbie. Practice of Social Research, 8th edition, New York: Wadsworth Publishing Company, 2001.

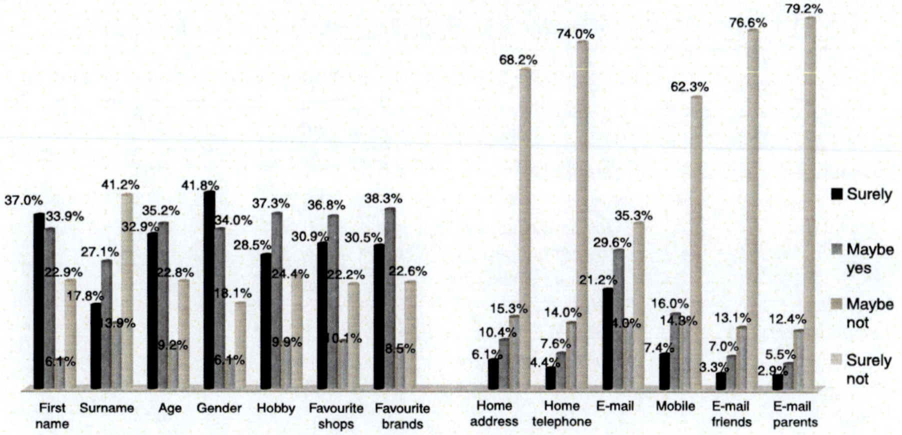

Fig. 14.1 Disclosure of personal data in websites ($N = 1{,}318$)

vast majority, all but one contact detail (e-mail) is surely not disclosed by a large majority of respondents. For example, 68.2% of teenagers would surely not divulge their home address to a commercial website and 79.2% of teenagers would not even think about forwarding the e-mail address of their parents. Within the two discerned categories of personal data some data are more readily disclosed than others. For profile data respondents seem to be less willing to disclose their surname than their gender. For contact data about 50% of teenagers would be certainly (21.2%) or potentially (29.6%) prepared to give out their e-mail address, while only about 8% of respondents would give their parents' e-mail address certainly (2.9%) or potentially (5.5%).

This study shows some interesting gender differences regarding privacy concern and data disclosing behaviour. Girls' mean level of privacy concern is significantly higher (M:8.93) than boys' privacy sensitivity (M:8.03) ($t = -7.69$, $p < 0.001$). Furthermore we observe that boys' mean score on the contact data scale is higher (M:38.59) than girls' score (M:34.46) ($t = 0.450$, $p < 0.001$) and that female pupils score significantly higher on the profile data scale (M:73.40) than boys (M:67.08) ($t = -4.74$, $p < 0.001$). These results indicate that boys are significantly more inclined to entrust contact data, whereas girls communicate more profile data.

14.4.2.2 Regression Results

To test the several hypotheses that were formulated based on the literature study, two multiple hierarchical regressions were performed.

Table 14.1 provides the reader with an overview of the estimated regression coefficients and the amount of variance that was explained by the variables included in the equation. The included variables explained 32% of the variance in willingness to provide contact data and 25% of the variance in willingness to disclose profile data. Our analyses found that the main explanatory variables for the first model on contact data, were privacy concerns and perceived benefits ($\Delta R^2 = 0.16$). Our

Table 14.1 Hierarchical multiple regression predicting willingness to provide contact data

		Willingness to provide contact data	
		Final β	*ΔR*^{2}
Demographics	Gender (1)	−0.160**	0.03
	Age	0.021	
Internet use	Online frequency	0.082**	0.02
Concerns/benefits	Privacy concerns	−0.155***	
	Disclosure benefits	0.250***	0.16
Disclosure	Profile data	0.345***	0.11
Parentalmediation	Parental active cosurfing	−0.110***	
	Parental restrictions	−0.074*	
	Parental monitoring	0.029	0.01
Total R^2			0.32***
		Willingness to provide profile data	
		Final β	*ΔR*^{2}
Demografics	Gender (1)	0.197**	0.03
	Age	0.034	
Internet use	Online frequency	0.012	0.00
Concerns/benefits	Privacy concerns	−0.082**	0.10
	Disclosure benefits	0.178***	
Disclosure	Contact data	0.380***	0.12
Parentalmediation	Parental active cosurfing	0.052	
	Parental restrictions	−0.047	
	Parental monitoring	0.055	0.00
Total R^2			0.25***

(1) coded as 0 = boy 1 = girl
*$p < .05$; **$p < .01$, ***$p < .001$

analyses show that these factors retain their importance in explaining respondents' willingness to disclose profile data ($\Delta R^2 = 0.10$). In exploring the interrelationship between disclosure of different types of personal data, our study found that 12% of variance in willingness to disclose profile data disclosure is explained by respondents' willingness to de disclose contact data ($\Delta R^2 = 0.12$) and 11% of willingness to disclose contact data is explained by willingness to disclose profile data.

Level of Privacy Concern and Perceived Benefits of Data Disclosure

H1 asserted a negative relationship between the level of privacy concern and the willingness to provide contact data (H1A) and profile data (H1B). The outcomes of our analyses indicate that there is a significant negative relationship between teenagers' level of privacy concern and the willingness to disclose both categories of personal data (contact data: $\beta = -0.155$; $p < 0.001$; profile data: $\beta = -0.082$; $p < 0.01$). Hence the present study finds statistical support for H1a and H1b.

H2 predicted a positive relationship between the benefits of information disclosure and the willingness to disclose contact data (H2a) and (H2b) profile data. As becomes clear when inspecting Table 14.1, the outcome of our analyses provide statistical support for this positive relationship with willingness to disclose contact data ($\beta = 0.250$; $p < 0.001$) as well as profile data ($\beta = 0.178$; $p < 0.001$).

Parental Mediation

H3, H4 and H5 asserted a negative relationship between on the one hand the three parental mediation styles found in the factor analysis (active cosurfing, restrictive mediation and parental monitoring) and on the other hand the willingness to disclose personal information.

Interpreting the results displayed in Table 14.1 we can see that parental mediation overall explains a rather small proportion of variance in willingness to disclose contact data ($\Delta R = 0.01$). The outcomes of the regression analyses further indicate that there is no statistical support for a potential influence any of the three mediation strategies (H3b, H4b, H5b) has on the willingness to disclose profile data. What our analyses did find, was statistical support for the hypothesized negative relation between active co-surfing (H3a) ($\beta = -0.110$; $p < 0.001$) and restrictive mediation (H4a) ($\beta = -0.074$; $p < 0.05$) and the willingness to disclose contact data. Although this was expected in H5a, our analyses did not find a significant negative relation between parental monitoring and teenagers' willingness to disclose contact data. In sum, only hypotheses H3a and H4a were statistically supported.

Other Variables: ICT-Use, Gender, Age

Consistent with the descriptive results above, our study found that female teenagers are less willing to disclose contact data than male teenagers ($\beta = -0.160$; $p < 0.01$). Conversely, our study unveils that male teenagers are less prepared to disclose profile data than females are ($\beta = 0.197$; $p < 0.01$). Age was not a significant factor in explaining disclosure of contact data and profile data. A significant, though rather weak relation was found between the amount of hours teenagers spend on the Internet and their willingness to disclose contact data ($\beta = 0.082$; $p < 0.01$). Overall, demographics and Internet, as control variables of the present study, explain a very limited amount of variance in the focal variables.

14.5 Conclusion

The increasing commercialization of youngsters' online environment has lead parents, consumer organizations and child advocates to voice their concern about minors' online privacy in the branded online marketplace. In the light of these concerns the present study aimed at examining which factors influence teenagers' willingness to disclose personal information in online marketing situations.

An important finding is that the majority of teenagers involved in this study does not take an uncommitted but a rather sceptical attitude towards marketeers' requests for personal data. Many are concerned about the further processing of disclosed data and look for more information in the privacy statement of a website before acceding to data requests. Furthermore, this study finds support for the "privacy paradox": despite the scepticism towards data processing performed by commercial websites, the outcome of our survey indicates that still a lot of teenagers disclose a considerable amount of commercially valuable information to marketeers. Not only profile data are easily disclosed (such as favourite shops and hobbies), but also e-mail address is being disclosed by about half of the teenage respondents involved in this study. Teens deal more cautiously with other contact data (phone number and home address) that constitute a higher direct privacy threat.

Evidence was found that the level of privacy concern negatively influences the willingness to provide contact data and profile data. Teenagers with higher privacy concern levels are less inclined to accede to requests for as well contact as profile data.

The outcomes of this study show that while girls are less inclined to disclose contact data than boys, they are more inclined to communicate profile data. This result seems to conflict with previous research and other findings of the present study, namely that girls' mean privacy concern is significantly higher. A possible explanation for these conflicting results may be that female Internet users consider profile data as being less risky and involving a smaller privacy threat than contact data and therefore do not bother too much about privacy risks associated with the disclosure of profile data.

In accordance with findings from previous research pointing out that heavy Internet users share lower levels of online risk perception, our data show that the more time teenagers spend on the Internet, the more they are prepared to disclose contact data for marketing purposes. Online frequency and the disclosure of profile data on the contrary were not significantly related.

Contrary to the researchers' expectations, the three parental mediation strategies explained only a very small proportion of variance in willingness to disclose personal data. A possible explanation is that parents' role as socialization agents decreases with their son or daughter aging toward adulthood. Socialization literature further indicates that teenagers perceive less severe consequences as they grow older if they do not comply with the rules and instructions made by their parents.[83] At the same time, however, the influence of peers increases as adolescents become older.[84] What parents say to their children is not longer taken for granted. The social norms and moral behaviours proposed and advised by parents are questioned. This "rebellion"-effect can possibly explain why the teenagers in this study were found

[83]Wendy S. Grolnick and Melanie Farkas, "Parenting and the Development of Children's Self-regulation," in Handbook of Parenting, ed. Marc H. Bornstein (Mahwah: Lawrence Erlbaum Associates, 2002), 89.

[84]Duane Buhrmester and Wyndol Furman, "The Development of Companionship and Intimacy," Child development 58(1967): 1101–1113.

not to obey the media consumption choices suggested by their parents. Another explanation is forwarded by Lwin[85] citing a study in which was found that parents themselves begin to exercise less control over children's leisure activities including computer usage, when the latter reach the age of 12. We see opportunities for future research to further explore the efficacy of parental mediation strategies among children and teenagers of different ages.

A strength of this study is the further differentiation in personal data categories by distinguishing contact data and profile data. This distinction is important because contact data contain more sensitive information and involve a higher privacy threat, as they allow marketeers or other Internet users to contact the minor. Another strength is the further differentiation made in parental involvement styles by distinguishing active cosurfing, restriction and monitoring and the possible influence on disclosing two major categories of personal data.

However, the present study has certain *limitations* that need to be acknowledged. Due to time restraints in administering the questionnaire certain interesting aspects related to online information disclosure could not be questioned. For instance an appropriate measurement instrument for teenagers' privacy protection is lacking and no measurement for risk perception was included. The distinction of the present study between two gross types of personal data, profile data and contact data, is still open for improvement. Future research could aim at identifying additional categories of personal information. Despite the fit of the regression models, a substantial amount of variance in teens' disclosure of personal data remains unexplained. In the present study only one concrete situation of data disclosure was presented, in this case a commercial website offering a gift in exchange for personal information. Future research could include a variety of data disclosure situations and incentives.

Finally, several implications of our study can be distinguished. In youngster's educational environment awareness raising initiatives concerning data protection could be further differentiated in accordance with the found gender differences in data disclosure and privacy concerns. Following the results of the present study marketeers will have to recognize that teenagers are concerned about their online privacy. Furthermore, the development of coping behaviour in response to excessive data requests is particularly important. In the future policy debate on children's and teenagers' online privacy rights, special attention should be devoted to the increasingly blurring boundaries between marketing communication and entertainment. This hybridization of commercials and entertainment ("promotainment" including "advergames" for instance) may undermine the often made assumption that teenagers are fully able to understand the marketing purpose of most teenage-oriented websites. This may complicate the identification of specific commercial purposes, when deciding on whether or not to disclose personal data in an online environment. This is important, since research on traditional marketing effects has

[85]May O. Lwin, Andrea J.S. Stanaland and Anthony D. Miyazaki, "Protecting children's privacy online: How parental mediation strategies affect website safeguard effectiveness," Journal of Retailing 84(2008): 205–217.

made clear that when children are not aware of the marketing purpose of commercial messages, they share a less sceptical attitude towards marketing initiatives.[86] This sceptical attitude however is necessary in order to make well-informed consumer decisions.

The United States Children's Online Privacy Protection Act and, for instance, the recommendation of, for instance, the Belgian Data Protection Authority stipulate a more rigorous protection of personal data, requiring parental consent, in those cases where children younger than thirteen are involved. Nevertheless, this study has made clear that teenagers (+14 years) are also (and even more) likely to disclose personal data. This raises questions concerning the lack of protection measures for minors in the later stages of teenage life. Moreover, privacy statements are often formulated in a specific jargon that is not adapted to young website visitors and therefore difficult to understand.[87] This induces us to express our doubts whether teenagers, 13 years or even older, fully understand the outcomes of personal data disclosure online.

Therefore, online data protection should be included systematically in school plans, according to pupils' age and the nature of the subjects taught. Parents and teachers could encourage teenagers to assess possible risks of data disclosure. However, under no circumstances children and teens can, for reasons of security, be confronted with over-surveillance that would restrain their autonomy. In this context, a balance should be kept between children's security and children's privacy.

Acknowledgements This survey was conducted as part of TIRO (Teens & ICT: Risks & Opportunities), a research project within the framework of and financed by the Belgian Federal Science Policy's programme on Society & Future. This research project was a collaboration between CRID and CITA FUNDP, SMIT VUB and MIOS University of Antwerp. The survey was prepared and conducted by assoc. prof. dr. Michel Walrave, Sunna Lenaerts, Sabine De Moor and Marijke De Bie (MIOS University of Antwerp).

References

Adler, R. P., L. Meringoff, T. S. Robertson, J. R. Rossiter, and S. Ward. *The Effects of Television Advertising on Children*. Lexington: Lexington Books, 1988.

Article 29 Working party. "Working document 1/2008 on the protection of children's personal data. General guidelines and the special case of schools." Article 29 Working Party. http://ec.europa.eu/justice_home/fsj/privacy/workinggroup/wpdocs/2008_en.htm (accessed August 3, 2009)

Austin, E. W. "Effects of family communication on children's interpretation of television." In Television and the American family, edited by Jennings Bryant and Alison J. Bryant, 377–392. Hillsdale, NJ: Lawrence Erlbaum Associates Inc., 1990.

Babbie, E. *Practice of Social Research*, 8th ed. New York, NY: Wadsworth Publishing Company, 2001.

---

[86]Richard P. Adler, Laurene Meringoff, Thomas S. Robertson, John R. Rossiter and Scott Ward. *The Effects of Television Advertising on Children*. (Lexington: Lexington Books, 1988), 26–30.

[87]Michel Walrave, Cyberkids' *e-privacy at stake? Data processing and privacy policies in websites aimed at minors. Privacy Paper N°4.s* (Antwerp: University of Antwerp, 2005), 12.

Buhrmester, D. and W. Furman. "The Development of Companionship and Intimacy." *Child development* 58 (1967): 1101–1113.

Cai, X., W. Gantz, N. Schwartz and X. Wang. "Children's website adherence to the ftc's online privacy protection rule." *Journal of Applied Communication Research* 5(2003): 346–362.

Carlson, L., S. Grossbart, K. J. Stuenkel. "The Role of Parental Socialization Types on Differential Family Communication Patterns Regarding Consumption." *Journal of Consumer Psychology* 1(1992): 31–52.

Castaneda, A. J. and F. J. Montoro. "The effect of Internet general privacy concern on customer behavior." *Electronic Commerce Research* 7(2007): 117–141.

Commission for the Protection of Privacy. "Advies Nr. 38/2002 van 16 september 2002: Advies uit eigen beweging betreffende de bescherming van de persoonlijke levenssfeer van minderjarigen op Internet." Commission for the Protection of the Privacy. http://www.privacycommission.be/nl/docs/Commission/2002/advies_38_2002.pdf (accessed on August 3, 2009)

Commission for the Protection of Privacy. "Direct Marketing en de Bescherming van Persoongegevens." Commission for the Protection of Privacy. http://www.privacycommission.be/nl/static/pdf/direct-marketing/20080805-nota-direct-marketing-nl.pdf (accessed on August 3, 2009)

Commission for the Protection of Privacy, "Direct Marketing en de Bescherming van Persoongegevens." Commission for the Protection of Privacy. http://www.privacycommission.be/nl/static/pdf/direct-marketing/20080805-nota-direct-marketing-nl.pdf (accessed on August 3, 2009)

De Souza, Z. and G. N. Dick. "Disclosure of information by children in social networking – Not just a case of 'you show me yours and I'll show you mine'." *International Journal of Information Management* 29(2009): 255–261.

Dinev, T., P. Hart. "Privacy concerns and levels of information exchange: an empirical investigation of intended e-services use." *E-Services Journal* 3(2006): 25–59.

Dinev, T., P. Hart, and M. R. Mullen. "Internet privacy concerns and beliefs about government surveillance – An empirical investigation." *Journal of Strategic Information Systems* 17 (2008): 214–233.

Dorr, A., P. Kovaric, and C. Doubleday. "Parent-child coviewing of television." *Journal of Broadcasting & Electronic Media* 1(1989): 35–51

Eastin, M. S., B. S. Greenberg, and L. Hofshire. "Parenting the Internet." *Journal of Communication* (2006): 486–504.

FEDMA. "European Code of Practice for the Use of Personal Data in Direct Marketing." FEDMA. http://img.custompublish.com/getfile.php/342991.1014.xacscqtseu/FEDMACodeEN.pdf?return=fedma.custompublish.com (accessed on August 3, 2009)

FEDMA. "Codes of Practice." FEDMA. http://fedma.custompublish.com/codes-of-practice.347966-59917.html (accessed on August 3, 2009)

Fogel, J., and E. Nehmad. "Internet social network communities: risk taking, trust and privacy concerns." *Computers in Human Behaviour* 1(2009): 153–160.

Fox, S. *Trust and privacy online: Why Americans want to rewrite the rules*. Washington, DC: Pew Internet & American Life Project, 2000.

Federal Trade Commission. "Children's Online Privacy protection Act of 1998." Federal Trade Commission. http://www.ftc.gov/ogc/coppa1.htm (accessed on August 3, 2009)

Graeff, T. R., and S. Harmon. "Collecting and using personal data: consumers' awareness and concerns." *Journal of Consumer Marketing* 4(2002): 302–318.

Grolnick, W. S., and M. Farkas. "Parenting and the Development of Children's Self-regulation." In *Handbook of Parenting*, edited by Marc H. Bornstein, 89–111. Mahwah, NJ: Lawrence Erlbaum Associates, 2002.

Horne, D. R., and D. A. Horne. "Domains of Privacy: Toward an Understanding of Underlying Factors" (paper presented at the Direct Marketing Educators' Conference, San Francisco, CA, October 11, 1998)

Kehoe, C. M. and J. E. Pitkow. "Surveying the Territory:GVU's Five WWW User Surveys Surveying the Territory: GVU's Five WWW User Surveys." WWW User Survey Home Page.http://www.cc.gatech.edu/gvu/user_surveys/papers/w3j.html (accessed on August 3, 2009)

Kerr, M., and H. Stattin. "What Parents Know, How They Know It, and Several Forms of Adolescent Adjustment: Further Support for a Reinterpretation of Monitoring." *Developmental Psychology* 3(2000): 366–380.

Krasnova, H., O. Günther, S. Spiekermann, and K. Koroleva. "Privacy concerns and identity in social networks." *Identity in the Information Society* 1(2009): 39–63.

Larose, R., and N. J. Rifon. "Promoting of Safety: Effects of Privacy Warnings and PrivacySeals on Risk Assessment and Online Privacy Behaviour." *Journal of Consumer Affairs* 1(2007): 127–149.

Lee, S.-J. and Y.-G. Chae. "Children's Internet Use in a Family Context: Influence on Family Relationships and Parental Mediation." *Cyberpsychology and Behavior* 5(2007): 640–644.

Lenhart, A., P. Hitlin, and M. Madden. *Teens and Technology*. Washington, DC: Pew Internet & American Life Project, 2005.

Lenhart, A., and M. Madden. *Teens, Privacy & Social Networks*. Washington, DC: Pew Internet & American Life Project, 2007.

Lewis, K., J. Kaufman, and N. Christakis. "The taste for privacy: An analysis of college student privacy settings in an online social network." *Journal of Computer-Mediated Communication* 14(2008): 79–100.

Liau, A. K., A. Khoo, and P. Hwaang. "Factors Influencing Adolescents Engagement in Risky Internet Behaviour." *Cyberpsychology and Behavior* 6(2005): 513–520.

Liebermann, Y., and S. Stashevsky. "Perceived risks as barrier to Internet and e-commerce usage." *Qualitative Market Research: An International Journal* 4(2002): 291–300.

Livingstone, S. and E. J. Helsper. "Parental Mediation of Children's Internet Use." *Journal of Broadcasting & Electronic Media* 4(2008):581–599.

Mahon, A., C. Glendinning, K. Clarke, and G. Craig. "Researching children: Methods and ethics." *Children and Society* 10(1996): 145–154.

May O. L., A. J. S. Stanaland, and A. D. Miyazaki. "Protecting children's privacy online: How parental mediation strategies affect website safeguard effectiveness." *Journal of Retailing* 84(2008): 205–217.

McNeal, J. U. *Children as Consumers*. Lexington, MA: Lexington Books, 1987, 211.

Milne, G. R., and A. J. Rohm. "Consumer privacy and name removal across direct marketing channels: exploring opt-in and opt-out alternatives." *Journal of Public Policy & Marketing* 2(2000): 238–249.

Milne, G. R., and M. J. Culnan. "Strategies for Reducing Online Privacy Risks: Why Consumers Read (Or Don't Read) Online Privacy Notices." *Journal of Public Policy & Marketing* 3(2004): 15–29.

Miyazaki, A. D., and A. Fernandez. "Consumer perceptions of privacy and security risks for online shopping." *The Journal of Consumer Affairs* 1(2001): 27–44.

Montgommery, K. C. "Digital kids: The new on-line children's consumer culture." In *Handbook of Children and the Media*, edited by Dorothy G. Singer and Jerome L. Singer, 635–650. Thousand Oaks, CA: Sage, 2001.

Moscardelli, D. M., and R. Divine. "Adolescents' Concern for Privacy When Using the Internet: An Empirical Analysis of Predictors and Relationships with Privacy-Protecting Behaviours." *Family and Consumer Sciences Research Journal* 3(2007): 232–252.

Moshis, G. P. "The Role of Family Communication in Consumer Socialization of Children and Adolescents." *Journal of Consumer Research* 4(1985): 898–913.

Nathanson, A. I. "Identifying and Explaining the Relationship Between Parental Mediation and Children's Aggression." *Communication Research* 2(1999): 124–143.

Nathanson, A. I. "Parent and Child Perspectives on the Presence and Meaning of Parental Television Mediation." *Journal of Broadcasting & Electronic Media* 2(2001): 201–220.

Neeley, S. M. "Internet Advertising and Children." In *Internet Advertising: Theory and Research*, edited by David W. Shumann and Esther Thorson, 343–362. Mahwah, NJ: Lawrence Erlbaum Associates, 2007.

Neuborne, E. "For kids on the Web, It's An ad, ad, ad, ad world." *Business Week* 3745(2001): 108–109.

Nowak, G., and J. Phelps. "Understanding Privacy Concerns." *Journal of Direct Marketing* 6(1992): 28–39.

Paine, C., U.-D. Reips, S. Stieger, A. Joinson, and T. Buchanan. "Internet users' perceptions of 'privacy concerns' and 'privacy actions'." *International Journal of Human Computer Studies* 65(2006): 526–536.

Phelps, J., G. Nowak, and E. Ferrell. "Privacy concerns and consumers' willing-ness to provide personal information." *Journal of Public Policy & Marketing* 1(2000): 27–41.

Rideout, V., D. F. Roberts and U. G. Foehr. *Generation M: Media in the Lives of 818 Year Olds*. Menlo Park, CA: The Henry J. Kaiser Family Foundation, 2005.

Sheehan, K. B. "An investigation of gender differences in online privacy concerns and resultant behaviors." *Journal of Interactive Marketing* 4(1999): 24–38.

Sheehan, K. B. and M. G. Hoy. "Flaming, complaining, abstaining: how online users respond to privacy concerns." *Journal of Advertising* 3(1999):37–51.

Shuman, D. W., and E. Thorson. *Internet Advertising: Theory and Research*. Mahwah, NJ: Lawrence Erlbaum Associates, 2007.

Staples, W. G. *Encyclopedia of Privacy*. Westport, CT: Greenwood Publishing Group, 2006.

Thorson, E., and D. W. Shumann. "The Internet waits for No One." In *InternetAdvertising: Theory and Research*, edited by David W. Shumann and Esther Thorson, 3–13. Mahwah, NJ: Lawrence Erlbaum Associates, 2007.

Turow, J., and L. Nir. *The Internet and the Family 2000: The View from Parents, The View from Kids*. Philadelphia, PA: The Annenberg Public Policy Center, 2000.

Valkenburg, P. M., M. Krcmar, A. L. Peeters, N. M. Marseille. "Developing a scale to assess three styles of television mediation: "instructive mediation," "restrictive mediation", and "social coviewing"." *Journal of Broadcasting & Electronic Media* 1(1999): 52–66.

Walrave, M. *Cyberkids' e-privacy at stake? Data processing and privacy policies in websites aimed at minors*. Privacy Paper No 4. Antwerp: University of Antwerp, 2005.

Walrave, M., S. Lenaerts, and S. Demoor. *Cyberteens @ risk? Tieners verknocht aan internet, maar ook waakzaam voor risico's ?* Brussels: BELSPO, 2008.

Wang, P., and L. A. Petrison. "Direct Marketing Activities and Personal Privacy: A Consumer Survey." *Journal of Direct Marketing* 1(1993): 193–220.

Warren, R., P. Gerke, and M. A. Kelly. "Is There Enough Time on the Clock? Parental In- volvement and Mediation of Children's Television Viewing." *Journal of Broadcasting & Electronic Media* 1(2002): 87–111.

White, T. B. "Consumer Disclosure and Disclosure Avoidance: A Motivational Framework." *Journal of Consumer Psychology* 1(2004): 41–51.

Youn, S. H. "Teenagers' Perception of Online Privacy and Coping Behaviors: A Risk- Benefit Appraisal Approach." *Journal of Broadcasting & Electronic Media* 1(2005): 86–110.

Youn, S. H. "Parental influences and teens' attitude toward online privacy protection." *Journal of Consumer Affairs* 3(2008): 362–388.

Youn, S. H., and K. Hall. "Gender and Online Privacy Among Teens; Risk Perception, Privacy Concerns and Protection Behaviours." *Cyberpsychology and Behavior* 6(2008): 763–765.

Chapter 15
Why Adopting Privacy Enhancing Technologies (PETs) Takes so Much Time

John J. Borking

15.1 About PETs and the Research Questions

Article 17 (1) of the Directive 95/46/EC (DPD) requires that the controller must implement appropriate technical and organizational measures to protect personal data against accidental or unlawful destruction or accidental loss, alteration, unauthorized disclosure or access, in particular where the processing involves the transmission of data over a network, and against all other unlawful forms of processing.

> The Directive states, that *(...) such measures shall ensure a level of security appropriate to the risks represented by the processing and the nature of the data to be protected.*[1]

The requirements referred to in the DPD must be implemented efficiently in the organization in order to give proper support to the citizen's right to privacy with respect to personal data. It is therefore important to devise a proper system of general processing measures and procedures that should be present in order to protect company processes and in connection with specific protective measures for the processing of personal data. Given the basic legal requirements for privacy protection and the risks of privacy incidents, it will be apparent that, if technical provisions are to be deemed adequate, they must go beyond the implementation of traditional security measures.

ICT offers solutions in the shape of privacy protection for users, consumers and citizens. The application of ICT to protect privacy has become widely known under the name Privacy-Enhancing Technologies (PET or PETs). PETs have been defined as a coherent system of ICT measures that protects privacy by eliminating

J.J. Borking (✉)
Borking Consultancy, Wassenaar, The Netherlands
e-mail: jborking@xs4all.nl

Dr. John J. Borking (1945) is owner/director of Borking Consultancy in Wassenaar The Netherlands and was a former privacy commissioner and board member of the Dutch Data Protection Authority. Address: Lange Kerkdam 27, 2242 BN Wassenaar, Netherlands; email: jborking@xs4all.nl

[1]Directive 95/46/EC, Official Journal L 281, 23/11/1995 P. 0031–0050.

S. Gutwirth et al. (eds.), *Computers, Privacy and Data Protection: an Element of Choice*, DOI 10.1007/978-94-007-0641-5_15, © Springer Science+Business Media B.V. 2011

or reducing personal data or by preventing unnecessary and/or undesired processing of personal data; all without losing the functionality of the data system (Borking, 2003; Hes and Borking, 2000; Van Blarkom et al., 2003; Communication from the Commission to the European Parliament and the Council on Promoting Data Protection, 2007). PETs are about technologies that *enhance* privacy and privacy protection is neither an equivalent of information security or confidentiality. The overlap and difference between privacy and information security and confidentiality, is visualised in the following Fig. 15.1.

PETs have to be used for implementing the legal specifications in the EU privacy directives 95/46/EC and 2002/58/EC, like data minimization, consent requirements, access rights of data subject, privacy safe construction of terminals in information systems (Borking, 2010).

PETs can guarantee data protection without making excessive demands on the processing of the data. By applying PETs and streamlining personal data processing, the organizations can continue to meet the high public expectations with respect to services and dealing with personal data (Koorn et al., 2004).

In Fig. 15.2, the different PETs options are positioned in relation to the effectiveness of the data protection. The diagram also shows the most important features of the different PETs options. The PETs staircase is not a growth model and does not have to be followed to the top. Once an organization has applied general PETs controls, it does not mean that it has to go on to "higher" levels of PETs. The suitability of the different PETs options depends on the individual situation.

The basic driver to invest in PETs is their potential to avoid privacy incidents and so to reduce the risks and subsequently the damage caused by privacy breaches.

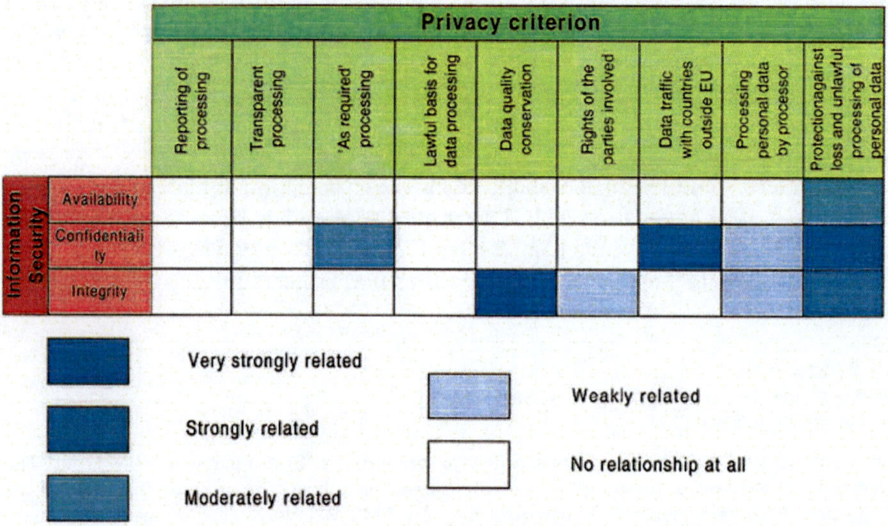

Fig. 15.1 Differences between privacy protection, information security and confidentiality (Borking, 2010)

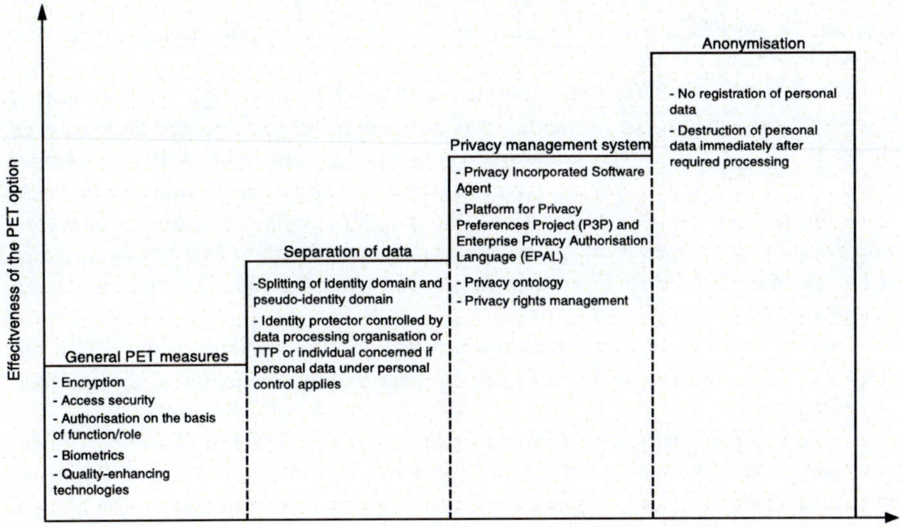

Fig 15.2 PETs staircase: the effectiveness of the different PETs options (Koorn et al., 2004)

In general terms a privacy incident can be defined as an event in which personal data are misused, because of the fact that personal data accompanied by a list with personal data constraints haven't been respected. The amended directive 2002/58/EC describes it as "*'a breach of security leading to the accidental or unlawful destruction, loss, alteration, unauthorised disclosure of or access to personal data, transmitted, stored or otherwise processed [. . .]*'.

Privacy breaches may impact an organization in different ways. Tsiakis and Stephanides distinguish direct, short-term, and long-term economic consequences (Tsiakis and Stephanides, 2005). Direct consequences are the costs for repairing or changing systems, costs of stopping or slowing down production or processes, costs of legal action. Short term consequences comprise the loss of existing customers, contractual relations, and the loss of reputation. Companies may loose business because of privacy breaches, which harm their trust relationships with customers and other business relations. Safeguarding privacy has been identified as a major component of building trust (Camp and Wolfram, 2000). Long term consequences include the loss of stock value and market value. An example of the latter is DoubleClick in 2000. After a serious violation of their existing privacy statement on their website and the lawsuit that was the result of this violation, their stock declined with 20% (Chapman and Dhillon, 2002). This also occurred with Choicepoint after their public announcement that they were hacked, and approximately 10 million data records were stolen. Their stock declined with 17% since the data breach (Privacy Rights Clearinghouse, 2007).

Cas and Hafskjold wrote in 2006: *So far PETs have not contributed as much as would be possible to the protection of privacy; partly because of a lack of availability of PETs, partly because of a lack of user friendliness.* (Cas and Hafskjold 2006)

Leisner and Cas in 2006 further pointed out that *PETs are insufficiently supported by current regulations; in particular it is not compulsory to provide the option of anonymous access to services or infrastructures.* (Leisner and Cas 2006) Sommer remarked *We still face major obstacles towards a deployment of such (PETs) technology in the field at a large scale (...) the part of convincing business to design their business processes in a way such that data minimization can be implemented as envisioned in PRIME will even be harder than has been the technological part.* (Sommer 2008)

However it isn't the user friendliness or the lack of availability of PETs, but there are other reasons why PETs aren't used by governmental or commercial organizations.

A group researchers (Bos, 2006), Borking, Dijkman, Fairchild, Hosein, Ribbers and Tseng have focused in the PRIME project, (Fairchild and Ribbers, 2008) on what business drivers lead organizations to adopt privacy enhancing technologies (PETs) for providing assurance for privacy.

The central research questions were:

- When starts an organization bothering about privacy?
- What factors impact the adoption of privacy-enhancing technologies tools in information systems as a measure to protect privacy sensitive data, and how do these factors affect the adoption decision?
- What are the drivers and inhibitors for adoption by organizations of PETs?

15.2 Technological Innovations

The capability of an organization to innovate or to apply an innovation is important in today's competitive environment (Tidd et al., 2005). If an organization lacks this capability it will fail to apply necessary transformations, to introduce innovation and as a result may create a competitive disadvantage.

An innovation is generally defined as the application of something new. According to Rogers (2003) the question whether something new is an innovation has to be considered from a relative point of view. Something that in a particular environment or by a particular person is subjectively perceived as new can be regarded as an innovation. An innovation can also be related to many things, like an idea, a method, a technology or a product. Each of these types of innovations has its characteristics, which play a role in the adoption process.

Given the innovative character of ICT, research of innovation in particular technical innovations, tends to focus on technological innovations like software or electronic services (Tidd et al., 2005). The OECD defines technological innovation as:

> a technological new product or process that includes a significant improvement and has been actually put into use. The technological new product or process consists of a variety of scientific, technical, organizational, financial and commercial aspects. OECD (2005)

PETs, given the relative recent introduction of the concept (Borking, 2003; Hes and Borking, 2000; Van Blarkom et al., 2003; Communication from the Commission to the European Parliament and the Council on Promoting Data Protection, 2007), the progress that is being realized with its application, and the new approach they offer with regard to privacy protection can be regarded as innovation.

The requirements referred to in the DPD must be implemented efficiently in the organization in order to give proper support to the citizen's right to privacy with respect to personal data. It is therefore important to devise a proper system of general processing measures and procedures that should be present in order to protect company processes and in connection with specific protective measures for the processing of personal data. The restrictions that the organization of information systems can impose on the possibility that their users can comply with privacy legislation are evident. One simple example is where a system contains an inescapable "date of birth" field, while analysis of the company's processes shows that recording the birth date of all persons included in the system is excessive. System design can just as easily ensure that users correctly observe the law. As a rule, privacy protection will constitute a supplementary system of measures and procedures in addition to the usual processing and security measures, but it should be assigned a significant place in management processes in order to implement and maintain a balanced processing policy for personal data.

When an organization is asked what it has done to protect privacy, it is apt to emphasize the personal data security measures it has in place. Although the use of safeguards to prevent unauthorized access to personal data is an important aspect of privacy protection, it is not sufficient in its own right. This is because such safeguards rarely involve the encryption of stored data; consequently, effective protection depends entirely on the security measures being correctly implemented and functioning properly.

It is therefore preferable to take technical measures that protect the individual's privacy at the point of data collection. Such measures may do away with the need to generate or record any personal data at all. Alternatively, they may minimize or even obviate the need to use or store identification data.

Given the basic legal requirements for privacy protection and the risks of privacy incidents, it will be apparent that, if technical provisions are to be deemed adequate, they must go beyond the implementation of traditional security measures.

15.3 Diffusion and Adoption of Technological Innovations

A central theme in the research on innovation is in particular the way technological innovations are spread in a specific environment and how subsequently these innovations are being accepted and utilized. This area is known as "diffusion and adoption" (Fichman 1992). Diffusion relates to how innovations are spread across a specific society or industry. Adoption is defined as the process through which a person or organization evolves from first getting acquainted with the innovation till its eventual utilization (Rogers 2003).

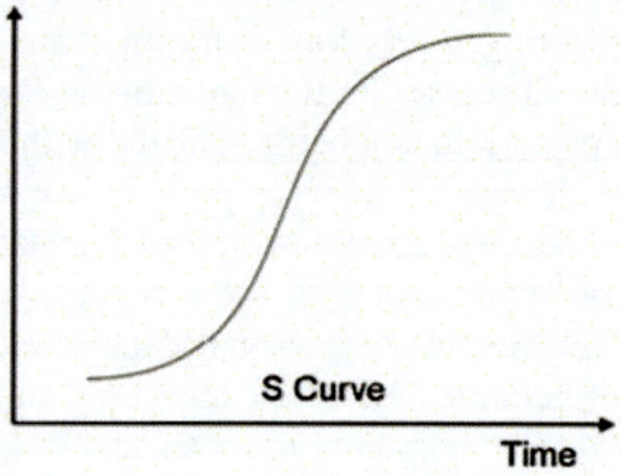

Fig. 15.3 Rate of adoption (Rogers, 2003)

In the study of diffusion and adoption many studies try to identify relevant impacting factors, so that predictive statements can be made (Jeyaraj et al., 2006).

Rogers (2003) considers adoption and diffusion as a process with a relatively known and constant pattern of evolution. He describes the rate of adoption as an S-shaped curve. See Fig. 15.3.

The idea of the S-shaped curve (limited interest for the innovation in the beginning, followed by an increased interest leading to an intensified use, which eventually will level off) applies to all types of adoption. Others, who state that also partial adoption, as a middle road between adoption and non-adoption, is a viable possibility, have supplemented Rogers' ideas; this reduces the contrast between adoption and non-adoption (Bayer and Melone, 1989).

15.4 Factors of Organizational Adoption of Technological Innovations

Rogers distinguishes various variables that influence the process of adoption of innovations. First he describes characteristics of the innovation itself: relative advantage or benefit, compatibility, complexity, testability, and visibility of the innovation. He also points their impact is determined by the perception of these factors by the potential adopter, and not so much by how they are in reality. Next he distinguishes various variables that characterize the organizations, which are open to adopt innovation: the general attitude of top management with regard to change, centralization, complexity, formalization, internal relatedness, organizational slack, size and openness of the organization to the environment.

Rogers' Diffusion of Innovation [DOI] Theory has gained quite a broad acceptance; the variables have been tested in multiple studies and found relevant. Also Fichman (1992) and Jeyarai et al. (2006) found that three clusters of factors explain the organizational adoption behavior: factors related to the technological innovation, to the adopting organization, and to the environment of both former factors. They investigated over a hundred variables that have been researched in different studies. They also performed an empirical test on the best predicting factors for the organizational adoption of IT-based innovations. Combined in clusters the dominant factors

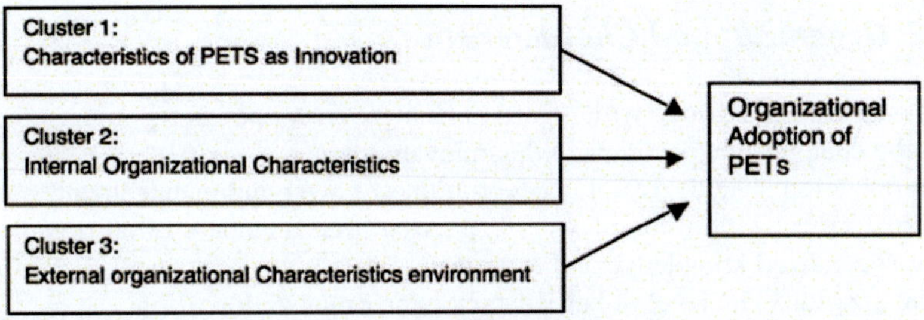

Fig. 15.4 Conceptual model (Fairchild and Ribbers, 2008; Bos, 2006)

appear to be those related to innovation characteristics, organizational characteristics, and environmental characteristics. Tung and Reck (2005) reach this conclusion in their study.

Others have emphasized other influences on the adoption process: Fichman (1992) argues that adoption of IT based innovations requires a different approach. Fichman (1992), Rivera and Rogers (2004) and Greenhalgh et al. (2004) point to specific effects of innovations in network organizations on inter-organizational relationships. The approaches of Jearay, Fichman and Rogers form the foundation for the Conceptual Model shown in Fig. 15.4.

The first cluster of factors encompasses those variables that are related to the technical innovation itself, and so to PETs. The second cluster looks at those variables that are related to the internal characteristics of the adopting organization. The third cluster contains factors related to the environment of the adopting organization and innovation. In case of PETs, in particular privacy policies and regulations, and level of enforcement seem to be particularly relevant.

15.5 Specific Characteristics

Rogers (2003) and Fichman (1992) distinguishes five innovation characteristics and eight organizational characteristics, which affect the organizational adoption of innovations.

15.5.1 Innovation Characteristics

Relative advantage or benefit (+): the advantage offered by the innovation, compared to the former practice or technology.

Compatibility (+): The extent that an innovation resembles its predecessor.

Complexity (–): The effort needed to learn how to use the innovation.

Testability (+): The extent that small-scale experiments with the innovation are possible.

Visibility (+): the extent to which the innovation is visible for the outside world.

15.5.2 Organizational Characteristics

Top Management's attitude with regard to change: How open is top management to accept the changes that accompany the innovation?

Centralization: The degree of concentration of power and management.

Internal Organization complexity: The extent that members of an organization possess specialized knowledge and expertise.

Formalization: The level of bureaucracy in an organization.

Internal relatedness: The extent that internal members of the organization are interrelated.

Organizational slack: The extent that an organization possesses uncommitted resources.

Size: The size of the organization.

Openness: The degree that organizations are in contact with other organizations.

15.6 Encompassing Model

Fichman (1992) compared different adoption studies and built an encompassing model that explains organizational adoption of complex information technology innovations. The model consists of three clusters, while each cluster contains a few groups of factors.

The three clusters are:

a. The Technology and Organization combination;
b. The Technologies and Diffusion environments;
c. The Organizations and Adoption environments.

The Technology and Organization Combination cluster stands for factors that describe the relationship between the innovation and a specific organization. This boils down to the fit between the innovation and the organization, the perception of organizational characteristics and factors that describe the possibilities for an organization to implement the innovation.

The Technologies and Diffusion environments cluster regards those factors that describe the innovation and the specific environment from which they emanate. These are in particular the innovation characteristics and possible roles of advising institutions.

The Organizations and Adoption environments cluster deals with factors that describe the adopting organization and their environment. These are organizational characteristics and characteristics of the environment and industry.

15.7 Interviews with Experts

In order to find variables that characterize each cluster, literature analysis has been combined with expert interviews (2006–2007). Factors that have been proposed to be relevant in the literature have been compared with the results of expert interviews, and vice versa. Five experts in the field of PETs have been interviewed (Bos, 2006) and 4 workshops have been held in Sweden, UK, Netherlands and Switzerland (Fairchild and Ribbers, 2008). In the workshops representatives of a broad range of industries participated. The results of the interviews are presented below. The variables mentioned by the experts and organizations have been grouped according to the categories innovation, internal organization and environment (Fairchild and Ribbers, 2005; Bos, 2006).

Factor: Innovation

Relative benefit:	Positive
Compatibility	Negative
Complexity:	Negative
Costs:	Negative[2]
Testability:	Positive
Role of advisory institutions:	Positive
Social recognition:	Positive
PETs woven into business processes:	Negative

Factor: Internal Organization

Top Management's attitude towards change caused by PETs:	Positive and Negative
Structure and Size of the organization:	Negative
Complexity of organizational processes:	Negative
Presence of key persons:	Positive
Ties with advisory institutions:	Positive
Perception and level of awareness of privacy regulations:	Positive
Diversity of information systems:	Negative
Type of the data processed:	Positive

Factor: Environment

External pressure by privacy laws:	Positive
Complexity of privacy laws:	Negative
Existing offer of PETs measures:	Positive
Visibility:	Positive

15.8 Explanations of the Terms

The terms mentioned under the factors innovation, internal organization and environment are explained hereunder.

[2]When showing the results of calculating the Return On Investment on PETs investments some participants showed a positive attitude change towards adoption (Borking, 2008).

15.8.1 Relative Benefit

The advantage of PETs is that it offers a clear privacy protection, which, when properly applied, is in line with legal requirements. The potential relative benefit compared to other protective measures is big. It however appears to be difficult to value in economic terms the relative benefits of PETs compared to other protective measures. This is caused by the existing ambiguity around PETs and privacy. As a result, often more conventional measures are chosen instead. Calculating the ROI on privacy/PETs investment leads to more clarity.

15.8.2 Compatibility

Only when PETs resembles its predecessor the effect is positive.

15.8.3 Complexity

PETs have been perceived as a complex innovation. The implementation of PETs requires specific expertise in different disciplines. Except IT expertise also legal expertise is needed; this combination of is very scarce and has to be acquired externally.

15.8.4 Costs

PETs have been considered as an expensive innovation (with unclear benefits). Much depends however on the moment that PETs is introduced. If the introduction is when a new information system is put into use and directly integrated into this I.S, then costs are generally at an acceptable level. This is also basically the only realistic option. PETs are simply too complex to apply to existing systems, costs are then being perceived as higher than those of traditional measures.

15.8.5 Testability

The extent that small-scale experiments with PETs are possible is perceived as positive

15.8.6 Role of Advisory Institutions

Some organizations can play a key role in the diffusion of innovations. The Dutch Data Protection Authority has assumed this role with regard to PETs in the past till

2002, especially when giving advices with regard to large projects. This role and the attention given to PETs have impacted its adoption. At the moment the Dutch DPA does not get actively involved in the design of information systems anymore advising the use of PETs, with a lower rate of adoption as a result.

15.8.7 Social Recognition

The use of PETs does not receive a lot social recognition, which is the result of its limited visibility. Also privacy protection is not an issue with which organizations try to differentiate themselves unless it is a USP. The market for privacy protection is not transparent.

15.8.8 PETs Woven into Business Processes

An important characteristic of PETs is that its implementation requires integration in information systems. This requires a combination of legal and technical (ICT) expertise, which is hard to find.

15.8.9 Top Management's Attitude Towards Change Caused by PETs

If management is open to accept the changes that accompany PETs then it is seen as positive.

15.8.10 Structure and Size of the Organization

Contrary to the literature the interviews showed that large organizations aren't more positive about PETs than smaller ones.

15.8.11 Complexity of Organizational Processes

PETs-measures usually have to be customized for a specific organization or process. The more complex this is, the more difficult it is to implement PETs.

15.8.12 Presence of Key Persons

The utilization of PETs often depends on specific key persons in an organization, who know the concept and take the lead in the adoption process. Such a person has a strong impact on the adoption of PETs.

15.8.13 Ties with Advisory Institutions

The use of PETs sometimes depends on the ties that an organization has with advisory institutions (e.g. DPA). An organization that has no links with such institutions is not likely to put PETs into use.

15.8.14 Perception and Level of Awareness of Privacy Regulations

Privacy standards (norms) are often not perceived as being very important for business processes; also the consequences of not complying with the law aren't considered generally as important as the change to be caught when violating the privacy legislation is considered to be very low. As a result the adoption of PETs is in most organizations not high on the management agenda. However in the interviews with multinationals in the field of consumer electronics, energy, banking and telecommunications the pressure of privacy legislation is considered as relevant.

15.8.15 Diversity of Information Systems

The more diversity of information systems in organizations, the less likely PETs will be introduced in the organization.

15.8.16 Type of Processed Data

When the level of risk associated with privacy breaches is high, then there is a bigger incentive to apply PETs.

15.8.17 Pressure by Privacy Laws

Privacy laws exert little pressure on organizations to really put PETs into use. Only in a few cases the law refers to PETs, however the decision makers are left free what to choose as protective measures.

Management of the interviewed organizations considers the EU privacy directives as of a too general and abstract character. In general there is little awareness of PETs. The focus of decision makers is on the key business processes; privacy is often a secondary issue. However the interest for privacy is increasing. The Commission's Communication on promoting Data Protection by Privacy Enhancing Technologies (PETs) (COM (2007) 228 Final, Brussels, 2.5.2007) is viewed as a positive stimulus. A mandatory requirement to use PETs is felt by the stakeholders as necessary. There is also very limited demand for privacy audits, because there is no felt need to have one unless the audit results in a visible result, like obtaining the EuroPrise[3] privacy certificate (seal). At its essence, EuroPrise (from the Independent Centre for Privacy Protection Schleswig-Holstein with Accredited EuroPriSe Legal and/or Technical Experts) is a voluntary certification program by which any company or individual could: (a) Gain assurance that its product or service is in compliance with EU data protection laws, and (b) Send a message to the marketplace and to consumers (end-users) stating: We take user's privacy seriously. EuroPrise states on its website www.european-privacy-seal.eu/ that this privacy certificate aims to facilitate an increase of market transparency for privacy relevant products and an enlargement of the market for Privacy Enhancing Technologies and finally an increase of trust in IT.

15.8.18 Complexity of Privacy Laws

Organizations often do not know/understand what privacy laws require them to do. Because privacy laws are overly complex and ambiguous, they do not use the right set of protective measures.

15.8.19 Existing Offer of PETs Measures

The lack of PETs-measures have a negative influence on the adoption of PETs, especially as many organizations are using standard package software in which PETs-measures haven't been foreseen. When PETs measures can be applied in an organization (like anonymization or privacy management systems) are available then it is a positive factor.

15.8.20 Visibility

When PETs in systems and services can be proven by privacy seals/certificates then it is a positive factor

15.9 Summary of the Results

A number of factors are perceived to have a negative impact on the adoption process. Decision makers assume is that PETs are difficult to implement efficiently and effectively. Also the internal organizational characteristics have a negative impact.

---

[3]EuroPrise (privacy seals) has been subsidized by EU Commission under the eTEN Programme. The EuroPrise project started op June 10 2007 and ended February 28, 2009. http://www.european-privacy-seal.eu/about-europrise/fact-sheet.

Although there is enough code developed, the limited offer of PETs tools by software suppliers appears to have a negative impact. Only the legal and regulatory pressure with regard to privacy protection has an undivided positive impact on the adoption process. However, the existing legislation provides too little reference to the concept of PETs, to make a difference in the adoption process. The promotion by advisory bodies appears to have a strong positive influence.

A conclusion of this study is that the adoption of PETs is problematic (Bos, 2006). There are only of a limited use. Looking at the conceptual model (Fig. 15.3), in particular those factors that are related to regulatory and legal compliance, to improved coordination and advice and information with regard to PETs, seem to help to solve this problem. The relative advantage of PETs is perceived by SMEs to be zero. However in interviews with large international organizations the use of PETs in relation to preventing reputation damage is seen as positive. Both educational activities and adaptation of the law seem to be necessary. Legal requirements are generally observed; however in privacy laws there is insufficient reference to PETs. Also the minimum level of privacy protection required by the law is perceived as insufficient for constituting an incentive to apply PETs (Fairchild and Ribbers, 2008).

15.10 Identity and Access Management (Iam) Maturity Model

To examine under what conditions an organization would adopt PETs into its business process, researchers explored how an IAM maturity model can be adapted to examine privacy adoption maturity in organizations. The hypothesis behind the choice for the IAM maturity model is that as protection of personal data is closely linked with identity issues, the increased attention for identity in the organizational processes must lead to the awareness of informational privacy.

> A maturity model is defined as *a staged structure of maturity levels, which defines the extent to which a specific process is defined, managed, measured, controlled and/or effective, assuming the organization develops and adopts new processes and practices, from which it learns, optimizes and moves on to the next level, until the desired level is reached.* Smit (2005)

During the last decade several maturity models have been developed in specific research areas such as business IT alignment, software development and information security. All of these models have one thing in common; they all describe the maturity of one or more processes within an organization. As a basis for this IAM maturity model, a number of existing models were examined. The descriptions of these maturity levels differ among the models, but are quite similar in general. Every model characterizes the first maturity phase as being chaotic and dealing with processes on an ad hoc basis. The second one is characterized by the planning of processes. The third maturity level is characterized by the implementation of standards aimed at particular processes and outputs for processes are defined. Quantitative management characterizes the fourth maturity level.

Processes and quality are controlled based on quantitative measures. Based on the measures taken out of the quantitative measures implemented in maturity level four, maturity level five improves the organization. These improvements are continuous, incremental and connected to the business objectives' measures (Bos, 2006; Fagerberg et al., 2005; Stanford Organizational Maturity Levels; Vandecasteele and Moerland, 2001). The following general phase descriptions can be discerned:

Phase 1: Only few processes have been defined and processes are conducted on an ad hoc base.
Phase 2: Processes that seem to work and be in order are repeated.
Phase 3: Processes are standardized and documented to review if they are executed accordingly.
Phase 4: Performance and success are measured and quality measures are done
Phase 5: Processes are systematically improved with the help of quantitative feedback of results, test results and innovative ideas.

Based on a KPMG (Vandecasteele and Moerland, 2001) model, researchers then integrated maturity phases into these processes, and developed an IAM maturity model shown below (Fig. 15.5):

The filled out maturity model can in turn be translated into a more general description of maturity phases for IAM in general. This means that the whole IAM situation is described per maturity phase. Describing the situation in general leads to a more practical and understandable image of the Identity and access management processes.

Through all of these five maturity phases the awareness and importance of IAM processes increases within the organization (Van Gestel, 2007). The organization

	Immature	Starting-up	Active	Pro-Active	Top Class
Authentication Management	No authentication means	Arbitrarily formulated authentication requirements (authentication means are provided, adjusted and deleted on user request)	Authentication Requirements based on a one time survey	Authentication Requirements based on continuous risk analysis	Authentication requirements based on continuous risk analysis and are continously adjusted
User Management	Double and inconsistent entries because of chaotic and ad hoc processes	Entries can be double but they are consistent	Central registration, Limited user group, manual procedures	Central registration, controlled authorization processes, manual procedures	Central real-time controlled authorization sources, automated procedures
Authorisation Management	No authorization matrixes, authorization is defined ad hoc	Authorization matrixes defined but are not updated	Authorization matrixes are updated periodically	Role Based Access Control used for critical applications	Role Based Access Control for all applications and continuous updated authorizations
Provisioning	Manual process locally	Limited Automated unreliable processed locally	Limited Automated but reliable processes locally	Limited Automated and reliable for multiple sources	Automated and reliable for multiple sources
Monitoring(Audit)	No responsibility delegated into a AO/IC organization	Sporadically delegated responsibility of AO/IC	Partial delegation of responsibility to AO/IC	Full responsibility to AO/IC	Full responsibility to AO/IC with periodic reporting

Fig. 15.5 Conceptual identity and access management (IAM) maturity model (Fairchild and Ribbers, 2008)

going through all these sequential phases not only needs to adjust its identity and access management processes, but also its own organizational structure and policies need to be adjusted. These adjustments like the adjustments to the IAM processes need to be evolutionary not revolutionary. Since IAM can entail the creation of roles or positions within the existing organizational structure, the impact of an IAM implementation can be quite significant. In order to deal with these changes the organization needs to be ready and willing to accept these changes or adjust the IAM project to suit the organizational structure, meaning that the organization and IAM need to be adjusted to each other for IAM to be successful after implementation. This could be an argument to introduce organizational structure as a part of the IAM maturity model. However there already exist organizational maturity models for organizations dealing with the questions of IT projects (Davenport, 1993). Introducing organizational maturity into the maturity would also introduce organizational facets that are not immediately related to Identity and access management. The development of IAM in organizations follows a S-curve as described by Rogers (2003), starting at the immature/monitoring level and ending at the top class/authentication management level. See Fig. 15.6.

In the White book on Privacy Enhancing Technologies by Koorn et al. (2004), is stated that PETs are composed out of several technologies divided in four different PETs categories (see Fig. 15.1):

1. General PETs controls (i.e. identity and access management);
2. Separation of data (identity and pseudo-identity domains);
3. Privacy management systems for personal data that can't be encrypted at the intake because many laws require the collection of clear (non- encrypted) data;
4. Anonymisation.

Authentication Management	No authentication means	Arbitrarily formulated authentication requirements (authentication means are provided, adjusted and deleted on user request)	Authentication requirements based on a one time survey	Authentication Requirements based on continuous risk analysis	Authentication requirements based on continuous risk analysis and are continously adjusted
User Management	Double and inconsistent entries because of chaotic and ad hoc processes	Entries can be double but they are consistent	Central registration, Limited user group, manual procedures	Central registration, controlled authorization processes, manual procedures	Central real-time controlled authorization sources, automated procedures
Authorisation Management	No authorization matrixes, authorization is defined ad hoc	Authorization matrixes defined but are not updated	Authorization matrixes are updated periodically	Role Based Access Control used for critical applications	Role Based Access Control for all applications and continuous updated authorizations
Provisioning	Manual process locally	Limited Automated unreliable processed locally	Limited Automated but reliable processes locally	Limited Automated and reliable for multiple sources	Automated and reliable for multiple sources
Monitoring(Audit)	No responsibility delegated into a AO/IC organization	Sporadically delegated responsibility of AO/IC	Partial delegation of responsibility to AO/IC	Full responsibility to AO/IC	Full responsibility to AO/IC with periodic reporting
	Immature	**Starting-up**	**Active**	**Pro-Active**	**Top Class**

Fig. 15.6 IAM topology with organization class segments for maturity

These technologies in turn require a certain IT infrastructure. It also becomes clear from the White book that implementing PETs requires a solid foundation in the form of Identity and Access Management in order to minimize the use and access to sensitive personal data. With the help of Identity and Access management, PETs tries to minimize the use of and access of sensitive personal data. Especially the use of the PETs that secure access makes this clear. Secured Access however is only the first step for PETs. Privacy Enhancing Technologies also strive to segregate sensitive information in order to secure a person's identity. Not only segregation however is used to achieve this goal. Depending on the organizational information needs, information can also be immediately removed after use or not even registered in the first place.

Along the maturity curve of IAM runs the S-shaped maturity curve of awareness for privacy protection (Hahn et al., 2008), although interviews with management of organizations indicate that this S- curve starts in a much later phase of the IAM S-curve.

If the rights to access can be bound to a certain group, profile, person or user within an organization then IAM can be used to make sure that the user or user group only gets access to the information for which they are authorized. IAM then can also be used to provide the means of identification to make sure that the right user gets access to the user profile that is authorized to access certain sensitive information. Next to user management, authentication management and authorization management, provisioning and monitoring and audit can also play an important part in a PETs implementation. For instance when a central database of information is accessed by different organizations provisioning (automated or not) can play an important to keep user accounts for that database up to date at the different locations. Monitoring and Audit plays an important role when reviewing the current status of user accounts and controlling if authorized users only are accessing data. Thus depending on the requirements of the organization on its PETs implementation a certain level of maturity is required for the relevant IAM processes.

For the implementation of PETs, certain maturity of the organization is required. It is highly unlikely that immature organizations will implement PETs, let alone that these organizations have any awareness of privacy protection. The level of maturity for IAM is a strong indication for the introduction of PETs in an organization (Fairchild and Ribbers, 2008).

Based on interviews of the management of large (multi-national) organizations it becomes clear that the choice of advanced PETs occur in the pro active and top class maturity segments (Borking, 2010) (See parallelogram in Fig. 15.7).

This leads to assume that there can be recognized three S-curves concerning the application of PETs: one for the adoption of PETs with as most important positive stimulating factor the pressure of the legislation which regulates the protection of personal data and the role of the recommending privacy supervisors (i.e. DPAs, Privacy Commissioners); one for the application of IAM processes where the maturity of the IAM processes must be high; and one for the integration of the protection of privacy with the company processes as reflected in GAP privacy level model running from the initial level till the optimal level (see Fig. 15.8).

Authentication Management	No authentication means	Arbitrarily formulated authentication requirements (authentication means are provided, adjusted and deleted on user request)	Authentication requirements based on a one time survey	Authentication Requirements based on continuous risk analysis	Authentication requirements based on continuous risk analysis and are continously adjusted
User Management	Double and inconsistent entries because of chaotic and ad hoc processes	Entries can be double but they are consistent	Central registration, Limited user group, manual procedures	Central registration, controlled authorization processes, manual procedures	Central real-time controlled authorization sources, automated procedures
Authorisation Management	No authorization matrixes, authorization is defined ad hoc	Authorization matrixes defined but are not updated	Authorization matrixes are updated periodically	Role Based Access Control used for critical applications	Role Based Access Control for all applications and continuous updated authorizations
Provisioning	Manual process locally	Limited Automated unreliable processed locally	Limited Automated but reliable processes locally	Limited Automated and reliable for multiple sources	Automated and reliable for multiple sources
Monitoring(Audit)	No responsibility delegated into a AO/IC organization	Sporadically delegated responsibility of AO/IC	Partial delegation of responsibility to AO/IC	Full responsibility to AO/IC	Full responsibility to AO/IC with periodic reporting
	Immature	**Starting-up**	**Active**	**Pro-Active**	**Top Class**

Fig. 15.7 Level of maturity for PETs in parallelogram

Initial	Activities are ad hoc, with: • No defined policies, rules, or procedures. • Eventually lower-level activities, not coordinated. • Redundancies and lack of teamwork and commitment.
Repeatable	The privacy policy is defined, with: • Some senior management commitment. • General awareness and commitment. • Specific plans in high-risk areas.
Defined	The privacy policy and organization are in place, with: • Risk assessments performed. • Priorities established and resources allocated accordingly. • Activities to coordinate and deploy effective privacy controls.
Managed	A consistently effective level of managing privacy, privacy requirements, and considerations is reflected in organization, with: • Early consideration of privacy in systems and process development. • Privacy integrated in functions and performance objectives. • Monitoring on an organizational and functional level. • Periodic risk-based reviews.
Optimizing	Continual improvement of privacy policies, practices, and controls, with: • Changes systematically scrutinized for privacy impact. • Dedicated resources allocated to achieve privacy objectives. • A high level of cross-functional integration and teamwork to meet privacy objectives.
	— Source: Hargraves et al 2003

Fig. 15.8 Generic privacy maturity levels

The three s-curves' combination results in Fig. 15.9:

As can be concluded from Fig. 15.9, the moment of decision for the adoption of PETs appears to be at the higher levels of the IAM maturity (organizations in the Top Class and Pro-Active maturity level, with the exception for organizations at the level: active that update authorization matrixes periodically) (Fairchild and Ribbers, 2008) and in the lower levels of privacy maturity, thus where IAM measures reach the level of PET measures. There are exemptions for those organizations that belong to the category of (micro/mini) SMEs where trust is a critical success factor, like in the medical profession, barristers, notaries etc. Although processes mentioned in

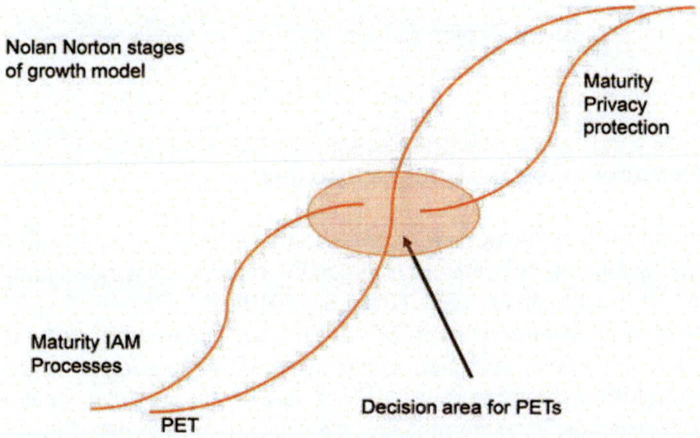

Fig. 15.9 Nolan Norton growth S-Curves concerning IAM-, Privacy en PETs (Borking, 2010)

the maturity model are non-existent in these organizations, it may be expected that these SMEs will protect personal information of their clients encrypted or will use rudimentary PETs tools.

15.11 Changing the Negative Adoption Factor of Costs into a Positive One

The perceived costs for PETs measures are an important negative adoption factor. However there isn't very much understanding about the business case for investment in PETs.

In order to best understand the likely adoption of PETs we must understand the challenges that privacy poses for organizations. This can be done best through engaging with experts and practitioners. To achieve this, the researchers conducted a number of consultations with industry experts, through direct discussions and by using a workshop-format (Fairchild and Ribbers, 2008).

Traditionally when the researchers put forward the question whether organizations have some inherent interest in privacy, a list of drivers emerges. These drivers include: compliance with legal obligation, fear of reputation damage from privacy failure, the need to generate trust with clientele, and the promotion of a good corporate practice. Yet if this was truly the case then privacy enhancing technologies would be already implemented everywhere across both industry and government organizations. Reality appears more complex (Fairchild and Ribbers, 2008).

But organizations need also to know what the business case of investments in PETs is (Economic motives for the use of PETs) in order to support the privacy protection required by the law and/or the policy of the firm. As many data are uncertain, scenarios have to be designed and assessed to give decision makers an understanding of the behavior of cost and benefit factors and their eventual effect on the business case outcome.

15.12 Business Case for Pets Investments

Investments in (risk reducing) PETs require insight into the costs and the quantitative and qualitative benefits. It is essential for the decision-making process concerning the investment for PETs (Borking, 2010).

The decision to spend money on privacy in any direction has to be financially justified. There is no point in implementing an expensive solution if a less expensive solution would offer the same risk reduction and because of that a better privacy protection. Beyond the legal compliance, it makes no sense to invest in a solution if its true costs are greater than the value it offers.

From the perspective of a business, privacy implies an investment to be measured in Euros saved as a result of reduced cost, or in additional revenues and profits from new activities that would not have occurred without an investment in privacy.

From the risk management literature a number of metrics have been identified to measure security risks; some of them apply to privacy risks as well (Fairchild and Ribbers, 2008).

15.13 Annual Loss Expectancy

One of the most common measures for the assessing the risk of a harmful event is Annual Loss Expectancy, or ALE. ALE is the product of the expected yearly rate of occurrence of the event times the expected loss resulting from the occurrence. Other yardsticks here are SLE and ARO. SLE stands for the Single Loss Exposure; this is the true cost of a security incident. ARO means annual rate of occurrence; this is the frequency in which a risk happens on a yearly basis. The annual loss expectancy foreseen from all of an organization's operations would be the sum of the expected yearly losses that could result from multiple (privacy) threats. Determining adequate inputs to this ALE equation is however very difficult, due to lack of statistical data.

For example if a bank estimates the probability of a serious security incident at one of its subsidiaries during 2008 as one in a million and the direct and indirect cost of such incident as 150 million Euros, the ALE created by the risk of this security incident for 2008 will be € 15 million times 1/1,000,000 = € 150. Of course the actual costs of this risk will never be that of the ALE, but it will be either € 0 or €150 million. In most cases the situation will be less certain and the probability or cost may range between one in five hundred thousand and one in a million and the cost may vary between € 100 million and € 200 million. The ALE would then be between: (€100 M or €200 M) × (1/5,000,001/1,000,000) = €100 or €400 (Blakley et al., 2002).

15.14 Return on Investment (ROI)

A metric that is quickly gaining in popularity is Return On Investments and specifically Return On Security Investments (ROSI) (Sonnenreich et al., 2006). Cardholm writes that: "Return on Investment (ROI) is a straightforward financial tool that measures the economic return of a project or investment. It is also known as return on capital employed. It measures the effectiveness of the investment by calculating the number of times the net benefits (benefits minus costs) recover the original investment. ROI has become one of the most popular metrics used to understand, evaluate, and compare the value of different investment options" (Cardholm, 2006)

> The equation is (Borking, 2010): ROI = [(Savings from safeguards) + (profits from new ventures)] / costs of safeguards = [ALE (baseline) – ALE (with safeguards) + (profits from new ventures)] / costs of safeguards.

Hereunder follows an example. Suppose an organization decides to implement a Privacy Management System (PMS). The business case could be substantiated as follows:

If PMS were not implemented, the minimum *annual* costs for a company employing 1,000 staff to comply with privacy policies are estimated as follows:

1. *Annual costs*

   Salary costs for Privacy Protection Officer (100% time allocation) Euro 100,000;
   Management and secretarial salary costs Euro 40,000;
   Costs for privacy audit Euro 30,000;
   Security costs with respect to privacy compliance (excluding essential information security) Euro 20,000;
   Report maintenance, regulations, settling registered people's rights, information, image and other damage, etc. Euro 20,000.
   This leads to the total annual costs of Euro 210,000.

   When comparing the situation where a PMS is used, the picture is as follows:

2. *Development and implementation of PMS*

   For the acquisition of PMS has to be paid: Euro 150,000;
   Consultancy for PMS implementation (60 days) costs Euro 80,000;
   Start-up costs after implementation Euro 20,000.
   The total one-off costs are Euro 250,000.

To these costs have to be added:

a. Annual costs PMS
b. PMS operational costs are Euro 30,000;
c. Maintenance costs are ± 15% of acquisition cost per annum: Euro 22,500;

d. Costs for privacy audit: Euro 10,000;
e. Salary costs for Privacy Protection Officer (50% time allocation) Euro 50,000; In this situation the total costs are Euro 112,500.

The saving per annum compared with the situation when there wasn't an investment in PMS is Euro 210,000–Euro 112,500–Euro 97,500. Thus the extra investment costs for PMS would be already fully recovered after approx. 2 years and 2 months.

15.15 Return on Security Investment (ROSI)

ROSI (Fig. 15.10) is a special application of ROI. The Return On Security Investments (ROSI) formula, developed by a team at the University of Idaho led by researcher HuaQiang Wei, is the most well known ROSI calculation in the security industry. They used what they found in the research area of information security investments and combined it with some of their own theories, assigning values to everything from tangible assets (measured in dollars with depreciation taken into account) to intangible assets (measured in relative value, for example, software A is three times as valuable as software B). Different types of attacks, or incidents, were assigned as individual costs. To verify the model, the team went about attacking an intrusion detection box they had built, to see if the costs the simulation produced matched the theoretical costs. They did. Determining the cost-benefit became the simple task of subtracting the security investment from the damage prevented. ROSI is an approach to look at the investment costs of security protection and the risk the investment removes. Assuming that the annual benefit of a security investment will be received throughout the lifetime of the investment, ROSI calculates the sum of the annual benefits over its cost. Benefits are calculated by adding expected cost savings to the new profit expected from new activities and sales.

Cardholm states that "it is basically a "saving" in Value-at-Risk; it comes by reducing the risk associated with losing some financial value" (Cardholm, 2006). Three core elements are determinative for the output calculation of the investment, namely: costs, turnovers and non-financial measurable elements. ROSI can be calculated using the equation below.

The earlier discussed ALE can also be written as: Risk Exposure multiplied with %RiskMitigated or Risk mitigated because of the investment in security (Borking, 2010).

The difficult parts in ROI method is determining ALE and SLE the risk-mitigating benefits of the security investment, since it is very difficult to know the

$$Rosi = \frac{(RiskExposure \bullet \%RiskMitigated) - SolutionCosts}{SolutionCost}$$

Fig. 15.10 ROSI equation (Sonnenreich, 2006)

true cost of a security incident. According to Sonnenreich et al. (2006) there is very little known about those costs, because very few companies track those incidents.

Cardholm has a better approach with less uncertainty. His calculation is as follows:

$$\text{ROSI} = \text{R} - (\text{R} - \text{E}) + \text{T},$$

or

$$\text{ROSI} = \text{R} - \text{ALE, where ALE} = (\text{R} - \text{E}) + \text{T}$$

The terms in Cardholm's equation can be described as:

- ALE: What we expect to lose in a year (Annual Loss Expectancy)
- R: The cost per year to recover from any number of incidents.
- E: These are the financial annual savings gained by mitigating any number of incidents through the introduction of the security solution.
- T: The annual cost of the security investment (Cardholm, 2006).

15.16 ROI for Privacy Protection

The ROI calculation methods can be applied also analyzing the return on investments that mitigates privacy risks, it means investments in PETs.

PETs investments differ from "normal" ICT investments, since the investment may not directly improve the workflow, or does not make a process more efficient. The costs from PETs are tangible and because of that are relatively easy to know. The benefits however are mostly intangible, because for example reputation improvement and a decreased risk for privacy incidents are not easy to quantify. However, these intangible benefits have the biggest value in a PETs investment.

Luckily, the value of risk mitigated can be calculated using the method of Darwin (2007). The Darwin Calculator can be found at www.tech-404.com/calculator.html.

The focus in this method will then be on the tangible benefits, the value of risk mitigated and the total costs, related to the PETs investments. This method will be named: Return on Privacy Investments (ROIPI).[4] How these figures will be calculated will be explained hereunder in more detail in the example of the Ixquick Europrise seal business case.

The formula is: ROIPI = {(Tangible Benefits +Value Of Risk Mitigated) – Total Costs} divided (/) by the total costs

When the ROIPI gives a positive result, it means that the investment is beneficial for the company since the benefits outweigh the costs. Note that if the value of risk

[4] Fritsch, 2008 and Dijkman, 2008 were the first that used the term ROPI. I prefer ROIPI preventing misunderstanding amongst auditors

mitigated is positive this also has a positive influence on the ROIPI. The strong point of this formula is that it is not necessary to derive at an accurate estimate. The ROIPI only has to be precise enough to support the decision-making.

ROIPI assumes that the organization will fully comply with the law. This isn't often the fact. Violation of privacy, i.e. the illegal use of personal data, generates a lot of revenue and the chance that violation will lead to a prosecution is almost nil, due to the lack of resources of the National Data Protection Authorities.

15.17 Ixquick

Ixquick is a meta search-machine. The website of Ixquick might be found at www.ixquick.com. Ixquick revenue model is the number of hits times the advertising benefits. The revenue is highly correlated to the search queries done through the site.

In 2003 and 2004, Internet traffic went down. In 2005, Internet traffic only went down with 5% and stabilized. In 2006 and 2007 the traffic increased again, due to the fact that Ixquick anonymized the IP addresses and search results in June 2006. Because of the anonymization, the traffic in 2006 and 2007 increased considerably. Due to the optimalization of the privacy protection of the users of the Ixquick meta search engine, triggered by the requirements for obtaining the EuroPrise privacy certificate,[5] the number of visitors of the website increased again substantialy in 2008, thanks to the investment in the PETs tool anonymization. With the increased traffic the revenue od Ixquick went up as well.

The reason of Ixquick for using PETs was that it is a unique selling point; Ixquick became and is still the first fully anonymized meta search engine. Besides this reason the other driver was privacy risk minimalisation.

The investment costs for the PETs tools were Euro 129.800, inclusive the extra investments needed for meeting the requirements of the EuroPrise certificate. The expenditure for the optimalized privacy protection amounted to € 37.000 for the technical and legal expertise. For press releases and communication costs announcing the Euro Prise privacy certificate award in July 2008 (Andriessen, 2008) € 8.000 was spent. The mentioned costs were non-recurrent one-off expenses.

Moreover there are also recurring costs for the maintenance and the further development of the system amounting to € 16.500 per year. The total costs for the whole PETs investment was: € 183.300.

The ROIPI equation can now be used for calculating whether Ixquick's privacy protection investment was the right decision of Ixquick's management.

ROIPI = {(Tangible Benefits + Value Of Risk Mitigated) – Total Costs} / (divided) by the total costs.

[5] http://www.european-privacy-seal.eu/about-europrise/fact-sheet

The total PETs costs are Euro 183.300. The tangible benefits of using PETs tools are the extra revenues in because of the increased data traffic. The directly tangible advantage for Ixquick due to the use of PETs for the period of PETs investments (2005–2008) is estimated by the author[6] at Euro 345.800. To estimate the factor "risk mitigated" the calculation tool of Darwin (2008) has been used. It will be assumed that in a privacy incident 10.000 records were stolen. Based on the daily users of the Ixquick search machine, the actual risk was much higher. The risk class of this data is of risk class II according to the guideline of the Duch Data Protection Authotity (CBP) (Borking, 2003; Hes and Borking, 2000; Van Blarkom et al., 2003; Communication from the Commission to the European Parliament and the Council on Promoting Data Protection, 2007) since the data consist of searches, these can consist of IP address, social security numbers and credit card numbers.

Based on the Darwin calculator (2008) the value of risk mitigated is Euro 1.050.300 on the 80% level (loss of 10.000 records) and the Dollar/Euro exchange rate in November 2008.

Using the values, the ROIPI equation produces as result:

Total Costs= Euro183.300
Tangible Benefits= Euro 345,800
Value Of Risk Mitigated= Euro 1.050.300
The intangible costs and benefits are appreciated as Euro 0.

Thus

$$\mathrm{ROIPI} = \{(345.800 + 1.050.300 + 0) - 183.300\}/183.300$$
$$= \mathrm{ROIPI} = 66,165 = \text{approx. } 662\% \text{ of the PETs investment.}$$

As this ROIPI value is very high, the conclusion is that the investment is very worthwhile. This number is also very high because of the value of risk mitigated. The ROIPI equation is especially preferable for SMEs because of its simplicity. This formula is a quick and reliable indicator whether the investment is worthwhile.

The intangible costs and benefits have been appreciated as zero euro, but if these intangible elements would be calculable, then the result would be even more favorable. However the ROIPI value is here significantly large enough to carry out the PETs investment and to justify the investment from a business economy point of view.

Others advocate rightfully that organizations should discard the above equations and instead use discounted cash flow methods for investments that have different costs and benefits in different years. The theoretical flaw in ROI (and so in ROSI, ROIPI and related approaches) is that it processes financial figures irrespective of the dates that will be received or paid. The value of 1 euro today is not the same as of 1 Euro in 2 years time. The Discounted Cash flow methods (DCF) encompass two

[6]The real financial figures are confidential

separate methods, the internal rate of return (IRR) and the Net Present Value (NPV). The allotted space for this chapter doesn't allow elaborating on the IRR method.

15.18 Net Present Value (NPV)

The Net Present Value (NPV) of a project or investment is defined as the sum of the present values of the annual cash flows minus the initial investment. The annual cash flows are the Net Benefits (revenues minus costs) generated from the investment during its lifetime. These cash flows are discounted or adjusted by incorporating the uncertainty and time value of money. NPV is one of the most robust financial evaluation tools to estimate the value of an investment (Cardholm, 2006).

The calculation of NPV involves three simple yet non trivial steps. The first step is to identify the size and timing of the expected future cash flows generated by the project or investment. The second step is to determine the discount rate or the estimated rate of return for the project. The third step is to calculate the NPV using the equations shown below:

$$\text{NPV} = \text{initial investment} + (\text{Cash flow year 1 divided by } (1 + \text{r})^1) \ldots (\text{Cash flow year 1 divided by } (1 + \text{r})^{\text{n}})$$

Or

$$\text{NPV} = \text{Initial investment} + \sum_{t=1}^{t=\text{end of project}} \frac{(\text{Cash Flows at Year } t)}{(1+r)^t}$$

The meaning of the terms is as follows:

- Initial investment: This is the investment made at the beginning of the project. The value is usually negative, since most projects involve an initial cash outflow. The initial investment can include hardware, software licensing fees, and start-up costs.
- Cash flow: The net cash flow for each year of the project: Benefits minus Costs.
- Rate of Return (r): The rate of return is calculated by looking at comparable investment alternatives having similar risks. The rate of return is often referred to as the discount, interest, hurdle rate, or company cost of capital. Companies frequently use a standard rate for the project, as they approximate the risk of the project to be on average the risk of the company as a whole.
- Time (t): This is the number of years representing the lifetime of the project.

Experts are convinced that a company should invest in a project only if the NPV is greater than or equal to zero. If the NPV is less than zero, the project will not provide enough financial benefits to justify the investment, since there are alternative investments that will earn at least the rate of return of the investment (Cardholm, 2006).

Ribbers developed a specific NPV equation for investments in PETs. Within the context of the NPV method, the following data have to be collected:

- the *initial investment in privacy protection* [I(p)], which encompasses cash outlays for Privacy Risk Analysis, process modeling, PETs, implementation of PETs, productivity loss, change management.
- the *yearly recurring cashflow*, which contains all yearly financial effects of the proposal. This calculation bears on an analysis of expected cashflow patterns that would occur with and without the investment; it reflects a difference between two defined situations. The so-called "without" situation will usually be the continuation of the current situation. This can for example be a situation with existing privacy protection in place, where the added value of PETs is considered. The "without" situation might also be a situation without any privacy protection. The definition of the "without" situation depends on the starting position of the decision-maker.

Ribbers proposes to take into account the following cash flow components: Annual Loss Exposure (ALE), Reputation Recovering Costs (RRC), Expected Revenue Accrual (ERA), Recurring Privacy Costs (RPC) (Fig. 15.11) (Fairchild and Ribbers, 2008).

ALE is the multiplied projected costs of a privacy incident and its annual rate of occurrence. Basically this encompasses revenue losses, legal claims, and productivity losses because of privacy breaches, repair costs and lost business.

RCC contain those expenses needed to restore the reputation of the company damaged by privacy breaches; examples are additional costs for PR and Marketing. Moreover if a privacy breaches affects the share price of the company (see ChoicePoint, Double Click), possibly breaches affects the share price of the company (see ChoicePoint, Double Click), banks and other financial institutions may require possibly additional financial guarantees.

ERA represents, on the positive side, possible marketing impacts on market share and revenue of publicized implementation of PETs.

RPC contains the yearly (additional) privacy costs caused by the proposal; this will encompass needed privacy threat or impact analyses, audits, privacy officers etc.

As said, the analysis compares the project situation with the situation without the project. Basically this comes down to analyze the cash flow differences between the two situations. This can be done either by applying a factor RM (Risk Mitigated) to the situation without the investment or by subtracting the full-expected cash flow of the two situations from one another.

The RM factor for the applied privacy risk reducing/protection solution indicates what part of ALE and RRC has been compensated by the solution. Mitigated Risk is expressed as a reduction of the expected number of privacy breaches per year.

The resulting NPV of a privacy protection solution is consequently as follows:

$$NPV = - I(p) + \sum_{j=1}^{n} \{(ALE + RRC)\, RM + ERA - RPC\} / (1+i)^{j}$$

Fig. 15.11 Privacy investment net present value (Fairchild and Ribbers, 2008)

15.19 The Case of the National Victim Tracking and Tracing System (ViTTS)

The nation-wide implementation in the Netherlands of the Victim Tracking and Tracing System (ViTTS) is an important contribution to an effective disaster management. The system provides regional medical officials with a concrete support to execute their tasks, through access to the required relevant contextual information; it supports the allocation of injured persons to local and regional hospitals, and it provides the relevant competent authorities with necessary information. Moreover, municipalities will be better placed to execute mandatory registration procedures under the municipal disaster plan, and hospitals will be provided with timely information about the numbers of victims and the nature of their injuries. Due to the fact that sensitive personal medical information is processed about victims, the DPD requires optimal protection of such sensitive personal data. Privacy issues with respect to the health sector are particularly sensitive.

The EU PRIME[7] research team[8] has applied the NPV calculation approach in several case studies. One of the case studies is ViTTS. The following data have been collected from ViTTS.

The initial investment in privacy protection *I(p)* comprises the following components:

– System analysis and design, prototyping, test runs: Euro 15,000
– Privacy audit and Privacy risk assessment: Euro 50,000
– Smart Cards for on line authentication and encryption: Euro 25,000
– Implementation costs of PETs measures: Euro 80,000

The total initial investment in reducing the risks of privacy incidents: Euro 170,000

Privacy breaches affecting the process of handling victims would have serious consequences and should be at all cost avoided. The privacy threat analysis showed that without privacy protection the VITTS system would undergo privacy breaches on a regular basis. The damage that would result from that can be estimated as follows.

---

[7]PRIME (Privacy and Identity Management for Europe) Contract No. 507591 Research periode 2004–2008

[8]The PRIME researchers were P.Ribbers (UoT), A.Fairchild (VUB), J.Tseng (EUR), R-J.Dijkman (UoT) and J.J.Borking (BC)

The direct consequence of a breach (SLE – Single Loss Exposure) would be loss of reputation of the national government, possible wrong allocation of victims to hospitals with ineffective treatment and possibly deceases as a consequence. This may lead to significant legal claims. Claims of Euro 100,000 per case are not exceptional.

Such a breach would necessitate a nation-wide roll out of system adaptations: for which is needed two man-months per designated preventive health care safety region at Euro 100 per hour:

Total costs	Euro 347,000
Test and Trials to prove effectiveness of the system:	
Euro 80,000 per region:	
Total cost	Euro 800,000
Training and education roll out:	Euro 50,000
The total recovering costs (RCC) would amount to:	Euro 1,197,000

The expected revenue accrual (ERA) can be estimated as follows. The most important reason for designated preventive health care safety regions to adopt the system is the built-in optimal privacy protection. So without privacy protection or with a much less rigid privacy protection there wouldn't have been developed such a system.

The estimated salary costs to replace the system by manual procedures would amount to 3 FTEs per region, which amounts to Euro 180,000 per region.

Nationwide this would result in a cost of:	Euro 1,800,000
The total benefits of protecting privacy and reducing the risks of privacy incidents can be estimated at:	Euro 2,277,000
(in this number legal claims are not included)	

Scenario

For the NPV calculation it is assumed:

1. a time horizon of 6 years
2. a serious privacy breach every 2 years
3. a cost of capital of 5%

Applying the equation results into the following:	
I(p):	Euro 170,000

Recurring cash flows:

– costs avoided every 2 years:	Euro 2,277,000
– yearly recurring privacy costs:	Euro 400,000
– privacy costs in year 3 (no costs in year 6 given the assumption):	Euro 25,000

Under this assumption the calculation would be as follows:

$$\text{NPV} = -170{,}000 + 2,277,000(0.9707029 + 0.822702 \\ +0.710681) - 25,000(0.863838) - 400,000\ (5.242137) = \text{Euro}+ \\ 3,268,368$$

This (positive) business case does not include possible legal claims.

The business case for the investment mitigating the risk of privacy incidents is positive. Other scenarios lead to a positive business case as well. The privacy protection will even be profitable under the unrealistic assumption of a privacy breach only occurring once (and taking legal claims into account).

15.20 Conclusion

In this contribution the causes have been discussed why PETs, compared to the millions of computer systems which process personal data, hardly is used and that organizations trust on rather traditional organizational and technical data protection measures. The adoption of PETs by an organization appears be influenced by a large number of factors and the level of maturity of that organization.

S-curves for Identity and Access Management, for the maturity of organizations, for privacy protection and for the application of PETs itself, give an explanation for the slow application of the PETs solutions to adequately protect personal data. When the positive adoption factors are exploited belonging to the general PETs S-curve for promoting PETs, then a faster adoption of PETs by organizations which a large intensity of information processing (thus more need for privacy protection) may be realized. Good education concerning the technical possibilities of PETs and concrete requirements in the legislation (such as a privacy impact (PIA) or threat analysis assessment is necessary for promoting the PETs applications. If the legislation would stipulate the option that users should be in the position to choose for approaching services anonymously, then the use of PETs measures would be stimulated. In summary only legal and regulatory pressure (and the promotion by such advisory or supervisory bodies as the data protection agencies (DPA)) with regard to privacy protection is perceived to-date as having an undivided positive impact on the adoption process.

Costs for investment in PETs is an important negative adoption factor. This negative adoption factor can be converted into a positive one. The ROI and NPV calculation methods are useful tools for management for assessing the (planned) investments in PETs, reducing the risks of privacy incidents considerably.

The ROI and NPV calculation methods are useful tools for management for assessing the (planned) investments in PET, reducing the risks of privacy incidents considerably.

ROI, ROSI and ROIPI provide useful insights. For a "quick and dirty" assessment of a PET investment ROIPI is useful especially for SMEs, like in the Ixquick business case. However ROIPI and other ROI methods are based on evaluating reductions in risks and do not take a time factor into account. The best approach would be to consider investments in PET as regular investments, characterized by cash flow patterns.

The Net Present Value approach is applied on the ViTTS case. This approach is effective in the context of assessing investments in PET, reducing privacy risks and enhancing privacy protection.

As many data are uncertain due to the lack of recording privacy incidents, scenarios have to be designed and assessed to give decision makers an understanding of the behavior of cost and benefit factors and their eventual effect on the business case outcome. A mandatory disclosure and registration of privacy incidents as foreseen in the modification of the EU Directive 2002/58/EC, will contribute to this end, provided these disclosures will be recorded in an European register accessible for every citizen (Bayer and Melone, 1989).

References

Andriessen, V. "Nederlandse Internetzoekmachine Ixquick ontvangt eerste Europese privacycertificaat." In *Het Financieele Dagblad*,15 juli 2008.

Bayer, J., and N. Melone. "A Critique of Diffusion Theory as a Managerial Framework for Understanding Adoption of Software Engineering Innovations." *The Journal of Systems and Software* 9, 2 (1989): 161–166.

Blakley, B., E. McDermott, and D. Geer. Information management is Information Risk Management. in *Proceedings NSPW'01*, Cloudecroft, New Mexico, 2002.

Borking, J. "The Status of Privacy Enhancing Technologies." In *Certification and Security in E-Services*. E. Nardelli, S. Posadziejewsji, and M. Talomo. Boston, MA: Kluwer Academic Publishers, 2003, p. 223.

Borking, J.J. "Assessing investments mitigating privacy risks." In *Het binnenste buiten, Liber Amicorum ter gelegenheid van het emeritaat van prof. dr. H.J. Schmidt, hoogleraar Recht en Informatica te Leiden*, edited by L. Mommers, H. Franken, J. Van den Herik, F. Vander Klauw, and G-J. Zwenne. Leiden: Leiden University Press, 2010.

Borking, J.J. "The Business Case for PET and the EuroPrise Seal," Report for EuroPrise EU Research project on Privacy Seals 2008; http://www.european-privacy-seal.eu/about-europrise/fact-sheet

Borking, J.J.F.M. *Privacyrecht is Code, Over Het Gebruik van Privacy Enhancing Technologies*. Deventer: Kluwer, 2010.

Bos, Tj. *Adoptie van privacy-enhancing technologies bij publiek/private instellingen*. Den Haag: Ministerie van Binnenlandse Zaken, 2006.

Camp, L.J., and C. Wolfram. Pricing Security. in *Proceedings of the CERT Information, Survivability Workshop*, Kluwer Academic Press, Boston MA, 2000.

Cardholm, L. Adding Value to Business Performance Through Cost Benefit Analyses of Information Security Management, Thesis, Gävle, 2006.

Cas, J., and Ch. Hafskjold. *Access in ICT and Privacy in Europe, Experiences from technology assessment of ICT and Privacy in seven different European countries*. Geneva: EPTA, 2006, p. 41.

Chapman, S., and G.S. Dhillon. Privacy and the Internet: The Case of DoubleClick, Inc. – Social Responsibility in the Information Age: Issues and Responsibilities. XXX, Fort Lauderdale-Davie, 2002.

Communication from the Commission to the European Parliament and the Council on Promoting Data Protection by Privacy Enhancing Technologies (PETs), COM (2007) 228 Final, Brussels, 2.5.2007.

Darwin. www.tech-404.com/calculator.html, 2007, 2008.

Davenport, Th.H. *Process Innovation – Reengineering work through Information Technology.* Boston, MA: Havard Business School Press, 1993.

Fagerberg, J. et al. *The Oxford Handbook of Innovation.* New York: Oxford University Press Oxford, 2005.

Fairchild, A., and P. Ribbers. "Privacy-Enhancing Identity Management in Business." In *Privacy and Identity Management for Europe*, edited by J. Camenish, R. Leenes, and Sommer, D. Report for the EU Commission X, Brussels, 2008, pp. 69–100.

Fichman, R.G. Information Technology Diffusion: A Review of Empirical Research. in *Proceedings of the Thirteenth International Conference on Information Systems (ICIS),* Dallas, (1992), pp. 195–206.

Greenhalgh, T., et al. "Diffusion of innovations in service organizations: Systematic review en recommendations." *The Milbank Quarterly* 4 (2004): 581–629.

Hahn, U., K. Askelson, and R. Stiles. *Managing and Auditing Privacy Risks*, ltamonte Springs, 2008 http://www.theiia.org/guidance/technology/gtag/gtag5/

Hes, R., and J. Borking. *Privacy Enhancing Technologies: The Path to Anonymity (2nd revised edition).* Report from the Dutch Data Protection Authority AV no. 11 Den Haag, 2000.

Jeyaraj, A., J.W. Rottman, and M.C. Lacity. "A review of the predictors, linkages, and biases in IT innovation adoption research." *Journal of Information Technology* 21, 1 (2006): 1–23.

Koorn, R., H. Van Gils, J. Ter Hart, P. Overbeek, P. Tellegen, and J. Borking. *Privacy Enhancing Technologies – Witboek voor Beslissers*; (Whitebook for Decision Makers – Ministry of Internal Affairs and Kingdom Relations) Den Haag, 2004.

Leisner, I., and J. Cas. *Convenience in ICT and Privacy in Europe, Experiences from technology assessment of ICT and Privacy in seven different European countries.* Geneva: EPTA, 2006, p. 50.

OECD Organization for Economic Co-operation and Development, *Oslo Manual: Guidelines for Collecting and Interpreting Innovation Data, 3rd edition*, 2005. Available at: http://www.oecd.org/

PRIME, acronym for project name: Privacy and Identity Management for Europe, Contract No. 507591 Research period 2004–2008

Privacy Rights Clearinghouse: A Chronology of Data Breaches. 2007. Available at: http://www.privacyrights.org/ar/ChronDataBreaches.htm#CP

Rivera, M.A., and E.M. Rogers. "Evaluating public sector innovation in networks: extending the reach of the national cancer institute's web bases health communication intervention research initiative." *The Innovation Journal: The public Sector Innovation Journal* 9, (2004): 1–5.

Rogers, E.M. *Diffusion of Innovations.* New York: Simon & Schuster, 2003.

Smit, N. *A maturity Model.* Zoetermeer: EUR, 2005.

Sommer, D. "The PRIME Architecture." In *Privacy and Identity Management for Europein*, edited by J. Camenish, R. Leenes, and D. Sommer. Report for the EU Commission, Brussels, 2008.

Sonnenreich, W., J. Albanese, and B. Stout. "Return on Security Investment (ROSI) A practical approach." *Journal of Research and Practice in Information Technology* 38, 1, (Feb. 2006).

Stanford Organizational Maturity Levels, http://www2.slac.stanford.edu/comp/winnt/systemadministration/Organizational%20Maturity%20Levels.doc

Tidd, J., et al. *Managing Innovation: Integrating Technological, Market and Organizational Change*, Chichester: Wiley, John & Sons, Incorporated, 2005.

Tsiakis, T., and G. Stephanides. "The Economic Approach of Information Security." *Computers & Security* 24, 2005.

Tung, L.L., and O. Reck. "Adoption of electronic government services among business organizations in Singapore." *Journal of Strategic Information Systems* 14, (2005): 417–440.

Van Blarkom, G.W, J.J Borking, and J.G.E Olk. *Handbook of Privacy and Privacy-Enhancing Technologies, The Case of Intelligent Software Agents*. The Hague: College Bescherming Persoonsgegevens, 2003, pp. 22–30.

Van Gestel, G.P.C. Creating an Identity and Access Management Maturity Model, Thesis, Universiteit van Tilburg, Tilburg, 2007.

Vandecasteele, J., and L. Moerland. *Groeimodel voor IV-functie – Het systematisch weergeven van een herinrichtingproces*. Amstelveen: KPMG Management Consulting, 2001.

Part IV
Privacy and Data Protection in the Cloud

Chapter 16
Can a Cloud Be Really Secure? A Socratic Dialogue

Gurpreet Dhillon and Ella Kolkowska

16.1 Prologue

Cloud Computing has indeed emerged as a new buzz-word. However the concepts therein are not new. In fact in 1961 John McCarthy, computer scientist famed for logic programming which resulted in LISP (List Processing Language),[1] publicly suggested at a MIT centennial lecture that computing power and specific applications could be sold through the utility business model, just like water or electricity. Cloud Computing in many ways is like Grid Computing that is characterized by large scale distributed computing where a pool of abstracted, virtualized, dynamically scalable, managed computing power, storage, platforms and services are delivered on demand to external customers.[2]

While in the literature a number of Cloud Computing related information security challenges have been postulated, there is little consensus or appreciation of their particular nature and scope. Some researchers[3] consider Cloud Computing to be inherently unreliable, others[4] appreciate the ability to centralize computing

G. Dhillon (✉)
Virginia Commonwealth University, Richmond, VA, 23284-2512, USA
e-mail: gdhillon@vcu.edu

[1]McCarthy, J. "Recursive Functions of Symbolic Expressions and their Computation by Machine, Part I," *Communications of the ACM* 3, 4 (1960): 184.

[2]Foster, I., at al., "Cloud Computing and Grid Computing 360-Degree Compared," in *Grid Computing Environments Workshop 2008, GCE '08* (2008);Luis, V.M., at al., "A Break in the Clouds: Towards a Cloud Definition," *SIGCOMM Computer Communication Review* 39, 1 (2009): 50.

[3]Balachandra, R.K., R.V. Paturi, and A. Rakshit, "Cloud Security Issues," in *IEEE International Conference on Services Computing* (IEEE, 2009).

[4]Beaty, K., et al., "Desktop to Cloud Transormation Planning," in *2009 IEEE International Symposium on Parallel & Distributed Processing* (IEEE, 2009) ; Descher, M., et al., "Retaining Data Control to the Client in Infrastructure Clouds," in *2009 International Conference on Availability, Reliability and Security* (2009).

S. Gutwirth et al. (eds.), *Computers, Privacy and Data Protection: an Element of Choice*, DOI 10.1007/978-94-007-0641-5_16, © Springer Science+Business Media B.V. 2011

resources as a boon. Nevertheless, issues such as data location, data segregation, recovery, and long term viability have often emerged as most contentious.[5]

In this paper we engage in a Socratic Dialogue to understand at least two viewpoints with respect to security in Cloud Computing. Socratic Dialogues have been used in the literature to uncover and address conflicting viewpoints.[6] The dialogue takes place between two scholars of information security – one a proponent of the socio-organizational perspective in managing information security and the other who believes more in the technical solutions for ensuring information security. The two scholars are also geographically dispersed – one who is contextualized in the North American psyche and the other in the Scandinavian school of thought. In many ways the opposing viewpoints represent the divergent "management cults". The ensuing dialogue helps in understanding the perspectives and perhaps in building alliances between the social and the technical.

16.2 The Dialogue

Gurpreet: My dear Ella, you have often talked about our ability to technically secure and manage the information resources. While I do believe that it may be possible to achieve some success through technical means, an exclusive reliance of technical solutions for security can actually be an impediment. Did you know of the recent Sidekick debacle? On Oct 1, 2009 T-Mobile announced that every user of Sidekick data services had lost their private personal records (emails, contacts etc). Later Microsoft/Danger announced that they had been able to recover some of the data from their servers and blamed system failure in a core database and backup as a reason for the failure. Clearly this incident suggests that there are issues with reliability of "far-off-servers" that operate in the "Cloud". I am not sure if it will ever be possible to ensure security of the Cloud.

Ella: I have heard about the incident and I understand its seriousness for the effected users. I am sure they would be really frustrated as well. However didn't we have data losses prior to Cloud Computing? For example in 2005 Citigroup lost control of back-up tapes containing names, account history, and social security numbers of more than 3.9 million customers while shipping the tapes via UPS.[7] Furthermore, in 2008 different governmental departments have lost nearly 30 million records

---

[5]Kaufman, L.M. "Data Security in the World of Cloud Computing." *IEEE Security & Privacy Magazine* 7, 4 (2009): 61.

[6]Oliva, T.A., and C.M. Capdevielle. "Can Systems Really Be Taught: (A Socratic Dialogue)." *Academy of Management Review* 5, 2 (1980): 277; Mitroff, I.I. "The Tally: A Dialogue on Feyerabend and Ford." *Theory and Society* 3, 4 (1976): 601. Among others.

[7]CNN, "Info on 3.9 M Citigroup Customers Lost Computer Tapes with Information About Consumer Lending Lost by UPS in transit to Credit Bureau," *CNNMoney.com*, 2005, http://money.cnn.com/2005/06/06/news/fortune500/security_citigroup/ (10 September 2010).

as a consequence of missing laptops, USB sticks and CDs.[8] Therefore I think the problem of data loss is not greater in the Cloud than it was before. In fact I think individual users and companies have the same risk to data loss and system crashes as before. However the difference is that if it happens in large enterprises it gets more publicity.

You can't stop servers from crashing. You can only provide high quality backups. However in the case of Sidekick the need of backups was apparently overlooked. The question is, who should be responsible for doing backups: the Cloud provider or the user?

Gurpreet: Ella, that is exactly the point. Someone somewhere becomes complacent and then everybody else has to suffer. Simply moving data to the Cloud does not preclude our responsibility towards our own data. While service providers do ensure safety of our data, they are also not solely responsible for it. Providing Cloud Computing services is after all a profit making business. Hence the utility derived from Cloud Computing (calculated in terms of user hours) the revenue has to be greater than that of managing data in a traditional datacenter.

Ella, Cloud Computing service providers work on a very simple formula:

$$UH_{cloud} \times (R - C_{cloud}) \geq UH_{datacenter} \times (R - (C_{datacenter} \div U))$$

Where, UH_{cloud} is the hours a user spends in the cloud, R is the revenue and C_{cloud} is the cost in providing the services. On the right hand side, the costing is for a fixed-capacity datacenter, where $UH_{datacenter}$ is the user hours in the datacenter, $C_{datacenter}$ is the cost of running the datacenter and U is the utilization. As Armbrust et al.[9] argue, it is more cost effective to provide services in a cloud. However there are significant costs associated with ensuring security, particularly for maintaining confidentiality of data. So the emphasis is always on defining the most efficient solution.

My feeling is that where profit maximization is the motive, security and confidentiality do not necessarily drive the investment objectives. And at times it is virtually impossible to enforce confidentiality rules. Remember the case of "Do Not Call Registries" in the US.[10] Because of the hidden interconnectivity of the personal data and service providers for a given individual, it became virtually impossible to compartmentalize service providers who could call a given household and those who could not. While there was initial success in curbing unwanted phone calls from telemarketers, the interconnected web of databases, service providers and access rules

[8] Best, J. "Lost Data Total Nears 30 million records," (2008), http://www.silicon.com/publicsector/0,3800010403,39295167,00.htm (10 September 2010).

[9] Armbrust, M., et al., *Above the Clouds: A Berkeley View of Cloud Computing.* Berkley, CA, 2009.

[10] In June 2003 the US Federal Trade Commission opened the "Do Not Call Registry" to comply with the Do-Not-Call Implementation Act of 2003. The Act allows for companies to make calls up to 18 months where there is an existing business relationship. This period can easily be extended for any amount of time with a range of merger and acquisition tricks and other loopholes.

to maintain confidentiality simply got overwhelmed to the point where maintaining the "Do Not Call Registry" simply became meaningless.

Ella: I think you have a point here, my dear Gurpreet, indeed data retention and data mining can be seen by the service providers as a profitable business. Conti[11] stresses that Cloud providers, like other companies mostly care about their profits and interest of their shareholders. As a result, user needs get ignored. This can certainly lead to reduction of data confidentiality.

However most of the Cloud providers ensure confidentiality of data in their security and privacy policies, for example Google states that they have extensive policies, procedures and technologies in place to ensure the highest levels of data protection in all their services and that they follow principles for protection stated in U.S. Department of Commerce Safe Harbor-program[12]. Confidentiality of data is often ensured by well-known and commonly used methods such as VPN, encryptions and access control. Of course confidentiality of data stored and transmitted in the Cloud varies with both the design of the system and how well the safety measures are implemented by the Cloud providers.[13] Some services encrypt information both in transit and in storage in such a way that only the owner can decrypt it,[14] while others focus on encryption of transmitted data and do not ensure encryption in storage.[15] The problem is that Cloud Computing solutions allow data to be sent and stored everywhere, even around the world. While this makes it possible to reduce costs and maximize performance, the risks are higher since sensitive data can be stored in places where data protection regulations are insufficient. There is no doubt that for most enterprises, data confidentiality and data protection is the biggest barrier in Cloud Computing, largely because of the sensitive and confidential nature of the data.[16] Thus, perhaps sensitive data shouldn't be sent to the Cloud!

Gurpreet: Yes Ella that is one of the biggest challenges in Cloud Computing. Did you know in a 2008 study, Saikat et al.[17] note that privacy in the Cloud is economically determined by who pays for the services? In many situations, such as Google, advertisers, rather than users pay for the Cloud. Hence typically advertiser

---

[11]Conti, G. *Googling Security: How Much Does Google Know About You?* Addison-Wesley Professional, 2009.

[12]Google privacy center, "Privacy Policy", Last modified: March 11, 2009, http://www.google.com/privacypolicy.html (10 September 2010).

[13]Kaufman, L.M. "Data Security in the World of Cloud Computing." *IEEE Security & Privacy Magazine* 7, 4 (2009): 61.

[14]Descher, M., et al., "Retaining Data Control to the Client in Infrastructure Clouds," in *2009 International Conference on Availability, Reliability and Security* (2009).

[15]Tian, X., X. Wang, and A. Zhou, "DSP RE-Encryption: A Flexible Mechanism for Access Control Enforcement Management in DaaS," in *2009 IEEE International Conference on Cloud Computing* (2009).

[16]Balachandra, R.K., R.V. Paturi, and A. Rakshit, "Cloud Security Issues," in *IEEE International Conference on Services Computing* (IEEE, 2009).

[17]Saikat G., K. Tang, and P. Francis. "NOYB: Privacy in Online Social Networks." in *Proceedings of the first workshop on Online social networks*, Seattle, WA, USA (2008).

interests take priority over conventional wisdom (for instance the principle of least privilege gets overlooked).

Ella: I understand what you mean, if the advertisers pay for the Cloud, privacy needs might be overlooked. In fact, ensuring privacy is of significant concern in Cloud Computing[18]. Some problems arise because of the very nature of the Cloud. Data is sent, divided and stored at different places around the world and privacy laws vary significantly.[19] Other problems are related to lack of reliable user authentication solutions for digital identity management. Experts claim that without proper identity management and trustworthy authentication solutions, it will be impossible to assure individual users that their private data are secure in the Cloud.[20]

I believe that in the future, societal pressures and individual users will be able to force Cloud providers to comply with privacy regulations. Did you hear that The Electronic Privacy Information Centre (EPIC) complained to the US Federal Trade Commission about Google's insufficient safeguarding of the users confidential information[21]? This accident could actually lead to Google online services being closed or at least modified. I hope that privacy will be improved in the Cloud in the future, but currently privacy can certainly not be guaranteed.

Gurpreet: I had heard about the EPIC complaint. While I still believe that individuals and organizations are better custodians of their data than someone else, there is a related problem of managing integrity of data, especially when there is loss of control as to who has access and who can change it. There is also a fine balance between security, usability and availability. A good example is Google Health, where users can manage their own health profiles. Google allows users to import details of doctor visits from participating health care providers. There is no doubt that the Google Health Cloud is an interesting technological advancement, giving users control of their own health records. However problems emerge when the custodian of data in the Google Health Cloud is able to share the records with others. Not only are we relying on the users to be security aware, but also be discerning enough to share information with people who matter. Proponents of Could Computing argue that centralization helps manage data confidentiality and integrity. I argue that the converse may also be the case.

Armstrong et al.,[22] for example have argued that categorizing data into geographic spaces makes data vulnerable. In the context of health care, they argue:

---

[18]Pearson, S. "Taking Account of Privacy when Designing Cloud Computing Services," in *2009 ICSE Workshop on Software Engineering Challenges of Cloud Computing* (2009).

[19]Balachandra, R.K., R.V. Paturi, and A. Rakshit, "Cloud Security Issues," in *IEEE International Conference on Services Computing* (IEEE, 2009).

[20]Europe's Information Society, "eHealth" (2005) http://ec.europa.eu/information_society/activities/eten/library/about/themes/ehealth/index_en.htm (10 September 2010).

[21]Nuttall, C. "US urged to probe Google's 'cloud' services," (2009), http://www.ft.com/cms/s/0/55572a2e-1425-11de-9e32-0000779fd2ac.html?nclick_check=1 (10 September 2010).

[22]Armstrong, M.P., G. Rushton, and D.L. Zimmerman. "Geographically Masking Health Data to Preserve Confidentiality." *Statistics in Medicine* 18, 5 (1999): 497.

> The greater the area of geographic space that we can eliminate as the source of any health events, the greater is the risk of disclosure. Conversely, the greater the number of other records that one can show to have possibly originated in the same area from which a particular record might have come, the less is the risk of disclosure (p. 521).

Services such as Google Health or Microsoft's Health Vault increase sources of health-related micro-data. And there is a demand by public health authorities and researchers to not only have access to this data for policy formulation and combating epidemics (e.g. H1N1 in 2009), but at the same time assuring integrity and confidentiality. This indeed is a challenge.

Ella: Indeed it is a challenge to ensure data integrity in the Cloud. However do we really have a choice in not using the possibilities offered by Cloud provides? For example there is an ongoing discussion within the EU parliament about an eHealth Platform for the deployment of internet-based technologies for patient monitoring and exchange of medical records, both nationally and internationally, between all member countries. Cloud Computing and the grid have been considered as an international Healthcare infrastructure[23]. It is said that such solutions will make the health sector more cost-effective and will ensure high quality[24]. Of course, with these solutions come risks, but don't you believe, my dear Gurpreet that there must be a way of reducing risks and gaining benefits?

You know, data security was earlier a major concern in any outsourcing arrangement, and today there are specific contracts that regulate and specify security requirements in outsourcing relationships. The outsourcing providers are also obligated to comply with the legal regulations related to the transfer and storage of a customer's data.[25] I believe that in the future data security will also be regulated in the Cloud, however today it is not the case. Clients of Cloud providers do not have the same possibility to negotiate terms and condition related to security. I think that so far data protection is not regulated in the Cloud. Customers have to be careful and check if the Cloud solutions keep them in regulatory compliance and ensure confidentiality and integrity of the data. The important thing is to evaluate the provider's solution regarding control of access to the data. However a precondition for ensuring access control is an effective and a reliable way of identifying people and companies in the Cloud. A number of scholars argue that finding digital identity services that realize a number of necessary requirements is actually of great concern in the Cloud.[26] Others claim that when proper digital identity services are adopted, sufficient access control can be implemented and privacy of the users can be ensured

[23] Europe's Information Society, "eHealth" (2005), http://ec.europa.eu/information_society/activities/eten/library/about/themes/ehealth/index_en.htm (10 September 2010)

[24] Europe's Information Society, "Information can save your life" (2007), http://ec.europa.eu/information_society/tl/qualif/health/index_en.htm (10 September 2010)

[25] Baker, R.K. "Offshore IT Outsourcing and the 8/sup thsup/Data Protection Principle – Legal and Regulatory Requirements – with Reference to Financial Services." *International Journal of Law and Information Technology* 14, 1 (2006): 1.

[26] Halperin, R., and J. Backhouse. "A Roadmap for Research on Identity in the Information Society." *Identity in the Information Society* 1, 1 (2008): 71.

and the full power of Cloud Computing can be utilized.[27] Finding a working digital identity in the information society is therefore a priority issue and a number of multidisciplinary research projects have been founded to address the concerns. For instance, within EU, PRIME – Privacy and Identity Management for Europe[28] and FIDIS Future of Identity in the Information Society[29] have made significant contributions.

Gurpreet: The more I think of Cloud Computing and the related security issues, the more scared I get. Recently I came across a paper by Nurmi et al.[30] where the authors present an open-source software framework for Cloud Computing. The authors describe the *Infrastructure as a Service* concept and suggest that it is possible to give users the ability to run and control entire virtual machine instances across various physical devices. While this seems like a rather interesting idea, my question is – who ensures that data is going to be made available for almost 100% of the time. Even Cloud Computing providers such as Google have experienced outage problems. For instance on July 2, 2009, Google reported[31].

> There was a serious issue in one of App Engine's datacenters with GFS, Google's low level storage system. GFS underlies Bigtable, which in turn underlies App Engine's Datastore. GFS also provides storage for our application serving infrastructure, so GFS unavailability caused problems for Datastore reads and writes, as well as application serving.

While it may seem prudent for companies such as Google to take upon a larger role, especially since they are setting themselves up as quasi-utility companies, in reality however such companies have shunned away from taking responsibility for their actions.

Ella: I cannot agree with you in this case, my dear Gurpreet. I argue that availability to the data and services has actually been improved with Cloud Computing. In the Cloud you have almost limitless flexibility in accessing different pieces of software and databases and the ability to combine them into customized services. You can also access your data wherever you can connect to the Internet and you can use different devices such as cell phones, PDAs, personal video recorders and many others. Then of course a service can be unavailable like the Google example you referred to. However Google promises its customers 99.9% uptime, which means that the e-mail service might be unavailable for about 9 hours in a year. According to Google's reports, its e-mail service were unavailable on an average of 10–15 min

[27] Cavoukian, A. "Privacy in the Clouds, A White Paper on Privacy and Digital Identity: Implications for the Internet," 2008, http://www.ipc.on.ca/images/Resources/privacyintheclouds.pdf (10 September 2010).

[28] PRIME. "Privacy and Identity Management for Europe," 2008, https://www.prime-project.eu/ (10 September 2010).

[29] http://identityproject.lse.ac.uk/identityreport.pdf (26 May 2010).

[30] Nurmi, D., et al., "The Eucalyptus Open-Source Cloud-Computing System," *Proceedings of the 2009 9th IEEE/ACM International Symposium on Cluster Computing and the Grid* (Volume 00, 2009).

[31] http://groups.google.com/group/google-appengine/browse_thread/thread/e9237fc7b0aa7df5?pli=1 (26 May 2010).

per month in 2008, which means that they kept the promise of 99.9% availability in 2008[32]. The fact is that most enterprises don't deliver higher reliability on their own systems; the difference is that outages on big services get more publicity.

Gurpreet: You certainly have a point Ella. I however feel that the onus resides with the data owners and/or companies that are using services of a Cloud provider. A provider may guarantee high availability of services, but it is more important to understand the internal functioning of the organization and assess as to how data is organized and organizational roles and responsibilities defined. It is unfortunate that in many cases companies are unable to differentiate between responsibilities associated with a role, accountable in performing the job function, and decision-making authority when things go wrong. Since Cloud Computing providers deal with significant amounts of data, definition of proper structures is paramount. However responsibility structures are poorly conceived within organizations. At best they are thought of as a kind of "coercive protection", which on the one hand aims to maintain the overall integrity in an organizational relationship and on the other aggressively seek to control actions through direct intervention.[33]

My problem, dear Ella, is that while it is important to establish proper structures of responsibility for ensuring security, there is a general lack of organizational competence in doing so. In terms of Weick and Roberts,[34] the "know-how" and "know-that" need to be integrated into "heedful purposeful interactions". This means that the knowledge about *what* to do has to be linked to the knowledge about *how* to do a certain task. In terms of Cloud Computing security, there is a need to develop understanding about the kinds of data that need to be secured, besides the mechanisms involved. Doing so would ensure that the Cloud Computing service provider has the requisite competence. As Dhillon[35] notes, this has to be by design, not by default.

I believe that data security in the cloud should be viewed along two dimensions. First the value of data, particularly if there were a data loss or non availability. Second the strategic importance of data. In situations where strategic importance of data is low, it may be relatively easy to suggest that such data can reside in a Cloud, however in cases where data is critical to the operation, care needs to be taken to ensure adequate availability. When the strategic importance of data is high, use of Cloud services should be avoided, but in case it is unavoidable, then special contractual arrangements need to be put in place – agreements that focus more on upside premiums and downside penalties. What ever be the case, I believe

---

[32]BBC News, "Gmail down again for some users," 2009, http://news.bbc.co.uk/2/hi/7934443.stm (10 September 2010).

[33]De Waal, A. "Darfur and the Failure of the Responsibility to Protect," *International Affairs* 83, 6 (2007): 1039.

[34]Weick, K.E., and K.H. Roberts. "Collective Mind in Organizations: Heedful Interrelating on Flight Decks." *Administrative Science Quarterly* 38, (1993): 357.

[35]Dhillon, G. "Organizational Competence in Harnessing IT: A Case Study." *Information & Management* 45, 5 (2008): 297.

responsibility structures need to be created and sustained. Only when this happens, would it be possible to ensure confidence in a Cloud Computing environment.

Ella: Absolutely, I totally agree with you in this respect. I think dealing with "responsibility" in the Cloud is extremely difficult. Just to understand and define all involved parties is a challenge as well. We have Cloud service providers, Cloud customers (companies and individuals) and those that regulate the Cloud. Today it is unclear how responsibilities should be divided between these different actors[36]. For example who should be responsible for compliance with data protection regulations such as the European Union Directive 95/46/EC? Is it the customers' responsibility to check if Cloud providers comply with data protection regulations and keep the data secure? Or, is it the responsibility of Cloud providers? This is just one example of ambiguousness in the Cloud world, which leads to many instances where organizations are unable to differentiate between who is responsible, who is accountable, who has authority and when things go wrong, who is to be blamed. Certainly there is a need for rules and contractual arrangements to regulate responsibilities and obligations in the Cloud.

Gurpreet: Ella, I do not think that technical solutions to security problems in the world of Cloud Computing are an answer. As the context changes, so do the security requirements. When I look at the evolving field, my sense is that we are using an old administrative logic to deal with newly emerging problems. Issues that seem to be more important today are the integrity of people occupying roles in organizations, trust and ethics. We may outsource our computing needs to a Cloud, but the Cloud Computing service providers need to have their act together. The "logical locks and keys" (i.e. passwords and encryption) can do only so much. The manner in which they hire, retain and compensate their employees becomes an important issue. Individuals need to identify themselves with the organization, which determines their ability to manage the information assets of a firm (as postulated by Ashforth and Mael,[37] 1989 in the Social Identity Theory). Failure to do so may result in a range of security challenges. Dhillon[38] noted that majority of information security problems occur because someone in the organization circumvents the controls. While the problem still exits, it gets compounded with loss of control by one organization to the other. Hence the motivation to protect data also gets transformed from largely being socially and ethically grounded to being commercially motivated.

Ella: Yes, I think that there are risks related to social aspects that are often ignored in Cloud Computing. Cloud providers collect enormous quantities of data of significant value and to make use of the data, employees in the online companies must have access to it. This fact causes a possibility for the employees to abuse

---

[36] http://www.trustguide.org/ (10 September 2010)

[37] Ashforth, B.E., and F. Mael, "Socia Identity Theory and the Organization." *Academy of Management Review* 14, 1 (1989): 20.

[38] Dhillon, G. "Violation of Safeguards by Trusted Personnel and Understanding Related Information Security Concerns." *Computers & Security* 20, 2 (2001): 165.

their trust for profits. Thus, even if strong safeguards are in place to protect information assets and control access to the data, they are insufficient if the trusted personnel abuse their trust.[39] A number of examples illustrating this phenomenon can be found in the literature.

Gurpreet: My dear Ella, I am glad that we share similar concerns. One aspect that we often overlook or do not pay as much attention relative to what we should, is that at the end of the day there are people organizing, storing, managing and ensuring access to data. An organization (Cloud Computing provider in this case) can at best define organizational structures and processes, hire the best people, but the manner in which people behave and operate is a function of (a) Individual motivation to be responsible (b) Organizational influences to comply with security policies (c) Working environment in an organization leading to trusting relationships (d) Workplace ethics.

I think in situations where there is no trust within or amongst organizational members, ethical principles end up taking a back seat. In a very interesting article, Shapiro brings to fore an interesting paradox – that guardians of trust are themselves trustees, which can cause significant problems. Even when there is prevalent trust, the identification of "sources of trust may provide the opportunity and means for its abuse"[40]. The nature and role of trust in organizations is an important one and links formal and informal responsibilities on the one hand with role identification on the other.

In the context of Cloud Computing, interpersonal trust becomes a rather critical construct. Granovetter[41] has argued that given a choice, individuals and organizations usually interact with people and institutions with whom they have had past dealings. This becomes evident if one were to see personal connections and cross listing of board members of some of the major Cloud Computing service providers and their clients[42].

Ella: It is certainly true. By placing applications and their data files on centralized servers the customers lose control of their data. Critical information resides on the servers of Cloud providers companies[43] and the customers need to trust the providers that their data is protected. I think that the importance of interpersonal relationships is often forgotten in the Cloud. The truth is that Cloud customers must trust both the technical systems and personnel at the Cloud providers they interact with. In the end there will be people (representatives from the different parties)

---

[39]Ibid.

[40]Shapiro, S.P. "The Social Control of Impersonal Trust." *The American Journal of Sociology* 93, 3 (1987): 623.

[41]Granovetter, M. "Economic Action and Social Structure: The Problem of Embeddedness." *The American Journal of Sociology* 91, 3 (1985): 481.

[42]We have had to make this assertion generic to maintain anonymity. It is however based on interview data collected by one of the authors in October 2009 of interpersonal relationships between Cloud Computing provides and their clients

[43]Conti, G. *Googling Security: How Much Does Google Know About You?* Addison-Wesley Professional, 2009.

who will take part in the interaction between the companies. There will be people who express their security objectives, recognize conflicting objectives and hopefully negotiate compromises. Any break in this chain of trust could defeat your best efforts at securing information assets. Both Cloud providers and customers are dependent on working relationships in the Cloud. Cloud providers that build their business on data-mining of the collected data and targeted advertising depend on customers who trust them to manage their data. On the other hand the customers must trust the online companies that the data is protected and maintained with respect to security and privacy.

Cloud providers offer extremely efficient and effective services that may be an interesting option for many companies. Instead of investments in expensive and inflexible IT-solutions, companies can use Cloud services for their processes and services. Denying access to these services and solutions due to the existing security risks is not a solution. There must be a way to ensure security in the Cloud.

16.3 Epilogue

Cloud Computing is a system of interrelated dependencies. While the technical dependencies are seemingly obvious, *Infrastructure as a Service* establishes relationships between organizational structures, processes, competencies and people within and across firms. Given the evolving nature of the structures, requirements for security have also emerged along the way.

For instance, in the 1970s it made perfect sense to simply focus on security requirements such as confidentiality, integrity and availability largely because computing resources were centralized and administratively managed in a top–down hierarchical manner. This changed in later years largely because of increased networking and the need to authenticate and focus on non repudiation requirements.[44] As systems development environments became more complex, the focus shifted to correctness in specification.[45] In later years other organizational issues emerged to be important – responsibility, integrity of people occupying roles, individual trust, ethicality.[46] The question is, what aspects should be focused upon in the future, particularly as new challenges emerge?

As indicated in Table 16.1, over the past several decades we have witnessed the emergent complexity of information security. This means that a fresh administrative logic is necessary for managing security in the ever so complex world of social and

---

[44]Parker, D. *Computer Security Management.* Reston, VA: Reston Publishing, 1981.

[45]Baskerville, R. "Information Systems Security Design Methods: Implications for Information Systems Development." *ACM Computing Surveys* 25, 4 (1993): 375; Wing, J.M. "A specifier's Introduction to Formal Methods." *Computer* 23, 9 (1990): 8.

[46]Dhillon, G., and J. "Backhouse. Information System Security Management in the New Millennium." *Communications of the ACM* 43, 7 (2000): 125.

Table 16.1 Chronological progression of information security issues

Year	Relationships
1970s	InfoSec = CIA where C is confidentiality; I is integrity; A is availability of data
1980s	InfoSec = CIA + $A_{u,\ n}R$) where A_u is authenticity; $_nR$ is non repudiation
1990s	InfoSec = CIA + ($A_{u,\ n}R$)) + C_{Spec} where C_{Spec} is correctness in specification
2000s	InfoSec = CIA + ($A_{u,\ n}R$)) + C_{Spec} + RITE (individual focus) where R is responsibility; I is integrity of people; T is trust and E is ethicality
2010s	?

technical interactions. As is obvious from the dialogue presented in this paper, information security in the Cloud tends to be technically oriented, with some emphasis on regulatory compliance. The value of human actors, business structures and processes is either overlooked or inadequately addressed. However, Cloud Computing security is socio-technical in nature. Scholars have argued that social and technical systems are strongly correlated, with one being dependent on the other.[47] This means that besides the technical and regulatory compliance aspects, issues such as responsibility, integrity of individuals, trust and ethics need to be addressed as well. We also believe that with increased virtualization, management of identity is going to emerge as a critical challenge, more so when individuals and organizations lose control of their data. With this in mind we believe that we can give a more comprehensive view of the chronological progression of information security issues illustrated in Table 16.1. In the years to come information security will be defined as followed:

2010s InfoSec = CIA + (($A_{u,\ n}R$) + C_{Spec} + RITE (albeit at organizational level as well) + I_{dn} where I_{dn} is Identity protection

Given the discourse and discussions related to the range of challenges, we make a call for a *Socio-Technical Perspective* in understanding and managing information security in an increasingly virtualized world. In order to ensure security of a Cloud, we postulate that the following condition be met:

1. *Forego legacy assumptions regarding the nature of information security.* While various researchers have argued that information security thinking needs to evolve as the context changes, in reality this has not happened. Continuous learning and awareness programs help in ensuring that all stakeholders stay current

[47]Hedberg, B., and E. Mumford. "The Design of Computer Systems: Man's Vision of Man as an Integral Part of the System Design Process. Human Choice and Computers," in *The IFIP Conference on Human Choice and Computers*. Amsterdam: North-Holland Publishing Company, 1975; Mumford, E. "The Impact of Systems Change in Organisations. Results and Conclusions from a Multinational Study of Information Systems Development in Banks." in *Systems Design and Human Needs,* edited by. N.-B. Andersen, B. Hedberg, D. Mercer, E. Mumford and A. Solé. Alphen aan den Rijn, Holland: Sijthoff & Noordhoff, 1979.

with the evolving needs. Typical response of the corporations is to train employees on the latest techniques and challenges. Such training in itself is problematic; since knowledge of fundamental principles related to emerging contexts (e.g. Cloud Computing) never gets taught.

2. *Shift focus from too much reliance on technical solutions.* In a hierarchical and an extremely structured environment, it made sense for more reliance on technical solutions relative to behavioral. However with services moving to the Cloud, it is imperative to ensure integrity, not just of the technical edifice, but also of the people involved. Many a times it becomes an ethical responsibility of various individuals to take necessary actions.
3. *Information security practices needs to be contextualized.* Until now information security, may it be in conventional organizations or in virtualized environments, has been handled in a rather reactive manner. Implementers generally have a series of checklists that form the basis for ensuring security. However each and every context is different. In some cases importance needs to be placed on confidentiality, while in others non-repudiation may be extremely important. Yet, in other cases identity management may be critical. So, depending on the context information security objectives need to be formulated.
4. *Responsibility and authority structures need to derive access rights.* Business sensitive information has great value and organizations need to consider whom they allow to access it. The access rights should be based on well-defined authority and responsibility structures. Information security literature emphasizes the importance of clarifying responsibility and authority structures. It is equally important to differentiate between who is responsible, who is accountable and who has authority. In the complicated structures in the Cloud, it is ever more important for the involved parties to understand what their respective roles and responsibilities should be. Responsibility in this context can be defined in terms of accountability, blameworthiness and obligation. However being responsible in this changing and ambiguous environment means not only accountability for blame after something has gone wrong; it also refers to handling of unexpected situations in the future. In the unregulated world of Cloud relationships it might be necessary for the involved parties (organizations and individuals) to develop their own work practices on a basis of a clear understanding of their responsibilities.
5. *People issues need to be adequately addressed.* As it has been pointed out in the information security literature, majority of security problems are directly or indirectly related to employees who violate or neglect policies and disobey rules.[48] While people management problems still exists in the information security field, they gets compounded in the Cloud when organizations lose control of their data to other parties and the motivation to protect data gets transformed from being socially and ethically grounded to being commercially motivated.

[48] Stanton, J.M., at al., "Analysis of End User Security Behaviors." *Computers & Security* 24, 2 (2005): 124.

6. *Trust in relationships needs to be inculcated.* In diffused environments when close supervision is impossible, trust is extremely important in managing information security. Relationships are built on trust rather than control. In such an environment, actors are expected to act according to accepted norms and patterns of behavior. By placing applications and data in the Cloud, companies (Cloud customers) lose control over their own data. Critical information exists on the servers of Cloud provider companies and the customers need to trust the providers that their data is protected. The Cloud relationships are so far diffused and unregulated that they have to be based on self-control, responsibility and trust. Two types of trust are important in Cloud relationships: trust within an organization and trust between organizations. In both cases levels of norms and patterns of behavior for involved parties (individuals and organizations) must be well defined and explained unambiguously in the policies.
7. *Encourage and define good ethical principles*. Ethicality in an organization refers to defining practices that should be followed by employees where rules do not exist.[49] In the context of the Cloud, defining ethical principles is especially important because the phenomenon is new and there are almost no rules governing how the Cloud should be used and regulated. Moreover new possibilities of using the Cloud are emerging as are the technical developments. This changing environment results in difficulty to define rules and regulations that can be applicable in all emergent situations. As a consequence the involved actors must act in accordance with some ethical principles.

References

Armstrong, M.P., G. Rushton, and D.L. Zimmerman. "Geographically Masking Health Data to Preserve Confidentiality." *Statistics in Medicine* 18, 5 (1999): 497–525.

Armbrust, M., A. Fox, R. Griffith, A.D. Joseph, R. Katz, A. Konwinski, G. Lee, D. Patterson, A. Rabkin, I. Stoica, and M. Zaharia. *Above the Clouds: A Berkeley View of Cloud Computing.* Berkley, CA, 2009.

Ashforth, B.E., and F. Mael. "Socia Identity Theory and the Organization." *Academy of Management Review* 14, 1 (1989): 20–39.

Baker, R.K. "Offshore IT Outsourcing and the 8/sup thsup/Data Protection Principle – Legal and Regulatory Requirements – with Reference to Financial Services." *International Journal of Law and Information Technology* 14, 1 (2006): 1–27.

Balachandra, R.K., R.V. Paturi, and A. Rakshit, "Cloud Security Issues." In *IEEE International Conference on Services Computing* (IEEE, 2009).

Baskerville, R. "Information Systems Security Design Methods: Implications for Information Systems Development." *ACM Computing Surveys* 25, 4 (1993): 375–414.

BBC News. Gmail Down Again for Some Users, 2009, http://news.bbc.co.uk/2/hi/7934443.stm. (10 September 2010).

Beaty, K., A. Kochut, and H. Shaikh. "Desktop to Cloud Transormation Planning." In *2009 IEEE International Symposium on Parallel & Distributed Processing* (IEEE, 2009).

[49]Dhillon, G., and J. Backhouse. Information System Security Management in the New Millennium. *Communications of the ACM* 43, 7 (2000): 125.

Best, J. "Lost Data Total Nears 30 million records," (2008), http://www.silicon.com/publicsector/0,3800010403,39295167,00.htm (10 September 2010).

Cavoukian, A. "Privacy in the Clouds, A White Paper on Privacy and Digital Identity: Implications for the Internet," 2008, http://www.ipc.on.ca/images/Resources/privacyintheclouds.pdf (10 September 2010).

CNN, "Info on 3.9M Citigroup Customers Lost Computer Tapes with Information About Consumer Lending Lost by UPS in transit to Credit Bureau," *CNNMoney.com*, 2005, http://money.cnn.com/2005/06/06/news/fortune500/security_citigroup/ (10 September 2010).

Conti, G. *Googling Security: How Much Does Google Know About You?* Addison-Wesley Professional, 2009.

Descher, M., P. Masser, T. Feilhauer, A.M. Tjoa and D. Huemer, "Retaining Data Control to the Client in Infrastructure Clouds." In *2009 International Conference on Availability, Reliability and Security* (2009).

De Waal, A. "Darfur and the Failure of the Responsibility to Protect." *International Affairs* 83, 6 (2007): 1039–1054.

Dhillon, G. "Organizational Competence in Harnessing IT: A Case Study." *Information & Management* 45, 5 (2008): 297–303.

Dhillon, G. "Violation of Safeguards by Trusted Personnel and Understanding Related Information Security Concerns." *Computers & Security* 20, 2 (2001): 165–72.

Dhillon, G., and J. Backhouse. "Information System Security Management in the New Millennium." *Communications of the ACM* 43, 7 (2000): 125–128.

Europe's Information Society, "eHealth" (2005) http://ec.europa.eu/information_society/activities/eten/library/about/themes/ehealth/index_en.htm (10 September 2010).

Foster, I., Z. Yong, I. Raicu, and S. Lu. "Cloud Computing and Grid Computing 360-Degree Compared." In *Grid Computing Environments Workshop 2008, GCE '08* (2008).

Google privacy center, "Privacy Policy", Last modified: March 11, 2009, http://www.google.com/privacypolicy.html (10 September 2010).

Granovetter, M. "Economic Action and Social Structure: The Problem of Embeddedness." *The American Journal of Sociology* 91, 3 (1985): 481–510.

Halperin, R., and J. Backhouse. "A Roadmap for Research on Identity in the Information Society." *Identity in the Information Society* 1, 1 (2008): 71–87.

Hedberg, B., and E. Mumford. "The Design of Computer Systems: Man's Vision of Man as an Integral Part of the System Design Process. Human Choice and Computers." In *The IFIP Conference on Human Choice and Computers*. Amsterdam: North-Holland Publishing Company, 1975.

Kaufman, L.M. "Data Security in the World of Cloud Computing." *IEEE Security & Privacy Magazine* 7, 4 (2009): 61–64.

Luis, V.M., R.M. Luis, C. Juan, and L. Maik. "A Break in the Clouds: Towards a Cloud Definition." *SIGCOMM Computer Communication Review* 39, 1 (2009): 50–55.

McCarthy, J. "Recursive Functions of Symbolic Expressions and their Computation by Machine, Part I." *Communications of the ACM* 3, 4 (1960): 184–195.

Mitroff, I.I. "The Tally: A Dialogue on Feyerabend and Ford." *Theory and Society* 3, 4 (1976): 601–609.

Mumford, E. "The Impact of Systems Change in Organisations. Results and Conclusions from a Multinational Study of Information Systems Development in Banks." In *Systems Design and Human Needs,* edited by. N.-B. Andersen, B. Hedberg, D. Mercer, E. Mumford and A. Solé. Alphen aan den Rijn, Holland: Sijthoff & Noordhoff, 1979.

Nurmi, D., R. Wolski, C. Grzegorczyk, G. Obertelli, S. Soman, L. Youseff and D. Zagorodnov, The Eucalyptus Open-Source Cloud-Computing System, *Proceedings of the 2009 9th IEEE/ACM International Symposium on Cluster Computing and the Grid* (Volume 00, 2009).

Nuttall, C. "US urged to probe Google's 'cloud' services" (2009), http://www.ft.com/cms/s/0/55572a2e-1425-11de-9e32-0000779fd2ac.html?nclick_check=1 (10 September 2010).

Oliva, T.A., and C.M. Capdevielle. "Can Systems Really Be Taught: (A Socratic Dialogue)." *Academy of Management Review* 5, 2 (1980): 277–279.

Parker, D. *Computer Security Management.* Reston, VA: Reston Publishing, 1981.
Pearson, S. "Taking Account of Privacy when Designing Cloud Computing Services." In *2009 ICSE Workshop on Software Engineering Challenges of Cloud Computing* (2009).
PRIME. "Privacy and Identity Management for Europe" 2008, https://www.prime-project.eu/ (10 September 2010).
Saikat G., K. Tang, and P. Francis. NOYB: Privacy in Online Social Networks. in *Proceedings of the first workshop on Online social networks*, Seattle, WA, USA (2008).
Shapiro, S.P. "The Social Control of Impersonal Trust." *The American Journal of Sociology* 93, 3 (1987): 623–658.
Stanton, J.M., K.R. Stam, P. Mastrangelo, and J. Jolton. "Analysis of End User Security Behaviors." *Computers & Security* 24, 2 (2005): 124–133.
Tian, X., X. Wang, and A. Zhou. "DSP RE-Encryption: A Flexible Mechanism for Access Control Enforcement Management in DaaS." In *2009 IEEE International Conference on Cloud Computing* (2009).
Weick, K.E., and K.H. Roberts. "Collective Mind in Organizations: Heedful Interrelating on Flight Decks." *Administrative Science Quarterly* 38, (1993): 357–381.
Wing, J.M. "A specifier's Introduction to Formal Methods." *Computer* 23, 9 (1990): 8–24.

Chapter 17
Privacy Regulations for Cloud Computing: Compliance and Implementation in Theory and Practice

Joep Ruiter and Martijn Warnier

17.1 Introduction

Privacy is considered to be a fundamental human right (Movius and Krup, 2009). Around the world this has led to a large amount of legislation in the area of privacy. Nearly all national governments have imposed local privacy legislation. In the United States several states have imposed their own privacy legislation. In order to maintain a manageable scope this paper only addresses European Union wide and federal United States laws. In addition several US industry (self) regulations are also considered.

Privacy regulations in emerging technologies are surrounded by uncertainty. This paper aims to clarify the uncertainty relating to privacy regulations with respect to Cloud Computing[1] and to identify the main open issues that need to be addressed for further research. This paper is based on existing literature and a series of interviews and questionnaires with various Cloud Service Providers (CSPs) that have been performed for the first author's MSc thesis (Ruiter, 2009). The interviews and questionnaires resulted in data on privacy and security procedures from ten CSPs and while this number is by no means large enough to make any definite conclusions the results are, in our opinion, interesting enough to publish in this paper.

The remainder of the paper is organized as follows: the next section gives some basic background on Cloud Computing. Section 17.3 provides an overview of several US and EU privacy regulations and Section 17.4 discusses the privacy regulations in relation to Cloud Computing. Next follows a more general discussion and the paper ends with conclusions.

J. Ruiter (✉)
Faculty of Sciences, VU University Amsterdam, The Netherlands
e-mail: jrr260@few.vu.nl

[1] Note that with regard to Cloud Computing, this paper is limited to Business to Business (B2B) Cloud Computing initiatives. Cloud Computing initiatives directed to consumers, such as Microsoft's Windows Live Mail or Google's Gmail are not part of this research.

S. Gutwirth et al. (eds.), *Computers, Privacy and Data Protection: an Element of Choice*, DOI 10.1007/978-94-007-0641-5_17,

17.2 Cloud Computing

Cloud Computing is a new paradigm in Information Technology (IT). In their research Vaquero et al. (2009) propose the following definition:

> Clouds are a large pool of easily usable and accessible virtualized resources (such as hardware, development platforms and/or services). These resources can be dynamically reconfigured to adjust to a variable load (scale), allowing also for an optimum resource utilization.

In traditional IT environments, clients connect to multiple servers located on company premises. Clients need to connect to each of the servers separately. In Cloud Computing clients connect to the Cloud. The Cloud contains all of the applications and infrastructure and appears as a single entity. Cloud Computing allows for dynamically reconfigurable resources to cater for changes in demand for load, allowing a more efficient use of the resources.

In Cloud Computing, end users are provided with dedicated hardware or a virtualized machine. To end users, this virtual machine appears as an isolated machine, where each user has isolated access. In Cloud Computing standardization has not yet emerged. Using software in a Cloud Computing environment therefore depends on the CSP. Virtualization in Cloud Computing allows distributing computing power to cater for load fluctuations. Standard web protocols provide access to Cloud Computing and control is centrally managed in various data centers.

Cloud Computing is offered through three types of services (Lin et al., 2009; Weinhardt et al., 2009). These services are Infrastructure as a Service (IaaS), Platform as a Service (PaaS) and Software as a Service (SaaS).

Infrastructure as a Service (IaaS), sometimes referred to as Hardware as a Service (Wang et al., 2008), allows the use of hardware through commonly available interfaces, such as web interfaces (Leavitt, 2009; Weinhardt et al., 2009) Due to the ubiquity of the web and the abstraction these interfaces provide, access to IaaS is claimed to be simple and easy. Although some researchers place storage as a separate service (e.g. Grossman, 2009), we will not do this and follows other researchers who define storage as a part of the IaaS concept.

Platform as a Service (PaaS) provides users with a platform to develop and execute software through similar interfaces as IaaS and SaaS. Developing software on PaaS allows users to collaboratively write the code and execute it in the Cloud.

Software as a Service (SaaS) provides users with applications that are easily accessible by providing common and ubiquitous interfaces. In contrast to normal applications, the applications in SaaS are installed on remote computers and not on the user's computer.

The three Cloud services of Cloud Computing are related. They can be consumed as separate services or can be combined. That is, PaaS can be installed on IaaS (Lederman et al., 2008; Lin et al., 2009). This relationship is visually represented in Fig. 17.1.

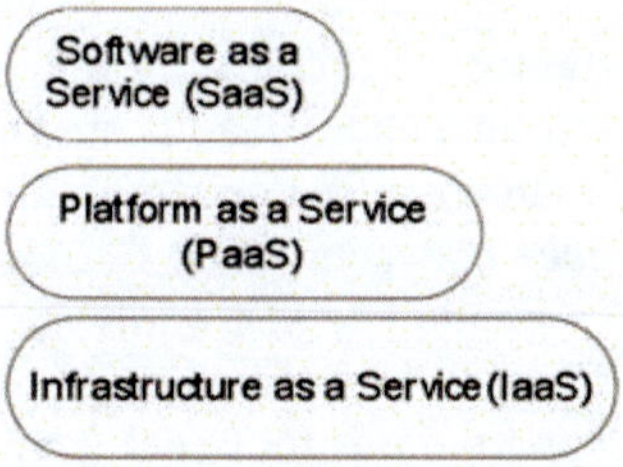

Fig. 17.1 The Cloud service layers (adapted from Grossman, 2009; Lin et al., 2009 and Weinhardt et al., 2009)

Portraying the Cloud services in layers resembles the OSI stack that comprises traditional computing. At the same time the layers represent the amount of control users have over their Cloud Computing initiative. Each layer provides further abstraction to users of Cloud Computing. IaaS hereby offers the least abstraction and SaaS the most. With more abstraction, more control of the technology stack is taken away by the Cloud Service Provider or IT organization.

These cloud services can be obtained from 3rd parties, referred to as Cloud Service Providers (CSPs) (Armbrust et al., 2009; Vaquero et al., 2009). Organizations can also opt for Cloud Computing technology within their own datacenter (Grossman and Gu, 2009; Leavitt, 2009).

Cloud Computing technologies can be classified into four different types: public Clouds, private external Clouds, private internal Clouds and hybrid Clouds. Security aspects, interoperability, pricing and benefits of Cloud Computing depend on the type of Cloud. Table 17.1 provides an overview of the classification and its characteristics.

In a public Cloud, organizations use Cloud Computing technologies through a CSP. The Cloud is physically located outside the premises of the organization. The Cloud is fully outsourced to the CSP, leaving the organization with little direct control over the hardware (Grossman, 2009). Public Clouds are typically offered

Table 17.1 Cloud type classification

	Managed by	Owner of infrastructure:	Dedicated hardware
Public	Cloud service Provider	Cloud service Provider	No
Private, external	Cloud service provider	Cloud service provider	Yes
Private, internal	Internal Organization	Internal Organization	Yes
Hybrid	Mixed	Mixed	Depends on contract with the CSP

through virtualization and distributed among various physical machines. Often multiple Clouds are hosted on the same hardware.

In private external Clouds, Cloud Computing is still offered by a CSP. The difference between public Clouds and private external Clouds is found in the hardware. In public Clouds, hardware is shared among different Clouds. In private external Clouds, the hardware only hosts the Cloud of one customer. This provides more opportunities for better (physical) security.

In private internal Clouds, organizations use Cloud Computing technologies within the organization's data center (Grossman, 2009). Private internal Clouds allow organizations to use the scaling of resources Cloud Computing provides, without handing over any control to a CSP. Private internal Clouds allow the organization full control over the Cloud. Organizational hardware, software and security standards can be used without the need for concessions to a CSP.

Hybrid Clouds are a combination of the other Cloud types. In a hybrid Cloud, organizations use a CSP in cases where additional resources are required.

17.3 Privacy Regulations

This section provides an overview of the most important privacy regulations in the United States and the European Union that are applicable to Cloud Computing. Policies on the creation of privacy legislation in the European Union and the United States differ. The United States favor a more laissez-faire approach. Industry self-regulation is favored over federal law (Baase, 2007; Movius and Krup, 2009; Steinke, 2002). It is believed that businesses shape their policies according to consumer preferences, following economic theory. This theory implies that consumer preferences determine market share, and that a higher market share leads to higher profits (Strauss and Rogerson, 2002). The Payment Card Industry Data Security Standards (PCI-DSS), discussed below, is an example of a self regulation policy. In situations where self regulation fails, sector specific laws are created so that other sectors are not hindered (Movius and Krup, 2009; Strauss and Rogerson, 2002). The sector specific laws only apply to a specific sector and do not oppose self-regulation initiatives in other sectors.

Privacy in the United States is dispersed among various different sector specific laws (Sarathy and Robertson, 2003). This paper is limited to a selection of sector specific laws and focuses on the privacy aspects of these laws. These sectors include the health care sector for the Health Insurance Portability and Accountability Act (HIPAA) and the financial services sector for the Gramm-Leach-Bliley Act, the Fair Credit Reporting Act and the Payment Card Industry Data Security Standards.

The European Union has a different approach concerning legislation. The European Union approach to legislation favors participation among businesses and governments as opposed to the US self-regulation approach (Movius and Krup, 2009). The European Union set privacy regulations up front as opposed to relying on industry self regulation (Baumer et al., 2004; Movius and Krup, 2009; Steinke, 2002).

17.3.1 EU Directive 95/46/EC

Directive 95/46/EC, commonly known as the Data Protection Directive (Birnhack, 2008), was implemented in October 1995 (EU Directive, 1995). The main purpose of the directive was to harmonize the privacy laws that existed in the different member states of the European Union and to provide a basic standard on privacy protection (Birnhack, 2008; EU Directive, 1995; Jentzsch, 2003).

Directive 95/46/EC addresses personal data, or personally identifiable information. Personal data is defined as any information relating to an identified or identifiable natural person ("data subject"). An identifiable person is one who can be identified, directly or indirectly, in particular by reference to an identification number or to one or more factors specific to his physical, physiological, mental, economic, cultural or social identity. Directive 95/46/EC consists of 32 articles setting requirements on handling personal data and mandating the countries of the EU to implement them (EU Directive, 1995).

The directive makes an implicit distinction between *data controller* and *data processor*. The data controller, the legal entity that chooses if and how data is processed, is responsible for compliance. It can choose to use a third party (the data processor) for data processing and should ensure that this is done in compliance with the directive. If the data processor resides in the EU and the data controller does not then it is the responsibility of the data processor to enforce the EU Directive 95/46/EC. Note that this is especially relevant in the context of cloud computing: all 'European' clouds, i.e., running on hardware located in a EU member state, have to ensure compliance with the EU Directive 95/45/EC – even if the data controller is not an EU company.

Directive 95/46/EC was written with the purpose of safeguarding the privacy of European Union inhabitants and to integrate different privacy legislation of EU member countries. There are several ways of complying with Directive 95/46/EC. European based organizations should adhere to its principles. Organizations outside the EU may use the Safe Harbor Agreement, Standard Contractual Clauses or Binding Corporate Rules.

17.3.2 The Safe Harbor Agreement

From a European point of view, the United States do not provide adequate privacy protection. This prevents data transfers between Europe and the United States. To address this problem, the European Commission and the United States Department of Commerce negotiated the Safe Harbor agreement (Bull, 2001; Fromholz, 2000). The agreement aims to align the process for US companies to comply with the EU Directive 95/46/EC.

The Safe Harbor agreement is only applicable to transfers between the United States and the European Union. Organizations outside the United States that have business operations within the European Union, have to rely on different mechanisms to adhere to the Transborder Transfer principle from Directive 95/46/EC. This principle requires that personal identifiable information can only be transferred to those countries that are deemed to provide adequate security. A US-based organization can adhere to the principles of the Safe Harbor agreement which guarantee (i) notice if data is collected (ii) choice for individuals to opt-out of the collection of data (iii) no transfer of collected data unless explicitly consented to by an individual (iv) security of the collected data (v) integrity of the collected data, i.e., the data should be factual and accurate (vi) individuals have the right to access data held about them and (vii) the above rules must be enforced. An example (Cloud Computing) company that is safe harbor compliant is Google.

The Safe Harbor agreement provides a substitute for adequate protection. In order to comply with the Safe Harbor agreement an organization must follow the Safe Harbor Privacy Principles, disclose their privacy policies, be subject to the statutory powers of the Federal Trade Commission, verify compliance with the Principles through self-or third-party assessment and register with the Department of Commerce. The Department of Commerce maintains a list with organizations adhering to the Safe Harbor agreement.

There is a substantial difference in European and US privacy regulations: European privacy laws apply only to personal data, i.e. data of a natural person whereas in the US, there is something like a privacy of a legal person.

Related to the Safe Harbor agreement are the Binding Corporate Rules (BCRs). These are sometimes presented as an alternative to the Safe Harbor Agreement. This is not the case (Bender and Ponemon, 2006). BCRs are used to ensure a form of complience to EU rules *inside* an organization for transfer from the EU to any other country (not just the USA). The rules do provide a form of certification for the complaince of a company to the EU data directive, and thus give an indication of safe harbor compliance.

17.3.3 The FTC Fair Information Practice

The FTC Fair Information Practice forms a set of guidelines concerning fair use of information about individuals. They originated in 1973 in the US Secretary's Advisory Committee on Automated Personal Data Systems. The Federal Trade Commission (FTC) first mentioned its Fair Information Principles in the 1998 report Privacy Online: A Report to Congress (Annecharico, 2002). The latest version has been published by the FTC on the 25th of June 2007. Organizations are encouraged to adhere to the Fair Information Practice but cannot be enforced to comply with the principles.

The FTC Fair Information Practice have their roots in privacy principles in the United States, Canada, and Europe, including Directive 95/46/EC. The FTC Fair Information Practice consist of the following five principles portrayed (i) Notice/Awareness (ii) Choice/Consent (iii) Access/Participation (iv) Integrity/Security and (v) Enforcement/Redress

These principles are basically the same as the principles in Directive 95/46/EC, with the exception of the Transborder Transfer principle (though similar principals could be added for other countries). The Integrity and Security principle are combined into a single principle. The Enforcement principle calls for self-regulation, organizations are not mandated to comply with the FTC Fair Information Practice.

17.3.4 Other Privacy Regulations

Some other American privacy regulations are sector specific, they include (i) the Health Insurance Portability and Accountability Act (HIPAA) which is created specifically for the health industry (ii) The Gramm-Leach-Bliley Act (GLBA) which is specifically designed for the financial services sector and applies to financial institutions (see Section 17.4 for a more thorough discussion of these acts) and (iii) the Fair Credit Reporting Act (FCRA) applies to consumer reports of United States citizens. The Act covers Credit Reporting Agencies (CRAs) and is enforced by the Federal Trade Commission. All these acts basically implement the Fair Information Principles discussed in Section 17.3.3.

Another act that is relevant in this context is the United Strengthening of America by Providing Appropriate Tools Required to Intercept and Obstruct Terrorism (USA-PATRIOT) Act. It differs from other legislation this paper addresses. The USA-PATRIOT can be seen as a law limiting privacy, opposed to the other privacy preserving regulations addressed (Baase, 2007).

The USA-PATRIOT Act is compliant with none of the Fair Information Practice Principles and the principles found in Directive 95/46/EC. The Act, particularly in sections 215 and 505, allows for the collection of information without consent of the individual. Reasons for information collection are not completely disclosed. Secondary use of the information is allowed under "domestic terrorism" reasons. There is a lack of clarity regarding the purpose of information collected, which makes it impossible to evaluate relevance and data quality. Accountability is absent on the part of those collecting and disclosing information (Regan, 2004). Note that, while there is no equivalent for the USA-PATRIOT act in a European context, individual police and secret services have similar possibilities for using wiretaps etc. thus also potentially hindering privacy is this context.

The Payment Card Industry – Data Security Standard (PCI-DSS) is an example of industry self regulation. The payment card industry has set compliance with the PCI-DSS as mandatory for organizations handling and processing payment card transactions (Wright, 2008).

The main privacy provisions of the PCI-DSS specifically address data related to card holders. These include (i) the requirement to protect cardholder data (ii) the requirement to encrypt transmission of cardholder data across open, public networks (iii) the development and maintenance of secure systems (iv) access restriction to cardholder data by businesses on a need to know basis (v) physical access restriction to cardholder data and (vi) the creation of a policy to increase employee awareness on compliance with the PCI-DSS.

Table 17.2 Common principles in privacy regulations

	FTC fair information practice principles	Directive 95/46/EC	The HIPAA	The Gramm-Leach-Bliley Act	The fair credit reporting Act	PCI-DSS
Notice	√	√	√	√	√	
Choice/consent	√	√	√	√	√	
Access	√	√	√		√	
Integrity	√	√	√	√	√	√
Security	√	√	√	√		√
Enforcement	√	√	√	√	√	√

17.3.5 Common Principles in Privacy Regulations

The privacy regulations discussed in this section have much in common, with the notable exception of the USA-PATRIOT act. The principles of Directive 95/46/EC and the FTC Fair Information Practice Principles are stated in similar terms. Additionally, it is said these principles are recognized worldwide as setting the standard for privacy (Movius and Krup, 2009; Regan, 2004). These principles therefore provide a standard in comparing privacy regulations. This comparison is shown in Table 17.2.

The horizontal axis states various privacy laws and regulations. A check means the principle is present in the regulation. The vertical axis portrays the common principles in the various privacy laws and regulations. This overview shows that the various privacy regulations are similar in nature. Nearly all the regulations provide individuals with a notice of the use of information, a form of consent for use of the information, require access to his/her data, require the integrity and security of the data and set demands for enforcement.

17.4 Privacy Issues for Cloud Service Providers

This section tries to identify the scope and applicability of the privacy regulations from the previous section regarding the Cloud Computing paradigm. An important aspect in enforcing privacy regulations is the physical location of an organization's Cloud Computing initiative. A CSP hosts an organization's Cloud Computing initiative in a distinct physical location. It is currently unknown what the consequences of local legislation on the Cloud's physical location are. Several researchers expect jurisdictional conflicts to arise (Jaeger et al., 2009; Mowbray, 2009). To place these jurisdictional conflicts in the scope of this paper; it is currently unknown if e.g. the Health Insurance Portability and Accountability Act (HIPAA) applies to a European health-care organization outsourcing health care data to a CSP located in the United States.

When organizations have a legal obligation to comply with legislation, these organizations are responsible and accountable for compliance (Eisenhauer, 2005; Lewis, 2009). Organizations can be held liable if a subcontractor breaches compliance with legislation. It is unknown if a CSP is legally considered the same as a subcontractor. Currently there is no jurisprudence on this matter. However, it is claimed that a CSP can be legally seen as a subcontractor (Gellman, 2009). This implies that organizations should ensure that a CSP is compliant with relevant privacy legislation. Various governments have posed laws, which require access to data stored in their jurisdiction for electronic discovery or anti-terrorism purposes (Gellman, 2009; Jaeger et al., 2008). An example of such a law can be found in the USA-PATRIOT Act. In most cases a form of subpoena or search warrant is required to provide a government legal authority to access stored data. The response to a search warrant or subpoena to a CSP differs per CSP (Soghoian, 2009). Some CSPs may object to the subpoena, others may comply without hesitation.

Cloud Computing offers the ability to dynamically reconfigure computing resources as demand for computing resources increases or decreases. A CSP needs to be capable of provisioning this demand. In cases where a CSP fails to provision this demand, the CSP itself may be forced to outsource organizational data to a different CSP, amplifying the location related privacy issues portrayed above.

With the exception of the USA-PATRIOT Act, all regulations addressed in Section 17.3 forbid secondary uses of covered data without consent from the data subject. A CSP may have the potential to use the data provided by an organization. Researchers have mentioned the potential of using this data for marketing and data mining purposes (Jaeger et al., 2008). When a CSP processes or transmits organizational data for purposes other than specified at the time an organization collected the data, the organization is no longer compliant to the corresponding privacy regulation.

In Infrastructure as a Service (IaaS) and Platform as a Service (PaaS) privacy issues pertaining to government access and secondary uses of data can be circumvented by use of encryption (Mowbray, 2009). For example, users can store their information in the cloud in encrypted form which prevents CSPs from accessing this data. In Software as a Service (SaaS) initiatives the use of encryption may not provide a solution to government access and secondary use of data. In SaaS the CSP actively processes organizational data in order to deliver its service. Encrypting data potentially renders the data unsuitable for processing by the CSP. State of the art encryption methods can counter this effect. For example, when using homomorphic encryption (Gentry, 2009) CSPs can still process the data without accessing the content, thus CSPs and governments will not be able to decrypt this data.[2]

[2]Note that in various jurisdictions it might be illegal to keep relevant encryption keys from law enforcers. For example see Part 3, Section 49 of the United Kingdom's Regulation of Investigatory Powers Act (RIPA, 2000).

17.4.1 The CSP and Privacy Regulations

None of the regulations described in Section 17.3 specifically mentions how using services offered by a CSP impacts compliance with the regulation. With the exception of the PCI-DSS, all regulations were created before the term Cloud Computing emerged. How Cloud Computing affects compliance with regulations is therefore subject to debate.

Solutions for compliance with pro-privacy regulations are given by several CSPs. A number of CSPs adhere to the Safe Harbor agreement. Adherence to the Safe Harbor agreement signifies compliance with Directive 95/46/EC. Examples of CSPs adhering to the Safe Harbor Agreement are Amazon, Google and Salesforce.com. Adherence to the Safe Harbor agreement obligates the CSP to adhere to several principles. These principles are the same principles as outlined in Section 17.3. The contents of the principles found in the Safe Harbor agreement resemble the FTC Fair Information Practice Principles.

The HIPAA act requires organizations subject to compliance to set up a business associate agreement with "Business Associates". A Business Associate is defined as "a person or organization, other than a member of a covered entity's workforce, that performs certain functions or activities on behalf of, or provides certain services to, a covered entity that involve the use or disclosure of individually identifiable health information" (HIPAA, 1996). In plain terms, a business associate is a third party, i.e., an employee of another company, performing services related to the organization that involved Public Health Information (PHI). When a third party merely acts as a conduit, for example a postal service, the third party is not categorized as a Business Associate. It is believed CSPs should be regarded as Business Associates. Gelmann states that: "A conduit transports information but does not access it except infrequently as necessary for the performance of the service, or as required by law. In theory, a cloud provider could possibly be a conduit for HIPAA purposes, but much depends on the terms of service. If the cloud provider reserves any rights to review, use, disclose, or post information submitted by a user, the provider will not qualify as a conduit" (Gellman, 2009).

The Financial Privacy Rule and Safeguards Rule of the Gramm-Leach-Bliley Act (GLBA) specifically mention service providers. Organizations striving for GLBA compliance need to take certain precautions when engaging in a business relationship with a service provider. The Financial Privacy Rule of the GLBA mandates organizations to hand out a notice to their clients stating the disclosure of the clients' Non-Public Information (NPI) to a service provider. The exchange of NPI with a service provider requires a contract. This contract should state the confidentiality of the NPI by guaranteeing the data is only used for the purpose for which it was shared. The Safeguards rule of the GLBA requires organizations to only select service providers capable of maintaining appropriate safeguards. A contract with the service provider should be established, requiring the service provider to maintain the appropriate safeguards. Organizations are required to ensure service providers comply with the contract. Furthermore organizations should oversee the handling of NPI by service providers.

The Fair Credit Reporting Act (FCRA) allows the sharing of consumer reports after the provision of the credit report to the related individual and an option to opt-out on the sharing (FTC, 2009).

Although the Payment Card Industry Data Security Standards (PCI-DSS) were created after the introduction of Cloud Computing, a CSP is not specifically mentioned in the PCI-DSS. The PCI-DSS gives a notion of general service providers. A service provider is defined as a "Business entity that is not a payment brand, directly involved in the processing, storage, or transmission of cardholder data. This also includes companies that provide services that control or could impact the security of cardholder data. Examples include managed service providers that provide managed firewalls, intrusion detection systems and other services as well as hosting providers and other entities. Entities such as telecommunications companies that only provide communication links without access to the application layer of the communication link are excluded" (PCI, 2009). If a CSP is seen as a service provider depends on the interpretation of the term "directly involved". When a CSP is directly involved in processing, storage, or transmission of cardholder data it is seen as a service provider. In this case, an organization engaging in a business relationship with a CSP needs to assure the CSP is compliant with the PCI-DSS. If a CSP is not "directly involved" in the processing, storage or transmission of cardholder data, the organization engaging in a business relationship with a CSP needs to clearly define which PCI-DSS requirements are handled by the CSP. The CSP then has two options to assure compliance with the PCI-DSS: undergo a PCI-DSS assessment themselves, or have their services reviewed during the course of each of their customer's PCI-DSS assessments. In general, it is assumed that CSPs adhere to the definition of service provider. Several CSPs assure PCI-DSS compliance by being compliant with the PCI-DSS themselves. Examples of PCI-DSS compliant CSPs are Aria Systems and OpSource. VISA Inc. maintains a list of PCI-DSS approved organizations (VISA, 2009). Organizations wishing to adhere to the PCI-DSS can confirm the compliance of their Cloud Computing initiative by verifying that the proposed CSP is mentioned on the VISA list and verifying the scope of the compliance.

This section addressed some of the privacy issues posed by Cloud Computing. The biggest threat to privacy in Cloud Computing is posed by outsourcing personal data to a CSP. The CSP is in physical control over data hosted within the Cloud Computing initiative while the accountability for non-compliance with privacy regulations lies with the CSPs client. The location in which the CSP physically hosts the data may pose issues with regulatory compliance. Another issue caused by the CSP is the disclosure of data to other non-affiliated third parties such as governments or marketing bureaus.

17.5 Privacy Regulations in Theory and Practice

This paper intends to provide clarity on the impact of Cloud Computing on privacy regulations and the impact of privacy regulations on Cloud Computing. An extensive literature study in combination with interviews with ten CSPs[3] (Ruiter, 2009) highlights several points of discussion:

- *Information security in Cloud Computing consists of established security solutions such as encryption, access management, firewalls and intrusion detection.* In internal Clouds the IT department has the ability to install all available security solutions it sees fit. In external Cloud Computing the security depends on the Cloud Service Provider (CSP). Some CSPs do not provide flexibility in the choice of security solutions, while others allow the implementation of client security requirements. The amount of control over the security depends on Cloud service. In IaaS, where clients are able to virtually manage an infrastructure, clients are usually able to implement more security measures than in SaaS, in which clients only use a software solution. Not all CSPs allow client-auditing of their security offerings. In these cases client organizations have to suffice with a CSP-provided audit statement, mostly SAS70 – Type II (SAS70), or have to take the CSPs word on the level of provided security.
- *Data storage, transmission and processing in Cloud Computing depends on the Cloud type, e.g. internal or external Cloud Computing, and the service, i.e., IaaS, PaaS, or SaaS.* In private, internal Cloud Computing the organization keeps all data within its own datacenter. Through techniques such as service-oriented computing and virtualization, the datacenter offers the benefits associated with Cloud Computing: faster and more efficient allocation of resources. In external Cloud Computing data is outsourced to a CSP. How the data is transmitted to the CSP depends on the CSP itself. Some CSPs allow encrypted data transmission, others do not. The storage of data dependents on the CSP as well. Some CSPs encrypt data or outsource data storage to a different CSP. It is unknown whether data is encrypted during the transfer between CSPs. Processing of data entirely depends on the CSP and the service. SaaS providers offer a specific processing service, whereas in IaaS the client organization determines to a large extent how the data is processed. CSPs may offer completely different services, thereby processing data in completely different ways.
- *The impact of privacy regulations is most dramatic between external Cloud Computing and traditional IT.* In external Cloud Computing, data gets outsourced to a CSP. The CSP has physical control over the Cloud Computing initiative while the accountability for non-compliance with privacy regulations lies with the CSPs client. Another issue in is related to the physical location where the CSP hosts the Cloud. The Transborder Transfer principle in Directive 95/46/EC requires organizations to exchange data only to countries that provide adequate protection. If an organization does not know where its data is hosted, this principle might be violated. Data location could also be an issue under local laws. It is unknown

[3]The size of the CSPs contacted ranged from startup companies to several large-scale service providers. The CSPs provided answers to these questions on the condition of anonymity. We realize the results are not conclusive (nor repeatable), but they give an indication of how CSPs currently address privacy issues.

if local regulations regarding data apply to the physical location of the Cloud. Another issue caused by the CSP is the potential disclosure of data to other non-affiliated third parties such as governments or marketing bureaus, i.e., use of data. Directive 95/46/EC, the FTC Fair Information Practice Principles, the Health Insurance Portability and Accountability Act (HIPAA) and Fair Credit Reporting Act (FCRA) all require that data only gets used in the purpose for which it was collected. In the Gramm-Leach-Bliley Act (GLBA) this requirements holds for data received indirectly; i.e. the CSP getting the data from its client organization.

- *The concept of Cloud Computing brings many uncertainties with respect to compliance with privacy regulations.* There are no clear answers on which privacy regulation requirements apply to Cloud Computing; none of the regulations from Section 17.3 explicitly mention Cloud Computing. The absence of cases in which an organization is accused of not being compliant with privacy regulations does not provide clarity either. There are only few case studies known in literature that describe HIPAA compliance in Cloud Computing. These cases leave several questions unanswered and do not provide enough information on the way certain regulatory requirements are implemented within the Cloud service. In general, the CSPs participating in the interviews from (Ruiter, 2009) do not know whether or not they are compliant with privacy regulations. A few notable exceptions are CSPs that are compliant with the Safe Harbor Agreement or have done PCI-DSS compliance audits themselves. This seems to be the only way in assuring compliance with privacy regulations: selecting CSPs which are compliant themselves. In those cases where CSPs are compliant with regulations, it is certain privacy regulations have affected the implementation of Cloud Computing: The services offered by the CSP are designed in such a way that compliance can be assured. In other cases the impacts of privacy regulations on Cloud Computing are not fully known.
- *Security is seen as a major issue in the adaptation of Cloud Computing, compliance to privacy regulations is not.* The interviews from (Ruiter, 2009) seem to indicate that CSPs in general do not know if they are compliant with privacy regulations. Customers seem to only inquire about the security of the Cloud, not about privacy regulations. One of the CSPs participating in the questionnaire stated privacy regulations did not influence the design of the security solutions at all. By combining these results, it seems privacy regulations have little influence on the security design of Cloud Computing. An exception to this rule are the CSPs' compliant with the PCI-DSS. The PCI-DSS sets standards on data security. PCI-DSS compliant CSPs have assured their security design/architecture is sufficient in adhering to the PCI-DSS.
- *It looks like many organizations are simply not aware of privacy issues in Cloud Computing (Ruiter,* 2009). Clients of the corresponding CSPs in general do not inquire about compliance with privacy regulations when establishing a business relationship with the CSP. Compliance with most regulations is mandatory. Therefore it seems organizations are not aware of the effects outsourcing data to a CSP may have with respect to compliance or that they think the issues are not important.

Privacy regulations are clearly not enough to solve all the privacy issues related to Cloud Computing. Raising awareness about both the issues and the existing regulations seems a good first step to remedy this.

17.6 Conclusions

There are still many uncertainties with respect to privacy regulations and Cloud Computing. There are very few case studies that concern compliance of privacy regulations with respect to Cloud computing in literature. The case studies that do exist do not clearly indicate how specific regulatory issues are handled. Non-compliant cases are nowhere to be found in literature. In general, CSPs seem to be unsure on how they should be handling privacy requirements. It is not possible to provide complete certainty on how organizations should implement Cloud Computing. The only way to provide certainty is when the CSP itself complies with the regulations. Adhering to the Safe Harbor agreement ensures compliance with Directive 95/46/EC. Several CSPs are adhering to the PCI-DSS as well. The ability to create HIPAA compliant Clouds is given by a couple of CSPs as well.

Even when the CSP is compliant with the privacy regulations, client organizations still need to make sure they adhere to the principles set in the various regulations themselves. More awareness amongst client organizations may eventually lead to more and better compliant CSPs.

We believe that research in the fields of privacy legislation and Cloud Computing would benefit substantially if future researchers could have access to more case studies addressing Cloud Computing. This could provide practical examples on how implementation of Cloud Computing affects the compliance of organizations with privacy regulations. Part of such research should be performed by people with sufficient knowledge of IT and legal practice. A legal background is required in order to determine the best way to interpret of regulations when applied to Cloud Computing. It is the opinion of the authors that due to the case law found in the United States, full clarity compliance can only be given when alleged non-compliance is brought to Court or when government officials create Cloud Computing specific legislation.

Acknowledgements The authors would like to thank Shay Uzery and Accenture for making this research possible. We also like to thank the anonymous reviewers for many suggestions that helped to improve the overall quality of the paper.

References

Annecharico, D. "Notes & Comments: V. Privacy after GLBA: Online Transactions: Squaring the Gramm-Leach-Bliley Act Privacy Provisions With the FTC Fair Information Practice Principles." *North Carolina Banking Institute* 6, (2002): 637–695.

Armbrust, M., A. Fox, R. Griffith, A. Joseph, R. Katz, A. Konwinski, G. Lee, D. Patterson, A. Rabkin, I. Stoica, et al. "Above the clouds: A berkeley view of cloud computing." *EECS Department, University of California, Berkeley, Tech. Rep. UCB/EECS-2009–28,* 2009.

Baase, S. *A Gift of Fire: Social, Legal, and Ethical Issues for Computing and the Internet*. Prentice Hall, 2007.

Baumer, D., J. Earp, and J. Poindexter. "Internet Privacy Law: A Comparison Between the United States and the European Union." *Computers & Security* 23, 5 (2004): 400–412.

Bender, D., and L. Ponemon. "Binding Corporate Rules for Cross-Border Data Transfer." *Rutgers Journal of Law & Public Policy*, (2006)

Birnhack, M. "The EU Data Protection Directive: An Engine of a Global Regime." *Computer Law & Security Report* 24, 6 (2008): 508–520.

Bull, G. "Data Protection – Safe Harbor, Transferring Personal Data To The USA." *Computer Law & Security Report* 17, 4 (2001): 239–243.

Eisenhauer, M. "Privacy and Security Law Issues in Off-shore Outsourcing Transactions." *Hunton & Williams, Atlanta Georgia* 15, (2005).

EU Directive. Directive 95/46/EC of the European Parliament and of the Council of 24 October 1995 on the protection of individuals with regard to the processing of personal data and on the free movement of such data, (1995).

Fromholz, J. "The European Union Data Privacy Directive." *Berkeley Technology Law Journal* 15, (2000): 461.

FTC. Federal Trade Commission, Fair Credit Reporting Act, (2009).

Gellman, R. "WPF REPORT: Privacy in the Clouds: Risks to Privacy and Confidentiality from Cloud Computing." *Released February* 23, (2009).

Gentry, C. *A Fully Homomorphic Encryption Scheme,* Phd Thesis, Standford University, (2009).

Grossman, R. "The Case for Cloud Computing." *IT Professional* 11, 2 (2009): 23–27.

Grossman, R., and Y. Gu. "On the Varieties of Clouds for Data Intensive Computing." *Data Engineering* 44, (2009).

HIPAA (1996). Health Insurance Portability and Accountability Act of 1996.

Jaeger, P., J. Lin, and J. Grimes. "Cloud Computing and Information Policy: Computing in a Policy Cloud?" *Journal of Information Technology & Politics* 5, 3 (2008): 269–283.

Jaeger, P., J. Lin, J. Grimes, and S. Simmons. "Where is the Cloud? Geography, Economics, Environment, and Jurisdiction in Cloud Computing." *First Monday* 14, 5–4, (2009).

Jentzsch, N. "The regulation of financial privacy: the United States Vs Europe." *ECRI Research Report* 5, (2003).

Leavitt, N. "Is Cloud Computing Really Ready for Prime Time?" *Computer* 42, 1 (2009): 15–20.

Lederman, L., B. Suri, J. Houston, and S. Itchhaporia. *The Next Stage of Computing*. William Blair & Company, (2008).

Lewis, S. "Cloud Computing Brings New Legal Challenges." *New York Law Journal*, (2009).

Lin, G., D. Fu, J. Zhu, and G. Dasmalchi. "Cloud Computing: IT as a Service." *IT Professional* 11, 2 (2009): 10–13.

Movius, L. and N. Krup. "U.S. and EU Privacy Policy: Comparison of Regulatory Approaches." *International Journal of Communication*, (2009):169–187.

Mowbray, M. "The Fog over the Grimpen Mire: Cloud Computing and the Law." *Scripted Journal of Law, Technology and Society* 6, 1 (2009).

PCI (2009). PCI Security Standards Council, Payment Card Industry (PCI) Data Security Standard – Requirements and Security Assessment Procedures version 1.2.

Regan, P. "Old Issues, New Context: Privacy, Information Collection, and Homeland Security." *Government Information Quarterly* 21, 4 (2004): 481–497.

RIPA (2000). United Kingdom. Regulation of Investigatory Powers Act.

Ruiter, J. *The Relationship between Privacy and Information Security in Cloud Computing Technologies.* Master's thesis, Vrije Universiteit Amsterdam, (2009).

Sarathy, R., and C. Robertson. "Strategic and ethical considerations in managing digital privacy." *Journal of Business ethics* 46, 2 (2003): 111–126.

SAS70. American Institute of Certified Public Accountants, Statement on Auditing Standard 70.

Soghoian, C. (2009). Caught in the Cloud: Privacy, Encryption, and Government Back Doors in the Web 2.0 Era.

Steinke, G. "Data privacy approaches from US and EU perspectives." *Telematics and Informatics* 19, 2 (2002): 193–200.
Strauss, J., and K. Rogerson. "Policies for online privacy in the United States and the European Union." *Telematics and Informatics* 19, 2 (2002): 173–192.
Vaquero, L., J. Caceres, M. Lindner, and L. Rodero-Merino. "A Break in the Clouds: Towards a Cloud Definition." *ACM SIGCOMM Computer Communication Review*, (2009): 50–55.
VISA (2009). VISA Inc, Global List of PCI DSS Validated Service Providers.
Wang, L., G. von Laszewski, M. Kunze, and J. Tao. "Cloud Computing: A Perspective Study." *Service Oriented Cyberinfrastruture Lab, Rochester Inst. of Tech–Dezembro de*, (2008).
Weinhardt, C., A. Anandasivam, B. Blau, and J. Stosser. "Business Models in the Service World." *IT Professional* 11, 2 (2009): 28–33.
Wright, S. *PCI DSS: A Practical Guide to Implementation*. IT Governance Ltd., (2008).

Chapter 18
Data Protection in the Clouds

Yves Poullet, Jean-Marc Van Gyseghem, Jean-Philippe Moiny, Jacques Gérard, and Claire Gayrel

18.1 Introduction

The Council of Europe (CoE) requested the Research Centre on IT and Law (CRID) to prepare a preliminary report identifying the main privacy issues related to cloud computing and the questions to be addressed in the future, in particular in the light of Council of Europe data protection standards and mainly of ETS 108 for the protection of individuals with regard to automatic processing of personal data (hereinafter referred to as ETS 108). This chapter is based on said report and as such, we chose to keep the report's structure, which is not the usual one for an article. In this perspective the conclusion is written under the form of questions which are open to discussion. This chapter does not aim to answer questions, but rather to raise them.

This chapter is structured as follows. It starts with a brief technical introduction illustrating the variety of services covered by the concept of "Cloud computing". As defined by the National Institute of Standards and Technology (NIST),[1]

> cloud computing is a model for enabling convenient, on-demand network access to a shared pool of configurable computing resources (e.g., networks, servers, storage, applications, and services) that can be rapidly provisioned and released with minimal management effort or service provider interaction.

Cloud computing services (hereinafter referred to as CCS's) include a large diversity of services going from those offered at the benefit of individuals – as the services

Y. Poullet (✉)
Research Centre on IT and Law (CRID), University of Namur, Namur, Belgium
e-mail: yves.poullet@fundp.ac.be

This chapter is based on the report written by the CRID in the context of the OCTOPUS Conference organised by the Council of Europe in order to strengthen the cooperation between Law Enforcement agencies under the Cybercrime Convention. The text of the report is available on the Council of Europe website. Even if the text of the article does represent a version deeply modified of the report, we have decided to limit the number of footnotes in order to keep the format of the report and to make reference not only but mainly to the text of the C.o.E Convention 108.

[1]Meil, P., and T. Grance. The NIST Definition of Cloud Computing, Version 15, 10-07-09, available on NIST (National Institute of Standards and Technology) web site.

S. Gutwirth et al. (eds.), *Computers, Privacy and Data Protection: an Element of Choice*, DOI 10.1007/978-94-007-0641-5_18, © Springer Science+Business Media B.V. 2011

offered by social networks – to those proposed at the benefit of companies in sharing a common software, or by using shared information infrastructures. Establishing a typology of cloud computing services is quite important because legal problems raised by each kind of computer services might be different to a certain extent. The second point is dedicated to the analysis of the adequacy of the ETS 108 definitions with the cloud computing reality. In particular, the status of the actors involved in the operations will be analysed. Thereinafter, our contribution analyses the duties of the persons subscribing to the cloud computing services or offering these services. Next, the crucial question of security is addressed. Finally, the chapter addresses the delicate questions of transborder data flows (hereinafter referred to as TBDF) and international private law, which are inherent to most of the cloud computing services.

Obviously cloud computing raises issues at many levels. Currently, cloud computing seems closer to fog than cloud and it might constitute a real danger for the users and data subjects whoever they are (legal entities, individuals).

Before analysing the different challenges raised by cloud computing services in views of the Data Protection legislations, let us try to explain why, with cloud computing, privacy issues have to be analysed in a deeply modified way. The first generation of data protection legislation took into account the sole risks linked to the processing by isolated information systems within a company or an administration. The risks of these systems were easily identifiable (sensitivity of data; processing purposes; etc). Nowadays, the terminals (PCs) are functioning in global and interactive networks with unprecedented options of exchanging more and more personal data. The network society has consequently raised new issues which are partly addressed by legislation like the EU 2002 e-privacy directive. The problems of confidentiality during the transmission, the need to regulate new types of data, such as traffic and location data, and the uptake of public communications services have significantly broadened the scope of the data protection. At the same time, terminals must be protected against illegal intrusion, and their functioning must be privacy compliant. With cloud computing, companies and administrations are invited to transmit their data, even their whole information assets, to the clouds by means of very user friendly web interfaces. The Cloud is, for obvious economic and security reasons, the answer to social pressure and provides individuals with the means to exist in the virtual society. The cloud computing service provider will use all the possibilities of the Net and of its information system, including those offered by other service providers for storing data, for ensuring their sharing amongst other users, and so on. What is quite noticeable is that, with cloud computing, the subscriber has lost the direct ownership/control of the information placed in the clouds because the data have left their own computer, or more generally their terminals, and are "somewhere in the cloud" in places determined by decisions, and notably the availability of the different elements of the Cloud Service operator's IT systems. These elements might be located in the same country as than subscriber's one, but often they are located somewhere else, even in non democratic countries. In other words, transborder data flows in the global network are becoming inherent to the essence of the CCS and the Internet is as such becoming the location of the processing of data. So starting from these premises, new challenges must be addressed. In our view, cloud computing issues will constitute a major data protection challenge in the next future with questions such as: How to ensure a certain ownership/control of the data by the data subject? How to solve the delicate problem of TBDF?

18.1.1 Some Technical Aspects and Specific Risks Linked with Cloud Computing Services

18.1.1.1 A Brief History

Users make use of applications with many functionalities which help them in their work or in their other activities. They reasonably expect that their data be stored in protected spaces in order to retrieve these data when needed. That constitutes the standard way.

In the sixties, the computer users used mainframes for running software. The data were stored on tapes, with no direct access for users. Everything was "online". The users did not know where, and on which media, their data were stored. They only knew that the data were in one splendid and large room in one specific building. Everybody has seen these ranks of tape machines on TV. The data access was controlled by the operators of the mainframe. And no external access was possible.

Later on external access to computing services and data were created by means of modems and controlled (for the rare persons that could try to it) by passwords. With the advent of the personal computer, data storage and computing facilities became local, everyone could have programs and data on their own computer. Users became responsible for the access control to their data. Nowadays, with the Internet, users can access the data stored on many machines from everywhere in the world. With this comes responsibility of end-users for their own and other people's data.

Thus, simple users can access data on "mainframes" located anywhere. They can also access data they manage on their own system (that is to say their local network). Finally, they can access data stored in computers from where they have access when connecting themselves on Internet. Four main components are needed in each of these cases:

- Hardware (processing, storage and memory)
- Operating system
- Applications
- Data

The use of external information systems might bring certain advantages because it implies the possibility for outsourcing processing or support to larger facilities. Another benefit might be found in the fact that all the expenses and efforts concerning the maintenance, upgrades and security of the information system shared as part of the cloud computing service are financially supported by the different users of these services, and technically supported by the company offering the cloud computing services. Cloud computing services in this sense do represent major scale economics for companies, particularly for small and medium-sized enterprises (SMEs). It should be underlined that this kind of benefit might be also offered in the context of a GRID. The main difference between GRID computing and the Cloud computing services mainly concerns the nature of the relationships between the users. GRID services concern users linked by a common professional interest and using the same information system (for instance, hospitals using the same datacenter or peculiar software in order to control their expenses). In the case of Cloud Computing services, services are not shared on equal footing by the users on the basis of individual agreements, but rather the selling of certain remote services, that we could describe as a commodity, by certain specialised (or not) companies. These commodities share the following characteristics:

- "On-demand self-service. A consumer can unilaterally provision computing capabilities, such as server time and network storage, as needed automatically without requiring human interaction with each service's provider.
- Broad network access. Capabilities are available over the network and accessed through standard mechanisms that promote use by heterogeneous thin or thick client platforms (e.g., mobile phones, laptops, and PDAs).
- Resource pooling. The provider's computing resources are pooled to serve multiple consumers using a multi-tenant model, with different physical and virtual resources dynamically assigned and reassigned according to consumer demand. There is a sense of location independence in that the customer generally has no control or knowledge over the exact location of the provided resources but may be able to specify location at a higher level of abstraction (e.g., country, state, or datacenter). Examples of resources include storage, processing, memory, network bandwidth, and virtual machines.
- Rapid elasticity. Capabilities can be rapidly and elastically provisioned, in some cases automatically, to quickly scale out and rapidly released to quickly scale in. To the consumer, the capabilities available for provisioning often appear to be unlimited and can be purchased in any quantity at any time.
- Measured Service. Cloud systems automatically control and optimize resource use by leveraging a metering capability at some level of abstraction appropriate to the type of service (e.g., storage, processing, bandwidth, and active user accounts). Resource usage can be monitored, controlled, and reported providing transparency for both the provider and consumer of the utilized service".[2]

This commodity can be offered through different deployment models. So, the NIST paper, already quoted, distinguishes:

[2]Meil, P., and T. Grance. The NIST Definition of Cloud Computing, Version 15, 10-07-09, available on NIST (National Institute of Standards and Technology) web site.

- "Private cloud. The cloud infrastructure is operated solely for an organization. It may be managed by the organization or a third party and may exist on premise or off premise.
- Community cloud. The cloud infrastructure is shared by several organizations and supports a specific community that has shared concerns (e.g., mission, security requirements, policy, and compliance considerations). It may be managed by the organizations or a third party and may exist on premise or off premise.
- Public cloud. The cloud infrastructure is made available to the general public or a large industry group and is owned by an organization selling cloud services.
- Hybrid cloud. The cloud infrastructure is a composition of two or more clouds (private, community, or public) that remain unique entities but are bound together by standardized or proprietary technology that enables data and application portability (e.g., cloud bursting for load-balancing between clouds)".[3]

18.1.1.2 Cloud Computing

The model of cloud computing, at least for forms (SaaS), implies that there is a simple computer that runs a browser that accesses a remote service. Users can use the browser to interact with applications in the "cloud" and stock their data in a folder in the cloud. In this case, an important question is access to these resources (application and data).

Cloud services can be distinguished in three types: "software as a service", "platform as a service" and "infrastructure as a service":

- "Software as a service" (SaaS) is easy to understand: users access applications on the Web, for example, a word processor, a spreadsheet or email software. The services offered by Google (e.g., Google Docs, Gmail) are well known examples of SaaS. Data are also stored on the Cloud providers' IT systems. In such context, the CCS provider (e.g. Google) is technically responsible for the application services and for the data of the users (secure storage and secure access).
- "Platform as a service" (PaaS) offers an operating system where users can install their own applications. The platform provides services such as application services and database services. The data are stored depending on the application, either on the provider's system or locally on the client's system.[4]
- "Infrastructure as a service" (IaaS) offers one "logical hardware" infrastructure.[5] Users have to install their own operating system, the applications they need and they have to decide which Storage provider to use and they have to decide how

[3]Meil, P., and T. Grance. The NIST Definition of Cloud Computing, Version 15, 10-07-09, available on NIST (National Institute of Standards and Technology) web site.

[4]See https://www.dropbox.com/ for a simple example of a cloud storage facility or http://msdn.microsoft.com/en-us/azure/default.aspx for a more complex example of a platform provider.

[5]See for example http://aws.amazon.com/ec2/.

to connect the different PaaS components they use. In general the user can not determine where the data is physically located because the Storage provider will store several copies of the data at possibly changing locations.

At a high level, these three services are carried out through two kinds of elements in the cloud: datacenters and clusters. The datacenters are specialised hardware where data are stored. They generally provide security for access and recovery services. Clusters offer high performance computing facilities.

For simple cases, customers can use simple infrastructures. Virtual servers are examples of IaaS. The virtual computer installed by a user can be moved from one location to another when needed. The segmentation of the infrastructure must be serious, because, if not, one instance can read or write in one other instance or virtual machine.[6] Hacking or destruction is then possible.[7]

In the case of the SaaS, only data are separate. Each user starts one instance of an application (e.g. word processor). The identification of the user is the only way to attribute data to the correct user. For this reason, the system must have proper authentication methods in place. In the two other cases, the problem is more complex, but access control and security are important issues.

18.1.2 Specific Risks Associated with Cloud Computing

This section briefly describes some risks related to, or accentuated by, the use of CCS which would justify a possible intervention of the CoE. The rest of the chapter will go into some of them in greater depth. At this point we already have to make clear that the legal issues may vary depending on whether services are directed to individuals or to companies or even to public administrations. Each risk may require a specific assessment depending on the actors involved on both sides (demand and offer of CCSs).

As regards services offered to individuals, such as social network sites,[8] or other large public available web 2.0 platforms, the following risks can rise:

- The possibility – for a third party or the cloud computing service provider itself – to *profile data subjects* by linking several databases/information related to an individual represented in the CCS's databases or result of their use of the service. This risk increases when consumers are invited to use CCSs free of charge

---

[6]IaaS services are typically used by multiple tenants at the same time, and hence multiple virtual machines will run simultaneously on the physical server.

[7]Segmentation is also an important requirement for the other types of CSS because they share vulnerabilities.

[8]As regards these services, see Article 29 Data Protection Working Party, Opinion 5/2009 on online social networking (WP163), adopted on 12 June 2009; Moiny, J.-P. "Facebook au regard des règles européennes concernant la protection des données", 2 E.C.J.L., 2010, pp. 235 and ff.

if they accept to receive one to one targeted adverts. They will be tempted to give privacy for "free" to CCSs.

- The concept of *consent*: when cloud users give up their privacy for services, their consent could not be free anymore. Beyond this risk, their consent can hardly be called sufficiently informed due to the opacity of many CCSs. Users are usually hardly aware of the true processing of their data, of cross-references made between different services, etc. Therefore, is consent still valid or aligned with the concepts of "free" and "explicit (informed consent)" which are its characteristics?
- For obvious reasons, *youngster's protection* might require additional measures. Is the consent by underage users valid? The US Children's Online Privacy Protection Act of 1998[9] requires parental consent for children aged under 13 years, and recently Article 29 Working Party has pleaded for forbidding the profiling of children[10].
- The problem of "*ownership*" of personal data: Consumers, once they have released their data in the cloud, might have difficulties not only to maintain access to these data (e.g., in cases of denial of access when they do not pay the service, or in case of bankruptcy of the CCS). But more fundamentally, they could have the difficulties exercise full control over the released data when they terminate their contractual relationship with the CCS. According to the general terms of many services, the provider could contractually reserve the right to keep the data even after such a termination (e.g., in the context of social network sites, where users commonly only have the option of deactivating their account instead of deleting it).
- Control over the data *after death*: when the subscriber of a service dies while his or her data are in a cloud computing system, who is then authorised to access these data (their heirs, the de cujus, the CCS)?

When a *company* subscribes to a CCS, additional questions might raise:

- The obvious need to distinguish clearly the concepts of user, subscriber and data subject, each of them referring to clearly different people involved into CCSs and being subject to different problems. So, the employee who is using the information system provided by his or her company might not be aware of the recourse made by his or her employer, the subscriber, to the cloud and to a CCS. As regards data located within the datacenters provided by the CCS provider, some are relating to customers, furnishers and so on, who are not necessarily aware of this fact. So to what extent can we consider that these persons are aware of the use of cloud computing services and is this recourse subject to possibilities of refusal or even of acceptance? Other specific questions relate to the distinction between

[9]See Sec. 1303, (b), 1, (a), ii of the Children's Online Privacy Protection Act of 1998, available on http://www.ftc.gov/ogc/coppa1.htm. Sec. 1303 (b), 2 however specifies some exceptions to the requirement of parental consent.

[10]Article 29 Data Protection Working Party, Opinion 2/2010 on online behavioural advertising (WP171), adopted on 22 June 2010, p. 17.

users (employees) and subscribers (employers). In case of death of an employee, who will have access to data stored in the datacenter? If the user dies, may the CCS provider erase the identifier and password of this user? Is he authorised to do so? In the negative, who has the authority to do so? Beyond this question, is it conceivable, to the benefit of the employees using the companies' information system, to make a difference between private and professional, excluding the former from the use of cloud computing services?

- The *protection of legal persons* and their know-how, industrial secrets, etc. On this matter, two different problems are identified. First, the company might place trade secrets concerning itself or a third party on the cloud servers. Such trade secrets might be compromised by a lack of security of the CCS. Second, the cloud provider might record certain transactional data generated by the use of the offered services, which could reveal substantive activities of the company and, as the case may be, sensitive information about it. For instance, the storage and analysis of communication of financial data between the subscribing company and a bank might reveal risks of bankruptcy.
- The *exclusion or subjection of the use of CCSs to strict conditions when some types of data are processed or when particular activities are at stake* (like the activities submitted to a professional secrecy). Certain legislations (see for instance the US Health Insurance Portability and Accountability Act of 1996[11] on Health data) regulate the disclosure of data to third parties. Insofar as the cloud provider might be considered a third party, they will be submitted to such regulations. In some cases, it could be deemed, due to the sensitive nature of the data and the risks inherent to CCSs, that the processing of these data is not to occur in the cloud because of the dispersion of data and, to a certain extent, the loss of control by the data controller on the data stored within the clouds. This leads to the question of the ban of cloud computing as regards specific processing of personal data or activities. One can consider that some matters, such as health, justice or administration are so sensitive that they cannot be reconcilable with the use of cloud computing because of the potential spreading of information over the Internet and major risk of disclosure.
- For the same reasons, one may wonder whether CCSs should be forbidden or subject to certain restrictions when specific processing of data driven by public administrations or authorities are concerned. In some case, the use of CCSs could threaten the confidentiality of data and, as the case may be, jeopardize State's sovereignty, particularly as regards States' Security concerns. The use of hybrid clouds operating only within the national borders might be a solution to this particular problem.
- The *bankruptcy or transfer of the cloud computing activities* might cause certain problems. The cloud provider's bankruptcy might lead to the sale of the cloud

[11] Available on https://www.cms.gov/HIPAAGenInfo/Downloads/HIPAALaw.pdf. About this example and others, see B. Gellman, Privacy in the clouds: Risks to Privacy and confidentiality from Cloud Computing, Report prepared for the World Privacy Forum, Feb. 23, 2009.

computing services to another company exercising competing activities with the subscriber's ones or having another privacy policy. The bankruptcy might in other cases lead to the termination of the activities. Anyway, the subscriber must be aware of the consequences of the disappearance or transfer of the cloud computing services on the data which are stored or put into circulation by it. So different questions would have to be analysed. Do we have to provide the continuity of the contract with its confidentiality or security guarantees, etc? Is it possible for the subscriber to unilaterally terminate the agreement for privacy or competition reasons and, if it is the case, to be sure to get back his or her data?

18.2 Personal Data Flows Within Any Cloud Computing System

Different personal data flows can be identified within any cloud computing system which involves several actors as the data controller, data processor, subscriber, user and data subject.

ETS 108 provides basic and useful definitions for the processing of personal data. However, this list does not take into account the peculiarities of cloud computing. Providing some additional definitions should clarify the understanding of the functions and duties of all the actors intervening in the cloud computing system. Any situation involving cloud computing can involve six major categories of actors – usually overlapping to some extent – and sometimes legally defined by ETS 108: a CCS provider, a subscriber to this service, data controllers, data processors, users and data subjects.

- *Cloud computing provider*: The natural or legal person providing a service (SaaS, PaaS and IaaS) in a Cloud computing system.
- *Subscriber*: The natural or legal person contracting with the cloud computing provider. It might be an individual (e.g., the average user of a social network site), a company or a public administration
- *User*: The natural or legal person actually using, in the context of his tasks, the CCSs. The user can be the same person as the subscriber, but also be a different person, such as an employee working in a company. This person would be the user, while the company would be the subscriber to the service (SaaS, PaaS and IaaS). In this respect, it could be assessed whether the cloud computing service provider should be subjected or not to specific obligations in favour of the user – only acting as a user? And which would be such obligations (e.g., a particular information duty)?
- *Data subject*: While ETS 108 already deals with the concept of "data subject", it doesn't give a complete definition. It appears important to precisely define this main actor in the personal data processing, whether in a Cloud computing system or not. To which extend, should we consider a legal body as a data subject to be protected on equal footing with individuals? Indeed, ETS 108 limits the concept

of data subject to individuals, thereby excluding legal bodies. Is such limitation still pertinent in a cloud computing environment?

- *Data processors and data controllers*: The distinction between data controller and data processor is at first glance quite clear according to the definition given by Directive 95/46/EC, but they are not defined by ETS 108[12]. The data controller processes data for his own purpose and defines the means to achieve this purpose; According to the article 2 (d) of the Directive 95/46, the data controller determines alone or jointly, both the purposes and the means of the processing of personal data, while the data processor operates data exclusively at the request of the data controller and does not pursue their own purpose[13].

In the context of cloud computing, the CCS provider might be considered in certain cases to be a data controller and in other cases as a data processor. It is quite clear, as recently asserted by the Article 29 working party,[14] that some CCS providers, for instance social networks, have to be qualified as data controller since they process personal data for their own purposes such as providing one to one marketing or transmitting data to third parties. The qualification might in other cases be quite difficult "since the cloud computing service provider could define in the broadest sense "means" of processing, that due to the characteristics of the service at stake, would justify some processing operations not directly requested by the subscriber – as the case may be, data controller". As an example, the provider of an IaaS, caring about the efficiency of its service, could automatically allocate processing and storage capacity between various facilities located worldwide. For instance, at a time "t", data centre and processing capabilities located in Germany are optimal. But, due to the increased use of these facilities at a time "t+1", it could be more sensible to have recourse to facilities located elsewhere in the world, for instance in India, in providing the service – which could involve a duplication of data, etc. In this respect, the technology at stake would automatically trigger a transborder data flow, the controller of which is not necessarily easy to determine. From another point of view, in a lot of cases, the cloud computing service provider might take advantage of storage or processing capacities offered by third parties, who could be considered as data processors of data processors.

In our opinion, it is difficult to qualify the CCS operator as data controller each time the processing operated by him are justified by the need to ensure the service proposed or to ameliorate it. It is quite obvious that the subscriber, by choosing a CCS operator, and by giving them certain latitude to define precisely how to achieve

---

[12]However, we can take the concept of data processor out of the article 7 of ETS 108.

[13]As regards the concepts of data controller and data processor of Directive 95/46/EC, see Article 29 Data Protection Working Party, Opinion 1/2010 on the concepts of "controller" and "processor" (WP169), adopted on 16 February 2010.

[14]WP169, Op. cit.

their tasks, determines the means of the processing. [15]The qualification of CCS as data controller or as data processor is a case by case decision. It depends from the possibility for the CCS to decide autonomously about the purposes of the data processing, knowing that the means might be trusted to a specialised service.

Even if we don't want to introduce "by force" the distinction between data controller and data processor in the context of a CoE amended Convention or by a specific recommendation, it would be necessary to specify the "legal" regime of this new actor and its specific duties. These duties concern, first, the subscriber having recourse to a data processor (obligation to have a written contract specifying the tasks given to the data processor, requirement as regards the quality of the data processor, etc) and, secondly, of the CCS operator independently of his qualification as data controller (prohibition or not of personal use of the data processed in the context of the tasks operated on behalf of the data controller, obligation to provide a high level of security, obligation to inform people in case of security breaches, etc.). It is quite clear that it would be wrong to assert that the only possibility to impose legal duties and obligations to CCS provider is linked to his qualification as data controller.

18.3 Domestic and Non Domestic Uses

Cloud computing serves the domestic and personal framework (e.g., social network sites, webmail, online blogs and word processors, etc), as well as professional environments (e.g., legal bodies decentralising their IT infrastructure to reduce costs, etc).

Bearing in mind that European Union has, voluntarily, limited the scope of Directive 95/46 to the non domestic processing of personal data, does this limitation remain relevant in the context of cloud computing? It is particularly relevant in the context of some cloud computing services such as social network sites. Here, individuals can make information concerning others available to the entire world, – rather than to a small circle of people who could qualify as household or domestic – which would make them a data controller.[16] A practical interpretation of domestic versus non domestic use has to be found which would not deprive data subjects of their rights enshrined in data protection legislation, and would not suffocate other individuals by heavy rules. As the case may be and depending on the cloud computing service at hand, it is necessary to think about the opportunity of establishing a softer data protection regime in spite of a wide application of an exemption to the scope of the legislation.

---

[15]In the same sense, we do not follow the Article 29 opinion referring to the SWIFT case, where the WP considers, a bit too rapidly and without appropriate nuances, that a furnisher of security services of data transmission become data controllers when they decide to answer to law enforcement agencies (LEA) requests. This would mean that a CCS providers would qualify as data controller each time they decide to answer positively to a lawful request issued by LEA.

[16]See the Lindqvist case which appeared before the European Court of Justice C 101/01 (2003).

This distinction might have harmful consequences for individuals as far as TBDF are concerned. Indeed and in some national laws, the rules dealing with such situations are applicable only to the non domestic use. This means that the data subject concerned by a non domestic process enjoys more protection than the others who could lack some protection in the context of cloud computing services.

18.4 The Protection of Legal Persons

Another issue resulting from cloud computing relates to the concept of personal data. Does this concept has to be confined to the definition given by the ETS 108 which says that personal data "means any information relating to an identified or identifiable individual ("data subject")[17]"?

In the context of, if need be, a specific regulation targeting cloud computing, wouldn't it be relevant to extend the concept of personal data to any information relating to an identified or identifiable *legal* person? In the surroundings of cloud computing, does the concept of personal data have to be extended – and how – to information such as industrial secrets, know-how, etc?

Most countries do not extend data protection scope to legal persons. The cloud computing system may change this conception because it will be used by the legal persons as a way to reduce their IT costs. And, depending on the relevant market, they could be deprived of any bargaining power (e.g., SMEs and non-profit organisations). This would compel them to contract under unfavourable conditions to stay competitive, having thereof less regards for data protection and privacy.

The extension of the scope of personal data from relating to persons to relating to legal persons is in line with decisions by the Strasbourg Court which has always asserted that article 8 ECHR protects not only the individuals but also legal persons notably their industrial secrets, know-how, etc.[18] Obviously, legal persons want to keep these safe from any disclosure to third parties without prior authorization. The concern is to determine to what extent a protection should be provided for by the law to legal persons, hearing that they can be economically and technically dependent on the use of CCSs. The information powers imbalance between individuals and companies or administrations created by IT use has been at the basis of data protection legislations.[19] Perhaps, it might be meaningful to extend data protection principles – at least some of them – to the protection of legal persons when it is clear that the same imbalance exists. And in so doing, the protection of the

[17] Article 2a.

[18] On that issue, see particularly, Bygrave, L. *Data Protection Law: Approaching Its Rationale, Logic and Limits*, The Hague/London/New York: Kluwer Law International, 2002, 448 pages.

[19] About the history of the privacy concept and the need to take fully into account the informational asymmetry between data subjects and data controllers, read notably Solove, D.J. "Conceptualizing Privacy", 90 California Law Review, 2002, 1085 et s.; Blok, P. Het recht op privacy, Boom Juridische uitgevers, 2003.

individual concerned by legal person's files or databases would also be improved. Furthermore it should be noticed that certain countries, members of the CoE, have already extended their data protection legislations to legal persons (see notably in Italy, Luxemburg, Norway and definitively in the E.U e-privacy Directive 2002/58 which was recently modified).

Therefore, these considerations coming from companies who may be major users of cloud computing systems, mainly on a B2B basis, have to be taken into account. Companies will store confidential information (such as the know-how, industrial secrets) on separate servers, or they will use cloud computing for internal communication (email, voice over IP, etc). Needless to say, they expect a reasonable protection of such information. And from an economic viewpoint, in case of lack of protection, they could be reluctant to use cloud computing systems.

A better protection of data related to legal persons may be necessary due to the economic pressures that could lead companies to adopt the cloud computing paradigm. Indeed, "meta-processing" is possible within such cloud systems because their providers may access information of various legal persons. With such a cross-source of data, CCSs providers could offer added value services (e.g. risk analysis on companies) to third parties. Such behaviour may constitute a major risk of disclosure of confidential or sensitive information to third parties.

Taking these concerns into account, it has to be determined if the concepts of personal data and data subject have to be extended to legal persons as regards cloud computing. Arguments might be drawn down from previous extension to legal persons as it has been the case as previously underlined under the EU e-Privacy Directive and under certain legislations of member states of the Council of Europe (Italy, Norway, Luxemburg, etc).

18.5 Liability of the Actors

Primary issues relate to the concepts of data controller and data processor. As has been argued earlier, the cloud computing system will involve both of them. The main question is to define who is who and who does what. For sure, the data controller *is the cornerstone of the data protection regulation.* The data controller has the responsibility of the main duties (information, security, etc). The determination of the data controller will have a huge impact on the legal structure of the cloud computing system.

In the identification of the data controllers involved in the cloud, we have to take into account the extraterritorial characteristic of actors and its consequences. Indeed, numerous CCS providers are set up out of the territories under the jurisdiction of the CoE's member States. Consequently, the control of the behaviour of the CCS provider can be difficult for both the authorities and the subscribers, as well as for the data subjects. Which then leads to difficulties as regards the control of the respect of the duties enshrined in data protection legislations, and as regards the sentence of the breach of these legislations.

As suggested, in some cases the CCS provider can be viewed as a data processor instead of as a data controller. According to this view, they can only act on behalf of the user (or subscriber) who himself processes personal data. But sometimes, the CCS operator pursues its own purposes for the processing of personal data. And as far as this processing is concerned, it is a data controller. Two issues result from this assessment. First, when a Cloud Computing service provider is data controller and data processor as regards a same user (or subscriber) could it be deemed appropriate and/or necessary to extend its quality of data controller for the whole processing operations? Secondly, if it is only a data processor, is it appropriate and/or necessary to establish, as the case may be, some specific duties (e.g., security, information, etc) and/or a specific rule of responsibility?

Of course, it appears preferable that data controllers be under CoE's member states' jurisdiction. The question directly relates to the right of protection of the users and data subjects. Then if the main actor is outside the scope of Europe's jurisdiction, how can the data subject or the subscriber or even the authorities control the processing of personal data and sue CCS providers if they breach their duties?

Consequently, from the data protection point of view, the following question raises: when is a CCS provider a data controller – or even a "joint-controller" – or a data processor?

A third option would be having the subscriber/user and the CCS considered jointly as data controller.

BUT the question raised by those three scenarios is whether it is possible for subscribers to require by contract with the CCS operator that the data generated or operated through their cloud computing services are located in the territories of the Member states and to forbid any onward transfer? What about the possibility for users to take benefit of this provision? A third party beneficiary provision ought to be included in cloud computing standard contract. At this point, we only lay down the question which will be elaborated in Section 18.9 when we will deal with the TBDF issues

However, it has to be pointed out that in practice, both models could, in some extent and in a same relationship between the CCS provider and its subscriber, overlap. Depending on the processing at stake, the provider could be data controller and data processor at the same time with regard to the same data or to the same data subjects. In this respect, it has to be determined if the subscriber – following a basic view, that is to say the data controller – could be a co-controller as regards the processing the controller of which is the provider – following the same basic view, that is to say the data processor. To this end, the following fundamental question can be raised: how to define "joint-controllers" and does such a definition have to be adapted in the context of the cloud computing? This is of course crucial due to the aforementioned scattered location of the actors of the cloud.

The following simple examples can illustrate the pertinence of the purpose. An employer decides to have recourse to encoding software offered by the cloud (SaaS) and designed to encode invoices from employees who seek refund for fees supported by them. The SaaS provider could offer its subscriber (employer) an additional – of course paid for – service to monitor the expenses of his employees. The service could consist of the sending of monthly reports detailing in descending order the total amounts of expenses per employee. In such a case, could – and should – the purpose of the processing – monitoring of employees in a specific field – being a complementary service, be deemed to be defined by the subscriber and the provider at once?

Another example comes from the social network sites context. The provider of such a network could offer a personalised advertisement service consisting of a SaaS enabling a company to choose a specific audience to deliver advertisements, without such company processing any personal data, the provider of the SaaS holding alone this task.[20] Could – and should – the company ordering the advertising campaign be deemed to be a co-controller of the processing at stake? In both cases, the providers of SaaS define means for the processing of personal data and suggest to subscribers a purpose they assigned to the means they created, purpose the subscriber chooses to appropriate, bearing processing of personal data. The very question is then the following: is it opportune to define – or redefine – a "joint-responsibility" of the actors in such cases, and how could and should it be done?

If the ETS 108 imposes duties on the data controller (controller of the file), there is nothing concerning the data processor since this latter is not considered by ETS 108[21] (even if we find this concept in embryonic form in the article 7 of ETS 108). Being a main actor of the cloud, it might be useful – and this has to be assessed – to impose on data processors themselves – or, as the case may be, on some data processors – specific duties by "law" instead of contract. Clearly, where a data controller does not have the bargaining power to impose their own warranties as regards data protection, the law could mitigate such an imbalance of powers. The specific duties of the data processors CCS providers could consist of security obligations, information obligations, a specific liability (e.g., as what exists as regards the responsibility of the intermediaries at the sense of the e-commerce Directive 2000/31/EC). As stated above, a particular liability could be established as regards joint-controllers. But this has to be further assessed. And the present considerations are of high importance since each time it is considered opportune to create new duties, the question of liability has of course to be studied.

18.6 Transparency and Duties of Information Including in Case of Security Breaches

We have to make distinction between the three situation drafted above. Depending on the role of each party, the duty of transparency/information towards the users and data subjects will be different. Nevertheless, this duty should be a fundamental objective of any cloud computing system. This objective involves the *information*

[20]See for example J.-P. Moiny, Op. cit., pp. 249–250.

[21]However, we can take the concept of data processor out of the article 7.

obligations definitively with regards to the users, but also perhaps more generally with regards to all the data subjects.

Providers are compelled by information duties if they are data controller. However, if the CCS provider is also a data controller pursuing his own purpose as regards data related to the users of the service, the subscriber has no duty, according to data protection rules, to inform the users of such a processing if he is not involved in the processing as data subject.

Moreover, if the CCS provider is only a data processor, the subscriber only outsourcing his IT infrastructure, in our view, users should be informed of the recourse to cloud computing technologies by the data controller. Finally, it needs also to be asked if a data controller relying on a CCS provider as data processor should not inform the data subject of this practice. Indeed, the use or not of a CCS could be decisive as regards the data subject's consent. This data subject does not necessarily want to send personal data to an unknown third party who is not his direct contractor, especially if he has no certainty about the final place of the processing.

Next to the general information duty enshrined in article 8 of ETS 108, article 5a of ETS 108 also concerns the transparency of the processing of personal data. It sets that the "*personal data undergoing automatic processing shall be obtained and processed fairly and lawfully*". The term "fairly" involves this concept of information. And it could be argued that it is unfair to rely on CCSs without informing users, as the case may be, even in the situation where the subscriber and the CCS provider are not data controllers. Therefore, it might be suggested to modify article 5a of Convention 108 in order to fit the specific transparency issues raised in any cloud computing system. In this respect, it needs to be determined to what extent the data subject has to be informed of the particular technology at stake and its technical implications, such as the relocation of the storage of information in another State, the chain of sub-processors, and, as the case may be, its legal implications such as the occurring of processing operations in a non Contracting States where even adequate – but different – data protection rules merits mention?

Still as regards the evolution of the ETS 108, or even in the framework of a CoE recommendation, it would be of high interest to consider the introduction of a *duty of information related to security breaches*.

Indeed, the concept of security breach is unknown by the CoE regulatory text, but has been introduced recently in European Union by the Amending Directive on e-privacy. This Directive defines "personal data breach" as a breach of security leading to the accidental or unlawful destruction, loss, alteration, unauthorised disclosure of, or access to, personal data transmitted, stored or otherwise processed in connection with the provision of a publicly available electronic communications service in the Community. The main idea is to put on the shoulders of certain communication services providers' new obligations provided that specific risks are linked with the nature of their services.

As regards, European Union Directive, the targeted services are limited to publicly available electronic communication services even if it has been recognised that in the future the concept must be extended to other services due to the risks existing in other services like banking on line services or electronic healthcare online services. Clearly the debate around the revision has asserted the need to re-open the debate about this limited scope and to follow the US example (see at the Federal level, the "Data Accountability and Trust Act", by extending certain obligations to any person engaged in interstate trade and who own or possesses electronic personal data shall notify a breach to individuals, if the breach leads to an unauthorised third person acquiring the data, and also to the Federal Trade Commission. So the first question is: "To what extent the specific nature of the risks linked with cloud computing services might justify the extension to these services?" Perhaps the U.S. extension or the extension to all cloud computing services is too broad since they will conduct to minimise the obligations to impose but considering the nature of the risks offered by cloud computing services acting or not as data controller and offering not a specific services like a service assisting people in order to fix meetings (like Doodle) but services including more sensitive processing, what remains to be defined. The main criterion must be the importance of risks incurred by the subscriber of the service but more generally by the concerned people.

The second question, having solved positively the first one, envisages the different obligations linked with the "Security Breach" regime. It consists of two kinds of additional obligations:

- First, all the legislation imposes a duty to inform the data subject through appropriate means, which might in case of cloud computing services go far beyond both the subscriber or the users and implies in cases of cloud computing offering purely technical or software facilities without having access to the data themselves a partition of the tasks between the service provider and the subscriber and that in order to afford them an opportunity to take the needed measures to avoiding or reducing the risk. As regards the list of the beneficiaries of this obligation, can we consider, on the basis of the previous remarks, that in certain cases this obligation to notify must be extended at the benefit not only of individuals but also of legal persons?
- Second point, does the legislator have to impose an obligation to alert at the same time the data protection authority? But in case of positive answer: which one (due to the global character of the provider)? Which information must be given? And through which channel?

Finally one pinpoints the idea for standardisation authorities of establishing in close connection with these independent agencies technical and security means.

18.7 Security

18.7.1 Introduction

The cloud computing pattern implies two main categories of data flows. A first one relates to the flows between users/subscribers and the cloud infrastructure. And a second one which groups the data transfers within the cloud system together. In such a context, two levels of security therefore have to be distinguished. The first one deals with the connection between the user and the cloud computing provider. And the second one relates to the cloud computing system itself.

Through this distinction, the cloud computing system is considered and promoted by their providers as a kind of safety deposit box which can be accessible only by authorised person. This is possible for SaaS services where the CCS provider assures that no other cloud services are used to provide the SaaS service or an IaaS, PaaS service where only one instance is used. In a PaaS, IaaS context complex systems can be realised. The single components and additional storage services are connected over the internet. The same is true for an SaaS service using other cloud services.

On the other side, the access to this safety deposit box must be secured to avoid any access to the transferred data by unauthorised persons. Such access should usually occur through the Internet as it should also most probably be the case of numerous data flows within the cloud. This clearly shows that Internet access providers (IAPs) also have a fundamental role in the cloud infrastructure as regards the conveyance of signals between users and the cloud system, but also within the cloud itself. IAPs offer an IP connection. Based on this connection secure environments such as VPNs can be realised between the endpoints of the communication, hence the system of the user and the system of the cloud provider.

Article 7 of ETS 108 imposes "*appropriate security measures*". It does not define who has to fulfil this obligation. It might be the data controller, the data processor or even the sub processor (even though these two last actors are not defined by the ETS 108 even though we could consider that the first one exists in an embryonic form). The concept of "security" is quite broad, even if not defined precisely by the article 7 of ETS 108. It means under article 17(1) of the Data Protection Directive, protection "*against accidental or unlawful destruction or accidental loss, alteration, unauthorized disclosure or access, in particular where the processing involves the transmission of data over a network, and against all other forms of unlawful processing*". So, for example, the risk of wiretapping by unauthorized third parties during the use of the services requires appropriate safeguards like the use of cryptography or secured lines (e.g., in case of electronic transmission of the credit card number). The possibility of intrusion within the provider's information system in order to collect all its customers' addresses or to manipulate certain data, imposes the necessity to install firewalls and other security measures. The sending of worms through the information systems of a communications service provider or the creation of a mirror site in order to lead astray certain communication are other specific risks linked with the use of communications services. The obligation is not limited to technical measures but encompasses also organizational measures which might be the nomination of a data security manager competent to ensure the compliance of the functioning of the service with all Data protection requirements.

Security is essential in case of CCSs since it is quite clear that by trusting a cloud computing service, the subscriber aims at being protected against all risks linked not only with confidentiality (disclosure or intrusion), but also with integrity and availability of the data stored somewhere in the cloud. In other words, because the cloud computing service provider is offering services founded on security in the broadest sense, it seems meaningful to impose them additional obligations as regards this obligation to security and more particularly in case of what is called: "security breach".

18.7.2 Specific Security Obligations

Regarding security and integrity of CCSs, according to the peculiar risks raised by such services due to the concentration of applications and or data used by different users and subscribers and the huge possibility for unauthorised people of aggregating all these data, it might be wise to impose new obligations on their providers. Amongst these appropriate security measures, three ones could be taken into consideration: The first one addresses the problem of unauthorised access by the provider's employees: providers of cloud computing services could be subject to an obligation to develop measures like identity management systems in order to fix and effectively control the respective privilege afforded to each member of personnel regarding access to personal data conveyed, stored or operated by the communications services. The second one would target the necessary protection of these data against any loss, destruction or illegal access or storage. This refers to various technological security measures such as the encryption of transmitted data, the adoption of automated control systems about the quality and integrity of stored or transmitted data, the setting up of log-in and log-out registries, etc. The last security measures would concern the adoption by the provider to express in clear language their security policy. This obligation contributes to an increasing accountability of the data controllers by compelling them to envisage the risks associated with the services they provide, to define exactly how they manage these risks and by making them responsible in case of non respect of their commitments. Furthermore it might be envisaged that the cloud computing services' provider would be required to cooperate with the competent data protection authority(ies) in case they would like to audit the security measures promised or implemented by the providers. In the same line, the possibility for these authorities or standardisation authorities to issue recommendations on best security practices ought to be assessed.

Some other organisational measures may be adopted in the context of the cloud computing matter as:

- Obligation to audit the system to put the risks and the lack of securities or confidentiality in an obvious place;
- Obligation to segregate the data stored by each subscriber in cases of multi-tenancy in order to avoid any accidental or unlawful access to these data by another subscriber;
- Obligation to have a person responsible for the security who will be in charge to warrant the security of the cloud computing system for the provider;

- Standardisation/normalisation of the sector to give to the user/subscriber a kind of security in its choice. This standardisation/normalisation goes hand in hand with the delivery of quality-labels available for cloud computing providers who insure the respect of several conditions/obligations of quality.

The Cloud Computing business model and architecture calls for a deeper examination of the relevance of non regulatory instruments. Indeed, cloud computing companies are mostly international and implemented in a great number of countries. Advantages and disadvantages of self-regulatory instruments, such as the European Union model of Binding Corporate Rules (BCR), whether as an alternative or complement to the existing legal framework, need to be assessed. Due to the globalised nature of cloud computing companies, we strongly believe that the European Union's experience with BCR could provide an interesting framework and point of departure for future debates.

18.8 Transborder Data Flows and Applicable Law to the Processing of Personal Data

Due to its highly virtualised architecture, CCSs involve great amount of data transfers, among which personal data as defined in the ETS 108, and by thus raise the issue of the applicability of the transborder data flows (TBDF) regime defined in the Additional Protocol 181. First, these transfers may occur between several actors: personal data may be transferred within the cloud provider's proprietary cloud, which can cover several countries; transfers may occur between cloud providers; transfers also occur between the cloud subscriber and his cloud provider, when he benefits from the cloud computing services wherever his location, such as when accessing, consulting or downloading personal data. Second, these transfers between actors may pursue different purposes: some transfers might be justified for purposes of transit or technical maintenance, while others are directly justified by the necessity to provide the CCSs requested by the user.

All these transfers may involve TBDF, since the cloud providers may resort to processing materials located in several countries to offer its services to subscribers/users soliciting cloud services from anyplace. Circulation of information, and as far as we are concerned, of personal data within and outside the cloud may occur in non State Parties to the ETS 108, among which most do not provide adequate level of protection. This state of fact raises the following issue.

18.8.1 Applicability of the Existing Legal Framework of Additional Protocol 181

The applicability of the existing legal framework to cloud computing technology requires deeper attention and assessment. Article 2 of additional protocol 181 basically prohibits international transfers of personal data toward states not party to the ETS 108 that would not ensure *adequate level of protection*. Any actor involved in cloud computing services, whether user, subscriber or cloud provider, should be fully aware of this prohibition and the legal risks associated with international transfers that would not satisfy the TBDF regime. It is obvious that in the context of contractual relationships between CCS providers and their subscribers, the last ones might impose certain restrictions to the first ones imposing for instance that the storage of data and their processing have to be operated in the country of the subscriber (certain governments impose that kind of restriction) or in specific countries where the adequate protection is obvious. This kind of "Zoning the Net"[22] could also be imposed through the design of the networks' infrastructure like SWIFT would have decided, according to its public statement, since January 2010. According to that decision, transfers concerning European citizens would be operated exclusively through the SWIFT European network. It would be possible to have, for certain types of data like sensitive ones, legislative mandatory rules prohibiting the use by CCS of their global networks in order to avoid risks of onward transfers to countries where no adequate protection is offered.

Derogations to this general prohibition as provided in additional protocol 181 need further examination. As provided in article 2 a), national laws may allow transfers of personal data toward non-adequate destinations in case of "*specific interests of the data subject*" or when legitimate interests, especially important public interests prevail. Rightly applied, these exemptions could constitute a basis for several international transfers in the cloud computing context. As a first instance, the data subject's consent to the transfers at stake could be solicited. As a second instance, international transfers could be justified by the necessity of the performance of the contract concluded in the interest of the data subject between the cloud provider and the cloud subscriber/controller. Public authorities resorting to cloud computing services in the framework of their tasks could justify international transfers in the name of legitimate important interests.

As far as the second set of exemptions is concerned, article 2, b) offers possibilities of international transfers "if safeguards, which can in particular result from contractual clauses, are provided by the controller responsible for the transfer and are found adequate by the competent authorities according to domestic law". Appropriate contractual clauses might constitute a relevant framework to ensure the legality of international transfers. However, such framework needs further assessment about its relevance in the cloud computing context, due to the necessity to take fully into account that the flows generated by the CCS'es often are concerning a lot of countries and a lot of companies as previously asserted. Perhaps the use of "Binding Corporate rules" adopted by large multinational companies offering cloud computing services can be at least a partial solution.

---

[22]The idea to come back notwithstanding the global character of the Internet to a certain "zoning" of the Net in order to ensure the sovereignty of the countries and national values, has been developed by Joel Reidenberg (Reidenberg, J. "Technology and Internet Jurisdiction", 153 UNIV. OF PENN. L. REV. 1951 (2005)).

In general, the applicability of these two sets of derogations to the cloud computing context needs further assessment from the point of view of the level of data protection aimed at by the CoE. In the context of unbalanced relationship between a cloud provider and its subscribers that could either be individuals or legal persons of little/medium influence, raising the data subject's consent or the necessity to perform a contract concluded between the cloud provider and the customer as primaries legitimate legal basis for international transfers could reveal wholly unsatisfactory. As regards the legitimacy of the TBDF based on the necessity of the performance of the contract, it might be questionable to ground flows as regards users or more generally data subjects having no contractual relationships with the CCS operators by the necessity of the performance of the contract concluded between the CCS and the subscriber author of these TBDF.

18.8.2 International Transfers of Personal Data/Storage of Personal Data and Law Enforcement Objectives

One of the most obvious and serious risks for data protection raised in the context of cloud computing architecture is a massive access by law enforcement authorities to the personal data and information stored in datacenters. Indeed, these datacenters can be established in countries that provide little or no protection of personal data in the framework of law enforcement activities. The development of datacenters might provide great opportunities to public authorities to access to great amount of information pertaining to its citizens or to foreign citizens.[23] Even considering democratic countries, the United States of America constitute a problematic example due to the very controversy third party data issue in the limited scope of the Fourth Amendment protection. We will come back on that issue.

18.8.3 Limitations to Transborder Flows and Applicable Law to the Processing of Personal Data

Cloud computing technologies involve countless TBDF. As regards the viewpoint of the CoE, these flows implicate Parties to the ETS 108 and its additional protocol (including European Union member States), as much as foreign States. A first set of rules is provided for in article 12 of the ETS 108 in consideration of TBDF *between*

---

[23]Except in cases where onion routing is used by cloud computing service. Onion routing is a technique allowing anonymous transactions within a computer network. The messages are encrypted repeatedly and sent through multiple networks nodes called onion routers. Each node decrypts the message in order to get the routing instruction and so encrypts and sends the message to the next onion router till the final destination. Intermediary nodes do not know the origin and the final destination of the message. In that case the national law enforcement agencies are unable to get access to the information if it is transmitted through onion router to a destination outside the national borders. On onion router example, see EFF'sTor: http://www.torproject.org.

Parties to this Convention – only indirectly taking into account TBDF intended for non contracting States (article 12.3, b). And a second one, provided for in article 2 of the additional protocol, directly addresses the issue of TBDF intended towards *non contracting States*. It has also to be underlined that the member States of the European Union have also to apply the TBDF regime provided for by Directive 95/46/EC (in particular, articles 25 and 26).

The aforementioned rules of ETS 108 pursue the specific aim of reconciling guarantying effective data protection and fundamental rights and liberties – even outside national borders – on the one hand, and on the other hand, ensuring the free international circulation of information between people, as the case may be, avoiding forms of protectionism. In this respect, TBDF *between Contracting States* should not be subject to any *special controls*; "in principle there shall not be permitted between Contracting States obstacles to transborder data follows in the form of *prohibitions* or *special authorisations* of data transfers"[24] (emphasis added by authors). Therefore, the ETS 108 prohibits what could be called an "administrative control" of data flows.

However, according to (article 12.3, a), a Party can disregard this rule if it has specific legislation for certain categories of personal data or for *automated personal data files*, because of the nature of those data or those files, except where the legislation of the other Party provide an equivalent protection. So, it can be asked whether CCS could – and should be, for instance due to the characteristics of the service at stake – deemed to constitute such a category of «automated personal *data file*» (e.g., health care online services) that needs to receive a specific treatment? In other words, in the context of CCS and its particular risks, the obligation imposed by the ETS 108 to the contracting States to adopt a particular regulation (e.g., concerning the processing of sensitive data through CCS) has to be assessed. As stated above, cloud computing covers various scenarios and it could require specific rules and particular treatment in some cases and not in others (e.g., depending on the public nature of the CCS, on the nature of the beneficiary of the service who can be a consumer, a children, a private corporation or a public administration).

As far as the additional protocol and the TBDF implying *non contracting States* are concerned, and except the exceptions provided for in article 2.2 of the additional protocol, article 2.1 of the latter compels contracting States to forbid these flows if the concerned non contracting State (or organisation) does not ensure an *adequate* level of protection for the intended data transfers. In this respect, the assessment of adequacy could be realised on a case by case basis. And it can relate to the processing of personal data for criminal investigation purposes by State agencies, or even for any "public" purpose (criminality, taxation, immigration, etc). In this respect, as the Directive 95/46/EC also forbids European States to authorise TBDF towards foreign States not ensuring an adequate protection, the adequacy assessment does not take into consideration the "processing operations concerning public security, defense, State security (including the economic well-being of the State when the

[24]Explanatory Report of the ETS 108, § 67.

processing operation relates to State security matters) and the activities of the State in areas of criminal law".[25] Therefore, to some extent, ETS 108 offers some added value. Anyway, due to the diversity of CCS, some distinctions could be drawn by the contracting States to ETS 108, and the protection offered by one foreign State could be adequate in one case and not in another. Which distinctions can and should/have to be drawn in this respect?

Two principal remarks can be made as regards TBDF directed towards non contracting States.

Firstly, the aforementioned rule should be without prejudice to an analogical – and *a fortiori* – interpretation of (article 12.3, a) of the ETS 108 in the present context of TBDF targeted to non contracting States. That is to say that the Convention should be interpreted in such a way that a contracting State can prohibit – or subject to authorisation – a TBDF related to a specific "automated personal data files" aforementioned if, for instance, the foreign State concerned does not offer an equivalent protection, *even though* it ensures an adequate level of protection. Restrictions to TBDF allowed between contracting States are *a fortiori* allowed between contracting and non contracting States.

Secondly and more generally, the additional protocol doesn't compel the contracting States to do anything else if the targeted foreign State offers an adequate level of protection; it only forbids allowing TBDF targeted to non contracting States. In this respect, despite the fact that the protocol also pursues the free flow of information, it does not explicitly prevent contracting State to forbid personal data flows targeted to a non contracting State offering an adequate protection. The same conclusions also apply as regards Directive 95/46/EC. So, the question in the context of cloud computing and TBDF to foreign States is also the following: could a contracting State deem that a particular processing involved in a CCS require an equivalent protection from the non contracting State, even if this particular processing is not deemed to constitute a particular "automated personal data files" under article 12.3, b) of the ETS 108, or to involve particular data? In other words, contracting States seems here to recover a larger margin of discretion than was the case under the ETS 108. But, on the one hand, how significant is this discretion? And, on the other hand, which CCS could and should/has to be specially treated through this potential margin?

Beyond what has been called an "administrative control" of TBDF, ETS 108 and its additional protocol, although they try to solve – in a certain manner – the issue of TBDF, do not provide for any rule related to the question of the applicable law to the processing of personal data. And this is also true regarding personal data flows between contracting States. As far as these latter are concerned, the explanatory report recognises that "it may not always be easy to determine which [...] national law applies", and it underlines that "the "common core" will result in a harmonization of the laws of the Contracting States and hence decrease the possibility of conflicts of law or jurisdiction". However, neither the Convention, nor the

[25] Article 3.2 of Directive 95/46/EC. These matters are outside the scope of Directive 95/46.

additional Protocol addresses the issue of applicable law. Moreover, the Explanatory Report also specifies that the principle of freedom of flow of personal data provided for in article 12.2 "does not mean that a Contracting State may not take certain measures to keep itself informed of data traffic between its territory and that of another Contracting State, for example by means of declarations to be submitted by controllers of data files". In the context of cloud computing, the scattered worldwide locations of the involved actors (i.e. CCS providers, subscribers, users and data subjects, controllers or processors) exacerbate conflict of laws concerns – that already existed – and have to be faced by national legislations; but how can they regulate and which constraints limit their margin? It is clear that harmonisation of the data protection rules is useful and that people (users and providers of the cloud) will benefit from such harmonisation. However, where no complete harmonisation exists in an inherently international context, a – common – conflict of law rule could bring some legal certainty.

European data protection law addresses, to some extent, the question of the applicable law through *Directive 95/46/EC*. This latter compels Member States to apply their national laws in the cases defined in article 4 of the directive.[26] This article marks the spatial boundaries of European data protection law. It seems that this rule needs to be implemented as a "unilateral conflict of law rule" defining the applicability of the national law at stake following the defined criterions. However, despite the fact that the directive also provides rules as regards TBDF targeted to a non Member State, it does not provide for a general "bilateral conflict of laws rule", that is to say a rule determining which law (of any State) apply to which situations. Therefore, the Member States could be deemed free to adopt their own conflict of law rules as far as they comply with article 4 of Directive 95/46/EC.

Contracting States (here, the legislator or the jurisdictions) have to define which law applies to which particular processing of personal data. And they have different ways to determine the applicable law. They can adopt a bilateral conflict of laws rule determining the applicable law in all instance (bilateral method), they can define the criteria of applicability of their law (for instance, taking into account the place of establishment of the data controller and/or the location of the equipments it uses for the purposes of a particular processing, see art. 4 of the directive 95/46/EC) with an unilateral rule (unilateral method), or they can also define a particular "public order

[26] As regards this rule, see notably Article 29 Data Protection Working Party, Working document on determining the international application of EU data protection law to personal data processing on the Internet by non-EU based web sites (WP56, adopted on 30 May 2002; Article 29 Data Protection Working Party, Opinion 1/2008 on data protection issues related to search engines (WP148), adopted on 4 April 2008, pp. 9–12; Article 29 Data Protection Working Party, Working document on Privacy on the Internet – An integrated EU Approach to On-line Data Protection – (WP37), adopted on 21st November 2000, p. 28; J.-P. Moiny, Op. cit., pp. 255–270. As regards data protection and jurisdiction, see in general C. Kuner, "Data Protection Law and International Jurisdiction on the Internet", Parts 1 and 2, 18 (2 and 3). *International Journal of Law and Information Technology*, 2010: 176–193, the second part will be published in a forthcoming number of the same review.

exception clause" compelling judges to apply national law in some specific cases or when the application of a foreign law leads to unwanted results.

In any case, cloud computing technologies require a reflection on part of contracting States to the Convention of the CoE on which criteria are the best ones to determine the applicability of their national data protection laws and to accommodate the particular issues arising from the above mentioned technologies. In this respect, for example, only some data protection rules could receive a particular territorial scope as regards cloud computing services in general or even some cloud computing services in particular. For instance, specific duties regarding information and right of access could have a more extended territorial scope if some data protection rules are extended to the processors, imposing them specific duties or responsibilities – if deemed necessary in the evolution of data protection law. And the applicability of these rules could depend on specific criteria differing from those applicable to the data controller according to already established general data protection rules. Needless to say, such a conflict of laws rule would gain in quality – from a practical point of view – if it would be discussed at an international level – for instance, under the auspices of the CoE. It should also be noted that directive 95/46/EC is under review. A discussion relating to conflicts of law seems to be of high interest and pressing to guarantee the practical enforcement of data subjects' protection, and to bring legal certainty to the emergent and promising market of cloud computing.

In such a reflection, it is required to take the technical peculiarities of CCS into consideration, for instance to avoid using irrelevant links with the territories of the State whose law has to apply. For instance, the place of the equipments used for the processing of personal data can be solely the result of efficiency considerations related to the working of the cloud. In this case, such a location seems less relevant as regards the identification of the applicable law, while Directive 95/46/EC links the determination of the applicable law to the location of the equipments used for the processing at stake. However, it has to be noted that the location of the processing capabilities can also help bringing legal certainty by localising the CCS offered in a specific geographic area to ensure the applicability of a specific legislation. It could be a convenient way to avoid conflict of law and to provide a wide range of services taking into account users wishes as regards data protection rules. The place of establishment of the CCS provider can also be of little relevance when this provider purposefully offers its services to consumers located in another country than the country where he is established.

A final point can be underlined as regards the applicable law to the processing of personal data and the TBDF's involved in the context of CCS: which influence would article 8 of the European Convention on Human Rights (ECHR) have on the international processing of personal data and on conflict of law?

Article 1 ECHR reads as follows: "The High Contracting Parties shall secure to everyone within their jurisdiction the rights and freedoms defined in Section I of this Convention". In this respect, the jurisdictions of these contracting Parties, applying the law of a non-contracting State of the ECHR, could have to ignore this foreign law if, in the particular case, it rises to a conflicting situation with the fundamental rights provided for by the ECHR.[27] For instance, the European Court of Human Rights has already approached the concern of the influence of the ECHR on private international law as regards article 6 ECHR and the exequatur of a foreign judgment.[28] Four questions need to be addressed. Firstly, which "rights" recognised under article 8 ECHR could influence the application, in a particular case, of conflict of laws rules? Secondly, which data protection rules fall within the scope of article 8 ECHR and these rights? For instance, which rules of the "common core" of the [ETS] 108"? It should be kept in mind to this respect that data protection rules proceed to the "horizontalization" of the human right to privacy in the information society. And finally, which "connections" an international case involving cloud computing technologies need to have with the CoE member States' territories to require the applicability of these identified rights? It has to be recalled that this would happen under the final control of the European Court of Human Rights.[29]

To sum up, closely regarding the specificities of cloud computing technologies, contracting States to the ETS 108 have to determine which applicability of which national data protection rule to international cases is desirable and permitted and/or required, avoiding, on the one hand, suffocating a new technology and, on the other hand, depriving people under their jurisdiction of rights they already have or of new rights it is deemed appropriate they have.

18.9 Law Enforcements Agencies and Data Retention

The fact that CCS operators are processing huge amounts of data including quite sensitive ones about the CCS customers or third parties, explains the interest of law enforcement agencies to have access to these data through the CCS provider cooperation or by imposing this latter similar obligations than to public communication services operators as regards data retention imposed by the EU 2006 Directive on Data retention.[30] In other words, we have to pay attention to the question of the extension of certain legal obligations for certain communications services' providers to retain data about the uses of their services or to cooperate with law enforcement authorities at their request or even at their own initiative.[31] That obligation would be more or less similar to the obligation imposed by the EU Directive to the IAPs and publicly available e-communication services' operators. Other questions might be

[27] Regarding the potential influence of the ECHR on conflict of laws, see notably Gannagé, L. *"A propos de l' "absolutisme" des droits fondamentaux", in Vers de nouveaux équilibres entre ordres juridiques – Liber amicorum Hélène Gaudemet-Tallon*. Paris: Dalloz, 2008, pp. 265–284.

[28] See European Court of Human Right, 20 July 2001, Pellegrini v. Italy.

[29] Mayer, P. "La Convention européenne des droits de l'homme et l'application des normes étrangères", *Revue Critique de droit international privé*, (1991): 664.

[30] Directive 2006/24/EC of the European Parliament and of the Council of 15 March 2006 on the retention of data generated or processed in connection with the provision of publicly available electronic communications services or of public communications networks.

[31] See the Council of Europe Cybercrime Convention, article 17.

raised. Cloud computing represents a shift of location which might have two consequences. The first one is the following: to what extent, can we consider that the LEA are authorised in case where data about a customer located in the LEA country are placed somewhere in the clouds to extend their searches to the foreign country where the data are located by the CCS and that using the online facilities offered to the customer? In line with this question we might pose the following: "Do the different national LEA'es have to cooperate together?"

Article 23 and ff of the Council of Europe Cybercrime Convention (art. 23 and ff) imposes such a duty to cooperate while fixing certain conditions of such cooperation and the means to ensure effectively that cooperation.

Within Europe, the 2009 Council Framework Decision on the European Evidence Warrant' "*for the purpose of obtaining objects, documents and data for use in proceedings in criminal matters*"[32] imposes cooperation, but according to article 10 of the Decision, requires regarding the transfers of personal data, the consent of the executing authority, meaning the LEA which is required to transfer data to a LEA of another country. Beyond this text, other multilateral agreements serve as the basis for mutual assistance in criminal matters in the EU. We quote the Council of Europe Convention on Mutual Assistance in Criminal Matters, 1959 ("the 1959 Convention"), and its protocols of 1978 and 2001 and the Schengen Agreement of 1985 and the EU Convention on Mutual Assistance in Criminal Matters 2000 (the "MLAC"[33]) and its protocol of 2001.[34]

The second question represents the other facet of the distinction between the place of the data and the customer. "*Is a national LEA investigating the computer of a suspected person located in its territory allowed to extend the research outside of the country to information systems connected with the investigated computer?*" According to their criminal procedure legislations, certain national LEA (for instance, Belgium, UK and France) have this power to assert jurisdiction over data stored in other countries, but accessible via electronic networks located in their own country. So, if LEA is investigating in Belgium in a computer located in Belgium, LEA might capture data stored in a third country but accessible via the local system. With the same argument it would be possible for LEA investigating in CCS premises located in their country to enter into the whole CCS information system. In other

---

[32]The Council's 2009 Framework Decision on the European Evidence Warrant ("EEW") applies the mutual recognition principle to judicial decisions for the purpose of obtaining evidence for use in proceedings in criminal matters. The EEW provides that Member States' law enforcement authorities should give immediate effect to judicial search and seizure orders emanating from other Member States. The EEW also provides standard forms for issuing orders, and fixed deadlines for executing orders.

[33]We underline that he MLAC only binds those States that choose to ratify it. To date, the MLAC has been ratified by 23 of the 27 EU Member States.

[34]About all these texts and for a detailed commentary, Spencer, J.R. "The Problems of Trans-Border Evidence and European Initiatives to Resolve Them" (2007) 9 Cambridge Yearbook of European Legal Studies 477, at 478.

words, cloud service providers may be subject to disclosure requests in countries outside of those where the data are stored.

A third question is more delicate: Does a CCS provider have a legal duty to cooperate with LEA'es? Do we need a legal framework for this cooperation? The GNI (Global Network Initiative) [35] has developed voluntary guidelines as regards the response to be given to governments' demands for access or blocking as solutions to the multinational dimension of the problem? The GNI pleads notably:

- for a global consensus as regards the governmental demands (Who? How? For which offences? etc);
- for prohibiting any overbroad demand and need for clear communications by writing;
- for a narrow interpretation of the demand (e.g. limitation in principle to data concerning data subjects located within the country).

Furthermore, the signatories clearly announce that they will challenge governmental demand before the courts when these demands seem inconsistent with the legal requirements and that they will take appropriate measures to make the Information services' users aware of the policies followed by the CCS providers and the governments.

The recent US-EU SWIFT agreement approved by the EU Parliament in July, 6 2010[36] could also be evoked in this context since one might imagine that a foreign LEA would like to obtain data stored in Europe in order to discover certain evidence of criminal infringements. On that point in the context of this SWIFT case, the European Union and the US have signed a revised agreement on sharing banking data to investigate suspected terrorist financing, moving the long-running negotiations over the deal a step closer to completion. Under the revised deal, an EU official would be posted in the US treasury in Washington to scrutinize the transfer of the European banking data to investigators. Information requests are also to be "tailored as narrowly as possible" and will be checked by Europol, the EU's police coordination agency. This solution might be extended to the access of official authority to data stored in Europe by a CCS operator.

---

[35] See the principles adopted in 2008 "Protecting and Advancing Freedom of Expression and Privacy in Information and Communications Technologies, available at the GNI website: www.globalnetworkinitiative.org.
These Principles on Freedom of Expression and Privacy ("the Principles") have been developed by companies, investors, civil society organizations and academics. "*They are based on internationally recognized laws and standards for human rights, including the Universal Declaration of Human Rights ("UDHR"), the International Covenant on Civil and Political Rights ("ICCPR") and the International Covenant on Economic, Social and Cultural Rights ("ICESCR")*".

[36] http://register.consilium.europa.eu/pdf/en/10/st11/st11222-re01.en10.pdf

18.10 Conclusions

The main and initial question raised by the considerations and questions set above is whether specific regulation on cloud computing is needed.

At this stage, following the considerations provided in this chapter, it is clear that, from a privacy legal point of view, different cloud computing services have different characteristics: Facebook does not raise the same problems than Microsoft's Azure or Amazon's EC2.

On the one hand, services have different natures – e.g. IaaS, PaaS or SaaS, private or public clouds, etc, – and various purposes – domestic, professional, public, etc. And on the other hand, the involved actors are also very different – individuals who are consumers or professionals, SMEs, NPOs, administrations, worldwide corporations, etc, and numerous imbalances could exist between them. Therefore, the questions identified above could receive varying answers according to the many facets of cloud computing technologies that will most probably continually evolve. In fact, these facets not necessarily raise the same concerns as regards data protection. Moreover, in the same sense, these questions could also vary according to the particular services and actors at stake, and they could not always have the same pertinence.

The questions raised in this chapter can be summarised as follows:

(1) Is the differentiation between domestic use/non domestic use pertinent, and do we need to extend the protection to the legal person and to change the concept of personal data?

- As seen before, such modification would be needed in the new environment of Cloud computing which concerns individuals as well as legal bodies.
- By maintaining the exclusion of domestic use from the protection – as Directive 95/46 does – would exclude many individuals using cloud computing in a domestic way (social network, Email services, etc) from any protection. Is this acceptable?
- The notion of personal data is very narrow and is not including the know-how, the commercial secret, etc.

(2) Who are the actors of cloud computing? Do they need to be legally defined if it is not already the case? If they are already legally defined, do the definitions at stake need to be modified? We identified five, sometimes overlapping, categories of actors: subscribers, users, data subjects, controllers (co-controllers) and data processors. This raises two principal questions:

- Does the concept of data processor need to be defined under ETS 108?
- Do legal persons need to be protected under the data protection rules of the ETS 108, with regard to which data (extension of the definition of the personal data and, therefore, of the data subject)?

(3) Which existing duties under ETS 108 need to be adapted? Which non-existing duties under ETS 108 need to be created? As the case may be, which actor has to bear these modifications or these creations? More precisely:

- Should data processors have to support specific duties provided for by the law, and which duties (e.g. in general as regards transparence and liability)?
- Should co-controllers to be targeted by specific liability rules and a particular allocation of duties under ETS 108?
- Should a specific duty regarding security breaches be established? Who would have to support this new duty (provider and/or subscriber), towards which actor (subscriber and/or data subjects) and in which cases?
- How to treat the distinction between non-domestic and domestic processing activities? When is it still relevant and how to improve the protection of data subjects when a domestic use exception could apply (total exclusion of data protection law or establishment of a softer legal regime)?
- Should data retention obligations have to be imposed on cloud computing services providers, when and how?
- Due to the possible imbalance between the actors of the cloud, is consent always an adequate basis of the legitimacy of the processing at stake or should data controllers – and if so when – have a duty to base the legitimacy of their processing on an additional basis?

(4) How could what call the "data protection continuity" be maintained? This question can be subdivided into the following concerns:

- When the cloud computing service provider or its user (data subject) terminates the contractual relationship at stake, how can it be guaranteed that the data subject (user) will recover the total "ownership" (control) of data relating to him?
- In cases of bankruptcies, mergers of corporations or sales of corporations, etc, how can it be guaranteed that the level of protection originally ensured to the data subject will remain at least equivalent?

(5) How to face the numerous concerns arising out of the international character inherent in cloud computing? This broad question also needs to be sliced into parts:

- Do some specific cloud computing services (e.g., involving sensitive data) need to be forbidden when they imply TBDF between contracting States and, a fortiori, non-contracting States ensuring an adequate level of protection?
- Which concerns can be solved by binding corporate rules?
- How to assess the adequacy of non-contracting States to ETS 108 as regards the processing of personal data for law enforcement purposes?
- How far could consent and contract authorise TBDF outside the territories of contracting States, towards non-contracting States not ensuring an adequate level of protection?

- How to resolve conflict of laws when actors involved in the cloud are located anywhere in the world and rules on conflict resolution do not yet exist. In other words, we should work out rules to solve conflicts of law at least in the context of Cloud computing.
- Does the "territoriality" of data protection rules have to be differently defined depending on the duties (e.g. security or transparence) and the actors (data controller or data processor) at stake, and if so, how?
- Finally when their data are in the clouds how to ensure the protection of the data subjects against the investigatory powers of the LEA? Is there an obligation for CCS providers to cooperate with the different LEA? Is that allowed for a national LEA investigating in the computer of a suspected person located in its territory to extend the research outside of the country to information systems connected with the investigated computer? Is the "zoning of the net" possible and if yes can we impose it through appropriate regulations? Are the recent EU-US agreement about SWIFT a good point of departure as regards the fixation of the limits of the cooperation between LEA?

(6) Do we have to ban or restrict the use of cloud computing services regarding sensitive matters, professions or activities (public or not)?

- This question raises the issue to impose on the CCS provider to limit its cloud or country of storage to a certain area such as the European Union. It would avoid to have sensitive data stocked in a non democratic regime who would nor guarantee the respect of privacy.

(7) On the security field, do we need to make special provisions for the cloud computing?

- What's about the role of standardisation bodies?
- Do we need to envisage security breach provisions in that context?

References

Article 29 Data Protection Working Party, Working document on Privacy on the Internet – An integrated EU Approach to On-line Data Protection – (WP37), adopted on 21st November 2000

Article 29 Data Protection Working Party, Working document on determining the international application of EU data protection law to personal data processing on the Internet by non-EU based web sites (WP56), adopted on 30 May 2002.

Article 29 Data Protection Working Party, Opinion 1/2008 on data protection issues related to search engines (WP148), adopted on 4 April 2008.

Article 29 Data Protection Working Party, Opinion 1/2010 on the concepts of "controller" and "processor" (WP169), adopted on 16 February 2010.

Article 29 Data Protection Working Party, Opinion 2/2010 on online behavioural advertising (WP171), adopted on 22 June 2010, p. 17.

Blok, P. Het recht op privacy, Boom Juridische uitgevers, 2003.

Bygrave, L. *Data Protection Law: Approaching Its Rationale, Logic and Limits*, The Hague/London/New York: Kluwer Law International, 2002.

Gannagé, L. *"A propos de l' "absolutisme" des droits fondamentaux", in Vers de nouveaux équilibres entre ordres juridiques – Liber amicorum Hélène Gaudemet-Tallon*. Paris: Dalloz, 2008.

Gellman, B. Privacy in the clouds: Risks to Privacy and confidentiality from Cloud Computing, Report prepared for the World Privacy Forum, Feb. 23, 2009.

Kuner, C. "Data Protection Law and International Jurisdiction on the Internet", Parts 1 & 2, 18 (2 & 3). *International Journal of Law and Information Technology*, (2010).

Mayer, P. "La Convention européenne des droits de l'homme et l'application des normes étrangères", *Revue Critique de droit international privé*, (1991): 651–665.

Meil, P., and T. Grance. The NIST Definition of Cloud Computing, Version 15, 10-07-09, available on NIST (National Institute of Standards and Technology) web site.

Moiny, J.-P. "Facebook au regard des règles européennes concernant la protection des données", 2 E.C.J.L., 2010, pp. 235 and ff.

Reidenberg, J. "Technology and Internet Jurisdiction", 153 UNIV. OF PENN. L. REV. 1951 (2005).

Solove D.J. "Conceptualizing Privacy", 90 California Law Review, 2002, 1085-.

Spencer, J.R. "The Problems of Trans-Border Evidence and European Initiatives to Resolve Them" (2007) 9 Cambridge Yearbook of European Legal Studies 477.

Chapter 19
Privacy-Preserving Data Mining from Outsourced Databases

Fosca Giannotti, Laks V.S. Lakshmanan, Anna Monreale, Dino Pedreschi, and Hui (Wendy) Wang

19.1 Introduction

In recent years, there has been considerable interest in the so-called data mining-as-service paradigm for enabling organizations with limited computational resources and/or data mining expertise to outsource their data mining needs to a third party service provider (Agrawal and Srikant, 2000; Bu et al., 2007; Chen et al., 2007; Lakshmanan et al., 2005; Wong et al., 2007). As an example, the operational transactional data from various stores of a supermarket chain can be shipped to a third party which provides mining services. The supermarket management need not employ an in-house team of data mining experts. Besides, they can cut down their local data management requirements because periodically data is shipped to the service provider who is in charge of maintaining it and conducting mining on it in response to requests from business analysts of the supermarket chain. It is generally expected that the paradigm of "mining and management of data as service" will grow with the advent and popularity of cloud computing (Buyya et al., 2008).

In the above example, the supermarket chain, the client, is a data owner and the service provider is referred to as the server. One of the main issues with this paradigm is that the server has access to valuable data of the owner and may learn or disclose sensitive information from it. For example, by looking at the transactions, the server (or an intruder who gains access to the server) can learn which products (items) are co-purchased, and in turn, the mined patterns that describe the supermarket customers' behavior. In this context, both the sale transactions and the mined patterns and all the information that can be extracted from the data are the property of the supermarket and should remain safe from the server and any other intruder. Indeed the knowledge extracted from the data can be used from the supermarket in important marketing decisions to improve their services.

This problem of protecting private information of organizations/companies is referred to as "corporate privacy" (Clifton et al., 2002). Unlike personal privacy,

F. Giannotti (✉)
ISTI-CNR, Pisa, Italy
e-mail: fosca.giannotti@isti.cnr.it

S. Gutwirth et al. (eds.), *Computers, Privacy and Data Protection: an Element of Choice*, DOI 10.1007/978-94-007-0641-5_19,

which only considers the protection of the personal information recorded about individuals, corporate privacy requires that both the individual items and the patterns of the collection of data items are regarded as corporate assets and thus must be protected.

In this position paper, we study the problem of outsourcing the *association rule mining* task within a corporate privacy-preserving framework. Association rule mining has the objective of discovering groups of products, or items, that are frequently purchased together by the supermarket's customers: the expected output of such task, given the sale transaction database as input, is the list of all possible groups of items, such as {milk, beer, diapers}, that occur together in a fraction of the market baskets that is statistically significant. The complexity of this task is evident: there are tens of thousands of distinct products in the assortment of a supermarket, and therefore the number of potential candidate groups of products quickly explodes with the size of the group. This computational complexity motivates the introduction of an outsourcing model, where the data owner, like our supermarket, gives the data in outsourcing to a service provider to obtain an association rule mining service from it, within a privacy-preserving framework, i.e., without disclosing neither the sale data nor the information deriving from the mining analysis.

In order to achieve a strong data protection, we need to assume an adversarial model where the attacker, who wants to acquire information on the sale data and the mined patterns, has a rich background information; to this aim, we assume that the attacker knows with precision the set of items in the original transaction database and their popularity, i.e., how many times each individual item is sold. This information can be obtained from a competing company or from published reports.

To counter this attack, we propose an encryption scheme that transforms the original database: (1) by replacing each item by a 1–1 substitution function and (2) adding fake transactions to the database in such a way that each item (itemset) becomes indistinguishable with at least *k–1* other items (itemsets); in other words, in the outsourced database, for each item (itemset) there are at least *k–1* other items (itemsets) that have the same number of occurrences into the database.

On the basis of this simple idea, our framework guarantees that not only individual items, but also any group of items has the property of being indistinguishable from at least *k* other groups in the worst case, and actually many more in the average case. This protection implies that the attacker has a very limited probability of guessing the actual items contained either in the sale data or in the mining results; on the contrary, the data owner can efficiently decrypt the mining results with limited computational resources, because the information that it maintains to this purpose is negligibly small; also, the initial encryption of the database can be done in an efficient way. In this paper we discuss the above privacy-preserving outsourcing model and show some preliminary results obtained applying this model over large-scale, real-life transaction databases donated by a large supermarket chain in Europe.

The architecture behind our model is illustrated in Fig. 19.1. The client/owner encrypts its transaction database (TDB) using an encrypt/decrypt module, which can be essentially treated as a "black box" from its perspective. This module is

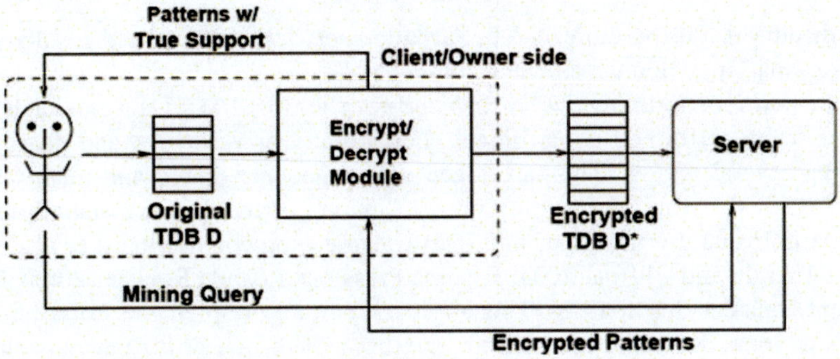

Fig. 19.1 Architecture of mining-as-service paradigm

responsible for transforming the TDB D into an encrypted database (TDB D^*). The server conducts data mining and sends the (encrypted) patterns to the owner. Our encryption scheme has the property that the returned number of occurrences of the patterns are not true. The encrypt/decrypt (ED) module recovers the true identity of the returned patterns as well their true number of occurrences.

19.2 Related Work

In this section, for the purpose of clarifying similarities and differences with our problem setting, we outline the work on privacy-preserving data publishing, mining, and outsourcing, which can be classified into the following six categories.

- *Privacy-preserving data publishing* (PPDP): The idea is that data is published by an owner for the common benefit of allowing analysts to mine patterns from it. Data is published with appropriate suppression, generalization, distortion, and/or decomposition such that individual privacy is not compromised and yet the published data is useful for mining (Fung et al., 2007; Machanavajjhala et al., 2006; Samarati, 2001; Xiao and Tao, 2006; Xu et al., 2008). Clearly, this approach can protect personal privacy, but not corporate privacy, i.e., privacy of the assets consisting of the transaction database and the patterns that can be mined from it.
- *Privacy-preserving data mining* (PPDM): The main model here is that private data is collected from a number of sources by a collector for the purpose of consolidating the data and conducting mining. The collector is not trusted with protecting the privacy, so data is subjected to a random perturbation as it is collected. Techniques have been developed for perturbing the data so as to preserve privacy while ensuring the mined patterns or other analytical properties are sufficiently close to the patterns mined from original data. This body of work was pioneered by Agrawal and Srikant (2000) and has been followed up by several papers since (Agrawal and Haritsa, 2005; Rizvi and Haritsa, 2002). Again, this approach is not suited for corporate privacy, in that some analytical properties

are disclosed. Furthermore, the perturbation restricts the returned results to an approximation, while we aim at accurate results.

- *Secure multiparty mining over distributed datasets* (SMPM): Data on which mining is to be performed is partitioned, horizontally or vertically, and distributed among several parties. The partitioned data cannot be shared and must remain private but the results of mining on the "union" of the data are shared among the participants, by means of multiparty secure protocols (Gilburd et al., 2005; Kantarcioglu and Clifton, 2004; Krishna Prasad and Pandu Rangan, 2006). They do not consider third parties. This approach partially implements corporate privacy, as local databases are kept private, but it is too weak for our outsourcing problem, as the resulting patterns are disclosed to multiple parties.
- *Privacy-preserving pattern publishing* (PPPP): The central question is how to publish results of mining such as frequent patterns without revealing any sensitive information about the underlying data (Atzori et al., 2008). Once again, the resulting patterns are disclosed.
- *Database outsourcing*: To protect the security of the outsourced data, they will be encrypted. Most of the work then focus on efficient query evaluation over encrypted databases (Agrawal et al., 2004; Hacigumus et al., 2002; Song et al., 2000), and cannot be applied to our model. Furthermore, the frequency-based attack in our work is seldom studied in the database outsourcing work.

The particular problem attacked in this paper is outsourcing of pattern mining within a corporate privacy-preserving framework. A key distinction between this problem and the above mentioned PPDM problems is that, in our setting, not only the underlying data but also the mined results are not intended for sharing and must remain private. The work that is most related to ours is Wong et al.'s system presented at VLDB 2007 (Wong et al., 2007). Similar to our work, first, they utilize a one-to-n item mapping together with non-deterministic addition of cipher items to protect the identification of individual items. Second, they assume that the adversary may possess some prior knowledge of frequency of the itemsets, which can be used to decipher the encrypted items. In contrast, our attack model focuses on single items with the assumption that the attacker knows the exact frequency of every single item. The major issue left open by (Wong et al., 2007) is a formal protection result: their security analysis is entirely conducted empirically on various synthetic datasets. The notion of (h, k, p)-coherence presented by Xu et al. at KDD 2008 (2006) achieves an effect similar to our approach of encrypting the database such that items fall into equivalence classes of size $\geq k$. However, this anonymization method is meant for data publishing purposes, and it is not applicable to privacy-preserving outsourcing.

19.3 Preliminaries: Pattern Mining

We now give some basic account on association rule mining. This mining task allows us to discover regularities between products in large-scale transaction data of supermarkets. For example, the rule $\{milk\} \rightarrow \{bread\}$ found in the database in

TDB
Bread
Milk Bread
Bread Milk
Water Milk
Bread Beer
Bread Eggs
Water

Fig. 19.2 An example of transaction database

Fig. 19.2 would indicate that if a customer buys *milk*, he/she is likely to also buy *bread*. Similar information can be used for marketing decisions such as promotional pricing or product placements.

Two major steps in mining association rules are: (i) finding frequent patterns or itemsets and (ii) computing the association rules from them. Step (i) is, by far, the computationally dominant step. We briefly review frequent pattern mining below. Let $I = \{i_1, \ldots, i_n\}$ be the set of items and $D = \{t_1, \ldots, t_m\}$ a transaction database (TDB), where each transaction is a set of items. We denote the *support* of an itemset $S \subseteq I$ as $supp_D(S)$ and the *frequency* or *relative support* by $freq_D(S)$. Recall, $supp_D(S)$ is the number of occurrences of the itemset S in the database D; in other words, it is the number of transactions containing the itemset S. While $freq_D(S) = supp_D(S)/\,|D|$. For each item i, $supp_D(i)$ and $freq_D(i)$ denote respectively the individual support and frequency of i. In the database in Fig. 19.2 the support of the item *Water* is equal to 2 and its frequency is 2/7.

The whole function $supp_D(i)$, for each item i of the database D, is also called the item support table of D.

It can be either represented in tabular form (see, e.g., table (a) in Fig. 19.6), or plotted as a histogram, as in Fig. 19.3, where the item support distribution of the

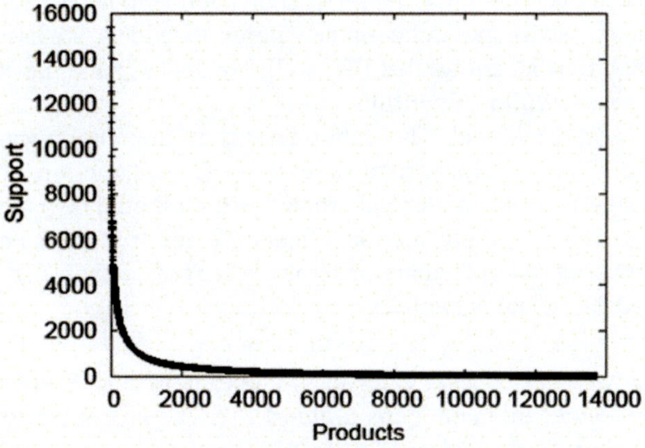

Fig. 19.3 Item support distribution of a real TDB analyzed

real-life TDB is reported; both in support tables and in histograms items are listed in decreasing order of their support.

The length of a transaction $t \in D$ is the number of items in t. We define the size of a TDB D as the sum of lengths of all its transactions, i.e., $\|D\| = \sum_{t \in D} |t|$. It can also be defined as $\|D\| = \sum_{i \in I} supp_D(i)$. This corresponds to the area under the support distribution graph (e.g., see Fig. 19.3).

The well-known frequent pattern mining problem (Agrawal and Srikant, 2000) is: given a TDB D and a support threshold σ, find all patterns (itemsets) whose support in D is at least σ. In this paper, we confine ourselves to the study of a (corporate) privacy-preserving outsourcing framework for frequent pattern mining.

19.4 Privacy Model

We let D denote the original TDB that the owner has. To protect the identification of individual items, the owner applies an encryption function to D and transforms it to D^*, the encrypted database. We refer to items in D as *plain items* and items in D^* as *cipher items*. The term *item* shall mean plain item by default. The notions of plain item sets, plain transactions, plain patterns, and their cipher counterparts are defined in the obvious way. We use I to denote the set of plain items and E to refer to the set of cipher items.

Adversary Knowledge. The server or an intruder (*attacker*) who gains access to it may possess some background knowledge which they can use to conduct attacks on the encrypted database D^* in order to make inferences.

We adopt a conservative model and assume that the attacker knows exactly the set of (plain) items I in the original transaction database D and their true supports in D, i.e., $supp_D(i)$, $i \in I$. The attacker may have access to similar data from a competing company, may read published reports, etc. Moreover we assume the attacker has access to the encrypted database D^*. Thus, he also knows the set of cipher items and their support in D^*, i.e., $supp_{D^*}(e)$, $e \in E$.

In this position paper we propose an encryption scheme based on: (i) replacing each plain item in D by a 1–1 substitution cipher (ii) adding fake transactions to the database. In particular, no new items are added. We assume the attacker knows this and thus he knows that $|E| = |I|$. We also assume the attacker knows the details of our encryption algorithm.

Attack Model. The data owner (i.e., the corporate) considers the true identity of: (1) every cipher item (2) every cipher transaction, and (3) every cipher frequent pattern as the intellectual property which should be protected. If the cipher items are broken, i.e., their true identification is inferred by the attacker, then clearly cipher transactions and cipher patterns are broken, so they also must remain protected. The attack model is twofold:

- *Item-based attack*: for each cipher item $e \in E$, the attacker constructs a set of candidate plain items $Cand(e) \subset I$. The probability that the cipher item e can be broken $prob(e) = 1/\,|Cand(e)|$.

- *Set-based attack*: Given a cipher itemset E, the attacker constructs a set of candidate plain itemsets *Cand(E)*, where for each $X \in Cand(E)$, $X \subset I$, and $|X| = |E|$. The probability that the cipher itemset E can be broken $prob(E) = 1/\,|Cand(E)|$.

We refer to *prob(e)* and *prob(E)* as *probabilities of disclosure*. From the point of view of the owner, minimizing the probabilities of disclosure is desirable. Intuitively, *Cand(e)* and *Cand(E)* should be as large as possible. Ideally, *Cand(e)* should be the whole set of plaintext items. This can be achieved if we bring each cipher item to the same level of support, e.g., to the support of the most frequent item in *D*. Unfortunately, this option is impractical, as this would lead to a large increase in the size of D^* compared to *D*, i.e., a large size of the fake transactions. This in turn leads to a dramatic explosion of the frequent patterns, making pattern mining at the server side computationally prohibitive. This is the motivation for relaxing the equal-support constraint and introducing k-anonymity as a compromise.

Definition 1 (Item k-anonymity). Let *D* be a transaction database and D^* its encrypted version. We say D^* satisfies the property of *item k-anonymity* provided for every cipher item $e \in E$, if there are at least $k-1$ other distinct cipher items $e_1, \ldots, e_{k-1} \in E$ such that $supp_{D^*}(e) = supp_{D^*}(e_i)$, $1 \le i \le k-1$.

Figure 19.4 shows the effect of grouping together cipher items into groups of k items. For a given value of *k*, the support distribution resembles a descending staircase. With small *k*, the graph tends to the original support distribution in *D*; while as *k* increases, the graph gets closer to the horizontal line discussed above. As the size of D^* is the area below the graph, we can control the size of D^* by an appropriate choice of *k*.

To quantify the privacy guarantee of an encrypted database, we define the following notion:

Definition 2 (k-Privacy). Given a database *D* and its encrypted version D^*, we say D^* is *k-private* if:

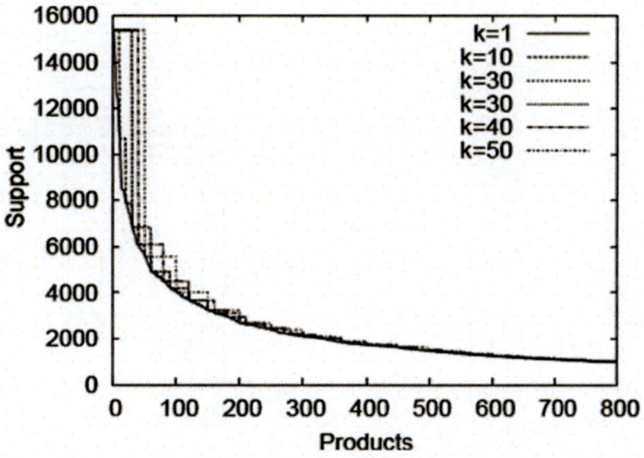

Fig. 19.4 Item support distribution on encrypted TDB with k=10, 20, …, 50

1. for each cipher item $e \in D^*$, $prob(e) \leq 1/k$; and
2. for each cipher itemset E with support $supp_{D^*}(E) > 0$, $prob(E) \leq 1/k$.

This definition does not constrain the disclosure probability of cipher itemsets which have no support in D^*. Intuitively, such cipher itemsets are not interesting. This will be exploited in the next section in designing effective k-private encryption schemes. Formally, the problem we study is the following:

Problem Studied Given a plain database D, construct a *k-private* cipher database D^* by using substitution ciphers and adding fake transactions such that from the set of frequent cipher patterns and their support in D^* sent to the owner by the server, the owner can reconstruct the true frequent patterns of D and their exact support. Additionally, we would like to reduce the space and time incurred by the owner in the process and the mining overhead incurred by the server.

19.5 Encryption/Decryption Scheme

In this section, we describe the ED module, responsible for the encryption of TDB and for the decryption of the cipher patterns coming from the server. The general idea of our Encryption/Decryption method is show with an example in Fig. 19.5.

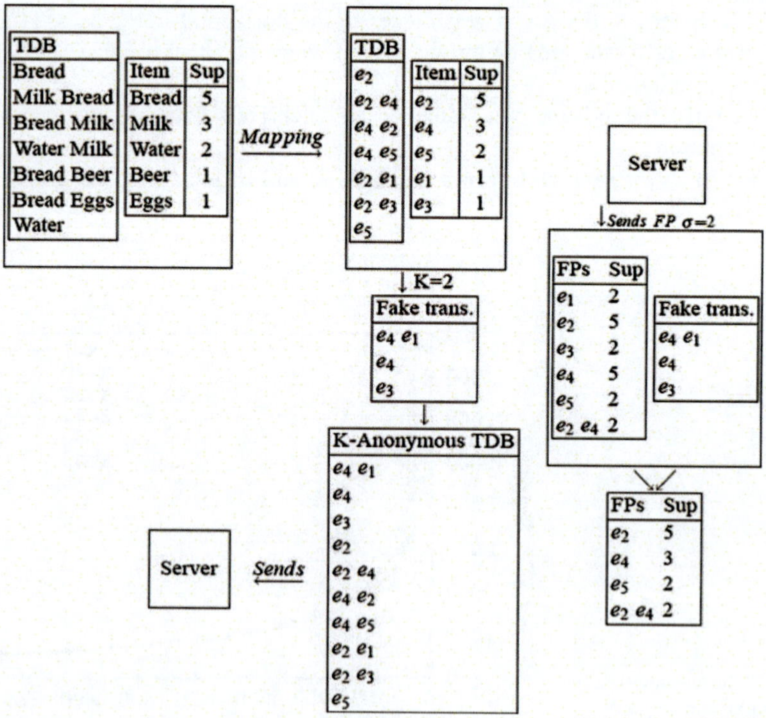

Fig. 19.5 Module for encryption (*left*) and decryption (*right*)

The client/owner replaces each item by a 1–1 substitution function and adds some fake transactions to the database in such a way that for each item (itemset) there are at least *k–1* other items (itemsets) that have the same number of occurrences into the database. Then, the client sends the modified database to the server. The server, when receives a mining query, computes the encrypted frequent patterns and sends them to the client that knows the fake transactions and so can recover the true patterns from the extracted patterns with the real support.

19.5.1 Encryption

In this section, we introduce the encryption scheme, which transforms a TDB D into its encrypted version D^*. Our scheme is parametric w.r.t. $k > 0$ and consists of three main steps: (1) using 1–1 substitution ciphers for each plain item; (2) using a specific item k-grouping method; (3) using a method for adding new fake transactions for achieving k-privacy. The encryption scheme is a countermeasure to the item-based and set-based attacks discussed in Section 19.4: since the attacker knows the exact support of each item, we create a k-private D^*, such that the cipher items cannot be broken based on their support.

k-Grouping Method. Given the items support table, several strategies can be adopted to cluster the items into groups of k. We assume the item support table is sorted in descending order of support and refer to cipher items in this order as e_1, e_2, etc. In order to obtain the formal protection that itemsets (or transactions) cannot be disclosed with a probability higher than 1/k, we need to use only grouping methods that yield groups of items that are unsupported in D. This means that if we consider Fig. 19.6 we cannot use a grouping method that generate the group $\{e_2, \ e_4\}$ as this two items appear together in a transaction in the original database showed in Fig. 19.5. We call such grouping methods robust:

Definition 3. Given a TDB D and a grouping G of the items occurring in D, G is called robust for D if and only if, for any group G_i of G, $supp_D(G_i) = 0$.

The above definition directly suggests a procedure for checking whether a given grouping G for a TDB D is robust: it is sufficient to check that the support in D of each group G_i in G is 0. If this is the case, the grouping can be safely used to

(a) IST

Item	Support
e_2	5
e_4	3
e_5	2
e_1	1
e_3	1

(b) Grouping

Item	Support
e_2	5
e_5	2
e_4	3
e_1	1
e_3	1

(c) Noise Table

Item	Support	Noise
e_2	5	0
e_5	2	3
e_4	3	0
e_1	1	2
e_3	1	2

(d) Hash Tables

	Table1
0	$\langle e_5, 1, 2\rangle$
1	$\langle e_3, 2, 0\rangle$

	Table2
0	$\langle e_1, 2, 0\rangle$

Fig. 19.6 Encryption with $k = 2$

obtain the maximum privacy protection guaranteed by our method. The following definition introduces our grouping method.

Definition 4. Given the TDB D and its item support table in decreasing order of support, our grouping method:

STEP1: groups together cipher items into groups of k adjacent items starting from the most frequent item e_1, obtaining the grouping $G = (G_1, \ldots, G_m)$ (i.e., $G_1 = \{e_{1 \ldots} e_k\}$, $G_2 = \{e_{k+1, \ldots} e_{2k}\} \ldots$).

STEP2: modifies the groups of G by repeating the following operations, until no group of items is supported in D:

- Select the smallest $j \geq 1$ such that $supp_D(G_j) > 0$,
- Find the most frequent item $i' \in G_j$ such that, for the least frequent item i of G_j we have: $supp_D(G_j \,|\{i\} \cup \{i'\}) = 0$,
- Swap i with i' in the grouping.

The output of grouping can be represented as the noise table. It extends the item support table with an extra column Noise indicating, for each cipher item e, the difference among the support of the most frequent cipher item in e's group and the support of e itself, as reported in the item support table. We denote the noise of a cipher item e as $N(e)$. The noise column indicates, for each cipher item e, the number of occurrences of e that are needed in D^* in order to bring e to the same support as the most frequent item of e's group. As such, the noise table represents the tool for generating the fake transactions to be added to D to obtain D^*. In particular, the total size of the needed fake transactions is exactly the summation of all the values in the Noise column of the noise table. The noise table provides a compact synopsis (using $O(n)$ space, where n is the number of items) that can be stored by the ED module, to support both the creation of the fake transaction and the decryption step.

For example, consider the example TDB in Fig. 19.5, and its associated (cipher) item support table in table (a) in Fig. 19.6. For $k = 2$, the grouping method generates two groups: $\{e_2, e_5\}$ and $\{e_4, e_1, e_3\}$ (table (b) in Fig. 19.6), that is robust: none of the two groups, considered as itemsets, is supported by any transaction in D.

Fake Transactions. Given a noise table specifying the noise $N(e)$ needed for each cipher item e, we generate the fake transactions as follows. First, we drop the rows with zero noise, corresponding to the most frequent items of each group or to other items with support equal to the maximum support of a group. Second, we sort the remaining rows in descending order of noise. Let $e'_1, \ldots, e'_m$ be the obtained ordering of (remaining) cipher items, with associated noise $N(e'_1), \ldots, N(e'_m)$. The following fake transactions are generated:

- $N(e'_1) - N(e'_2)$ instances of the transaction $\{e'_1\}$
- $N(e'_2) - N(e'_3)$ instances of the transaction $\{e'_1, e'_2\}$
- $\ldots$

- $N(e'_{m-1}) - N(e'_m)$ instances of the transaction $\{e'_1, \ldots, e'_{m-1}\}$
- $N(e'_m)$ instances of the transaction $\{e'_1, \ldots, e'_m\}$

Continuing the example, we consider cipher items of non-zero noise in table (c) in Fig. 19.6. The following two fake transactions are generated: 2 instances of the transaction $\{e_5, e_3, e_1\}$ and 1 instance of the transaction $\{e_5\}$. We observe that fake transactions introduced by this method may be longer than any transactions in the original TDB *D*, as in the examples above, where the maximum transaction length *lmax* in *D* is 2 and there are fake transactions of length 3. So, we consider shortening the lengths of the added fake transactions so that they are in line with the transaction lengths in *D*. In our running examples above, as *D* only contains the instances of length 1 and 2, we split the two instances of the transaction $\{e_5, e_3, e_1\}$ into two instances of fake transactions $\{e_5, e_3\}$ and 2 instances of $\{e_1\}$. So, we obtain 2 instances of $\{e_5, e_3\}$, 2 of $\{e_1\}$ and 1 instance of $\{e_5\}$.

In order to implement the synopsis efficiently we use a hash table generated with a minimal perfect hash function. Minimal perfect hash functions are widely used for memory efficient storage and fast retrieval of items from static sets. In our scheme, the items of the noise table e_i with $N(e_i) > 0$ are the keys of the minimal perfect hash function. Given e_i, function h computes an integer in $[0, \ldots, n-1]$, denoting the position of the hash table storing the triple of values $<e_i, times_i, occ_i>$ where:

- $times_i$ represents the number of times the fake transaction $\{e_1, e_2, \ldots, e_i\}$ occurs in the set of fake transactions
- occ_i is the number of times that e_i occurs altogether in the future fake transactions after the transaction $\{e_1, e_2, \ldots, e_i\}$.

Given a noise table with m items with non-null noise, our approach generates hash tables for the group of items. In general, the *i-th* entry of a hash table *HT* containing the item e_i has $times_i = N(e_i) - N(e_{i+1})$, $occ_i = \sum_{j=i+1, \ldots, g} N(e_j)$, where g is the number of items in the current group. Notice that each hash table *HT* represents concisely the fake transactions involving all and only the items in a group of $g \leq lmax$ items. The hash tables for the items of non-zero noise in table (c) (Fig. 19.6) are shown in table (d) (Fig. 19.6). Given that in our example, $lmax = 2$, we need to split the 3 items e_5, e_3, and e_1 of non-zero noise in Fig. 19.6 into two sets, $\{e_5, e_3\}$ and $\{e_1\}$, each with associated fake transactions, coded by the two hash tables. Notice that any pattern consisting of items from different hash tables will not be put into a fake transaction.

Finally, we use a (second-level) ordinary hash function *H* to map each item *e* to the hash table *HT* containing *e*.

The constructed fake transactions are added to *D* (once items are replaced by cipher items) to form D^*, and transmitted to the server. All the fake transactions, i.e., $DF = D^* \setminus D$, are stored by the ED module, by the compact synopsis described above.

19.5.2 Decryption

When the client requests the execution of a pattern mining query to the server, specifying a minimum support threshold σ, the server returns the computed frequent patterns from D^*. Clearly, for every itemset S and its corresponding cipher itemset E, we have $supp_D(S) \leq supp_{D^*}(E)$. Therefore, our encryption scheme guarantees that all itemsets frequent in D will be returned, in cipher version, by the server. But additional patterns frequent in D^*, but not in D, are returned as well. For each cipher pattern E returned by the server together with $supp_{D*}(E)$, the ED module trivially recovers the corresponding plain pattern S as follows:

$$supp_D(S) = supp_{D^*}(E) - supp_{D^* \setminus D}(E) \tag{1}$$

This calculation is efficiently performed by the ED module using the synopsis of the fake transactions in $D^* \setminus D$ described above.

19.6 Preliminary Experimental Results

Data Sets: We empirically assess our encryption method with respect to a real-life transaction database donated by one of the largest supermarket chain in Europe. We selected 300,000 transactions occurring during a period of time in a subset of stores; the transactions involve 13,730 different products.

We implemented our encryption scheme, as well as the decryption scheme in Java. All experiments were performed on an intel Core2 Duo processor with a 2.66 GHz CPU and 6 GB RAM on a Linux platform (Ubuntu 8.10). We adopted the Apriori implementation by Christian Borgelt (http://www.borgelt.net) written in C and one of the most highly optimized implementations.

Encryption overhead: we assess the size of fake transactions added to *TDB** after encryption; Fig. 19.7 reports the sizes of fake transactions for different *k* values. We observe that the size of fake transactions increases linearly with *k*.

Mining overhead: We study the overhead at server side in the pattern mining task over *TDB** w.r.t. *TDB*. Instead of measuring performance in run time, we measure the increase in the number of frequent patterns (FP) obtained from mining the encrypted TDB, considering different support thresholds. Results are plotted in Fig. 19.8, for different values of *k*; notice that k=*1* means that the original and encrypted TDB are the same. The x-axis shows the relative support threshold in the mining query, w.r.t. the total number of original transactions; the number of frequent patterns obtained is reported on the y-axis. We observe that the number of frequent patterns, at a given support threshold, increases with *k*, as expected. However, mining over *TDB** exhibits a small overhead even for very small support thresholds, e.g., a support threshold of about 1% for k=*10* and 1.5% for k=*20*. We found that, for reasonably small values of the support threshold, the incurred overhead at server side is kept under control; clearly, a trade-off exists between the level of privacy,

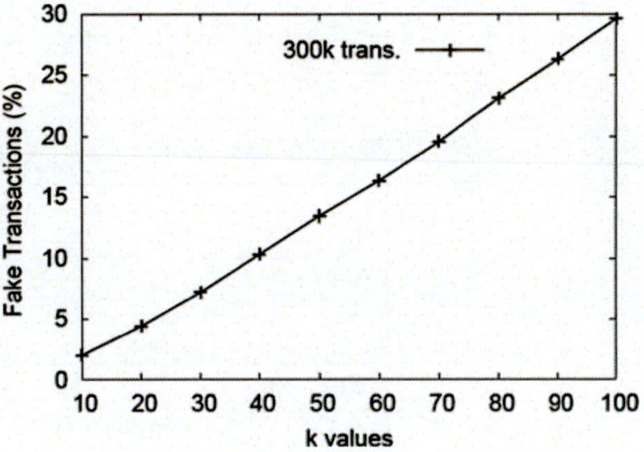

Fig. 19.7 Fraction of fake transactions

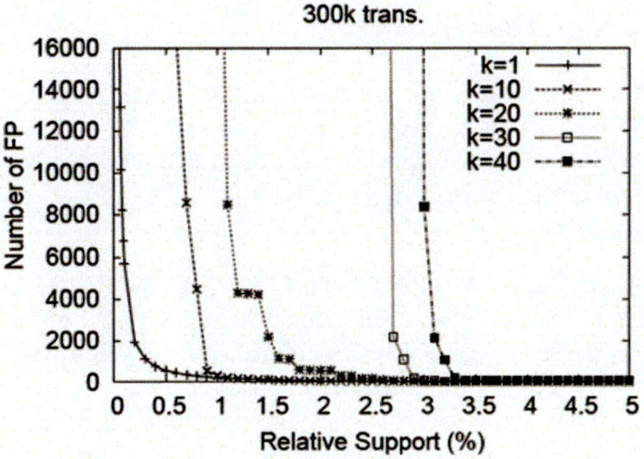

Fig. 19.8 Mining overhead at server side

which increases with k, and the minimum affordable support threshold for mining, which also increases with k.

Decryption overhead by the ED module: We now consider the feasibility of the proposed outsourcing model. The ED module encrypts the TDB once which is sent to the server. Mining is conducted repeatedly at the server side and decrypted every time by the ED module. Thus, we need to compare the decryption time with the time of directly executing Apriori over the original database. As shown in Fig. 19.9, the decryption time is about one order of magnitude smaller than the mining time; for higher support threshold the gap increases to about two orders of magnitude. Indeed, for support equal to 2.6% the mining time by Apriori is about 1.10 sec while the decryption time for k=10 is about 0.01.

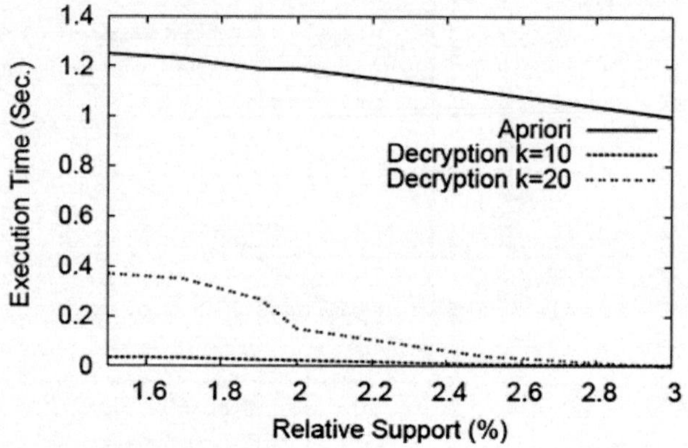

Fig. 19.9 Decryption time vs. mining time

19.7 Future Work

Our first results are encouraging, and open the way for the definition a general framework for privacy-preserving outsourcing of association rule mining. In future, we intend to investigate further the problem because many interesting problems are still open. The next steps include:

1. The study of a formal analysis based on our attack model and the proof that the probability that an individual item, a transaction, or a pattern can be broken by the server can always be controlled to be below a threshold chosen by the owner, by setting the anonymity threshold k.
2. The complexity analysis in space and in time of the encryption/decryption scheme proposed in this position paper, to better understand the real applicability of our approach.
3. The definition of a strategy for incrementally maintaining the synopsis at the client side against updates in the form of appends.
4. A detailed experimental analysis of our scheme using large real data set with different sparsity/density properties. This is important to understand how our scheme works in different settings. Another interesting investigation could be the analysis of the scalability of the proposed approach.

19.8 Summary

We studied the problem of (corporate) privacy-preserving outsourcing of association rule mining.

Our encryption scheme is based on 1–1 substitutions together with addition of fake transactions such that the transformed database satisfies k-anonymity w.r.t. items and itemsets.

This framework allows to a data owner, like a supermarket, to give its data in outsourcing to a service provider and to obtain an association rule mining service from it, without disclosing important information, deriving from the mining analysis, describing for example the customers' behavior. The effort and computational resources required to the data owner for the encryption are negligible respect to those required to mining the data. We showed some preliminary results obtained by experiments based on a large real data set. Our first results are encouraging; naturally, there are many interesting open issues to be investigated.

References

Agrawal, R., and R. Srikant. "Fast Algorithms for Mining Association Rules." (Paper Presented at the Annual International Conference on Very Large Data Bases, 1994).

Agrawal, R., and R. Srikant. "Privacy-Preserving Data Mining." (Paper Presented at the Annual International Conference SIGMOD, 2000).

Agrawal, R., J. Kiernan, R. Srikant, Y. Xu. "Order Preserving Encryption for Numeric Data." (Paper Presented at the Annual International Conference SIGMOD, 2004).

Agrawal, S., and J.R. Haritsa. "A Framework for High-Accuracy Privacy-Preserving Mining." (Paper Presented at the Annual International Conference on Data Engineering, 2005).

Atzori, M., F. Bonchi, F. Giannotti, and D. Pedreschi. "Anonymity Preserving Pattern Discoery." *VLDB Journal* 17, (2008): 703.

Bu, S., L.V.S. Lakshmanan, R.T. Ng, and G. Ramesh. "Preservation of Patterns and Input-Output Privacy." (Paper Presented at the Annual International Conference on Data Engineering, 2007).

Buyya, R., C.S. Yeo and S. Venugopal. "Market-Oriented Cloud Computing: Vision, Hype, and Reality for Delivering IT Services as Computing Utilities." (Paper presented at the annual International Conference on High Performance Computing and Communications, 2008).

Chen, K., G. Sun, and L. Liu. "Toward Attack-Resilient Geometric Data Perturbation." (Paper Presented at the Annual SIAM International Conference on Data Mining, 2007).

Clifton, C., M. Kantarcioglu, and J. Vaidya. "Defining Privacy for Data Mining." (Paper Presented at the International NSF Workshop on Next Generation Data Mining, 2002).

Fung, B.C.M., K. Wang, and P.S. Yu. "Anonymizing Classification Data for Privacy Preservation." *TKDE* 19, (2007): 711.

Gilburd, B., A. Schuster and R. Wolff. "A New Privacy Model and Association-Rule Mining Algorithm for Large-Scale Distributed Environments." (Paper Presented at the Annual International Conference on Very Large Data Bases, 2005).

Hacigumus, H., B. Iyer, and S. Mehrotra. "Providing database as a service." (Paper Presented at the Annual International Conference ICDE, 2002).

Kantarcioglu, M., and C. Clifton. "Privacy-Preserving Distributed Mining of Association Rules on Horizontally Partitioned Data." *TKDE* 16, (2004): 1026.

Krishna Prasad, P., and C. Pandu Rangan. "Privacy Preserving BIRCH Algorithm for Clustering Over Vertically Partitioned Databases." *Secure Data Management*, (2006).

Lakshmanan, L.V.S., R.T. Ng, and G. Ramesh. "To Do or Not To Do: The Dilemma of Disclosing Anonymized Data." (Paper Presented at the Annual International Conference on SIGMOD, 2005).

Machanavajjhala, A., J. Gehrke, and D. Kifer. "l-diversity: Privacy Beyond k-anonymity." (Paper Presented at the Annual International Conference on Data Engineering, 2006).

Rizvi, S., and J.R. Haritsa. "Maintaining Data Privacy in Association Rule Mining." (Paper Presented at the Annual International Conference on Very Large Data Bases, 2002).
Samarati, P. "Protecting Respondents' Identities in Microdata Release." *TKDE* 13, (2001): 1010.
Song, D.X., D. Wagner, and A. Perrig. "Practical Techniques for Searches on Encrypted Data." (Paper Presented at the IEEE Symposium on Security and Privacy, 2000).
Wong, W.K., D.W. Cheung, E. Hung, B. Kao, and N. Mamoulis. "Security in Outsourcing of Association Rule Mining." (Paper Presented at the Annual International Conference on Very Large Data Bases, 2007).
Xiao, X., and Y. Tao. "Anatomy: Simple and Effective Privacy Preservation." (Paper Presented at the Annual International Conference on Very Large Data Bases, 2006).
Xu, Y., K. Wang, A.W. Fu, P.S. Yu. "Anonymizing Transaction Databases for Publication." (Paper Presented at the Annual International Conference on Knowledge Discovery and Data Mining 2008).

Chapter 20
Access Control in Cloud-on-Grid Systems: The *PerfCloud* Case Study

Valentina Casola, Raffaele Lettiero, Massimiliano Rak, and Umberto Villano

20.1 Introduction

According to the definition by NIST[1], cloud computing is a pay-per-use model for enabling available, convenient, on-demand network access to a shared pool of configurable computing resources (e.g., networks, servers, storage, applications, services). These resources can be rapidly provisioned and released with minimal management effort, without interaction with the service provider. Stated another way, the cloud computing paradigm is based on the idea of delegating to the network the provision of almost all functionalities of present-day computer systems, from a high-availability and fast hardware infrastructure to complex applications tailored to user needs. Distinctive features of clouds are the extensive use of the SOA model[2] and of virtualization techniques[3,4]. Owing to the latter, the users can even have full administrative control on the virtualized computing resources received from the cloud. However, the control of the underlying physical resources is left in the cloud, which is owned and managed by the cloud provider.

In order to classify the different contexts in which the cloud paradigm can be applied, the recent literature proposes a taxonomy that takes into account the possible delivery models (*Infrastructure as a Service* (IaaS), *Platform as a Service* (PaaS) and *Software as a Service* (SaaS)) and deployment modalities (Private, Public, Managed, Hybrid)[5]. Our research interests lie essentially in the provision of high performance computing infrastructures. Hence, from here onwards, in this paper

V. Casola (✉)
Dipartimento di Informatica e Sistemistica, Università degli studi di Napoli Federico II, Napoli, Italy
e-mail: casolav@unina.it

[1]Mell, P., and T. Grance. *The NIST Definition of Cloud Computing*. 2009.

[2]W3C Working Group. *Web Services Architecture* (2004), http://www.w3.org/TR/ws-arch/.

[3]Barham, P., et al., "Xen and the Art of Virtualization." *SIGOPS Operating Systems Review* 37, (2003): 164–177.

[4]VMWare Staff, *Virtualization overview*. (White Paper) http://www.vmware.com/pdf/virtualization.pdf.

[5]Cloud Security Alliance. *Security Guidance for Critical Areas of Focus in Cloud Computing*. (2009).

S. Gutwirth et al. (eds.), *Computers, Privacy and Data Protection: an Element of Choice*, DOI 10.1007/978-94-007-0641-5_20,

the stress will be on the IaaS delivery model, with any of the above-mentioned deployment models.

Besides being a means for demand-driven provision of relatively low-parallelism computing resources, cloud computing is also an incredibly powerful solution for offering large distributed computational facilities in a simple and effective way. The obvious reference in this field is grid computing, a paradigm that aims at enabling access to high performance distributed resources in a simple and standard way. As such, it is widely diffused in the e-science world. In practice, grid is born with the Globus project, and currently the Globus toolkit and gLite are the most relevant implementations available. In grids, users can compose complex stateful services in order to build up complex and computation-intensive tasks. This is obtained by means of a middleware paradigm: every host has a grid interface, and developers adopt middleware-dependent APIs for building up their applications.

As a matter of fact, currently the scientific community is willing to spot the differences between clouds and grid systems. At least in principle, the two have obvious analogies. However, even if the cloud and grid computing paradigms are intended for different classes of users and applications, it is a fact that the differences between them are not yet widely recognized[6,7]. To complicate the matter, it is possible to implement clouds on the top of grid systems (cloud-on-grid approach)[7,8], or even the opposite (grid-on-cloud approach)[9].

The integration of the two technologies (grid and cloud) is currently of great interest, as it allows addressing different problems with the same technological platform, unifying the two user communities. The cloud-on-grid approach has gained large interest in the scientific community, as it helps to manage some of the most common problems with parallel programming: the incredible variety of different software (and software versions), configurations, operating systems and hardware layers that often should coexist, but are not mutually compatible. Thanks to the adoption of clouds and of their underlying virtualization techniques, it is possible to provide to the grid user and to the parallel application developers a "clean" environment, completely and freely customizable.

In previous papers, the authors have presented *PerfCloud*[10,11], a framework that integrates cloud and grid paradigms, adopting a cloud-on-grid approach. In terms

---

[6]Jha, S., A. Merzky, and G. Fox. "Using Clouds to Provide Grids Higher-Levels of Abstraction and Explicit Support for Usage Modes." *Concurrency and Computation: Practice & Experience* 21, 8 (2009): 1087–1108.

[7]Foster, I., et al., "Virtual Clusters for Grid Communities." In: CCGRID 2006, 513–520. IEEE Computer Society Press, 2006.

[8]Keahey, K., et al., "Virtual Workspaces: Achieving Quality of Service and Quality of Life in the Grid." *Scientific Programming* 13 (2005): 265–275.

[9]Cherkasova, L., et al., "Optimizing Grid Site Manager Performance with Virtual Machines." in *Proc. of the 3rd USENIX Workshop on Real Large Distributed Systems (WORLDS06)*, (2006).

[10]Mancini, E.P., et al., "*PerfCloud*: Grid Services for Performance-Oriented Development of Cloud Computing Applications." in *Proc. of Emerging Technologies for Next generation GRID (ETNGRID-2009/WETICE-2009)* (2009).

[11]Casola, V., et al., "*PerfCloud*: Performance-Oriented Integration of Cloud and Grid." in *Proc. of CloudComp 2009, Munich (DE)* (2010).

of functionality, *PerfCloud* is a system able to build up virtual clusters (i.e., clusters of virtual machines) using grid services, offering also additional services for performance evaluation. The rationale for the cloud-on-grid integration is the possibility to exploit the existing complex but robust distributed infrastructure of grids for supporting the more user-friendly cloud paradigm.

One of the key grid subsystems is undoubtedly the security infrastructure[12], which is worth to be re-used for a cloud implementation. Security in cloud environments is an incredibly wide field of research. The Cloud Security Alliance[5] points out a set of 15 different security domains related to the cloud paradigm, each of which involves a great number of open issues. Our work is concerned with identity and access management (domain 13). In particular, in this paper we will focus on fine grained access control. Access Management in cloud environments has a number of distinctive characteristics, which makes it different from the analogous problem in classical distributed architectures (and in particular, in grids). Typically, in a distributed system we have two main kinds of actors: administrator and users. The first one has full access rights to the physical machines, and may configure the system from the hardware, up to the operating systems and the main services offered from the system. On the other hand, users have limited rights, may only configure their applications and have controlled access to available resources. In gridS, the security infrastructure enables the set up of Virtual Organizations, in which the administrators of different physical machines expose a set of services to the users. Moreover, the administrator may define complex policies for the access of the system to resources.

In a virtualized cloud environment, instead, the final user can obtain resources on which he has full administrative control. The resources obtained (in the IaaS model, a virtual machine, or even a cluster of virtual machines) are indistinguishable from physical ones. Among other things, he can offer them as components of a larger grid, or simply manage the services exposed (deploy or un-deploy services, start-up a new certification authority, . . .).

Security requirements are strongly related to both delivery and deployment model of the cloud, as architectural choices and service provision activities imply the adoption of proper security policies to guarantee data integrity, privacy and user confidentiality. In this paper the stress will be on security issues in IaaS systems and, in particular, on the implications linked to the use of a cloud-on-grid approach. We will discuss these topics using our framework *PerfCloud* as a case study.

We will identify and classify the different kind of users (actors) in an IaaS cloud scenario, discussing the role they play and the respective security implications. Then we will show how we have implemented fine-grain access control mechanisms in *PerfCloud*. In particular, after defining specific security access control policies for all the actors, we will enforce these policies by exploiting the underlying Globus Toolkit 4 (GT4) security infrastructure.

---

[12]The Globus Security Team. *Globus Toolkit Version 4 Grid Security Infrastructure: A Standards Perspective,* www.globus.org/toolkit/docs/4.0/security/GT4-GSI-Overview.pdf (2005).

The paper will go on as follows. Section 20.2 sketches the *PerfCloud* architecture. Section 20.3 deals with the problem of access control in cloud-on-grid systems, recalling briefly the authentication and authorization (AA) facilities provided by the GT4. The next section outlines the roles of the cloud users as far as access control is concerned. The implementation of fine-grain access control mechanisms in *PerfCloud* is thoroughly dealt with in Section 20.5. The paper closes with a discussion on related work, followed by the conclusions and by a summary of our future research.

20.2 *PerfCloud* Architecture

PerfCloud is a complete framework that provides virtual cluster creation and management, along with performance prediction services in an e-science cloud (a cloud suitable for high-performance computing applications). The design relies on the adoption of a set of grid services able both to create a Virtual Cluster (VC) and to predict the performance of a given target application on that particular VC.

As mentioned in the introduction, *PerfCloud* builds an IaaS (Infrastructure as a Service) cloud environment upon a Globus GT4 grid infrastructure. The *PerfCloud* model of the infrastructure is a collection of clusters, each of which is composed of a front-end node (FE) and a set of computing nodes exchanging messages on a private network. Both the nodes and the network can be physical or virtual. In its current implementation, *PerfCloud* does not support hybrid clusters, i.e., clusters made up of a mixture of physical and virtual nodes.

In a "pure" cloud-on-grid system, the virtualized environments assigned to users are mutually insulated, and do not know of each other. Given two scientific communities working on different but related problems, the users of each group have access to a freely customizable computing platform, but cannot invoke the software services developed by the other group. Even if their computing environments exploit physically the same grid, they are not interoperable. In *PerfCloud* this problem is solved by integrating the virtual resources offered by the cloud in the underlying grid. Doing so, all the leased virtual environments become part of a single grid including virtual and physical computing resources and, if necessary, can cooperate using customary grid–based interactions. In other words, given an existing computing grid, users can gain access to virtualized resources (e.g., to virtual clusters) through a cloud interface, and these virtual resources are integrated in the existing grid and can cooperate with its component systems. Stated another way, the virtual clusters provided by the cloud are automatically part of the underlying grid infrastructure.

The clusters managed by *PerfCloud* participate in the underlying grid and offer their computational resources to the grid infrastructure. Their FEs host a Globus container and are certified within the grid Virtual organization. The FEs also host

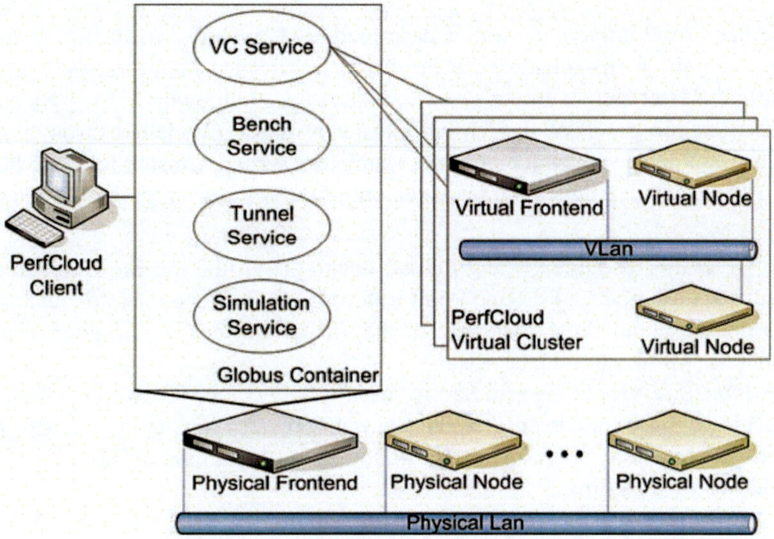

Fig. 20.1 The *PerfCloud* architecture

job schedulers (such as PBS[13] and Condor[14]) that distribute the workload on their computing nodes.

Figure 20.1 describes the overall architecture of *PerfCloud*. The *PerfCloud* application client resides on a user machine (which has access to the grid environment) and interacts with the *PerfCloud* system by invoking grid services. Furthermore, it manages grid connections, also providing utilities for end-users (notably, the performance analysis services). The architecture provides different grid services that enable the user to build up a new cluster as a grid Virtual Workspace[7,8] with full access rights.

The grid services of *PerfCloud* also include performance evaluation services (simulation, tuning and benchmarking) that can be invoked to simulate and to predict the performance of the environment just built. In order to help user interaction with the clusters, *PerfCloud* provides a tunnelling grid service, which lets the users execute commands on the target clusters. Moreover, *PerfCloud* offers a set of virtual-machine preconfigured images, which can be used to set-up virtual clusters. The images are ready-to-use cluster configuration enriched with all the software needed to execute High Performance Computing (HPC) applications (compilers, MPI and OpenMP platforms, Globus containers, job schedulers, . . .).

---

[13]Thain, D., et al., "Distributed Computing in Practice: The Condor Experience." *Concurrency – Practice and Experience* 17, (2005): 323–356.

[14]Henderson, R. "Job Scheduling Under the Portable Batch System." in *Job Scheduling Strategies for Parallel Processing, Lecture Notes in Computer Science 949*. Springer, (1995): 279–294.

The *PerfCloud* implementation is described in (Mancini et al., 2009). In the following, we will focus on the core *PerfCloud* services, i.e., the services adopted for virtual cluster management (creation, start-up, shutdown, . . .). The core of *PerfCloud* is composed of the Virtual Cluster resource (modelled through a Web Service Resource Framework – `WSRF`) and the Virtual Cluster to Grid Service (`VC2GS`)[10], which enables the management of the VC resource; in particular:

- The VC `WSRF` resource is the grid interface to the virtual machines. It is directly identified through an EPR (End Point Reference), the Globus pointer, and exports information about the actual state of the virtual cluster (i.e., number of nodes, number of CPUs per node, . . .).
- The `VC2GS` service is the grid Service used to access the VC resource. For example, it is possible for a user to create a new VC (i.e., to start-up the images and to obtain the resource EPR), to change its structure (the number of nodes, . . .) or to shutdown the machines.

The first version of the *PerfCloud* prototype offered a basic set of services, but did not enforce any access control mechanism. These have successively added to guarantee, for example, that if a cloud user starts a virtual cluster, no one else can turn it off, and that a generic grid user cannot start a virtual cluster. The details of the access control architecture and implementation are the object of the next sections.

20.3 Access Control in Cloud-on-grid Architectures

Currently two different aspects of security in cloud computing are being actively investigated. The first focuses on problems related to the "security management" in distributed and open systems. In this case, a black-box approach based on Service Level Agreement (SLA), security policies, security guidelines and agreements is adopted: the expected behaviour of the system is studied considering the way service provider and requestors interact. On the other hand, the second aspect of security in clouds focuses on the security mechanisms adopted in the infrastructure. In this case, the system is not considered a black box, but a layered system. In this paper, we will focus on the latter, as our targets are systems that build infrastructures as a service.

As already mentioned, security requirements are strongly related to both the delivery and deployment model of the cloud, because architectural choices and service provision activities imply the adoption of proper security policies to guarantee data integrity, privacy and user confidentiality. In the remainder of this paper, we will consider access control mechanisms for cloud services implemented on the top of grid platforms. As shown in Fig. 20.2, from the security point of view, a generic cloud-on-grid system can be viewed as made of three different layers: hardware layer, grid layer and cloud layer. The hardware layer security issues are primarily related to securing the operating system (trusted O.S., reliable disks,

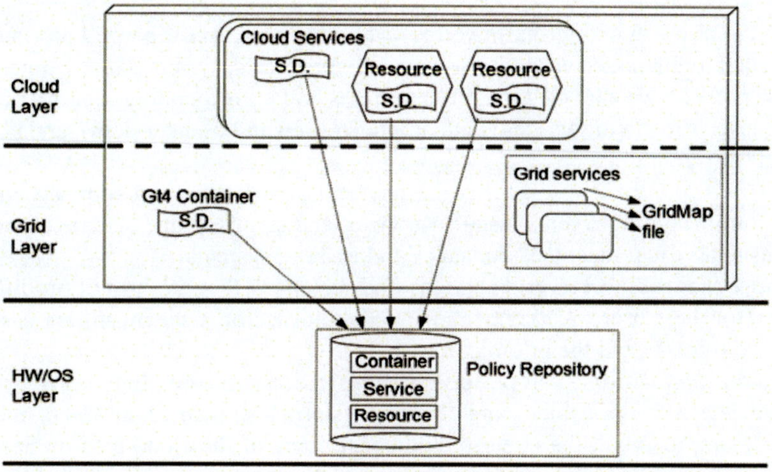

Fig. 20.2 Cloud-on-grid architecture

security patches, software updates and maintenance ...). System administrators usually perform these activities.

As for the security issues in the grid layer, the security mechanisms are primarily concerned with user authentication and authorization to run a job on machines in the grid. The grid Security Infrastructure (GSI) adopts digital certificates issued by a Public Key Infrastructure (PKI), and a delegation model (myProxy) in order to authenticate users[15,16]. These mechanisms are adopted also for authorization purposes, by combining the adoption of the digital certificate subject field and the gridmap file to map such a subject on a local operating system account12. The Globus Toolkit 4 offers innovative mechanisms to enforce policy-based access control through the adoption of security descriptors and security policies12. These can be very useful when implementing fine-grained access control mechanisms[17] at service and resource levels[18,19] and hence for the implementation of cloud services.

The security policies at the cloud layer involve many different users who can play different roles within the cloud platform. They need different privileges on system

[15] Welch, V., et al., "X.509 proxy certificates for dynamic delegation." in *Proc. of the 3rd Annual PKI R&D Workshop*, (2004).

[16] Welch, V., et al., "Security for Grid Services." in *Proc. of the 12th International Symposium on High Performance Distributed Computing (HPDC-12)*, (2003).

[17] Ferraiolo, D.F., and D. Richard Kuhn. "Role-based access control." in *Proc. of the 15th National Computer Security Conference*, (1992).

[18] Lang, B., et al., "A Multipolicy Authorization Framework for Grid Security." in *Proc. of the Fifth IEEE Symposium on Network Computing and Application*. IEEE Computer Society Press, (2006).

[19] Keahey, K., and V. Welch. "Fine-Grain Authorization for Resource Management in the Grid Environment." in *Proc. of the Grid2002 Workshop, Lecture Notes In Computer Science 2536*. Springer, (2002).

resources (physical and virtual). At this level, fine-grain access control mechanisms are needed to guarantee access to service administrators, system administrators and end-users. We have analyzed the different roles and corresponding security policies for the access to cloud services (both administration and user services), and pointed out four different roles:

System Administrator: manages the physical architecture and software components (turns on/off servers, installs services and applications, secures operating systems and virtual machines, manages and updates systems, . . .).

Grid User: can access grid services and resources that are hosted on different servers (for these users, Authentication and Authorization mechanisms are based on digital certificates and the gridmap file).

Cloud Administrator: can perform administration tasks on cloud services (e.g., create/delete virtual machines and clusters, assign clusters to Cloud Users, manage Cloud Users, maintain the integrity and consistency of the cloud environment . . .) but is not the owner of the created resources, as they are assigned to Cloud Users.

Cloud User: can access and manage cloud resources (virtual services and resources) and administrate his own virtual cluster(s). He is authorized by passing a fine-grain access control.

Summarizing, in the grid there are system administrators who manage physical machines from the hardware up to the Globus layer, and grid users who just access and use the grid-exposed resources. In the cloud, instead, there are different classes of resources to protect (physical and virtual ones). They are "assigned" to external users and must be independently managed and protected at different architectural levels (container, service, resource).

Furthermore, according to the cloud NIST definition of "on-demand self-service"[1], the cloud administration tasks should be performed by automated procedures or may be supervised by humans in "free" environments as, for example, in the academic context; since the grid platforms are widely adopted in academic context, we decided to include the cloud administrator role in our analysis.

As *PerfCloud* services and their security are based on the security mechanisms of Globus, in the following we will sketch the security solutions in GT4.

GT4 offers a flexible mechanism to enforce authentication and authorization. It is based on the adoption of security description files that describe both authentication requirements, along with authorization policy decision points where role-based access control policies are evaluated[17]. The authorization decision can be performed by standard GSI components or by external authorization services as XACML[20], PerMIS[21] or gridShib[22].

[20]The OASIS technical committee. *Xacml: extensible access control markup language* (2005), http://www.oasisopen.org/committees/xacml/repository/.

[21]Chadwick, D.W., et al., "Permis: A Modular Authorization Infrastructure." *Concurrency and Computation: Practice and Experience* 20, (2008).

[22]Barton, T., et al., "Identity Federation and Attribute-Based Authorization Through the Globus Toolkit, Shibboleth, Gridshib, and Myproxy." in *Proc. of 5th Annual PKI R&D Workshop*, (2006).

In particular, GT4 uses the concept of security descriptors as standard method for configuring the security policies of clients and services. GT4 provides four different types of security descriptors:

- *Container security descriptors* specify the container level security requirements that need to be enforced, i.e., the authentication and authorization mechanisms adopted to let a user access services and resources in the container;
- *Service security descriptors* specify the service level security requirements that need to be enforced, i.e., the authentication and authorization mechanisms adopted to let a user access a given service;
- *Resource security descriptors* specify the resource level security requirements that need to be enforced, i.e., the authentication and authorization mechanisms adopted to let a user access a given resource;
- *Client security descriptors* specify in the GT4 clients the security mechanisms to be used on service invocations.

Container, service and resource security descriptors have different priority levels. The most restrictive policy is applied at the resource level, and overrides the others. Service and container security descriptors are provided as XML files, defined in the deployment descriptor and locally stored. On the other hand, resource security descriptors can be created only dynamically, either programmatically or by means of a descriptor file. GT4 offers APIs to define the security descriptor or in the client code or through an XML file.

The typical structure of an XML file containing the configuration of the security descriptor of a container, service or resource type, is made of two main elements: `auth-method` and `authz`, as in the following:

```
<securityConfig xmlns="http://www.globus.org">
<auth-method>
    <GSISecureConversation>
         <protection-level>
            <integrity/>
         </protection-level>
   </GSISecureConversation>
</auth-method>
</method>
<authz value="pdp1:org.foo.PDP1 pdp2:org.foo.PDP2
foo1:org.<foo.authzMechanism bar1:org.bar.barMechanism"/>
</securityConfig>
```

As regards authentication, GT4 uses digital certificates to authenticate and to delegate users. Furthermore, GSI allows enabling security at transport level and at message level. Transport-level security means that the complete communication (all the information exchanged between a client and a server) is encrypted. With message-level security, only the contents of the SOAP message are encrypted. GSI offers two message-level protection schemes (*GSISecureMessage* and *GSISecureConversation*), and one transport-level scheme (*GSITransport*).

Summarizing, four types of authentication methods are provided by GT4:

– *none*: no authentication is performed;
– *GSISecureMessage*: each individual message is encrypted;
– *GSISecureConversation*: a secure context is first established between client and server; all the following messages can reuse that context;
– *GSITransport*: transport-level security is provided by using TLS.

Moreover, by means of *GSISecureMessage*, *GSISecureConversation* and *GSITransport*, the security administrator can specify the integrity (data are signed) and privacy (data are encrypted and signed) ... protection level. Clients must be configured to adopt a compliant authentication mechanism.

As regards authorization, container, services and resources can also be protected by different authorization mechanisms (enforcing different Policy Decision Points – PDP) with different mechanisms for collecting attributes (Policy Information Points – PIP).

20.4 Access Control and Roles in *PerfCloud*

As previously discussed, the security requirements of the cloud environments are very different from the grid ones. In the fact, there are different classes of resources to protect (physical and virtual ones), which are independently managed and which can be protected at different architectural levels (container, service, resource). In particular, in *PerfCloud* the virtual clusters should be managed by a restricted set of users with administration roles but privileged users can use them. As *PerfCloud* is built using the cloud-on-grid approach, we exploited the available grid security infrastructure. The access control mechanism we have implemented is able to enforce a role-based access control model[17] thanks to the adoption of the GT4 security infrastructure. We defined the following security policy for the *PerfCloud* roles:

– *System Administrators* can manage the physical machines from HW up to the operating system level. They are responsible for installing, configuring and starting the Globus platform and its Certification Authority, and for managing grid identities and accounts (issuing digital certificates, enabling users); furthermore, they are responsible for updating the security policies on the system;
– *Grid Users* can create and use grid resources;
– *Cloud Administrators* are grid Users with additional rights. They can supervise the cloud environment creating/maintaining new Virtual Clusters and managing Cloud User rights. In particular, they can enable/disable a Cloud User for the access to one or several Virtual Clusters.

– *Cloud Users* are grid users with additional rights. They can turn on/off, access and use, configure Virtual Clusters previously assigned to them by the Cloud Administrator.

Figure 20.3 illustrates the use cases for the *PerfCloud* System Administrator role, and Fig. 20.4 illustrates the main use cases for the other roles.

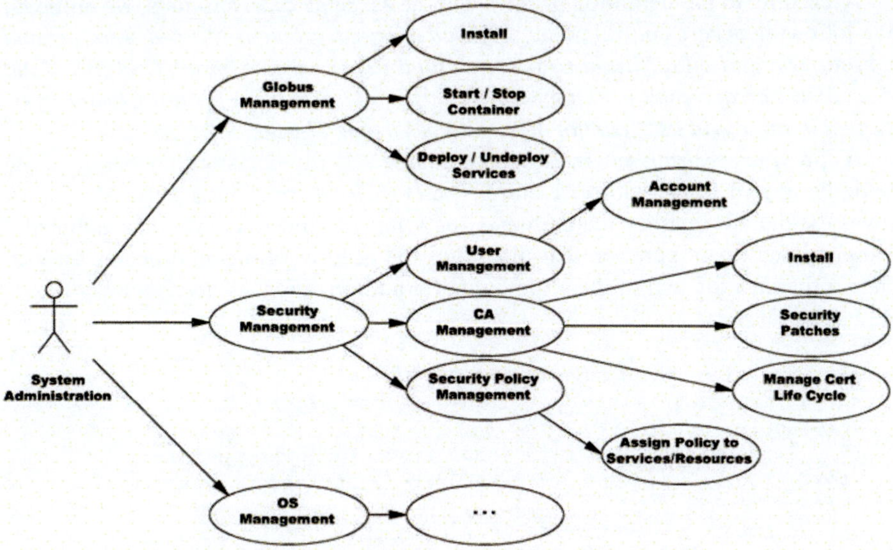

Fig. 20.3 System administrator use cases

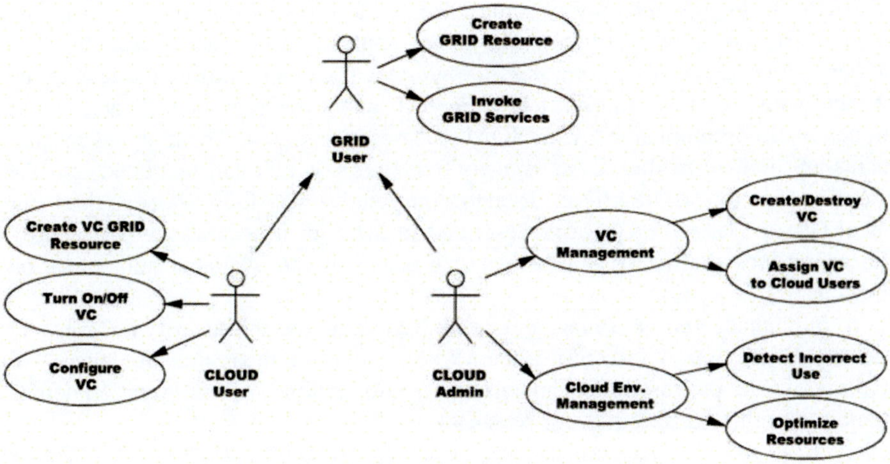

Fig. 20.4 Cloud roles and use cases

20.5 The Implementation of Access Control Mechanisms in *PerfCloud*

In this section we will illustrate the implementation of the *PerfCloud* Authentication and Authorization mechanisms, showing how they enforce an access control policy. We have created three different accounts for our tests: *Max*, *Raffaele* (Role = Cloud User) and *Valentina* (Role = Grid User).

According to the definition of roles and the associated permissions, we enforced the following access control policy: *a cloud user can create a VC grid resource and can turn on and off a Virtual Cluster; a cloud user can turn on a VC iff the Cloud Administrator assigned it to him; a cloud user can turn off a Virtual Cluster iff he turned it on; a grid user cannot manage any Virtual Cluster.*

As previously discussed, in *PerfCloud* all the security decisions are based on the underlying grid security infrastructure. The *PerfCloud* container security descriptor gives the default security requirements for a generic grid User. We have adopted a simple default configuration that just states the need of digital certificates for grid user authentication and of the gridmapfile for authorization, as described below:

```
<?xml version="1.0" encoding="UTF-8"?>
<securityConfig xmlns="http://www.globus.org">
<credential>
<key-file value="/etc/grid-security/containerkey.pem"/>
<cert-file value="/etc/grid-security/containercert.pem"/>
</credential>
<gridmap value="/etc/grid-security/grid-mapfile"/>
</securityConfig>
```

As for *Authentication*, any user must create its proxy certificate before accessing the grid and cloud services. The client application can adopt few different authentication mechanisms to access the services (i.e., can delegate the task to the channel using a secure transport layer as SSL using the proxy certificate, or use higher-level communication with SOAP secured messages). The adopted mechanisms are defined in the *Client security descriptor*, which can be coded into the client application or described through additional XML files. We enriched our client (`PerfCloudTrayIcon`, described in[11]) to let it access external security descriptor files that make it possible to use any of the available authentication methods.

To implement *Authorization* rules, we have configured the security policies and security description files to enforce them within the different architecture layers. We will present the proposed mechanisms and the corresponding configuration details by means of three different usage scenarios:

- *Scenario 1*: the user invokes the `VC2GS` service to `CREATE` a new grid resource associated to his VC;

- *Scenario 2*: the user invokes VC2GS to TURN-ON the Virtual Cluster and use it;
- *Scenario 3*: the user ends its session and invokes VC2GS again to TURN-OFF the Virtual Cluster previously started.

These scenarios are illustrated in Fig. 20.5 through a communication diagram; in particular, the CREATE scenario entails the steps from 1 to 6, the TURN-ON scenario the steps from 7 to 12, and the TURN-OFF scenario the steps from 13 to 18. The diagram illustrates the different components of *PerfCloud* that are involved in the process. In the following we will present them, dealing also with some configuration/implementation details.

Scenario 1: in this scenario a Cloud User invokes the VC2GS method for the creation of the Virtual Cluster resource (step 1 in Fig. 20.4). VC2GS being a cloud service, we enriched it with a service security descriptor, which overrides the default container security descriptor. Before granting access, VC2GS enforces the policy evaluation (steps 2 and 3 in Fig. 20.4) through the *Policy Decision Point* in the *service security descriptor*, to grant or to deny the creation of the resource to the cloud user. The proposed service security descriptor defines the authentication methods required (in the file security-config-first) and the authorization module (local_PDP_policy) to be used both for service method invocation and for resource creation.

The security-config-first.xml file is used to indicate the required authentication methods (in our case, *GSISecureConversation* or *GSISecureMessage*) and which type of Authorization Service must be used for authorization (in our case, LocalConfigPDP).

The localPDP_policy.xml file contains the authorization policy, written in the policy language that is compliant with the indicated authorization service; in particular, in the current implementation, the policy is per-account based (see the example below). Nevertheless the localPDP is not the only authorization service that

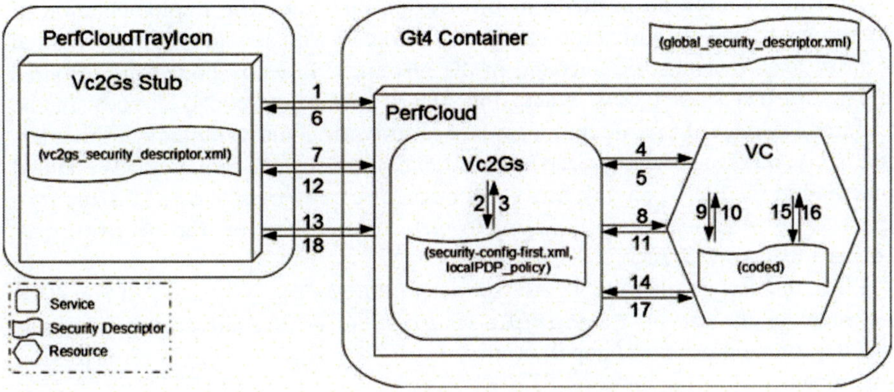

Fig. 20.5 The *PerfCloud* Authentication and Authorization architecture

can be adopted in the Globus toolkit. In fact, we are already experimenting external services as XACML[20] and Shibolet[22] that provide more flexible policy languages.

For example, in this scenario, the contents of the `localPDP_policy` file may be:

```
/O\=Grid/OU\=GlobusTest/OU\=simpleCA-vega.dii.unina2.it/
OU\=dii.unina2.it/CN\=Max={http://www.globus.org/namespaces/
virtual/core/FactoryService}createResource

/O\=Grid/OU\=GlobusTest/OU\=simpleCA-vega.dii.unina2.it/
OU\=dii.unina2.it/CN\=Raffaele={http://www.globus.org/
namespaces/virtual/core/FactoryService}createResource
```

Each line of the file contains the binding between the user and the allowed methods. In this example, only *Max* and *Raffaele* can create the resource. As result of the enforcement of such a policy (step 3), when a Cloud User (*Max* or *Raffaele*) invokes the service, he is authenticated and the service authorizes him to proceed. If a grid user (*Valentina*) tries to invoke the service the authorization is instead denied.

Once the `CREATE` action is authorized, the `VC2GS` service creates the resource (step 4), and returns an EPR (End Point Reference) to the client (steps 5 and 6). It should be noted that, up to this step, the resource security descriptor (which is coded in the resource grid interface) has never been involved in the procedure. In fact, any Cloud User may create a resource from existing images.

Scenarios 2 and 3: In the second scenario, the client sends a request to `VC2GS` to `TURN-ON` the Virtual Cluster (i.e., to perform an action at resource level) (step 7 in Fig. 20.4).

So far, we have only defined the policy at service level. This makes it possible to authorize a user to execute a service, but at service level we cannot enforce a rule such as the following: *the only user authorized to start a VC is the user who created the resource.* This kind of rules may be applied only at resource level, taking in consideration an attribute of the created resource. The implementation of the resource level authorization cannot be enforced with the adoption of external security files, but must be encoded in the resource. So we re-coded the resource into a `SecureResource`, where the security rules are coded directly in the resource interface. In particular, a `SecureResource` must implement the interface `org.globus.wsrf.impl.security.SecureResource`. We added a method in the file `org.globus.virtual.servi-ces.core.factory.impl.VirtualResource.java` to get an instance of the `Resource-SecurityDescriptor`.

When the user invokes the `VC2GS` service to `TURN-ON` a VC (step 7 in Fig. 20.4) that involves the use of a `SecureResource`, the `VC2GS` directly refers to the coded resource descriptor (step 9), as the security descriptors of the lower layers are overridden.

To enforce the desired security rule, a `SecureResource` must include methods to obtain the identity of the resource creator (the `getCaller()` method).

Thus it builds an authorization chain and instantiates a PDP that contains only one rule: it requires that the identity of the invoker is the same of the creator. Moreover, the `SecureResource` must indicate the supported authentication method (which is the only method added to the list of Authentication Methods: `desc.setAuthMethods(list)`). Once authorized (step 10), the VC is turned on and a positive response is returned to the Cloud User (steps 11 and 12).

For clarity's sake, the following code contains the secured resource just described:

```
String identity = SecurityManager.getManager().getCaller();
ResourceSecurityDescriptor desc=new ResourceSecurityDescriptor();
String authzChain = "identity";
ResourcePDPConfig PDPconfig = new ResourcePDPConfig(authzChain);

PDPconfig.setProperty("idenAuthz", "identity", identity);
java.util.ArrayList list= new java.util.ArrayList();
list.add(org.globus.wsrf.impl.security.descriptor.
GSISecureConvAuthMethod.BOTH);
desc.setAuthMethods(list);
desc.setAuthzChain(authzChain, PDPconfig,
"Name of Chain", "Some id");
this.config=new ResourceSecurityConfig(desc);
this.config.init();
```

As a result of these embedded security rules, if a Cloud User, say *Max*, creates the VC, only Max is authorized to start the resource. *Raffaele* or *Valentina* cannot be authorized, independently of the fact that they are Cloud or grid users.

So, summarizing the steps from 7 to 12, the client invokes the `VC2GS`, which enforces the resource level authorization policy; the coded decision policy checks the caller identity, verifies that it is correctly authenticated and that he is the resource owner. If so, the virtual machine is started up and the service returns control to the client.

The last scenario is analogous to the previous one. The components involved are exactly the same, and the sequence of steps from 13 to 18 is the same as in the previous scenario. The only difference is in the coded rules that must be enforced; in this case, the `SecureResource` must check that the Cloud User requesting the action is the same that turned on the virtual cluster. For brevity's sake, we do not present here the corresponding code.

20.6 Related Work

In the last years, a great interest on cloud computing has been manifested from both academic and private research centers; as a result, a large number of projects from industry and academia have been proposed. The concept of cloud computing is born with the Amazon Elastic Compute Cloud (EC2)[23], which is based on a simple idea: to offer a set of web services and a command line interface to let the users manage virtual machine images on the Amazon datacenter. In commercial contexts, it is worth mentioning the IBMs Blue Cloud[24] the Sun Microsystems Network.com[25] the Microsoft Azure Services Platform[26], The Google App Engine[27], The Dell Cloud computing solutions.[28]. Most of these commercial systems adopt proprietary solutions (such as the virtualization engine by VMWare[4]) and relatively few details are available on the adopted architectures and on the enforced security solutions.

Even if the cloud concept is born in the commercial environment, it is just an evolution of the virtualization techniques that have been object of research in the last years. At the state of the art, in this context the most advanced research project is the Reservoir project[29], which includes technologies as OpenNebula[30]. The most widely adopted virtualization engine is Xen[31], even if alternative solution do exist (Virtualbox[32], KVM[33], ...). In the academic world, and especially in the HPC area, cloud computing is in "competition" with the grid model, as outlined in many scientific papers6[,34].

The idea of grid-Cloud integration originated in a scientific context. The idea is to exploit an existing grid infrastructure as basis for the cloud environment. This solution is the one chosen for the Virtual Workspaces8, adopted by the Nimbus project[35], and used for building many e-science clouds. All the above-presented solutions have architectures similar to the one presented in this paper and delegate the security problems to the underlying grid infrastructure, but, at the best of authors' knowledge, the details of the choices done for the management of security in cloud services are not available.

---

[23] Amazon Inc., "Elastic Compute Cloud," (2008), http://aws.amazon.com/ec2.

[24] IBM Inc., "Blue Cloud Project," (2008), http://www03.ibm.com/press/us/en/pressrelease/22613.wss.

[25] Sun Microsystems, "Network.com," http://www.network.com.

[26] Microsoft Co., "Azure Services Platform," http://www.microsoft.com/azure/default.mspx.

[27] Google Inc., "Google Application Engine," http://code.google.com/intl/it-IT/appengine.

[28] Dell Co., "Dell Cloud Computing Solutions," http://www.dell.com/cloudcomputing.

[29] Reservoir Consortium: Reservoir Project, http://www03.ibm.com/press/us/en/pressrelease/23448.wss, (2009).

[30] Distributed Systems Architecture Research Group, Opennebula project. Technical report, Universidad Complutense de Madrid. http://www.opennebula.org, (2009).

[31] Barham, P., et al., "Xen and the Art of Virtualization." *SIGOPS Operating Systems Review* 37, 5 (2003): 164–177.

[32] Sun Inc., "VirtualBox," http://www.virtualbox.org/.

[33] Qumranet, "KVM," http://www.linux-kvm.org/page/Main Page.

[34] Foster, I., et al., "Cloud Computing and Grid Computing 360-Degree Compared." in *Proc. of 2008 Grid Computing Environments Workshop*. IEEE, (2008): 1–10.

[35] University of Chicago, "Nimbus Project," http://workspace.globus.org/clouds/nimbus.html, (2009).

20.7 Conclusions and Future Work

In this paper we have presented an analysis of security issues in cloud architectures and, in particular, in *PerfCloud*, a framework that offers a set of cloud-on-grid functionalities. The architecture of *PerfCloud* makes use of existing grid and virtualization technologies to manage at low-level the virtual clusters, and integrates them in the existing grid.

Cloud services can be deployed and delivered in many different ways, and many security issues are still open. In this paper, we have focused our attention on access control mechanisms, one of the main priorities identified by the Cloud Security Alliance. In particular, we have analyzed the need of a fine-grain access control in the cloud, where services, resources and physical/virtual machines should be properly managed and configured to guarantee a given degree of trust to end-users. We have pointed out the actors that play a role in this context, and defined specific security access control policies that we are able to enforce by adopting the security components of the Globus Toolkit security infrastructure.

As regards the future evolution of our work, we will integrate *PerfCloud* with different authorization services (as XACML, PERMIS or ShibGRID) to evaluate the trade-off between the security level provided and the introduced overhead. This will make it possible for the cloud to offer guarantees on the quality of service and to negotiate Service Level Agreements in an automatic way.

References

Amazon Inc., "Elastic Compute Cloud," (2008), http://aws.amazon.com/ec2.

Barham, P., B. Dragovic, K. Fraser, S. Hand, T. Harris, A. Ho, R. Neugebauer, I. Pratt and A. Warfield. "Xen and the Art of Virtualization." *SIGOPS Operating Systems Review* 37, 5 (2003): 164–177.

Barton, T., J. Basney, T. Freeman, T. Scavo, F. Siebenlist, V. Welch, R. Ananthakrishnan, B. Baker, M. Goode, and K. Keahey. "Identity Federation and Attribute-Based Authorization Through the Globus Toolkit, Shibboleth, Gridshib, and Myproxy." In *Proc. of 5th Annual PKI R&D Workshop*, (2006).

Casola, V., M. Rak, and U. Villano. "PerfCloud: Performance-Oriented Integration of Cloud and Grid." In *Proc. of CloudComp 2009, Munich (DE)*, Springer, (2010).

Chadwick, D.W., G. Zhao, S. Otenko, R. Laborde, L. Su, and T.A. Nguyen. "Permis: A Modular Authorization Infrastructure." *Concurrency and Computation: Practice and Experience* 20, (2008): 1341–1357.

Cherkasova, L., D. Gupta, and A. Vahdat. "Optimizing Grid Site Manager Performance with Virtual Machines." In *Proc. of the 3rd USENIX Workshop on Real Large Distributed Systems (WORLDS06)*, (2006).

Cloud Security Alliance. *Security Guidance for Critical Areas of Focus in Cloud Computing*. 2009.

Dell Co., "Dell Cloud Computing Solutions," http://www.dell.com/cloudcomputing.

Distributed Systems Architecture Research Group. *Opennebula Project.* Technical report, Universidad Complutense de Madrid (2009). http://www.opennebula.org.

Ferraiolo, D.F., and D. Richard Kuhn. "Role-based access control." In *Proc. of the 15th National Computer Security Conference*, (1992): 554–563.

Foster, I., T. Freeman, K. Keahey, D. Scheftner, B. Sotomayor, and X. Zhang. "Virtual Clusters for Grid Communities." In CCGRID 2006, 513–520. IEEE Computer Society Press, 2006.

Foster, I., Y. Zhao, I. Raicu, S. Lu. "Cloud Computing and Grid Computing 360-Degree Compared." In *Proc. of 2008 Grid Computing Environments Workshop*. IEEE, (2008): 1–10.

Google Inc., "Google Application Engine," http://code.google.com/intl/it-IT/appengine.

Henderson, R. "Job Scheduling Under the Portable Batch System." In *Job Scheduling Strategies for Parallel Processing, Lecture Notes in Computer Science 949*. Springer, (1995): 279–294.

IBM Inc., "Blue Cloud Project," (2008), http://www03.ibm.com/press/us/en/pressrelease/22613.wss.

Jha, S., A. Merzky, and G. Fox. "Using Clouds to Provide Grids Higher-Levels of Abstraction and Explicit Support for Usage Modes." *Concurrency and Computation: Practice & Experience* 21, 8 (2009): 1087–1108.

Keahey, K., and V. Welch. "Fine-Grain Authorization for Resource Management in the Grid Environment." In *Proc. of the Grid2002 Workshop, Lecture Notes In Computer Science 2536*. Springer, (2002): 199–206.

Keahey, K., I. Foster, T. Freeman, and X. Zhang. "Virtual Workspaces: Achieving Quality of Service and Quality of Life in the Grid." *Scientific Programming* 13 (2005): 265–27.

Lang, B., I. Foster, F. Siebenlist, R. Ananthakrishnan, and T. Freeman. "A Multipolicy Authorization Framework for Grid Security." In *Proc. of the Fifth IEEE Symposium on Network Computing and Application*. IEEE Computer Society Press, (2006): 269–272.

Mancini, E.P., M. Rak, and U. Villano. "PerfCloud: Grid Services for Performance-Oriented Development of Cloud Computing Applications." In *Proc. of Emerging Technologies for Next generation GRID (ETNGRID-2009/WETICE-2009), 201-6*. IEEE Computer Society Press, (2009).

Mell, P., and T. Grance. *The NIST Definition of Cloud Computing*. 2009.

Microsoft Co., "Azure Services Platform," http://www.microsoft.com/azure/default.mspx.

Qumranet, "KVM," http://www.linux-kvm.org/page/Main Page.

Reservoir Consortium. Reservoir Project (2009), http://www03.ibm.com/press/us/en/pressrelease/23448.wss.

Sun Inc., "VirtualBox," http://www.virtualbox.org/.

Sun Microsystems, Network.com, http://www.network.com.

Thain, D., T. Tannenbaum, and M. Livny. "Distributed Computing in Practice: The Condor Experience." *Concurrency – Practice and Experience* 17, (2005): 323–356.

The Globus Security Team. *Globus Toolkit Version 4 Grid Security Infrastructure: A Standards Perspective* (2005), http://www.globus.org/toolkit/docs/4.0/security/GT4-GSI-Overview.pdf.

The OASIS technical committee. *Xacml: extensible access control markup language* (2005), http://www.oasisopen.org/committees/xacml/repository/.

University of Chicago: Nimbus Project (2009) http://workspace.globus.org/clouds/nimbus.html.

VMWare Staff. *Virtualization Overview*, http://www.vmware.com/pdf/virtualization.pdf.

W3C Working Group. *Web Services Architecture* (2004), http://www.w3.org/TR/ws-arch/.

Welch, V., F. Siebenlist, I. Foster, J. Bresnahan, K. Czajkowski, J. Gawor, C. Kesselman, S. Meder, L. Pearlman, and S. Tuecke. "Security for Grid Services." In *Proc. of the 12th International Symposium on High Performance Distributed Computing (HPDC-12), 48*. IEEE Computer Society Press, (2003).

Welch, V., I. Foster, C. Kesselman, O. Mulmo, L. Pearlman, S. Tuecke, J. Gawor, and F. Siebenlist. "X.509 proxy certificates for dynamic delegation." In *Proc. of the 3rd Annual PKI R&D Workshop*, (2004).

Chapter 21
Security and Privacy in the Clouds: A Bird's Eye View

Wolter Pieters

21.1 Introduction

Over the last years, something called "cloud computing" has become a major theme in computer science and information security. Essentially, it concerns delivering information technology as a service, by enabling the renting of software, computing power and storage. Some argue that the phenomenon is essentially nothing new; others state that gradual developments have given rise to a qualitatively different computing architecture.

In this contribution, we describe the major security and privacy challenges in cloud computing. In doing so, we take a bird's eye view. The view is partly based on the discussions at the Workshop on Security and Privacy in Cloud Computing (SPCC) at the 2010 Computers, Privacy and Data Protection Conference. However, rather than summarising the results of the workshop, we focus on the fundamental questions that underlie the lively discussions.

Firstly, we will discuss the phenomenon of cloud computing in more detail, including its relation to de-perimeterisation, or disappearing boundaries. Secondly, the ideal of encrypted processing is presented, which could be a "holy grail" for cloud computing, together with its major limitations and practical implications. Thirdly, we focus on the physical security properties that are lost when virtualising information infrastructure, and on the question how to put these back in. Fourthly, we discuss the relation between cloud computing and the changing use of information technology by individuals as a means to establish their identity, notably by means of social networking services. Finally, we discuss ethical implications of disappearing boundaries in information technology, by proposing the concept of informational precaution.

W. Pieters (✉)
University of Twente, Enschede, The Netherlands
e-mail: w.pieters@utwente.nl

S. Gutwirth et al. (eds.), *Computers, Privacy and Data Protection: an Element of Choice*, DOI 10.1007/978-94-007-0641-5_21, © Springer Science+Business Media B.V. 2011

21.2 Cloud Computing

21.2.1 Foundations

The foundations of cloud computing lie in the outsourcing of computing tasks to third parties. In the old days of computing, this entailed renting a mainframe for one's computations. With the rise of the personal computer, these rentals became largely obsolete, except for costly scientific calculations.

When computers were connected to the Internet, a new option for renting computing power became available. It was possible to use the combined power of connected computers for performing one's calculations. So-called "grids" were established for scientific tasks. A well-known example is the SETI project,[1] where cosmic signals are analysed in the processor's idle time of computers around the world, in order to find signs of extraterrestrial intelligence.

Due to the development of the Internet, and its associated clustering of processing in grids and server rooms, computing has gradually become an "infrastructure", like the electricity network. Just as companies do not generate their own electricity, they increasingly see computing as something that has to be provided to them. Therefore, they outsource the maintenance of their IT infrastructure, bypassing the need for an in-house IT department. More and more often they will also rent the infrastructure from their provider, relieving them from buying the hardware themselves.

In classical outsourcing, companies can negotiate with their provider to establish the terms of service, including security aspects of the data processing. More and more providers, however, offer IT infrastructure as a commodity, with standard contracts and little room for smaller companies and individuals to negotiate. The combination of a) rental of information processing as-a-service and b) this service having the characteristic of a commodity, is what we call cloud computing.[2] According to Gartner, cloud computing is defined as "a style of computing where massively scalable IT-related capabilities are provided 'as a service' using Internet technologies to multiple external customers".[3]

Cloud computing (and outsourcing in general) is one of the main causes of the broader development called *de-perimeterisation*.[4] This term denotes the fading

---

[1] "SETI Institute Homepage," accessed May 10, 2010, http://www.seti.org/.

[2] As discussed in Paolo Balboni's presentation at SPCC.

[3] *Gartner Says Security Delivered as a Cloud-Based Service Will More Than Triple in Many Segments by 2013*, Gartner, 2008, accessed April 29, 2010, http://www.gartner.com/it/page.jsp?id=722307.

[4] *Jericho whitepaper*, Jericho Forum, The Open Group, 2005, accessed May 10, 2010, http://www.opengroup.org/jericho/vision_wp.pdf.

van Cleeff, A., and R.J. Wieringa. "Rethinking De-Perimeterisation: Problem Analysis And Solutions." in *Proceedings of the IADIS International Conference Information Systems 2009, 25–27 Feb 2009, Barcelona*. IADIS press, (2009): 105–112.

Pieters, W. "Converging Technologies and De-perimeterisation: Towards Risky Active Insulation." in *Proceedings of SPT 2009: Converging technologies, changing societies*. Enschede: CEPTES, University of Twente, (2009): 58–60.

of the boundaries of organisations and their information infrastructure, thereby invalidating a security approach that focuses on those boundaries. In traditional IT security, many organisations employed such a boundary-based approach, for example based on firewalls. It is then assumed that everything within the boundary is trusted, and everything outside is not. When the organisation's data is hosted elsewhere, such approaches are not adequate anymore.

Outsourcing and cloud-computing thus lead to de-perimeterisation. Other drives in this direction are the use of mobile devices by employees, and the hiring of consultants from third parties, who have to work within the organisation's boundaries. All these developments challenge a containment-based approach to information security, and force organisations to implement data-level security instead. Already in 1996, challenges to the existing security paradigm were discussed, and many of those considerations have become only more valid since.[5] Cloud computing thus reinforces existing challenges to current security paradigms.

21.2.2 Implementations

The implementation of cloud computing is typically realised by (a) invoking the Internet browser to use software, platforms and infrastructure online, and (b) virtualising the underlying infrastructure to cope with flexible demand.[6] Virtualisation here refers to the implementation of so-called "virtual machines", with properties more or less independent of the capacities of the underlying physical machines. This means that a single virtual machine can make use of several physical computers to increase its capabilities, but also that multiple virtual machines can run on a single physical computer. The advantages lie in not having to reserve a single physical machine for a particular task, thereby reducing hardware and operational costs.

Several distinctions have been proposed with respect to cloud services. The first of these has to do with the type of service offered. What is offered can be either software (software-as-a-service, or SaaS, for example Salesforce online bookkeeping or Gmail), a platform for developing and running applications (platform-as-a-service, or PaaS, for example Force.com) or an infrastructure that can be rented for processing or storing data (infrastructure-as-a-service, or IaaS, for example Amazon EC2).

Another distinction involves the control over the infrastructure. The infrastructure can be public, private, hybrid, managed, or community-owned. Each of these types involves particular decisions about who *owns* the infrastructure and who *controls* it. For example, in a managed cloud, a company owns its own IT infrastructure, but outsources the management to a third party. Obviously, public clouds – both

---

[5]Blakley, B. "The emperor's old armor," in *Proceedings of the 1996 workshop on new security paradigms*. ACM, (1997): 2–16.

[6]As discussed in Jean-Pierre Seifert's presentation at SPCC.

owned and managed by third parties, and possibly accessible to anyone including competitors – are the trickiest ones security-wise. Companies therefore need to evaluate carefully which types of cloud services are suitable for their needs.[7] Privacy legislation can play an important role here.[8]

21.2.3 Security

In general, the transfer of information-related tasks to other parties obviously entails security risks, in terms of confidentiality, integrity and availability of the data and services. There are basically two ways to solve these: trust the provider, or put technical guarantees in place that establish security properties even if the provider is not trustworthy. Usually, only a combination of these is feasible, making cloud security an inherently socio-technical problem,[9] of which legal developments form an essential part.

Tasks that can be performed in the cloud include storage, transfer and processing of information. With respect to storage and transfer, fairly standard security mechanisms can be applied. This does not mean that there cannot be security vulnerabilities, but the question on how to address those is mostly answered within the existing paradigm, called public key infrastructure. Data that is stored or transferred between parties is encrypted with the public key of the intended receiver, who decrypts it using her private key, when required. This means that the service provider that transmits or stores the data will not learn its contents, under the assumptions of the underlying cryptographic system.

For processing, there are no such standard solutions. In order to process data, it needs to be decrypted. Thus, whereas the communication between the data owner and the service provider can be considered secure, the service provider needs to access the plaintext data to do any meaningful processing. Therefore, in order to assume control over the security of the data, the owner needs to trust the service provider not to store the plaintext data, or transmit it to other parties.

In the following sections, we discuss a few noteworthy issues in cloud security and privacy, as apparent from SPCC and other cloud security venues.

21.3 The Ideal of Encrypted Processing

Scientists have for a while sought for the ultimate solution to the security of information processing in an untrusted environment. The idea is that if we can build programs that operate on encrypted data, and produce an encrypted version of the

[7] As discussed in Filip Schepers's presentation at SPCC.

[8] Ruiter, J., and M. Warnier. "Privacy Regulations for Cloud Computing, Compliance and Implementation in Theory and Practice," *this volume*.

[9] Dhillon, G., and E. Kolkowska, "Can a Cloud be Really Secure? A Socratic Dialogue," *this volume*.

correct output, then a service provider can perform calculations without having to possess the data in the clear. Such a system already exists for the simple operation of addition, and it is called homomorphic encryption. This means that if I provide encrypted versions of some numbers, then the system can produce the encrypted sum of the results without having to decrypt the data. Applications of this technique include counting votes in electronic elections.[10] Before 2009, the same trick could be done for multiplication, *but not with the same system.* When we can do both in the same system, we have so-called *fully homomorphic encryption*, and it can be shown that such a system can perform arbitrary operations on encrypted data.[11]

In 2009, the first fully homomorphic encryption system was proposed by Craig Gentry from IBM.[12] Unfortunately, the efficiency of the system is so low that it cannot be used for any practical purpose. Whether scientific progress can yield a workable solution in the near future is doubtful. Also, there are security disadvantages to homomorphic encryption, notably the possibility to calculate encrypted versions of certain plaintexts without knowing the associated key. This has consequences for integrity and authenticity of data.

For now, we are stuck with solutions to secure cloud processing for limited situations. An example of this approach is searching in encrypted data.[13] This means that we can store an encrypted database with a cloud provider, and we can search in the database without having to download the full database and decrypt it. We need only download and decrypt the results of the search. Similar solutions may be applicable for other specific cases of processing.

If encrypting is not feasible, we may at least wish to anonymise data before processing, in order to comply with privacy requirements. Techniques in this direction are being developed.[14]

21.4 Putting Physical Limitations Back in Place

Another principal question in cloud computing relates to the technology of virtualisation. When physical machines are replaced by virtual machines, what does this

---

[10] Schoenmakers, B., "A Simple Publicly Verifiable Secret Sharing Scheme and Its Application to Electronic Voting." in *CRYPTO '99*, volume 1666 of LNCS. Springer, (1999): 148–164. Hirt, M., and K. Sako. "Efficient Receipt-Free Voting Based on Homomorphic Encryption." in *Proc. EUROCRYPT 2000*, volume 1807 of LNCS. Springer, (2000): 539–556.

[11] Ishai, Y., and A. Paskin. "Evaluating Branching Programs on Encrypted Data." in *Proc. 4th Theory of Cryptography Conference (TCC)*, volume 4392 of LNCS. Springer, (2007): 575–594.

[12] Gentry, C. "On Homomorphic Encryption Over Circuits of Arbitrary Depth." in *The 41st ACM Symposium on Theory of Computing (STOC)*. ACM, (2009): 169–178.

[13] Brinkman, R. *Searching in Encrypted Data*. PhD thesis, University of Twente, 2007. Accessed April 29, 2010. http://doc.utwente.nl/57852.

[14] Giannotti, F., L.V.S. Lakshmanan, A. Monreale, D. Pedreschi, and H. (Wendy) Wang. "Privacy-preserving Mining of Association Rules from Outsourced Transaction Databases," *this volume.*

mean for security?[15] Similar questions have appeared in electronic voting (a physical ballot box versus a digital ballot box), and led to a lot of controversy about whether digital voting systems would be able to meet the same criteria as paper-based ones. Especially, the scale of fraud has been a source of concern, as hacking a digital ballot box would potentially allow one to steal all the votes in an election, whereas one would have to physically manipulate the ballot box in each district in case of paper voting. Moreover, in case of Internet voting, a digital attack could be launched from any location.

The technology of virtualisation has raised similar concerns in relation to cloud computing. If a vulnerability would exist in the software that creates and manages virtual machines, the so-called hypervisor, this would enable an attacker to compromise virtual machines around the world. In particular, the question has often been raised whether it would be possible to access the data of another virtual machine from one's own virtual machine, given that they run on the same physical computer.

Another type of concerns are those about location and time. On a privately owned device, one can be sure *where* the data is stored, from where it can be accessed, and *when* it will be made available, or made unavailable by deletion. These constraints are no longer present when data is stored in the cloud, and policies concerning location and time may need to be explicitly enforced. Such physical properties turn from *inherent* into *imposed* properties.[16]

For location, research is done into location-based or location-aware access control.[17] In such an approach, requirements can be put in place for the location of the *user* as well as the location of the *data*. This means that someone may not be able to access the data if she is in the Netherlands, but it may also mean that someone may not be able to access the data if *the data* resides in the Netherlands. Such a policy can thus enforce accessibility only within certain jurisdictions. How to implement the appropriate mechanisms is another question, and for most systems reliance is necessary on secure sensing of location, as one can always try to fake the signal representing location information. Other approaches can measure the time of certain communications and thereby determine distance.[18]

For the property of time, the most notable development is that of secure deletion. Secure wiping of privately owned storage devices has existed for a while, but such mechanisms can of course not be enforced if one does not own the storage, as in cloud computing. In that case, one would have to trust that the provider would do a

---

[15]van Cleeff, A., W. Pieters, and R.J. Wieringa. "Security Implications of Virtualization: A Literature Study." in *2009 IEEE International Conference on Computational Science and Engineering (CSE09)*. IEEE Computer Society, (2009): 353–358.

[16]Blakley, "The Emperor's Old Armor."

[17]See e.g. Ardagna, C.A., M. Cremonini, E. Damiani, S. De Capitani di Vimercati, and P. Samarati. "Supporting Location-Based Conditions in Access Control Policies." in *Proc. of the ACM Symposium on Information, Computer and Communications Security (ASIACCS'06)*. ACM, (2006): 212–222.

[18]Pavlovic, D., and C. Meadows. *Quantifying pervasive authentication: the case of the Hancke-Kuhn protocol.* Technical Report No. RR-09-09. OUCL, 2009. Accessed May 10, 2010. http://www.comlab.ox.ac.uk/files/2437/RR-09-09.pdf.

secure wipe upon request, which will typically not be the case if storage is offered as a commodity.

The idea that has been developed is to encrypt the data, and make sure that the decryption key is destroyed after a certain time period. This destruction then has to be delegated to a trusted party,[19] or based on some inherent property, like changing nodes in a peer-to-peer network.[20] The most obvious limitation of the approach is that, in order to use the data, it needs to be decrypted. It can then also be stored in decrypted form, which would make the destruction of the key useless. Therefore, storage of the decrypted data should be discouraged to make the approach work, just like in the prevention of copying music with digital rights management.

The transition from inherent to imposed security for these physical properties introduces new risks, as the mechanisms now need to be designed and implemented rather than using e.g. physical distance as a safeguard.[21] Vulnerabilities in the systems proposed are likely to be found. It remains therefore to be seen whether such approaches will really be used in practice.

21.5 Outsourced Identity

One of the apparent challenges in offering IT as a service is the identification of users and administrators. In the past, it could be enforced that administrator access would only be granted with a local login. In the cloud, there is no such thing as "local", and the administrator of a virtual machine is forced to log on remotely. Meanwhile, administrators of the provider will manage the physical machines, giving rise to multiple levels of administrator roles.[22]

Security risks have always been for a significant part due to employees of the organisation itself, with administrators as a prominent example. The entanglement of their responsibilities in the cloud also leads to new types of insiders, who have or can obtain the necessary credentials to endanger information security.[23] Insiders

---

[19] Perlman, R. *The ephemerizer: Making data disappear*. Technical Report TR-2005-140, Sun Microsystems, (2005). Tang, Q. "Timed-Ephemerizer: Make Assured Data Appear and Disappear," in *Sixth European Workshop on Public Key Services, Applications and Infrastructures*. Springer, (2009).

[20] Geambasu, R., T. Kohno, A.A. Levy, and H.M. Levy. "Vanish: Increasing Data Privacy with Self-Destructing Data." in *Proceedings of the USENIX Security Symposium*. USENIX association, (2009): 299–350.

[21] Pieters, W. "Converging Technologies and De-perimeterisation: Towards Risky Active Insulation." in *Proceedings of SPT 2009: Converging technologies, changing societies*. Enschede: CEPTES, University of Twente, (2009): 58–60.

[22] Casola, V., R. Lettiero, M. Rak and U. Villano. "Access Control in Cloud-on-GRID systems: the PerfCloud Case Study," *this volume*.

[23] Probst, C.W., R.R. Hansen, and F. Nielson. "Where can an insider attack?" in *Workshop on Formal Aspects in Security and Trust (FAST2006)*, volume 4691 of LNCS. Springer, (2007): 127–142.

are no longer only found within a single organisation, but spread among business partners and service providers, such as those of cloud services.[24]

It is therefore important to develop access control models and accountability mechanisms that can establish strong links between identity and activity, when necessary. We see the development of such architectures for example in Electronic Health Record (EHR) systems. Here, the right kind of authorisation is in principle necessary to obtain sensitive data. However, in case of an emergency, such a mechanism needs to be overridden, and accountability is established after the fact.

Apart from issues of authentication, cloud computing also raises questions concerning the nature of identity itself. Cloud use is not limited to companies. Increasingly, individuals store their information online instead of on their own devices. Identities are managed in Facebook, phone companies store users' SMS messages online and different forms of online cooperation are offered by Google and others. Considering the high availability rates of the Internet nowadays and the professional backup system usually connected to online storage, these services may indeed be quite attractive.

All this comes at a price, though, which is called loss of control. Unlike in businesses, control may not be such an explicit consideration for individuals. Indeed, not being in control may relieve the individual of unnecessary burdens. However, it is generally not the case that individuals would allow service providers to do with their data what they like. There are limitations based on privacy laws, but even then, users may wish to explicitly weigh the advantages of online storage and processing against the exposure of their data.

Some systems have indeed triggered user action. For example, many Dutch citizens have objected to the online transmission of their health data via the electronic patient file system. Facebook users have forced the provider to adapt default privacy settings and even shut down a new service that allowed friends to track one's shopping behaviour. And Google has explained that the target advertisements popping up with users' e-mails do not actually involve human reading of the e-mail contents – for what it's worth, for who has access to the machines actually processing these e-mails to match the ads? These examples illustrate that, in order to understand individual cloud use, it needs to be distinguished from company use.

One the one hand, individual use of cloud services seems to be merely a matter of convenience. If I can more reliably and more easily work with my information online, then why wouldn't I do so? On the other hand, the tendency to put *personal* information online for other purposes than processing it oneself is certainly new, and has to be evaluated more carefully. The use of Facebook, Twitter, LinkedIn is not comparable to other forms of cloud computing by individuals, as it concerns deliberate publication of personal information for social reasons.

[24]Nunes Leal Franqueira, V., A. van Cleeff, P.A.T. van Eck, and R.J. Wieringa. "External Insider Threat: A Real Security Challenge in Enterprise Value Webs." in *Proceedings of the Fifth International Conference on Availability, Reliability and Security (ARES'2010)*. IEEE Computer Society Press, (2010): 446–453.

Two trends have come together to produce the latter phenomenon. The first is the demise of traditional group memberships in society (called "ontzuiling" in Dutch, meaning "de-pillarisation"). Where individuals were born into or easily led into certain groups, based on family, gender, class, et cetera, this is now much less straightforward. The second is the rise of the Internet and associated cloud services. Combining the two, online services like Facebook have emerged to provide *online management of one's identity*. One now needs to explicitly group oneself, and social networking services are a tool to do precisely this. We call this phenomenon *outsourced identity*.

Three combined characteristics of outsourced identity distinguish it from earlier forms of external identity (diaries, village gossip, etc.).[25] Firstly, the outsourcing is intentional; secondly, the information is public or semi-public; and thirdly, the form of the information makes it easy to de-contextualise it.[26]

The consequences of this phenomenon are that the problems of profiling associated with online information about individuals cannot only be discussed in terms of social sorting. In outsourced identity, individuals *want* to get sorted. It is therefore necessary to distinguish good from bad sorting. The question is how we could make this distinction.

Informed consent may be the key concept here: if individuals are allowed to deny group membership, they maintain their autonomy in the face of undesirable assignments to groups. This of course implies that (1) they know about the group membership and (2) the membership can (at least theoretically) be denied. The former condition does not hold if decisions are being made in secret or based on secret profiling information; the latter does not hold if it concerns conditions such as age, race or gender, or probabilistic groups like "those who are likely not to pay their bills".

There is thus a difference in protecting privacy in the cloud from a company perspective and from an individual's perspective. From a company perspective, privacy-sensitive data needs to be processed in accordance with privacy laws, and the confidentiality of this data should be guaranteed as far as possible. This should then make sure that people do not suffer from undesirable consequences of loss of privacy, in particular in terms of discrimination. From an individual perspective, people use the cloud and its applications precisely to discriminate themselves; to make themselves stand out from the crowd and be assigned to groups simultaneously. This does not mean that individuals do not need privacy; rather, this outsourced identity cannot be addressed from the same perspective as privacy in traditional applications, and care is needed when judging the phenomenon from an ethical point of view. In the end, different architectures, less centralised

[25] See e.g. Clark, A., and D. Chalmers. "The Extended Mind." *Analysis* 58, 1 (1998): 7–19.

[26] Dumortier, F. "Facebook and Risks of 'De-Contextualization' of Information." In *Data Protection in a Profiled World*, edited by S. Gutwirth, Y. Poullet, and P. de Hert. Springer, 2010: 119–138.

and with more user control, may be more appropriate for applications like social networking.[27]

21.6 Informational Precaution

In the face of the development towards cloud computing, the need to separate pieces of information becomes more and more profound. If information is not properly secured, some parties may become too powerful, and trust relations may be broken. Unfortunately, the role of information security in providing this separation is poorly understood. On the one hand, technical solutions are being developed to secure information; on the other hand, policies are developed without much regard to what is technically possible, and sometimes policies seem to run ahead of the possibility of function creep, by allowing all kinds of extended usage scenarios on forehand. Substantial future research is needed to clarify the relation between technical forms of information security and societal goals, both from a social science and from a philosophical perspective.

In the meantime, we need a means to convince policy makers, both in government and in industry, of the real ethical dimensions of the issues. Implementing another distributed IT application is not just an increase in convenience; we are really changing the world here. Where environmental concerns are high on the agenda, the data protection community has not yet succeeded in providing a clear vision on the need of protecting something like information, with a view to social effects of information processing.

One approach to bridge this gap would be the translation of norms from environmental ethics to information technology. Earlier, we have proposed to do this for the precautionary principle.[28] This principle states that parties should refrain from actions in the face of scientific uncertainties about serious or irreversible harm to public health or the environment. It further holds that the burden of proof for assuring the safety of an action falls on those who propose it.[29] The precautionary principle has been applied successfully in the European Union, but is less popular in the United States, as it obviously implies government interference with what is desirable and what is not.

A similar baseline for computer ethics *as something that contributes to sustainability* does not exist yet. Even if the precautionary principle is not universally

---

[27]Cf. Jacobs, B. "Architecture is Politics: Security and Privacy Issues in Transport and Beyond." in *Data Protection in a Profiled World*, edited by S. Gutwirth, et al. Springer, 2010: 289–299.

[28]Pieters, W., and A. van Cleeff. "The Precautionary Principle in a World of Digital Dependencies." *IEEE Computer* 42, 6 (2009): 50–56.

[29]Raffensperger, C., and J.A. Tickner, editors. *Protecting Public Health and the Environment: Implementing the Precautionary Principle*. Washington, DC: Island Press, 1999. Rogers, M.D. "Scientific and Technological Uncertainty, the Precautionary Principle, Scenarios and Risk Management." *Journal of Risk Research* 4, 1 (2001): 1–15.

accepted in environmental ethics, it does provide a basis from where to start discussions on new technologies. Generalised to information technology, it can serve as a trigger for governments to at least consider the social implications of IT developments.

Whereas the traditional precautionary principle targets environmental sustainability, informational precaution would target social sustainability.[30] Clear definitions of this concept are however lacking, and more precise definitions are necessary. One could say that social sustainability relies on maintaining stable trust and power relations in society, and information is a key asset there. With that in mind, information security can indeed contribute to social sustainability, and is therefore an indispensable feature in ethical cloud scenarios – including government initiatives.

These considerations are part of larger-scale developments, and disappearing boundaries do not only occur in information security. The precautionary principle may play a role for society in general in dealing with these dependencies. As such, the cloud is an instance of our self-created dependence upon large-scale infrastructures, and thereby on the ethical behaviour of those who manage them. Precaution can serve to enforce this ethical behaviour, by designing technology such that it stimulates the right kinds of actions.[31]

However, also the precautionary principle itself needs to be treated with precaution. It easily becomes another tool in the hand of a few to control the many, and some authors even criticise what they call a "precaution state".[32] Therefore, informational precaution deserves our further attention and discussion in the cloud era.

21.7 Conclusions

In cloud computing, information storage, transmission and processing are purchased as a commodity from a service provider. Although security issues in storage and transmission can be addressed to a reasonable extent using standard tools, protecting data being processed is another story. Although fully homomorphic encryption would enable processing of encrypted data, the theoretical breakthrough of 2009 is far from practical. Therefore, efforts need to be put into securing data processing in limited cases, such as searching in encrypted data.

---

[30] McKenzie, S. *Social Sustainability: Towards some definitions*. Hawke Research Institute Working Paper Series No 27, 2004. Accessed May 10, 2010. https://www.sapo.org.au/binary/binary141/Social.pdf.

[31] Verbeek, P.P. *What Things Do: Philosophical Reflections on Technology, Agency, and Design*. University Park, PA: Pennsylvania State University Press, 2005.

[32] "Voorzorgstaat", see van Ooijen, C., and S. Soeparman. "Toezicht in de voorzorgstaat: Kennis en informatiegebruik tussen staatscontrole en sociabiliteit." in *Inzicht en Toezicht: Controle in de Kennissamenleving*, volume 6 of Jaarboek Kennissamenleving. Amsterdam: Aksant, (2010): 161–181.

Another fundamental issue in cloud computing is the abolishment of physical constraints that helped securing the data in the past. Like in a transition from paper voting in polling stations to Internet voting, this brings additional security challenges, and it is not even clear beforehand that all of these can be solved. In order to simulate physical constraints in a virtualised infrastructure, proposals like location-based access control and secure deletion have been put forward.

Meanwhile, the increasing outsourcing and specialisation not only affects companies, but also individuals. In this case, cloud computing not only means a different way for individuals to store their data, but also a different way to manage the external representations of their identity. In this "outsourced identity", people *intend* to get themselves socially sorted, and social sorting cannot be seen as problematic in and of itself. Instead, rules should be put in place to distinguish good from bad sorting, notably by requiring informed consent.

In order to provide ethical foundations for the complex interaction of governments, industries and the social environment in the age of cloud computing, research is needed in the relation between information technology and social sustainability. The precautionary principle may serve as a tool to translate successful approaches from environmental ethics to information ethics in the cloud.

Acknowledgments This research is supported by the research program Sentinels (www.sentinels.nl). Sentinels is being financed by Technology Foundation STW, the Netherlands Organization for Scientific Research (NWO), and the Dutch Ministry of Economic Affairs.

References

Ardagna, C.A., M. Cremonini, E. Damiani, S. De Capitani di Vimercati, and P. Samarati. "Supporting Location-Based Conditions in Access Control Policies." In *Proc. of the ACM Symposium on Information, Computer and Communications Security (ASIACCS'06)*. ACM, (2006): 212–222.

Blakley, B. "The Emperor's Old Armor." In *Proc. New Security Paradigms '96*. ACM, |(1997): 2–16.

Brinkman, R. *Searching in Encrypted Data*. PhD thesis, University of Twente, 2007. Accessed April 29, 2010. http://doc.utwente.nl/57852.

Clark, A., and D. Chalmers. "The Extended Mind." *Analysis* 58, 1 (1998): 7–19.

van Cleeff, A., W. Pieters, and R.J. Wieringa. "Security Implications of Virtualization: A Literature Study." In *2009 IEEE International Conference on Computational Science and Engineering (CSE09)*. IEEE Computer Society, (2009): 353–358.

van Cleeff, A., and R.J. Wieringa. "Rethinking De-Perimeterisation: Problem Analysis And Solutions." In *Proceedings of the IADIS International Conference Information Systems 2009, 25-27 Feb 2009, Barcelona*. IADIS press, (2009): 105–112.

Dumortier, F. "Facebook and Risks of 'De-Contextualization' of Information." In *Data Protection in a Profiled World*, edited by S. Gutwirth, Y. Poullet, and P. de Hert. Springer, 2010: 119–138.

Gartner, *Gartner Says Security Delivered as a Cloud-Based Service Will More Than Triple in Many Segments by 2013*. Press release, Gartner, 2008. Accessed April 29, 2010. http://www.gartner.com/it/page.jsp?id=722307.

Geambasu, R., T. Kohno, A.A. Levy, and H.M. Levy. "Vanish: Increasing Data Privacy with Self-Destructing Data." In *Proceedings of the USENIX Security Symposium*. USENIX association, (2009).

Gentry, C. "On Homomorphic Encryption Over Circuits of Arbitrary Depth." In *The 41st ACM Symposium on Theory of Computing (STOC)*. ACM, (2009): 169–178.

Hirt, M., and K. Sako. "Efficient Receipt-Free Voting Based on Homomorphic Encryption." In *Proc. EUROCRYPT 2000*, volume 1807 of LNCS. Springer, (2000): 539–556.

Ishai, Y., and A. Paskin. "Evaluating Branching Programs on Encrypted Data." In *Proc. 4th Theory of Cryptography Conference (TCC)*, volume 4392 of LNCS. Springer, (2007): 575–594.

Jacobs, B. "Architecture is Politics: Security and Privacy Issues in Transport and Beyond." In *Data Protection in a Profiled World*, edited by S. Gutwirth, Y. Poullet, and P. de Hert. Springer, 2010: 289–299.

Jericho Forum, *Jericho whitepaper*. Jericho Forum, The Open Group, 2005. Accessed May 10, 2010. http://www.opengroup.org/jericho/vision_wp.pdf.

McKenzie, S. *Social Sustainability: Towards some definitions*. Hawke Research Institute Working Paper Series No 27, 2004. Accessed May 10, 2010. https://www.sapo.org.au/binary/binary141/Social.pdf.

Nunes Leal Franqueira, V., A. van Cleeff, P.A.T. van Eck, and R.J. Wieringa. "External Insider Threat: A Real Security Challenge in Enterprise Value Webs." In *Proceedings of the Fifth International Conference on Availability, Reliability and Security (ARES'2010)*. IEEE Computer Society Press, (2010): 446–453.

van Ooijen, C., and S. Soeparman. "Toezicht in de voorzorgstaat: Kennis en informatiegebruik tussen staatscontrole en sociabiliteit." In *Inzicht en Toezicht: Controle in de Kennissamenleving*, volume 6 of Jaarboek Kennissamenleving. Amsterdam: Aksant, (2010): 161–181.

Pavlovic, D., and C. Meadows. *Quantifying pervasive authentication: the case of the Hancke-Kuhn protocol*. Technical Report No. RR-09-09. OUCL, 2009. Accessed May 10, 2010. http://www.comlab.ox.ac.uk/files/2437/RR-09-09.pdf.

Perlman, R. *The ephemerizer: Making data disappear*. Technical Report TR-2005-140, Sun Microsystems, (2005).

Pieters, W. "Converging Technologies and De-perimeterisation: Towards Risky Active Insulation." In *Proceedings of SPT 2009: Converging technologies, changing societies*. Enschede: CEPTES, University of Twente, (2009): 58–60.

Pieters, W., and A. van Cleeff. "The Precautionary Principle in a World of Digital Dependencies." *IEEE Computer* 42, 6 (2009): 50–56.

Probst, C.W., R.R. Hansen, and F. Nielson. "Where can an insider attack?" In *Workshop on Formal Aspects in Security and Trust (FAST2006)*, volume 4691 of LNCS. Springer, (2007): 127–142.

Raffensperger, C., and J.A. Tickner, editors. *Protecting Public Health and the Environment: Implementing the Precautionary Principle*. Washington, DC: Island Press, 1999.

Rogers, M.D. "Scientific and Technological Uncertainty, the Precautionary Principle, Scenarios and Risk Management." *Journal of Risk Research* 4, 1 (2001): 1–15.

Schoenmakers, B. "A Simple Publicly Verifiable Secret Sharing Scheme and Its Application to Electronic Voting." In *CRYPTO '99*, volume 1666 of LNCS. Springer, (1999): 148–164.

Tang, Q. "Timed-Ephemerizer: Make Assured Data Appear and Disappear." In *Sixth European Workshop on Public Key Services, Applications and Infrastructures*. Springer, (2009).

Verbeek, P.P. *What Things Do: Philosophical Reflections on Technology, Agency, and Design*. University Park, PA: Pennsylvania State University Press, 2005.